MARXISM AND THE
PHILOSOPHY OF SCIENCE

MARXISM AND THE PHILOSOPHY OF SCIENCE

A Critical History

The First Hundred Years

Helena Sheehan

VERSO

This edition published by Verso 2017
Based on the paperback edition published by Humanities Press International 1993
First published by Humanities Press International 1985
© Helena Sheehan 1985, 1993, 2017

1 3 5 7 9 10 8 6 4 2

Verso

UK: 6 Meard Street, London W1F 0EG
US: 20 Jay Street, Suite 1010, Brooklyn, NY 11201
versobooks.com

Verso is the imprint of New Left Books

ISBN-13: 978-1-78663-426-9
ISBN-13: 978-1-78663-428-3 (UK EBK)
ISBN-13: 978-1-78663-427-6 (US EBK)

British Library Cataloguing in Publication Data
A catalogue record for this book is available from the British Library

Library of Congress Cataloging-in-Publication Data
A catalog record for this book is available from the Library of Congress

Printed in the UK by CPI Group

CONTENTS

Dedication

In memory of those who made their mark upon the history set out here;
who symbolized in their very lives the ways in which
science, philosophy and politics have come together for Marxism;
who died tragically and before their time:

Christopher Caudwell (1907 - 1937)
David Guest (1911 - 1938)
Antonio Gramsci (1891 - 1937)
Georges Politzer (1903 - 1942)
Jacques Solomon (1908 - 1942)
Boris Mikhailovich Hessen (1883 - 1937)
Nikolai Ivanovich Vavilov (1887 - 1943)

ACKNOWLEDGMENTS

Because of the extraordinary scope and complexity of this project, the organization of my research has been a highly complicated affair and in it I have had an enormous amount of help. I have also benefited greatly from my discussions with philosophers, historians, and scientists of a wide range of views and from the criticisms of those who have read my manuscript.

My research took me abroad on a number of occasions: to the Soviet Union, the German Democratic Republic, Czechoslovakia, Yugoslavia, Romania, the Federal Republic of Germany, France, Austria, Great Britain, the United States of America, and Canada. Aside from access to sources and interviews with philosophers, historians, and scientists, my travels, in taking me to many of the places where the history I was writing had unfolded, made events far more vivid to me than they would have been otherwise.

I went twice to the Soviet Union. The second time, I spent four months of 1978 in Moscow, working in Soviet libraries, giving lectures, and having discussions with Soviet philosophers. During this time, I achieved an insight into the character of Soviet philosophical life and a vivid sense of the sociohistorical atmosphere in which Soviet philosophy developed in its own distinctive way that I could not have achieved otherwise. I was often at the Institute of Philosophy of the USSR Academy of Sciences in Moscow, where I gave a paper outlining my research and, in the lively discussion from the floor which followed, received many useful comments.

Among the Soviet philosophers who assisted me in my work, I would especially like to acknowledge the following: from the Institute of Philosophy in Moscow: Professors Y.V. Sachkov, T.I. Oiserman, A.D. Ursul, M.E. Omelyanovsky, I.A. Akchurin, L.G. Antipenko, K.K. Delakarov, Y.B. Molchanov, E.P. Pomagaeva, I.I. Petrov, B. Bogdanov, S. Brayović, E. Stepanov; from Moscow University: Professors S.I. Melyukin, J. Vogeler, Y.A. Petrov; from the Institute for the History of Science and Technology: Professor B.G. Kuznetsov; from other Moscow institutes: Professors E.P.

Sitkovsky, Y.A. Zamoshkin, A.V. Shestapol, V.P. Terin, A. Grigorian. My interview with Academician M.B. Mitin concerning events in which he was a central figure was particularly useful. Professor N.V. Matkovsky, a historian, also showed a warm interest in my work and did much to assist it.

I went twice to the German Democratic Republic in 1978 and 1979 under the auspices of the Central Institute for Philosophy of the GDR Academy of Sciences. On both occasions, I gave an account of my research, followed by extremely stimulating discussion. I also had long and searching discussions with a number of their philosophers and scientists. I am particularly grateful to Academician Herbert Hörz, who invited me and who has shown an extraordinary willingness both to help me with research arrangements and to discuss the philosophical problems involved. Other members of his department, the Sector for the Philosophical Problems of the Natural Sciences, have done much for me, particularly Dr. Ulrich Röseberg, and also Dr. Siegfried Paul, Dr. Eberhard Thomas, Dr. Rüdiger Simon, Dr. John Erpenbeck, Dr. Nina Hager. From Humboldt University, I came to know Professor Herman Ley, Professor Dieter Schultze, Professor Franz Löser, and Professor Klaus Fuchs-Kittowski and I am grateful to them for a series of very fine and useful discussions. Professor Alfred Kosing of the Academy of Social Sciences in Berlin also graciously assisted me, as did Academician Helmut Böhme, Director of the Central Institute for Genetics of the Academy of Sciences, who has shown a sustained interest in my work, discussing with me particularly those aspects of it relating to biology and genetics.

A trip to Prague in March 1978, to participate in an international symposium on Dialectical Materialism and Modern Sciences, sponsored by the Czechoslovak Academy of Sciences, and to read a paper, brought me into contact with a number of philosophers and natural scientists in Eastern Europe and was the occasion of a number of discussions relevant to my work. I am grateful to Academician Radovan Richta, Director of the Institute of Philosophy and Sociology of the Czechoslovak Academy of Sciences.

Those from Eastern Europe with whom I had discussions away from their home base were: from Poland, Professor Wladyslaw Krajewski of University of Warsaw whom I met in Dubrovnik, Professor Piotr Sztompka of Jagiellonian University of Krakow, whom I met in Hannover, and Professor Adam Schaff, whom I met in Düsseldorf and Vienna; from Bulgaria, Professor Azaria Polikarov of the University of Sofia, whom I met in Bucharest; from the German Democratic Republic, Dr. Wolfgang Harich whom I met in Vienna; from Czechoslovakia, Dr. Julius Tomin, whom I met in Dublin.

A trip to Vienna to meet with Professor Adam Schaff of the Polish Academy of Sciences and of the European Center for Coordination of Social Science Research in Vienna was especially important to me. His ruthless honesty in discussing many difficult matters with me was something for which I shall always be grateful, however painful it was at the time. Not in Vienna at this time, but extremely helpful to me in correspondence, has been Professor Walter Hollitscher of Vienna and of Karl Marx University in Leipzig.

A somewhat idyllic sojourn in Dubrovnik brought me the opportunity to discuss in some depth the development of Yugoslav philosophy and philosophy of science with quite a number of Yugoslav philosophers and scientists. I spoke at greatest length with Professor Svetozar Stojanović, Professor Mihailo Marković, Professor Ivan Supek and Dr. Srdjan Lelas. Dubrovnik was also the setting of some extremely stimulating discussions of the epistemology of science involving Professor Ernan McMullin, Dr. William Newton-Smith and Mr. Rom Harré which pushed me to probe certain issues further myself. Discussions of the current state of Marxist thought with Professor Joachim Israel and Professor Marx Wartofsky were also helpful.

Various international conferences I attended brought similar opportunities, particularly the 16th World Philosophical Congress in Düsseldorf in August 1978, the 6th International Congress of Logic, Methodology and Philosophy of Science in Hannover in August 1979, and the 16th International Congress of the History of Science in Bucharest in August 1981.

My numerous trips to Britain were extremely important to my work, not only for my research at the British Museum and the Marx Memorial Library, but for my contact with British Marxists. I am particularly indebted to Mr. Maurice Cornforth, Mr. Gordon McLennan, Mr. James Klugmann, Mr. Monty Johnstone, Mr. Desmond Greaves, Dr. John Hoffman, Dr. Julian Cooper, Mr. David Pavett, Dr. David Margolies, Mr. Jonathan Rée, Dr. E.P. Thompson, Professor Roy Edgley, Professor Joseph Needham, and Professor Ralph Miliband. I am also grateful to Ms. Rosemary Sprigg, executor of the estate of Christopher St. John Sprigg, for access to unpublished material relevant to this study. Interviews with Mr. Paul Beard, Mrs. Elizabeth Beard, Mr. Nick Cox, and Mr. William Sedley were useful to me in my research on Caudwell. So too were unpublished manuscripts of works on Caudwell by Dr. Jean Duparc of Nanterre and Dr. George Moberg to which they graciously gave me access. I am indebted to Dr. Duparc for lengthy and challenging discussions and generous hospitality when I came to Paris to read his thesis for the doctorat d'état. While in Paris, I also spoke with Dr. Dominique Lecourt about the various issues raised here.

My trips to the United States and Canada in 1979 and 1980 advanced my work in two ways. I gave lectures based on my research at the Center for Philosophy and History of Science of Boston University, in the Marxism and Science seminar in the History of Science Department of Harvard University, at the Philosophical Colloquium of the University of Toronto, at public lectures sponsored by the Departments of Philosophy of Queens College of the City University of New York, of Fairfield University, and of the University of Bridgeport and by the journal *Science and Nature*, and these raised challenging questions and points of discussion. Also, while in the United States, I had discussions with philosophers and historians of science that were extremely important to me as I got my work into its final form. I am especially grateful to Professor Robert Cohen of Boston University not only for challenging discussions and criticisms, but for organizing many things for me. I am also grateful to Professor Erwin Hiebert and Professor Everett Mendelsohn of Harvard University; Professor Dirk Struik and Professor Loren Graham of MIT; Professor Marx Wartofsky, Professor Joseph Agassi and Professor Diana Long Hall of Boston University; Professor Howard Parsons of University of Bridgeport; Professor Eugene Fontinell and Professor George Kryzwicki-Herbert of Queens College, CUNY; Professor Ernan McMullin of the University of Notre Dame; Professor Mihailo Marković, Professor Alexander Vucinich, and Professor Mark Adams of the University of Pennsylvania; Professor Robert Tucker of Princeton University; Professor John Stachel, Professor Imre Toth, and Professor Dudley Shapere of the Institute for Advanced Study; Professor Richard Bernstein of Haverford College; Professor Mario Bunge of McGill University; Dr. Dan Goldstick and Dr. Frank Cunningham of the University of Toronto. Also my talks with those outside of the university setting with a lively interest in Marxist philosophy of science, particularly with those gathered around the journal *Science and Nature* and with Mr. George Novack, author of many books on Marxist philosophy, were very encouraging.

I have been extraordinarily fortunate in the number of outstanding scholars who have shown interest in my work and have taken great care in reading and commenting upon what I have written. This manuscript, or at least substantial sections of it, has been read by the following: Marx Wartofsky, Robert Cohen, Ernan McMullin, Mario Bunge, Nicholas Rescher, Stephen Toulmin, Erwin Hiebert, Loren Graham, Mihailo Marković, Adam Schaff, Maurice Cornforth, Walter Hollitscher, Ulrich Röseberg, Helmut Böhme, Jean Duparc, John Stachel, Imre Toth, Tomas Brody, Roy Edgley, Jonathan Rée, Monty Johnstone and Peter Mew. I am much indebted to them for their discerning judgements and challenging criticism.

Closer to home, I must mention the general encouragement in my work given to me by colleagues in Irish universities and in the Irish Philosophical Society, particularly those in the departments in which I have been based while working on this book: the Department of Philosophy of Trinity College, Dublin, the Department of Politics of University College, Dublin, and the School of Communications of the National Institute for Higher Education in Dublin.

Many other people helped me as well, too many people, in too many different ways to be able to name them all here: librarians, interpreters, administrators, apparatchiks of all sorts, and simply friends. Of these, I should single out Mr. Séamus O Brógáin, who helped me with various technical matters and arrangements and Mr. Sam Nolan who encouraged me in getting through the final hurdles and seeing the task through to completion.

Above all, I wish to thank Mr. Eoin O Murchú, who not only typed the manuscript, despite his vehement disagreement with many of the views expressed in it, but did much more to facilitate my work than I can possibly recount here. My debt to him is really quite incomparable.

Although it is not customary to acknowledge publishers, I wish to break such a custom to express the appreciation I feel for the enthusiastic encouragement and robust joie de vivre of Mr. Simon Silverman.

However, acknowledging my intellectual sources is really a much more complicated matter than listing persons with whom I discussed my work and giving references to books or papers cited at the end of chapters. Larger movements and deeper processes at work, the weight of various intellectual traditions, and also the whole tempo of the times have made my orientation to this book what it has been.

From the time I began to engage in serious and systematic study of Marxism, many things fell into place for me. I cannot say that I am a Marxist in any unambiguous sense, for I have arguments with virtually all contending schools of thought within Marxism, both orthodox and revisionist, both past and present. I also believe that my view of the world has been forged in the convergence of a number of intellectual traditions, of which Marxism is only one. However, I must acknowledge that Marxism has left its mark on me and decisively shaped my modes of understanding as no other tradition has. As I worked through the history set out here, I was forced to come to terms with many issues in a sharper way than I would have otherwise. There were times when I was quite shaken by what I realized I had to write. My views on various questions evolved, reached points of crisis, and then resolution. My own relationship to Marxism became more and more complicated. I believe that this has been a good thing as far as this book is concerned, for Marxism is a very complex and controversial phenomenon and commentary on it has for too long been far too polarized in terms of simple and one-sided extremes. The waters have for far too long been muddied both by the shallow jargon and hollow self-praise and by the stereotyped polemics emerging in response to them. I have come strongly to believe it deserves much better. I have tried to look at it freshly, neither as apologist nor as prosecutor, but as someone who could recognize it as a formidable intellectual tradition and at the same time be free to subject it to critical assessment. I have, however, never believed that such openness of mind required detachment or lack of commitment, as the still-prevailing academic ethos would have it. No matter how critical or complicated my position in relation to Marxism and the various issues it posed for me became, I was at all times passionately involved with it. For this, I make no apologies, for I believe that the very intensity of my involvement has enhanced rather than distracted from my understanding.

Moreover, I believe that my active involvement with political organizations of the left, both old and new, at various times and in various places, has been epistemologically important for me. It is not simply that it brought me to know things I would not otherwise know, but it has involved a way of knowing that would not have been open to me in any other way. As is relevant to this book, the history of Marxism in the period under discussion is very much tied to the history of the communist movement. The fact that, while writing this book, I was coming to terms with the communist movement "for real" gave me many troubling experiences, but it also gave me a particular sort of insight that has, no doubt, colored these pages.

The concept of the "unity of theory and practice" has been a notion much vulgarized and much abused, but I think it would be unwise to throw it over. I have long been convinced of the correctness of Dewey's critique of the spectator theory of truth and have long since opted for a participational theory of truth. And so I believe that the world is *known* best by those who most actively take hold of it, interact with it, participate in it. It has been my privilege to have encountered people who have *known* in this way in the course of my research for this book. These include those still alive, whom I have interviewed, and those now dead, whom I have discovered through their texts. I am much indebted to them.

In the end, of course, I had to do my own thinking and make my own mistakes. For special reasons, I must insist that *only* I can be held to account for the views set out here, for better or for worse. I have been exposed to many and conflicting viewpoints and I have weighed seriously what others have said, but I had to decide for myself where I stood. And I have. This has sometimes meant taking issue with people who have been very kind to me, and if their assistance has been turned to ends neither they nor I foresaw, I ask them to try to understand. If they cannot accept the position I have taken, I hope they will at least accept my good faith in taking it. I am well aware of the controversial character of my conclusions, some of which may disturb others as much as they have disturbed me, but I have come to them gravely and could not do otherwise.

Helena Sheehan

INTRODUCTION

In these pages, there unfolds a complex and intricate story. Although it is a story of the most vital significance, it has long been left untold. In some quarters this has been because of a total ignorance that there was anything to tell, and in other quarters because of a fear of the telling of it. It is the story of the shifting nexus of science, philosophy, and politics within Marxism. It is astonishing to discover in how many different ways and on how many different levels these have converged within the development of the Marxist tradition, both for better and for worse, with a multiplicity of factors coming into play, including the impact of new scientific discoveries, new philosophical trends and new political formations upon the overall process. While it is a story of a progressive and audacious enterprise, to which intelligent and discerning minds have applied themselves, it is also a story with its dark side, with retrogressive episodes, unscrupulous characters, hasty and ill-conceived projects, and superficial solutions, sometimes with dire consequences.

This work attempts to give a historical account of the development of Marxism as a philosophy of science, as well as a philosophical analysis of the issues involved. This volume encompasses the first hundred years of the existence of Marxism, beginning with the mid-1840s when the philosophical ideas of Marx and Engels began to emerge in mature form, and ending with the mid-1940s with the dissolution of the Comintern and the end of the Second World War. It deals both with the mainstream of the Marxist tradition in the development of dialectical materialism as a philosophy of science and with the diverging currents advocating alternative philosophical positions. It shows the Marxist tradition to be far more complex and differentiated than is usually imagined, characterized by sharp and lively controversies for contending paths of development at every step of the way.

The debates about *how* to develop Marxism as a philosophy of science have taken a number of forms. One recurring theme has been that of the precise relationship between philosophy and the actual results of the empirical sciences, debated in the sharpest form in the controversy between the mechanists and dialecticians in the Soviet 1920s. Another recurring theme has been the relationship of specific scientific discoveries to the philosophy of dialectical materialism, such as the controversies over relativity theory or genetics (the much cited "Lysenko affair"). Yet another theme has been whether dialectical materialism or alternative philosophies of science, based on neo-Kantianism, Machism, or mechanistic materialism, are the most appropriate development of the Marxist tradition, such alternatives having been put forward by the Austro-Marxists, the Russian empirio-criticists and the Soviet mechanists. There have also been debates about *whether* to develop Marxism as a philosophy of science, disputes exemplified most sharply by the emergence of the neo-Hegelian trend within Marxism and the dialectics of nature debate set off by the Hungarian Marxist Lukács in the Comintern in the 1920s and continuing into the 1970s.

It all adds up to a formidable intellectual tradition in the philosophy of science which is virtually ignored by academic philosophy of science outside Eastern Europe. Certainly in the world of Anglo-American philosophy departments, insofar as there is any thought given to the *history* of the philosophy of science, in this era in which philosophy has become so woefully ahistorical, it is fixated on one line of development. The consensus undoubtedly is that the main dramatis personnae of nineteenth- and twentieth-century philosophy of science are such as Mach, Carnap, Popper, Kuhn, Lakatos, Feyerabend. This line of development has constituted the point of reference to which all commentators are expected to orient themselves, no matter how fundamental their criticisms of it, no matter how deep their commitment to charting a new way forward. I do not doubt that this line of development has been a vitally important one and that anyone working in this discipline without a thorough knowledge of the history of its major shifts and its present-day twists and turns would deserve harsh judgement from his or her colleagues. My point, however, is to call attention to the fact, too often neglected in this milieu, that this is not in fact the only major line of development in nineteenth- and twentieth-century philosophy of science.

Another line of development, stemming from the very bold and original work of Marx and Engels, has been one with a very different relationship to both philosophy and science, but with, nevertheless, its own full-blown tradition in the philosophy of science—its own history, its own shifts, its own very fascinating twists and turns. However, no one is judged badly by his or her colleagues for being utterly oblivious of it and those who bother to think of it at all consider the sideward glances at Marxism on the part of Popper and

Lakatos to be as much as it deserves, despite the fact that these forays have not been characterized by very full knowledge or very high standards of argument. Its history is known hardly at all and only the vaguest of caricatures of it prevail, with the name most readily connected with it almost inevitably being that of Lysenko. Lysenko has been part of it, to be sure, but so indeed have Marx, Engels, Lenin, Bukharin, Deborin, Bernal, Haldane, Caudwell, Langevin, Prenant, Geymonat, Hessen, Oparin, Fok, Ambartsumian, Kedrov, Hollitscher, and many others. Nor is it, as such names should indicate, simply an Eastern European phenomenon. It made, for example, significant inroads even into the hallowed Royal Society in 1930s Britain.

Although these two traditions have struggled with many of the same issues, the basic rhythms of their development have been very different. The tradition stemming from Mach and the Vienna Circle arose out of the impulse to defend scientific rationality in the face of the challenges posed to it by new developments in science. In an atmosphere of crisis in the epistemological foundations of science, with all forms of rampant obscurantism feeding off this crisis, they strove to set science upon secure foundations. They sought to purify and cleanse the intellectual inheritance of the ages of all superfluous accretions, to clear out the slag of the centuries, to subject all belief to the clear light of reason and the rigor of experiment. They did so, however, from a base that was too narrow, employing criteria that were too restricted, leaving out of the picture too much that was all too real. Rigidly separating the context of discovery from the context of justification, the logical positivist and logical empiricist schools took only the latter to be the proper concern of the philosophy of science. The process by which a theory came to be was irrelevant to judgements on its validity. And, if psychological, sociological, and historical considerations were irrelevant, metaphysical considerations were worse than irrelevant. The striving for a comprehensive world view, giving rise to the great basic questions of the history of philosophy, was renounced in no uncertain terms.

The trajectory of this tradition, from positivism to the current variety of postpositivist philosophies of science, has reflected the pressure of a complex reality upon conceptions too restricted to give an adequate account of it. The successive modifications of the tradition over the years, from verificationism to falsification, to the historicism of paradigm shifts, to the methodology of scientific research programs, to methodological dadaism, have been impressive but still inadequate attempts to come to terms with the metaphysical and historical dimensions of science. Despite such significant departures from the antimetaphysical and ahistorical heritage of the Vienna Circle, it still cannot be said that philosophers of science have yet brought to bear the full weight of the implications of metaphysics or historicity for science. Moreover, insofar as these dimensions have come into play, they have tended to do so in a negative way, in that they have been perceived as undermining the rationality of science. Many of the current debates are rooted in a persistent inability to reconcile the rationality of science with the metaphysical and sociohistorical character of science. In a strange way, the residues of positivism linger on and color the views of even the most radical of antipositivists. And the flames of the crisis in the epistemological foundations of science burn more wildly than ever. There is no consensus, indeed there is exceedingly sharp polarization, regarding the relationship of science either to philosophy or to history.

It has always seemed somewhat ironic to me that the most influential line of development in breaking with classical empiricism in the direction of a more contextual, sociohistorical, metaphysical view of science has come via Wittgenstein and Kuhn, when there were earlier bodies of thought already there, which had long since put forward far deeper and more radical critiques of the received view of science and far richer alternative versions. Both the radical empiricist tradition of James and Dewey and the dialectical materialist tradition of generations of Marxists have embodied atternative versions of empiricism (in the sense of seeing the origins of knowledge in experience), which were based on much richer notions of experience, which allowed the metaphysical and historical dimensions of knowledge to come more fully into play. Both rejected the formalist, individualist, particularist, passivist model of knowledge in favor of a more historicist, social, contextualist, activist model.

Long before Wittgenstein, these earlier thinkers understood that experience came already clothed in language; that meaning could only be understood in context; that no logical formalism could substitute for the real flow of actual experience. Long before Kuhn, they knew that science was a complex, social, human activity; that is, was a process characterized by both evolutionary development and revolutionary upheavals. Long before Popper, they spoke of the role of guessing in science; they criticized the view that science was a matter of straightforward induction and saw the part played by hypothesis and deduction. Long before the emergence of the Edinburgh School, they made very strong claims indeed about the role of social interests in knowledge; they believed sociological explanation to be relevant to all theories, whether in literature or science, whether true or false, rational or irrational. Indeed, so many contemporary themes, such as the role of theory in observation, the attention to the process of discovery as well as to the process of justification, the impossibility of strict verification, and many others, turn out to be anticipated in these now dusty tomes on yellowed page after yellowed page.*

---

* Ludwig Fleck's *Genesis and Development of a Scientific Fact* (1979) is one such previously neglected text that has contemporary themes. Fleck used historical-epistemological categories like "thought-style" and "thought-collective," and presented Wasserman's test for syphilis to emphasize the plasticity of scientific "facts."

The differences, however, are just as striking. Both of these earlier traditions placed science within a much wider sociohistorical context than did Kuhn and his successors. They understood far better the relationship of the history of science to the history of everything else, which undercut the possibility of such protected worlds-unto-themselves as Wittgensteinian language games, Kuhnian paradigms or Popperian third worlds. Their field of vision opened to a much wider world. Their understanding penetrated to a much deeper historicity. Their modes of thought were much more integral and so they were not prone to the one-sidedness of later views. Their realization of the importance of hypothesis and deduction did not lead them to deny any role whatsoever to induction. Their awareness of complexity did not lead them to conclude that there were no unifying patterns. Their historicism did not entail discontinuity. Their acceptance of relativity did not imply incommensurability. Their renunciation of the quest for certainty did not bring them to despair, to announcements of the "end of epistemology," to declarations that "anything goes." What is really most striking about these earlier thinkers is how much more robust they were; how much more able they were to live in an open, uncertain, unfinished universe with many risks and no guarantees; how much more willing they were to stake their lot with uncertified possibilities. Nothing in all that they knew about what a complicated and uncertain activity science was stopped them from taking their stand with science and even drawing far-reaching conclusions from it.

Here I leave to someone else or to another day the task of making the case further for the radical empiricist tradition of American philosophy. This book and the one to follow will hopefully make the case for Marxism. Given the tensions wracking contemporary philosophy of science, it could well be of value to look back on a tradition that has proceeded with the same quite crucial matters, but in such a different way. The most significant features of Marxism in respect to these problems are: (1) that it has seen scientific theories as inextricably woven into world views; (2) that it has made extraordinarily strong claims regarding the sociohistorical character of scientific knowledge; (3) that it has not tended to perceive these aspects as being in any way in conflict with the rationality of science. Marx and Engels saw the history of science as unfolding in such a way that science was a cognitive activity carried on within the framework of a whole world view, which was in turn shaped by the nature of the socioeconomic order within which it emerged. Such a characterization of science took nothing away from science in their eyes. The science of the past, grounded in the world views of the past, grounded by the relations of production of the past, had all been necessary stages in the evolution of human understanding. It was necessary to unmask the superseded ideologies of the past and present that distorted the development of science. Even more, it was necessary to move the process onward to the next stage: the further development of science, in the context of a new world view, in the context of a struggle for a new social order and new relations of production.

From the beginning, the Marxist tradition bravely set itself the task of elaborating the philosophical implications of the sciences of its times with a view to working out a scientific *Weltanschauung* adequate for its epoch. Engels's antipositivist materialism was an extraordinarily impressive achievement. He did not shrink from the great basic questions that have perplexed the philosophers of the ages, but he did insist that attempts at answers be grounded in the best empirical knowledge of the time. In so doing, he not only laid the foundations of a scientifically grounded world view, but set forth views on many issues, such as reductionism, the history of science, and the logic of scientific discovery, that not only anticipated certain contemporary theories, but are still in advance of them. Throughout its subsequent history, there were new challenges to meet, arising out of the revolutionary advances in the natural sciences as well as out of the emergence of new philosophical trends and new political formations, and these giving rise at every step along the way to new controversies, new arguments for contending paths of development. It is a complex tradition, with Marcuse or Colletti as far from Engels as Feyerabend from Carnap.

Some of these contending paths involved explicit renunciation of Engels's enterprise of striving for a comprehensive world view grounded in science and in continuity with the history of philosophy. There have been, for example, the Soviet mechanists who believed that science could do very well without philosophy, as well as Marxists of the neo-Hegelian variety who have tended to think that philosophy should keep its distance from science. Today, the interpretation of Marxism as a *Weltanschauung* tied to the development of science is by no means an uncontested position. It is no secret that there are today many contending schools of thought, all claiming to embody the correct interpretation of Marxism, some of which explicitly renounce the ideal of a *Weltanschauung* and display a marked hostility to the natural sciences, while others make a fetish out of an eccentric concept of science and tend to take a derogatory attitude to philosophy. Nevertheless, the mainstream of the Marxist tradition has defended the sort of synthesis of science and philosophy that Engels proposed, even it has not always proceeded in the best possible way to bring the project to fruition.

From the beginning as well, this whole project has been placed firmly within a wider socio historical context. Philosophy of science was no free-floating set of theories spontaneously thought up by free-floating philosophers of science. Nor was science a straightforward piling up of facts about the world unproblematically discovered by autonomous scientists in their sealed-off laboratories. Marxists have seen philosophy, science, and philosophy of science, and indeed all aspects of the intellectual culture, as all inextricably interwoven with each other and tied to a common social matrix. It is no accident that a whole distinct tradition in philosophy of science has emerged and developed in relation to such a distinct social theory and political movement as Marxism. Marxism made its entrance on the historical stage not only with its critique of capitalism, its labor theory of value, its materialist interpretation of history as class struggle, its call for socialist revolution. It also formulated in the same act the classical premises of the whole discipline of sociology of knowledge and the externalist tradition in the historiography of science.*

The Marxist tradition has come forward with the fullest and most highly integrated claims to date regarding the socioeconomic basis of the rhythms of intellectual history. It is one of the most striking features of the story told in these pages just how tight its participants perceived the relation of science, philosophy, and politics to be, as far as both their own and their opponents' theories were concerned. Marxism is therefore a tradition in which the relation of science to philosophy and of both to politics and economics is conceived of differently than in alternative traditions. The other traditions too have their socioeconomic basis, Marxists have argued, even if they have been unable to be fully self-conscious and explicit about it. Indeed, even the inability to rise to such self-consciousness and explicitness in this realm has a socioeconomic basis on this account.

There can be no doubt that Marxism has given rise to a multifaceted and protracted struggle, both creative and destructive, to confront the philosophical problems raised by probing into the role of ideology in relation to science and philosophy of science. There is much to be learned both from the profound insights and from the tragic disasters born of passionate debates revolving around the tensions between partisanship and truth, between "proletarian science" and laws of nature. It still cannot be said, however, that the fundamental philosophical problem has yet been satisfactorily and explicitly resolved by Marxists or by anyone else.

The underlying question, still to be sharply answered, is: How can science be the complex, uncertain, precarious, human process that it is—inextricably bound up with all sorts of philosophical assumptions and with all sorts of wider sociohistorical processes—and still be reliable knowledge of nature? The story

---

* The Marxist tradition has been seminal in the development of a whole cluster of disciplines, such as the sociology of knowledge, sociology of science, history of science, and history of technology, and has had a crucial influence even on their non-Marxist practitioners, such as Karl Mannheim and Robert Merton.

set out here does not give that question the sharp, clear, fully argued answer that we feel that it needs, but it does give much in the way of promise and indicates that the resources for that sort of answer may well be there.

In proceeding with this question ourselves, of course, we need to do our own thinking, but we should do it with a grounding in the best insights that all available intellectual traditions have to offer. The polarization, for or against Marxism, has been a formidable barrier to the sort of reciprocal interaction that could bear fruit.

This raises a recurring problem in the history of Marxism, i.e., the relation of Marxism to non-Marxist trends. This, it must be said, is a problem that Marxists have not always handled very well. Most have unfortunately gone to the one extreme or the other, *either* accommodating themselves too far in the direction of recurrences of modes of thought superseded by Marxism and compromising the very distinctiveness of Marxism beyond recognition without sufficient reason for doing so *or* considering Marxism a closed world, with nothing to learn from other schools of thought and heaping abuse and invective upon anyone who has suggested otherwise.

The traditions of Russian Marxism, with its tendency to identify opposition with treachery, and with its decided inability to be fair-minded towards opponents, has exercised a disproportionate influence in the latter direction. Even Lenin, who learned far more from non-Marxist ideas and was far more broad-minded than the subsequent generation of Russian Marxists, was notoriously unfair to Mach and Bogdanov. With lesser men, the exaggeration of the worst aspects of this tradition, backed by the powers of the NKVD,* has had disastrous and harrowing consequences. This has resulted in a hardening of the lines of debate between a fixed "orthodoxy" and a traitorous or martyred (depending on one's point of view) "revisionism."

This has led, on the one hand, to an unhealthy pressure against creative thinking and to a sterile adherence to fixed formulations. A nearly pathological fear of revisionism has obscured the fact that every tradition with the vitality to endure has revised and must continually revise itself with the onward march of history and with the progressive achievements of human knowledge. Marxists must judge matters on the basis of the evidence, on the basis of truth criteria established by the highest level of development of scientific method at any given time, and not on the basis of conformity or nonconformity to established Marxist premises, no matter how fundamental. Only if its most basic premises

---

* The Peoples Commissariat of Internal Affairs, the Soviet security police from 1934 to 1953, the predecessor of the KGB.

are continually scrutinized can the continued affirmation of them be meaningful. Of course, if really basic premises could no longer be affirmed in this way, then it would be legitimate to query whether the new position should still be considered Marxist. But only by being open to this possibility, by following Marx's own advice to question everything, can Marxism be adhered to and developed in a healthy way.

On the other hand, this rigidity has fanned the flames of opposition and polarized debate along unconstructive lines, resulting in the tendency of those who wanted to be creative and were repulsed by the rigidity to close their minds to any of the "orthodox" premises and to grasp eclectically and regressively at contradictory premises, sometimes just because they did contradict the orthodox ones. Political factors, which can either enhance or distort philosophical thinking, have often had such a distorting effect, insofar as theoretical judgements have been overshadowed both by administrative measures and antipathy to administrative measures. This being so, it is sad but true to say, as Marx Wartofsky has, that "communist politics, as well as anticommunist politics, left the tradition of Marxist scholarship enfeebled."[1]

Nevertheless, the various "revisionist" positions bear a closer look than they have tended to get, either from those who have dismissed them out of hand or from those who have glorified them uncritically. Those, such as the early Lukács or the various Marxists of the Second International, who sought to bring back into Marxism the neo-Kantian dichotomy between history and nature, were , it could be argued, reverting to an antithesis already transcended by a higher synthesis. It meant dropping something crucial to Marxism as a higher and more integrated mode of thought. It did, however, highlight certain problems within Marxist thought, that is, the relationship between freedom and necessity, the relationship between the role of critical thought, ethical norms, and revolutionary activity on the one hand and the role of science and causal laws on the other. These were aspects of Marxist thought that needed to be clarified or further developed with the decline of the liberal idea of progress, the more problematic character of science, the turn from materialism with a new balance of class forces as the bourgeoisie turned from an alliance with the proletariat against the feudal aristocracy to making their peace with the right against the forces rising against them on the left. In the same way, the Machist interpretation of Marxism may have brought back a subjectivism superseded by the Marxist unity of subject and object, but it did reflect very real epistemological problems raised by the new level of development of scientific knowledge. It called attention to questions that needed to be posed and answered in a new way, on the basis of new knowledge and at a new level of sophistication.

This in turn raises many problems of historiography with which I had to grapple in writing this book and I have come to believe that it is vital to break from received patterns of interpretation in thinking about the history of Marxism:

> 1. from seeing it as developing in an unproblematic straight line from its founders to today's Soviet textbooks of dialectical materialism, with every query along the way branded as an assault on the integrity of the tradition;
> 2. from thinking it would have been an unproblematic straight line but for the "cult of personality";
> 3. from grasping at the "classics of heresy,"[2] reading their propositions back into the works of Marx and writing off the rest as sheer dogmatism;
> 4. from taking selected texts, which are given an exceedingly forced "reading,"[3] and making these entirely constitutive of what counts as Marxism, and dismissing any consideration outside that framework as "historicist";
> 5. from throwing the whole lot together as the "illusion of the epoch."[4]

The history of Marxism insofar as it is written at all is, for the most part, written along the above lines. There is, to be sure, a healthy tradition in Marxist historiography, represented by such writers as Eric Hobsbawm, E.P. Thompson, and Christopher Hill, though little in the field of philosophy or philosophy of science. Bearing on this problematic under study here, there is very little that is relatively free of the above patterns of interpretation, the most notable exception being Loren Graham's *Science and Philosophy in the Soviet Union* (London, 1973). Graham, although not a Marxist, has given an extremely fair assessment of Soviet philosophy of science in a sympathetic but critical vein.

But almost all comprehensive or partial histories of Marxism bearing on philosophy and philosophy of science fit into the above categories, mostly into categories three and five. George Lichtheim's *Marxism: A Historical and Critical Study* (London, 1961) fits into the third category and only really covers the period to 1918, with only a sketchy and heavily caricatured concluding section dealing with the subsequent period. Gustav Wetter's *Dialectical Materialism: A Historical and Systematic Study of Philosophy in the Soviet Union*, complete with *imprimatur* and *nihil obstat*, falls into the fifth, as do a number of other such works written by neo-Thomist sovietologists, though it must be said that these are valuable as source material, the standards of scholarship being quite high, despite premises that rule out the possibility of sympathetic analysis. The most ambitious and seemingly comprehensive work to appear has been Leszek Kolakowski's three volume *Main Currents of Marxism* (London, 1978), a curious blend of the thinking associated with the third and fifth categories, reflecting the author's own growing dissociation with Marxism. The perspective from which he now seems to judge it all has been aptly described by Wartofsky in his review: "Between God, despair, and peaceful meliorism, Kolakowski offers no decisive choices, though he flirts with all three."[5] Whatever the merits of Kolakowski's work, and merits there are, it suffers severely from his inability to give a fair account of positions with which he disagrees. There are also significant gaps that make it far from comprehensive. So too with David McLellan's *Marxism After Marx* (London, 1979) which falls roughly into the third category and which is lacking in proportion and perspective and full of omissions and interpretations somewhat off the mark.

As for those connected with categories one, two and four, there are inhibiting factors against writing such histories. In the case of the first and second, there is simply too much material in existence contradicting these interpretations and such efforts that forge ahead despite this are a combination of evasion or blatant fabrication. The most crass example was the famous Stalinist *History of the CPSU(B): Short Course* (Moscow, 1939). A more recent, toned-down, example is Mikhail Iovchuk's *Philosophical Traditions Today* (Moscow, 1973).

The fact is that the history of Marxism has taken certain turns that present special problems in the matter of Marxist historiography. The history of Marxism is for many reasons dominated by the history of the communist movement. While noncommunists, for the most part anticommunists, have found this history almost impossible to comprehend, producing extremely distorted accounts of it, communists themselves have not helped much. For the stark truth of the matter is that communists have found it exceedingly difficult to be honest about their own history.* Chapters four and five will try to give some indication why. In all events, precedents were set in the most deplorable degradation of intellectual life: Stalin's denunciation of "archive rats," followed by a wave of denunciation of "rotten liberalism" and "bourgeois objectivism," purges, false accusations, insincere self-criticisms, arrests, executions, books disappearing overnight from bookshops and library shelves, names becoming unmentionable, photographs being doctored, NKVD requisitioning of archives, blatant lying. The days of such massive purges, arrests and executions may be over, the days of NKVD destruction of archives may be over, but the legacy of which these were a part is with us still. There is still much evasion and deceit. Casting its shadow over the academic life of the socialist countries and the intellectual life of some of the communist parties, there is the official history of Marxism, shallow, schematic, triumphalist, full

---

* It must be said, however, that Western European communists have made serious efforts to overcome this in recent years.

of semiceremonial phrases, empty jargon, hollow self-praise. Below it, the real history of Marxism, in so far as it is known, leads a subterranean existence. Names, unmentioned in public, are whispered in quiet places. The story can be told in bits and pieces, but it cannot be published.

Such books as Roy Medvedev's *Let History Judge* (London, 1976) and his brother Zhores Medvedev's *The Rise and Fall of T.D. Lysenko* (NY, 1969) have come as a breath of fresh air, given prevailing conditions, in that they have spoken loudly of much that is only supposed to be whispered and have faced the difficult personal consequences of doing so. Although admirable books in many respects, they do not escape the assumptions of the second category, putting an exaggerated emphasis on the personality of Stalin and failing to give due attention to other operative factors.

As to the fourth category, the Althusserian trend is very ambiguous about the nature or the value of history, in so far as it recognizes it exists. Upon unraveling a complex historical argument, I was once answered by an Althusserian: "There is no such thing as history; there are only books on shelves." At the time it left me speechless, though I must admit that all the arguments I thought of on my way home to justify what I had thought needed no justification deepened my sense of historicity. Even analytic philosophers, whom I had long since judged shockingly lacking in a sense of historicity, at least (if pressed), will admit that history *exists* and might even be nice to know (however inessential to *philosophy* in their sense of the term). As to Louis Althusser himself, he cannot outline the history of Marxist philosophy, because of a "symptomatic difficulty," for philosophy has no history:

> Ultimately, philosophy has no history. Philosophy is that strange theoretical site where nothing really happens, nothing but this repetition of nothing. To say that nothing happens in philosophy is to say that philosophy leads nowhere because it is going nowhere.[6]

He does admit that there is a history of the sciences and that the lines of the philosophical front are displaced according to the transformations of the scientific conjuncture, but this is "a history of the displacement of the indefinite repetition of a null trace whose effects are real."[7] It is a trend of wide influence today, with Barry Hindness and Paul Hirst starkly declaring:

> Marxism, as a theoretical and political practice, gains nothing from its association with historical writing and historical research. The study of history is not only scientifically but also politically valueless.[8]

And this, may I remark, in a book about precapitalist modes of production. My advice to anyone wanting to know about precapitalist modes of production or the history of Marxist philosophy or anything else for that matter is to look elsewhere. What attempts have been made by Althusserians to look at what the normal run of the species considers to be historical matters have, not surprisingly, been singularly unilluminating. An example bearing on the problematic of this book is Althusser's treatment of the debates in pre-revolutionary Russian Marxism in *Lenin and Philosophy* (London, 1971). Unfortunately, the history of Marxism is as little known by today's Marxists as by non-Marxists, with one Marxist recently finding it necessary to make the point that pre-1960s Marxism was no "pre-Althusserian dark age."[9]

As to my own approach to the historiography of Marxism, I have already revealed much and my chapters reveal more, but let me state clearly with regard to the interpretative categories set out above that:

1. While I feel I can trace a certain mainstream flow of the Marxist tradition, I find it to be far from an unproblematic straight line and I believe the various currents diverging from it must be analyzed respectfully and seriously as 'highlighting very real problems.* Moreover, I have defied the existing conventions, mentioned the unmentionable names, delved into delicate and difficult matters that others believed should be let lie. I have done so regretfully, even sorrowfully, for I could take no joy in the self-inflicted tragedies of the communist movement, as do anticommunist writers who are, for the most part, the only ones who write about such things. But, shaken though I sometimes found myself, I could not turn the other way, for I disagree totally with the premises underlying the tradition of sacrificing truth to "partisanship," in the name of which so many crimes against science and against humanity have been committed. The only justification for socialism can lie in arguing for its truth and its humanism. If so, truth, morality, and partisanship should coincide. Indeed, there have been truthful and moral communists, such as Gramsci and Caudwell, and no less partisan (in fact more so) for that.

2. I do not think this history would be an unproblematic straight line but for Stalin. While I admit his shadow looms very large over it and I have tried to assess his role in it, he was after all only one man and there was much happening.

---

* Parenthetically, let me explain that I have used the word *Marxist* in a fairly liberal sense, at least on the historiographical level, including virtually everyone who called themselves Marxist and situated their problematic within the Marxist tradition. I have, however, on the philosophical level, tried to show the inconsistencies of many such positions and the discontinuity of certain of their premises with certain features essential to Marxism as a distinctive intellectual tradition.

3. I take exception to the tendency to draw a sharp line between Marx, Lukács, Korsch, and Gramsci on the one side, representing "creative" Marxism, and Engels, Lenin, Stalin, and a mostly nameless coterie as embodying "dogmatic" Marxism on the other. I have argued against the tenuous interpretations of Marx and the caricatures of Engels and Lenin upon which this trend is based. I have dealt respectfully, but critically, with Marx, Engels, and Lenin, and I have tried to set the assessment of their philosophical contribution upon a firmer basis. As far as subsequent Marxist thinkers are concerned, I think that good and not-so-good philosophers can be found on both sides of this arbitrary divide. To look at the Comintern period, for example, there can be no doubt that Gramsci towers over Stalin and Mitin. However, it is also my opinion that Caudwell towers over Korsch, yet there is a great fuss made over Korsch while Caudwell is forgotten. Korsch's political dissent, more than the philosophical merit of his writings, is, in my opinion, responsible for his greater prominence than Caudwell's, who died as a Communist Party member.*

4. To the Althusserians, I must confess to being an unrepentant historicist.** My argument is quite simply that we cannot separate human thought from the movement, the flow, the context of human thinking without thoroughly distorting what it actually is. The unfolding of history has formed us as what we are and shaped our modes of thought and there is no understanding any of it without a sense of the process that has brought it into being.

5. Far from believing Marxism to be the illusion of the epoch, I believe that, however problematic it has become and whatever criticisms may legitimately be laid at its door, there is still a point to Sartre's statement that it remains the unsurpassed philosophy of our time.[10] Even if it be surpassed, and I am prepared to think that it may be, we shall still owe it much. For the time being, such features as its comprehensiveness, its coherence and its orientation towards science still recommend it beyond any of its contenders so far.

It is my view that the history of Marxism needs to be seen as a complex and intricate story of persons, ideas, and events; as a process in which philosophical ideas have emerged and contended with one another in a complex and

---

* There are other factors, I admit. After all, Gramsci too died a Communist Party member. Other factors also require analysis, such as why Italian Marxism has made far more of its own history than British Marxism.

** I realize this term is a slippery one, with a wide range of usage. I use it to designate thinking pervaded by a deep sense of temporal process, which insists upon the value of genetic, historical, and causal explanation, though not in opposition to structural, logical, or systemic explanation. So specified, it is a way of thinking antithetical to the Althusserian way, although it is not Althusser's own rather eccentric use of the term.

multifaceted interaction with the social, political, economic, and scientific forces at work at any given time. It needs to be viewed with a sensitivity to the passionate striving, to the astute analyses, to the intelligent and enduringly valid syntheses it has brought forth and, at the same time, with a willingness to look full in the face upon the dark side of it and to realize how and why it is that those who love wisdom need to walk warily, even among those who are supposedly of their own kind.

Regretfully, I am hard put to cite examples of work in the history of Marxist philosophy that meet these criteria—works that are comprehensive and sweeping, striving to bring into play the full network of interconnections, alert to the overall patterns of development, yet without sacrificing concreteness, thoroughness of research, or rigor of analysis; works that are sympathetic but critical. This, at any rate, is the challenge I have set for myself. It is for others to judge the extent to which I have met it. No doubt, I have fallen far short of what could have been done, given the complexity of the problematic with the immense possibilities it opens up and given the fact that it is nearly uncharted territory.

Not altogether uncharted, however. Although there is no overall history of Marxist philosophy of science, there are some rather good histories of certain episodes of it. The only area in which major works are already in existence, however, is in the history of Soviet philosophy of science, the most notable works being David Joravsky's *Soviet Marxism and Natural Science* (London, 1961), which covers the period from 1917 to 1932, and *The Lysenko Affair* (Cambridge, Mass., 1970); and Loren Graham's *Science and Philosophy in the Soviet Union* (New York, 1973) the weight of which is on the postwar period, but which nevertheless contains much valuable material on the prewar period. These works are invaluable sources, adhering to the highest standards of scholarship, and I have not hesitated to draw upon them in my fourth chapter. However, even in the field of Soviet philosophy of science, where such significant work has already been done, there are gaps to be filled and interpretative issues to be clarified. I believe that many of the facts and ideas set out by Joravsky should be seen in a somewhat different light. While basically I agree with Graham, I prefer to give greater weight to the political context in dealing with philosophy of science than he does.

Regarding my other chapters, there are no such major parallel works. However, there are, in every case, in addition to the primary sources, a variety of types of secondary sources that touch on certain aspects of the chapter. About the work of certain authors, there is a voluminous body of secondary literature, but little or nothing about others. I have undertaken to review this secondary literature where it exists and to take a specific interpretative position. In cases of an extensive body of literature, where an exhaustive survey would be impossible, I have attempted to select in a representative manner, so as to indicate the overall range of interpretative patterns. In matters of ongoing controversy, such as that of the Marx-Engels relationship and that of the relationship of the Oparin-Haldane hypothesis to Marxist philosophy, I have weighed the evidence and argued a very definite position.

In every chapter, my general method has been to sketch the relevant historical background, to outline the terms of the philosophical debates, to show their connections to other debates, to evaluate the various philosophical arguments put forward, to determine their significance within the history of Marxist philosophy of science as a whole, and to review critically the secondary literature. But that is only the roughest and most superficial characterization of the book's method. Working through it in detail and in depth has forced me to wrestle with a host of exceedingly intricate and thorny methodological issues.

Perhaps the most important of these underlying issues concerns the relationship between the philosophical and the historical dimensions of this work. This is not an easy matter to discuss. It would be difficult enough if I only had to explain myself to those who believe as I do in the historicity of philosophy, and I mean a deep, internal, and constitutive historicity, rooted in the historicity of rationality itself. To those who know, and know deeply, why the history of the philosophy of science is essential to the philosophy of science, I would have problem enough defending what I have done and how I have done it. However, living as I do in a world full of analytic philosophers and Althusserians, nothing can be taken for granted, and if I wish to defend my method, my task is far more complicated. I fear, however, that I may be talking across insuperable barriers in trying to explain myself to those to whom the slightest hint of historical scholarship automatically lays one open to the charge of being "insufficiently philosophical" or to those for whom a "symptomatic difficulty" already declares the end result to be a nullity.

Most philosophers today are utterly oblivious of the fact that philosophy or science is historical, except in the most trivial and superficial sense. Even when they do look at the history of philosophy or science they do so in such a thoroughly ahistorical and noncontextual way, that anybody could virtually have said anything at any time. In philosophy, the ideas of Plato, Aristotle, Descartes, Hume, Kant, Hegel, Carnap, and Quine are treated as discrete and interchangeable units, virtually independent of time and place, and generated in an autonomous activity with no necessary or integral connection to anything else. If temporal sequence or economic, social, political, or scientific developments are mentioned, it is only in an accidental or circumstantial way. It is exceedingly rare today to find a philosoher with a real sense of the flow, the movement, the *process* of the history of philosophy or science and with a vivid sense of its integral connection with economics, culture, politics, and science.

This being so, the whole debate over the relationship of the philosophy of science to the history of science among philosophers of science has been an extremely constructive one, though its inadequacies are painfully apparent in the strained attempts to set up awkward interactions between an ahistorical philosophy of science and a nonphilosophical history of science. Even while making a stab at overcoming this syndrome, Kuhn declares:

> Subversion is not, I think, too strong a term for the likely result of an attempt to make the two fields into one. . . . The final product of most historical research is a narrative, a story, about particulars of the past. . . . The philosopher, on the other hand, aims principally at explicit generalizations and at those with universal scope. He is no teller of stories, true or false. His goal is to discover and state what is true at all times and places rather than to impart understanding of what occurred at a particular time and place.[11]

The contrast is falsely drawn and the subversion he fears would be, in my opinion, no bad thing. I see no reason why the philosopher should not be a teller of stories and be fully a philosopher in doing so. Certainly coming to terms with the history of philosophy and the history of science is no diversion from being a philosopher and writing such history is a thoroughly philosophical task.

Actually, there is far greater philosophical rigor called into play in the real struggle to make sense of the turbulent complexity of the world of historical experience than in shallow formal schemes of the hothouse world in which the analytic philosophers build their careers. At any rate, in the practical task of writing this history, I felt fully a philosopher in doing so. So many times I found that historiographical problems came down to epistemological problems, which in turn came down to ontological ones. While on the surface of it, I sometimes experienced a tension between narrative and thematic considerations and I often found the task of keeping the thread of the story going, while doing full justice to the theoretical issues involved, a bit tricky to organize, it was a creative tension. On a deeper level, as I tried both to tell the story and to assess its philosophical significance, two tasks that I found to be thoroughly intertwined at every step along the way, I came to realize more fully just how deep and multifaceted are the interconnections between considerations involved in giving a historical account and those involved in putting forward a philosophical analysis. Lukács once expressed one dimension of this quite starkly:

> Without a *Weltanschauung*, it is impossible to narrate properly or to achieve a composition which would reflect the differentiated and especially complete variety of life.[12]

It is true. In any case, it is quite clear to me that my own world view has in deep, complex and decisive ways shaped my telling of this story. I acknowledge this straight out. Although I have tried to be fair even in relation to positions with which I disagree, I have been no detached observer. Although I have not used the first person in my chapters, except in footnotes, or given an explicit *declaration de foi* at every turn, I have nevertheless made my position, which in many ways cuts across the lines of historical debate, plain. Because I believe my philosophical assumptions to be justified, I believe this has enhanced my historical account. Those who do not share my assumptions will not think so. So be it, for it cannot be otherwise. Nevertheless, those who share my assumptions will inevitably see ways I could have fulfilled my task better and hopefully those who do not will not find me too wanting in standards of scholarship and argumentation.

It is, of course, one thing to analyze the integral connection of philosophy to history and context, and another to implement such an analysis in terms of philosophical and historiographical technique. Within my own terms of reference, this has meant a constant alertness, not only to all the forms of interaction in the shifting nexus of science, philosophy, and politics, but to all the levels of interaction within each of these forms. It has meant the need to show in each period and set of circumstances the effects of science on philosophy, of philosophy on science, of science and philosophy on politics, of politics on science and philosophy. But within each of these categories, there are many levels of effects.

For example, as to the effects of politics on philosophy of science:

There is first of all the most obvious level—the effect of overt political events on the fate of individual philosophers and scientists. Thus, the tradition was deprived of any further work on the part of Guest or Caudwell when they died fighting in the Spanish Civil War, of Politzer or Solomon when they were executed by the Gestapo, of Gramsci when he died in prison, of Hessen or Uranovsky or Vavilov when they were swept up in the purges.

On a slightly different level, a bit more complicated to assess, there are the effects of political transformations on the development of the disciplines of philosophy and science and the relations between them. Thus, the October Revolution opened to Marxism the resources of institutional power, bringing into the discussion of philosophy of science a great vigor and enormous scale that was highly valued as integral to the task of building a new social order, and then crushing it through shortcircuiting rational procedures with clumsy or vicious attempts to settle philosophical and scientific questions by administrative measures. There are many factors to be taken into account to comprehend the shift in Soviet intellectual life as the 1920s passed into the 1930s, such as the impact of the inauguration of the first Five Year Plan upon the changeover from bourgeois to "red" specialists, the function of the concept of "proletarian science," with many new complications arising; but it was even then a society which at one and the same time put an unprecedentedly high social value upon philosophy of science, even while they brought to bear the dark and destructive forces that would overpower it. Also, roughly on the same level, are the various shifts in the whole style of the intellectual life of the communist movement corresponding to the various shifts in Comintern policy, especially the shift from the sectarian third-period war on "social fascism" to the more expansive period of the Popular Front.

There are many further levels, with the various primary and secondary sources drawing all sorts of connecting links between political factors and philosophical and scientific ones, sometimes quite to the point and sometimes far from it, but all of them showing the truth of Haldane's discovery that, while the professors may leave politics alone, politics will not leave the professors alone.

Finally at the deepest level, there comes the problem of discerning just how the prevailing political milieu enters into the philosophical thought process itself and into the scientific thought process itself. Politics, itself rooted in and connected to economics in exceedingly intricate ways, mediates in the formation of ideologies* that bind fundamental assumptions to class interests in exceedingly intricate (and hidden) ways. I do not believe that intellectual history follows upon the development of economic and political factors in simple and uncomplicated ways. The connections can be of an extraordinarily complex and subtle nature, with all sorts of complications arising from overlapping and countervailing tendencies. It is an area beset with difficulties, for it has been so littered with facile and even malicious tendencies to assert direct and simple one-to-one correspondences between ideas and classes. The left's habit of writing off people and premises as "bourgeois" or "petty bourgeois" has all too often provided a refuge for those who have had neither arguments nor conscience against those who have and it has not been grounded in a proper understanding of the real class basis of patterns of thinking. It is an area in which it is necessary to proceed with caution, but at the same time it is an area that must not be avoided.

At all events, I have tried to be sensitive to all the interconnections I could discover, both overt and subtle, both direct and indirect, both immediate and epochal. I have probed to uncover the relationship of various thought patterns to various class forces and to various stages of development. I have tried to

---

* I use the term *ideology* to indicate a system of interrelated attitudes, based on a specific system of values and ties to the interests of a specific social force. I am aware of the fact that Marxists are split on the use of the term *ideology* and I believe the more general, nonpejorative sense of the term is preferable to the one that associates it with false consciousness.

understand what it is about the mode of existence of the bourgeoisie as a class that ties the social order under its hegemony to patterns of thought that swing between reductionism, dualism, and idealism. I have also tried to understand what it is about identifying with the cause of the proletariat that puts the stress on totality and makes materialism seem most appropriate philosophy. Marxism itself embodies its own explanation of this, being not only an integrated world view, but a critique of a social order that structurally inhibits the formation of an integrated world view and a theory of a new social order that is the necessary social matrix for an integrated world view. It renounces any disjunction of philosophy and politics as symptomatic of the intellectual fragmentation endemic to the capitalist mode of production and holds that the integration of all spheres can be achieved only in a new consciousness tied to a new mode of production.

This being the case, no history of Marxist philosophy of science could be anything but superficial if it could not give some account of the historical convergence between the critique of dualism, reductionism, and idealism on the one hand and the critique of capitalism on the other; between a tradition striving to set out an integrated world view and a movement towards a classless society. The Marxist argument that the historical convergence is grounded in an integral and logical connection is a forceful and fascinating one. But the logic can only manifest itself historically and the argument can only be filled out in and through the telling of the story.

NOTES

    1. Marx Wartofsky, "Politics, Political Philosophy and the Politics of Philosophy," in *Revolutions, Systems and Theories*, ed. H.J. Johnson, J.J. Leach, and R.G. Muehlmann (Dordrecht, 1979), p. 144.
    2. The older Lukacs used this phrase in objecting to the tendency to take his own early work and that of others as offering the solution to the present-day problems of Marxism. Cf. the interview with Lukacs in *New Left Review*, no. 68 (July–August 1971), pp. 55–56.
    3. The term is given a special meaning by Althusser and his followers. Cf. Althusser and Balibar *Reading Capital* (London, 1970).
    4. H.B. Acton, *The Illusion of the Epoch* (London, 1972). (Originally published, 1949)
    5. Marx Wartofsky, "The Unhappy Consciousness," *Praxis International* (October 1981): p. 291.
    6. Louis Althusser, *Lenin and Philosophy* (London, 1971), p. 56.
    7. Ibid., pp. 62–63.
    8. B. Hindness and P. Hirst, *Pre-Capitalist Modes of Production* (London, 1975), p. 312.
    9. Robert Gray, Review of *Rebels and Their Causes: Essays in Honour of A.L. Morton*, ed. Maurice Cornforth, in *Comment*, 23 June 1979.
    10. Jean-Paul Sartre, *Search for a Method* (New York, 1968), p. 30.
    11. Thomas Kuhn, *The Essential Tension* (Chicago, 1977), p. 5.
    12. György Lukács, *Marxism and Human Liberation* (New York, 1973), p. 126.

CHAPTER 1

THE FOUNDERS: Engels, Marx, and the Dialectics of Nature

The Nineteenth Century

The nineteenth century was a time of magnificent turbulence, marking a crucial turning point in the history of human thought about nature. The whole context, within which the problematic had until then been set, was to be dramatically transposed.

As the century opened, minds were still reeling beneath the blows to traditional patterns of thinking that had been struck during the great upsurge of revolutionary France. Paving the way, the French encyclopedists had fought to force the murky superstitions of the ages under the pure light of reason and thus to shatter forever the basis of human servility. The growth of science had strengthened their efforts immeasurably by giving natural answers to natural questions, undermining the foundations of all supernatural ones. It seemed that religion and all forms of idealism had been given the deathblow by this powerful and exuberant materialism.

But idealism proved to be more resilient than had been anticipated and indeed it reached unprecedented new heights in the grandiose system of Hegel, with a new emphasis on temporality, on development, on process. This idea of evolution, "in the air" so to speak long before the voyage of the Beagle, this totality and all-embracing historicity, answered a need not met by the materialism of the previous century, a materialism that relied on analysis at the expense of synthesis, that had focused on substance to the neglect of process. However, unlike other post-Enlightenment philosophies, which went the way of irrationalist reaction against the Age of Reason, Hegel was rationalist in his way and affirmed his continuity with the "magnificent sunrise" that had gone before. As Hegel saw it: "A sublime emotion swayed men at that time, an enthusiasm of reason pervaded the world."[1]

In time, Hegel's philosophy gave rise to two discordant reactions. The right-wing Hegelians seized upon his dictum that "what is real is rational" as an apologia for the existing order of things and for a time it became a force for the most oppressive forms of conservativism and traditionalism in the hands of the official defenders of the Prussian state. The left-wing Hegelians, however, emphasized the radical negativity of the dialectical process, seeing it as opening the way for a naturalistic and historical critique of origins, especially in religion and in politics, and issuing in the atheistic humanism of Feuerbach and the communism of Hess. Left-wing Hegelianism, most especially as it was taken up by Marx and Engels, led back again to materialism and converged with the new stage being reached by the natural sciences.

It seemed at this time that the boundaries of human knowledge were being pushed back further and further by the hour. Every day brought fresh news of new breakthroughs, and there was a new wave of excitement about science. It seemed omnipotent, bringing mankind to the threshold of a golden age. It was also proliferating and changing at a dizzying pace, transforming itself from a wide-ranging activity, engaging perhaps a hundred or so amateurs and polymaths, to an increasingly professionalized and specialized enterprise, with a mushrooming apparatus of laboratories, journals, learned societies, sources of finance, technical applications, and administrative infrastructures. Not that it brought only light and glory. There were some shocks in store, and there was much confusion and bewilderment as the clash between the old and the new played itself out.

Sciences such as biology and geology were pushing ahead with startling results. In the fossils and the rocks, the secrets of the past were being revealed and the story of man was turning out to be much longer and much stranger than anyone had been led to expect. It all came to a head with Darwin and the furor over the theory of evolution. In terms of historical impact, it was Darwinism that most forcibly and explicitly placed man and all of his ideas and institutions squarely within the setting of the natural world, giving them a natural origin and a natural history. It brought a glimpse of a time so far back as to be beyond all previous imagination, a time of darkness and terror, a time when our own kind wandered blindly. It was a startling experience to trace backwards the steps of one's ancestors and to see them suddenly transmuted into the steps of a beast, to discover them in the very process of becoming men, to learn of them stumbling around and beginning to name stones and gods. It became clear, despite all who burst into a frenzy denying it, that human origins were in the world of matter, that man did not spring full-blown from the hands of a creator, but emerged gropingly and painfully in the course of a long evolutionary history.

The Emergence of Marxism

It was no accident that Marxism made its entry onto the historical stage at the same historical moment as Darwinism. At the funeral of Marx, Engels made much of the connection between Darwinism and Marxism, stating that just as Darwin had discovered the law of development of organic nature, so Marx discovered the law of development of human history. This was one line of connection to be sure, but there was another, more directly connected with Engels's own work than with Marx's. It was not simply a matter of doing in the sphere of history what Darwin had done in the sphere of nature, for even in the sphere of nature Marxism extended and complemented Darwinism, pushing the conclusions of Darwin further in the direction of a new philosophy of nature. Some were inclined to shy away from elaborating on the philosophical implications of the theory of evolution, but Marx and Engels were not. They insisted that the new discoveries not only forced any intellectually honest person to be materialist, but demanded a dialectical interpretation of this material reality as well.

More than any of their contemporaries, Marx and Engels strove to comprehend the deeper implications of the forces that were stirring in their times. They were in tune with the prevailing materialism inspired by confidence in the great advances of the natural sciences, but they also saw that the natural sciences had reached a new stage that demanded that materialism be brought to a new stage as well. Evolution needed to be integrated into the very structure of materialism. Whereas the eighteenth century had seen the world as a timeless order of nature, the nineteenth century perceived the world as a temporal and developing process. The spirit of the age was pervaded by a deep sense of time and historical process, a new sense of the mutability of all that existed. But the *Zeitgeist* had not yet achieved philosophical coherence.

Nevertheless, it was coming. Looking back into history, Marx and Engels believed they had discovered the underlying pattern, the inner dynamic that relentlessly pushed history forward. The key was in the labor process, in the mode of production, in the means men employed to procure their material existence. From the economic basis of a given society, there arose a legal and political superstructure and corresponding forms of social consciousness. All of these elements were inextricably bound together within the historical whole. The belief that the economic, political, cultural and intellectual spheres were separate and independent of each other was the most fundamental of illusions. At crucial turning points in the historical process, the existing relations of production became a fetter upon the development of the forces of production, necessitating a revolution in the mode of production and throwing existing forms of political life and intellectual consciousness into crisis. Such a time was coming for capitalism, they believed, with the growing contradiction between the private ownership of the means of production and the social organization of labor, playing itself out in its two great classes. Its dominant class, the bourgeoisie, having overthrown the feudal aristocracy, had to create the proletariat, whose destiny it was in turn to overthrow it. Although history had been heretofore a history of class struggle, the mission of the proletariat was to create a classless society: in liberating themselves, they would liberate all. The communist society of the future would usher in the reign of universal human emancipation: the overthrow of private ownership of the means of production, the withering away of the coercive state, the end of the alienating division of labor, the abolition of antagonistic social relations, the transcendence of all forms of philosophical dualism.

Essential to this was the formulation of an integrated world view. Although the pivotal insight of the communist vision, the materialist interpretation of history, was articulated by Marx and Engels in the 1840s, it took the remaining decades of their lives to fill it out, to ponder its full implications, to weave all the strands together in the right way. Both Marx and Engels considered science a crucial factor in this synthesis. The post of May 30, 1873, gave some indication of the process at work in the development of their thinking. On that day, Friedrich Engels wrote to his friend Karl Marx:

> This morning while I lay in bed the following dialectical points about the natural sciences occurred to me: The subject matter of natural science—matter in motion, bodies. Bodies cannot be separated from motion. . . . One cannot say anything about bodies without motion, without relation to other bodies. Only in motion does a body reveal what it is. Natural science, therefore, knows bodies by examining them in their relation to one another, and in motion . . .

There ensued a series of reflections on current scientific theories, followed by a humorous postscript:

> If you think there is anything in it, don't say anyting about it just yet, so that no lousy Englishman may steal it on me. It may take a long time yet to get it into shape.[2]

These reflections, giving birth to a new conception, that of the materialist "dialectics of nature," came as a result of Engels's previous studies both of the natural sciences and of Hegelian philosophy. It did not arise in a vacuum, but it was new, and a complex and rich tradition in the philosophy of science stems from it.

It is a tradition that has given rise to enormous controversy. The story opens amidst fierce polemics and fierce polemics characterize its development at every step of the way. Moreover, the polemics emerge on every possible level. Engels's ideas first became known through a series of polemics against other trends of his day. All through its subsequent history, the tradition has wrestled with arguments for conflicting lines of development, as well as with arguments rejecting its basic premises, the latter coming both from some claiming to be within the Marxist tradition, as well as from those clearly criticizing it from without.

The Contribution of Engels

But all of the complex controversy refers back at some point to the work of Engels. There has been, over the years, much debate revolving around the assessment of the contribution of Engels to the early formulation of Marxist philosophy, with a fair degree of misrepresentation clouding the discussion in many quarters.

Frequently, Engels has been presented as a secondary figure, always in the shadow of Karl Marx with whom he worked in extraordinarily close collaboration over a period of forty years. Engels has often been pictured as a somewhat shallow, second-rate thinker, whose only real value was in providing considerable financial assistance to the Marx family and in editing Marx's unfinished *magnum opus, Capital*. This picture is, in fact, a crude and cruel caricature, for a serious examination of the work of Engels reveals him to be a serious and original thinker. The process has admittedly not been helped much by Engels's own excessive modesty in evaluating his own role in relation to Marx's. He spoke of himself as having been glad to play "second fiddle" to the "wonderful first violin" that was Marx, and summed up the matter this way:

> I cannot deny that both before and during my forty years partnership with Marx I had a certain independent share in working out the theory. But Marx was responsible for the leading basic ideas—particularly as far as economics and history were concerned—and he put those ideas in their final classic form. What I achieved—apart from work in a few specialized fields of study—Marx could have achieved without me. But what Marx achieved, I could not have achieved. Marx stood higher, saw further and took a wider and quicker view than all the rest of us. Marx was a genius; we others were at best talented. Without him the theory would not be by far what it is today. It therefore rightly bears his name.[3]

There is no reason, however, to think that Marx saw it this way. He seemed to have thought of Engels as in every way his equal and often acknowledged the decisive role Engels had played in formulating the leading basic ideas of their common outlook. Marx, too, engaged in a somewhat affectionate exaggeration, putting it the other way around:

> You know that (1) I get around to everything late; and (2) I always follow in your footsteps.[4]

But there is much more involved in the discussion than the assertion of the secondary status of Engels in relation to Marx. Many commentators indeed go further and insist on drawing the sharpest possible dividing line between Marx and Engels, maintaining that the conception of the dialectics of nature is a vulgar distortion of the ideas of Marx and that Engels alone is responsible for the philosophy that has become known as dialectical materialism. There have been two distinct lines of approach in the voluminous anti-Engels literature of recent years. Although they are contradictory, both are sometimes found in works by the same author. The first approach accuses Engels of positivism, scientism, reductionism, a crude and mechanistic form of materialism. The second castigates him for archaic Hegelianism and obsolete idealism. This literature has come to constitute something of an academic orthodoxy in certain quarters, though it has by no means gone unchallenged. The emergence of this literature is itself a phenomenon worth analyzing.

But before doing so, and before going into the various aspects of these interpretations of Engels and the problems of the Marx-Engels relationship as it bears on questions of philosophy and the natural sciences, it is first necessary to look at the historical context in which the ideas of Engels evolved and to present the basic concepts pervading his writings on these questions.

Biographical Background

The views Friedrich Engels developed as his outlook matured took him a long way from the world from which he had come. Born into a prosperous business family in Barmen in the Rhineland region of Germany in 1820, the prevailing ethos of his time and place was that of the strict fideist pietism, full of contempt for reason and science, that had spread in Germany as a reaction to the French Revolution. Because of this, Engels was not encouraged to finish his schooling and was constantly under pressure to turn his mind to the devotions of the church and the fortunes of the family business. Resisting both of these pressures, he gave free rein to his active, searching mind and became determined to accept only what could justify itself at the bar of reason. He naturally turned to philosophy, and indeed all his life retained the intensely philosophical frame of mind that often characterizes those who have had to struggle free of a strongly religious background. During the period of his military service in Berlin, he sought the company of the Young Hegelians and identified fully with the radical interpretation of Hegel that was causing such a stir in those days. He found this rationalist, antiauthoritarian milieu a healthy antidote to the irrationalist and authoritarian environment of his home. The ideas of Feuerbach, proclaiming a radical naturalistic humanism, were the rage and Engels and his contemporaries became Feuerbachians. He always acknowledged the crucial role that Feuerbach had played for his generation during their "period of storm and stress." After completing his military service, Engels moved on to Cologne, where he came under the influence of the moving spirit of the radical newspaper *Rheinische Zeitung*, Moses Hess, and at this time began to consider himself a communist.

At this time, he also had a rather unpromising first meeting with Karl Marx, at that time editor of the paper, who was at the point of breaking free from his own Young Hegelian fling. Although they had come to maturity in the Rhineland in the same years, Marx and Engels were from dissimilar backgrounds. Born in Trier in 1818, Marx was descended from a long line of rabbis on both sides and his family atmosphere was one permeated by the values of the French Enlightenment. He was a doctor of philosophy; Promethean in spirit, he saw man as creator, and from his youth proclaimed his hatred of all gods. When the two met again in Paris in 1844, they recognized themselves to be somewhat kindred spirits and began their lifelong collaboration. Together, in the early years, they produced *The Holy Family, The German Ideology,* and the *Communist Manifesto.* Their political and theoretical activities at this time took them from one European center to another: London, Paris, Brussels, and then back to Germany for the 1848 revolutions in which both played an active part. Upon defeat and the onset of counterrevolution, both Marx and Engels settled in England, where they both lived until their deaths. Through their political activities, including the formation of the International Workingmen's Association in 1864*, through their extensive correspondence and their personal contacts, but pre-eminently through their writing, they fought for their ideas within the emerging labor movement and guided the formation of a new generation to take up their cause.

Engels's intellectual interests were extremely wide-ranging, embracing politics, economics, linguistics, military strategy, anthropology, Irish history and many other fields as well. He had throughout his life an extraordinary interest in the natural sciences and was extremely well informed and up to date regarding the state of development of the natural sciences in his time. The British biochemist, J.B.S. Haldane, once said that Engels was "probably the most widely educated man of his day." As he put it:

> Not only had he a profound knowledge of economics and history, but he knew enough to discuss the meaning of an obscure Latin phrase concerning Roman marriage law, or the processes taking place when a piece of impure zinc was dipped into sulphuric acid. And he contrived to accumulate this immense knowledge, not by leading a life of cloistered learning, but while playing an active part in politics, running a business, and even fox hunting.[5]

---

* This became known in time as the First International.

One might be inclined to think, of course, that one who diverted his attention so widely might have the mind of a dilettante, but quite remarkably, whatever Engels did, he seems to have done well. In so many fields, he not only knew well the current state of the question, but shed new light on the matters at issue and actually advanced the field further. Although some of his actual conclusions can now be criticized on the basis of subsequent discoveries, his method of approach was never a superficial one and many of his ideas in many fields have shown themselves to be of enduring value. What also makes Engels an impressive figure is that he was an extremely gifted writer, his style being bright, colorful, and lucid, with his robust sense of humor constantly breaking through, though never detracting from the seriousness of purpose pervading his writings. It was quite in contrast to the opacity and heaviness that has so often characterized German philosophical writing.

Engels's most intense studies of the natural sciences began in the 1850s, during the period in which he was working in the family business in Manchester, though he only brought these studies into play in any systematic way in his writings of the 1870s after retiring and moving to London. The Marx-Engels correspondence shows him over a long period of time engaged in studies of physics, physiology, chemistry, comparative anatomy, and geology. He expressed great enthusiasm, for example, after reading Darwin's *Origin of Species* in 1858 and wrote to Marx, "Darwin, whom I am just reading, is magnificent—there has never been until now so splendid an attempt to prove historical development in nature."[6] It was an enthusiasm Marx fully shared: "Darwin's book is very important and it suits me very well that it supports the class struggle in history from the viewpoint of natural science . . . it not only deals the deathblow to teleology in the natural sciences for the first time, but also sets forth the rational meaning in an empirical way."[7]

Aside from his correspondence, there are three main sources that reveal the development of Engels's ideas on the philosophical implications of the natural sciences.

Anti-Dühring

The first and most influential source in its time was *Herr Eugen Dühring's Revolution in Science,* later known as *Anti-Dühring,* published as a book in 1878 after being serialized in *Vorwärts*, the German social democratic newspaper, the year before. As it was the fullest and most systematic exposition of the Marxist world outlook embracing not only philosophy but political economy and the theory of scientific socialism, it was the book that more than any other played the decisive rôle in forming the ideas of the up-and-coming generation of Marxists. Kautsky later recalled that it was only after the publication of *Anti-Dühring* that they began to think and act like Marxists.

The book was a polemic against Eugen Dühring, a lecturer at the University of Berlin, who had clamorously announced his conversion to socialism and had come to the fore as an exponent of a new brand of socialism in opposition to Marx's. He considered his attack on Marx to be "from the left" and mounted it in a highly pretentious and abusive manner. He delivered himself of his highly speculative theories with an authoritarian air that contemptuously swept aside all previous theories. He scoffed at Marx's "barren misconceptions," "buffoonery," and "philosophical and scientific backwardness," and dismissed his work as "a contemptible mass of language" that was "without any permanent significance."[8] Dühring was an extremely facile writer and became quite the rage in German radical circles. The fact that he was blind and had achieved his position through heroic determination, together with his outspoken radicalism in the face of the university authorities, made him a highly popular figure. In the beginning, even Marx and Engels had been favorably disposed towards him, as he was the only academic to break the "conspiracy of silence" against the first volume of Marx's *Capital* at the time of its publication. But it would seem that his popularity went to his head and he was swept away by a developing megalomania.

At first Marx and Engels thought him not worth answering. Marx referred to him as a "minor scribbler" and called him and his admirers "halfway elements . . . half-mature students and superwise Doctors of Philosophy who want to give socialism a 'superior' idealistic orientation, that is to say, to replace its materialistic basis . . . by modern mythology."[9] Engels was equally scornful and wrote of the "semi-ignorance of half-baked literati" whose eclectic views scantily clad in socialist garb were in effect inimical to scientific socialism.[10]

But as time went on, Marx and Engels became more worried. As various leaders of the German Social Democratic Party fell under his spell, not only Johann Most, but August Bebel, Wilhelm Bracke, and Eduard Bernstein, Wilhelm Liebknecht became increasingly alarmed and repeatedly urged Marx and Engels to embark on a systematic refutation of his theories. Eventually, Marx became convinced that Dühring was posing an serious threat to them in the struggle for intellectual leadership of the party, as well as a threat to the fragile unity that the party had just achieved in adopting the Gotha Program. He wrote to Engels that Dühring's views must be ruthlessly exposed. Engels was reluctant to interrupt the work that he was then doing—a more constructive effort to elaborate Marxist philosophy in relation to the development of the natural sciences. Finally he agreed, though not without complaining to Marx: "It is all very well for you to talk. You can lie warm in bed and study Russian agrarian conditions in particular and rent in general with nothing to disturb you—but I am to sit on the hard bench, swill the cold wine, suddenly interrupt everything again and tackle the boring Dühring."[11]

Dühring was putting forward elaborate grandiloquent theories on everything under the sun, including the natural sciences; or as Engels put it, he "speaks of all possible things and some others as well."[12] Engels considered his system building to be "bumptious pseudo-science" based on extremely defective knowledge and steeped in philistinism, which he was incapable of elaborating "without dragging in his repugnance to tobacco, cats and Jews."[13] Nevertheless, he used Dühring as a foil against which to elaborate the positive features of the communist world outlook, including the developing conception of the dialectics of nature. He claimed to have found his studies in the natural sciences to have been of great service to him in this task, for he had come to the point where he could move on this terrain with reasonable freedom and safety, though he still had to exercise caution. It is for this more constructive element that *Anti-Dühring* retains its significance, for Dühring and his ideas have long since faded into oblivion.

Dialectics of Nature

The work Engels was forced to interrupt in order to take on Dühring was never finished and the manuscript was published posthumously as *Dialectics of Nature*. It was to have been a major work elucidating the philosophical implications of the natural sciences, drawing on Engels's studies of various sciences over a number of years. He began work on it in 1873 and continued with it for three years until breaking off to write *Anti-Dühring*. He took it up again in 1878 and worked at it until the death of Marx in 1883. After this, Engels devoted himself to the mammoth task of editing the remaing volumes of *Capital* and he was never able to get back to his own projected work. After his own death in 1895, the manuscript was kept in the archives of the German Social Democratic Party. Bernstein showed the manuscript, or part of it, to Einstein, who felt it merited publication for its historical value, although he was highly critical of Engels's outdated notions of electricity. In the end, it was acquired from the German Social Democratic Party by the Marx-Engels Institute in the Soviet Union in 1925, was edited by David Riazanov, the director of the institute, and was published in the Soviet Union in both German and Russian in 1927.

The book was a remarkable one, though, as with any unfinished work, it was of extremely uneven quality. Some of it consisted of finished chapters, such as the brilliantly written one on the history of science. Other parts were only fragments, and sometimes the ideas were very sketchy indeed. Some parts of the book are quite dated, discussing controversies that have long been settled, or adhering to theories that have now been superseded. Some statements are most definitely in error, though it must be said that many of Engels's errors were those of the best science of his day. He was hardly alone in accepting the notion of ether, the Kant-Laplace theory of cosmogony, or the belief in the inheritance of acquired characteristics. The fact remains that he supported some of the most promising and progressive scientific theories of this period, such as Mendeleyev's periodic system of the elements, and opposed theories that were obstructing the real advance of science, such as the hypothesis of the heat-death of the universe.

Moreover, Engels sometimes anticipated scientific discoveries of the future. He pointed to the possibility of the existence of matter with no rest mass, and advanced the theory of the decisive role of labor in molding the physical and social forms of human existence. After reviewing the various points on which Engels's ideas are somewhat dated, Haldane observed: "When all such criticisms have been made, it is astonishing how Engels anticipated the progress of science in the sixty years which have elapsed since he wrote."[14] What is, however, of the greatest significance is the methodology and overall conception of the book. Looking back on the years that had passed between Engels's time and his own, Haldane further remarked:

> Had Engels's method of thinking been more familiar, the transformations of our ideas on physics which have occurred during the last thirty years would have been smoother.[15]

Ludwig Feuerbach and the End of Classical German Philosophy

The other major source for unraveling the philosophical thinking of Engels in relation to the natural sciences is *Ludwig Feuerbach and the End of Classical German Philosophy*. After the Danish philosopher, Carl Starcke, had produced a book on Ludwig Feuerbach in 1885, the editors of *Neue Zeit* asked Engels to write a critical review. Engels welcomed the opportunity to give a coherent analysis of the relation of himself and Marx to the philosophy of Hegel and Feuerbach and to settle accounts with the philosophy of their youth.

Although the emphasis in these diverse sources was somewhat different, there was a common outlook underlying them all. This is not to say that Engels had a completely worked-out philosophical system, with no gaps and full answers to every question, but he did have a reasonably coherent method of thinking, a definite philosophical orientation, based on certain fundamental presuppositions that give it a continuing relevance to contemporary philosophy of science.

Contending Trends of the Time

In unfolding his own philosophical orientation, Engels naturally was conscious of the need to situate it within the spectrum of ideas prevailing at the time and to differentiate it from contending views. In the course of doing so, Engels, in fact, specifically addressed himself to the very trends he is today accused by contemporary critics of embracing.

At the time Engels was writing, the enormous achievements of such scientists as Helmholtz, Kirchhoff, Hertz, Maxwell, and Thompson were bringing forth emphatic proclamations that the highest objective of science was the reduction of all laws of nature to the laws of mechanics. It was widely believed that all the rich qualitative complexity encountered in our experience was ultimately reducible to quantitative terms. Engels stood resolutely against the craze to reduce everything to mechanical motion and to quantitative analysis—a retreat to Pythagoras in his opinion—and he held views on reductionism that were extraordinarily enlightened for the time.

These views were elaborated with particular clarity in the course of his criticism of other materialist trends on the scene and were contrasted with those of such popularizers of materialism as Karl Vogt, Jakob Moleschott, and Ludwig Buchner. These were natural scientists whom Engels considered to be reflections of the German "would-be Enlightenment," which missed the spirit and movement of the real Enlightenment. Because they flourished at the time of deepest degradation of bourgeois Germany and official science, he thought them praiseworthy for advocating atheism and respect for science. However, he did not take very kindly to the abuse they directed against philosophy and he criticized the crudity and vulgarity of such conceptions as the famous one that the brain produces thought as the liver secretes bile. Engels argued for the qualitative distinctness of each level of being, while at the same time maintaining the continuity of levels; he admitted that thought would one day be explicable in biochemical terms, but doubted that this would exhaust the essence of thought. He attacked the rigidity of these scientists' thought and the fixity of their concepts, and called for a more flexible and fluid method of thinking.

Engels thought more highly of the French materialists of the eighteenth century. He appreciated them for their vigorous assault on the superstitions of the ages, and though he criticized their mechanism and their sweeping reductionism, he nevertheless realized that this was a progressive philosophy that took materialism as far as it could be taken at the time. (With physics, chemistry, and biology still in their infancy, and therefore very far from being able to offer the basis for a general outlook on nature, mechanics was the most highly developed science at the time.) Engels felt that it was the subsequent advance of science that had made it possible to put materialism on a firmer foundation than was possible in the previous century.

Engels was not so appreciative, however, of the tradition of English empiricism, which he saw as a reactionary trend that had gone to the furthest extremes in emptiness of thought, and had in fact led in the end to the most extreme credulity, fantasy and superstition. Beginning by exalting experience and treating thought with sovereign disdain, it bred a contempt for theory that in the final analysis brought the most sober of empiricists to the most barren of all superstitions. As Engels saw it, proponents of this school of thought imagined that they operated only with undeniable fact, whereas in reality they were bound by obsolete thought and were as credulous towards the outdated notions of their predecessors as they were sceptical towards the progressive ideas of their contemporaries. He felt that philosophical thinking was becoming more and more rigid under the English influence.*

The materialism of Feuerbach was also a decisive influence on the philosophical development of Marx and Engels. In the Hegelian system, nature had become the alienation of the abstract idea. Feuerbach brought nature once again back into its own and put materialism back on the throne. Eventually, however, Marx and Engels came to believe that this too was an inadequate form of materialism, as it was too abstract and too quiescent. Engels stated: "Feuerbach, who on every page preaches sensuousness, absorption in the concrete, in actuality becomes thoroughly abstract."[16] According to Engels, Feuerbach not only remained aloof from all the great political upheavals of his time, but all the epoch-making advances in the natural sciences passed him by as well.**

On the other hand, Engels set his thinking against the various idealist tendencies as well. Of primary importance, of course, was the natural philosophy of Hegel, which he criticized for trying to force the objective world into the framework of subjective thought and for not conceding to nature any development in time. The Hegelian philosophy of nature, however, not only contained "a great deal of nonsense and fantasy," but also "many fruitful seeds."[17] He opposed those like Vogt who simply scoffed at it without acknowledging its historical significance. The great merit of the Hegelian system, Engels said, was that it represented, for the first time, the whole world—natural, historical, intellectual—as a process, and that it traced out the internal connections that make it a continuous whole. This made Hegel, in

---

* Here Engels was far too dismissive of English empiricism and failed to take into account the progressive impulses of its origins. He was, however, right to argue that its antirationalist bias often led to more extreme forms of irrationalism.

** Engels's criticism of Feuerbach was not entirely justified. While not so politically engaged as Engels, nor so close to the natural sciences as Engels, he did join the SPD in 1870 and exhibited a sustained interest in the natural sciences. (For a fuller treatment of the philosophy of Feuerbach that takes these factors into account, cf. Marx Wartofsky, *Feuerbach*, Cambridge, 1977).

Engels's eyes, far superior to his empiricist contemporaries. He saw the natural philosophers as standing in the same relation to his "consciously dialectical natural science" as the utopians stood to modern communism. Nevertheless, however great was his acknowledged debt to Hegel, Engels was irreconcilably opposed to any philosophy that asserted the primacy of spirit to nature, for it inevitably assumed world creation in one form or another. And Engels was insistent on the need to explain the world in terms of the world itself.

Coming in the wake of the decline of Hegelianism were other philosophies creating in Engels's opinion an atmosphere of incoherence and confusion. The views of these philosophers were part of what Engels considered to be a relapse, especially "the vapid reflections of Schopenhauer and Hartmann."[18] Neo-Kantianism,* which was all that was left of classical German philosophy and "whose last word was the eternally unknowable thing-in-itself,"[19] he thought a regression—"merely a shamefaced way of surreptitiously accepting materialism while denying it before the world."[20]

So Engels was opposed to vulgar materialism, mechanistic materialism, reductionism, and empiricism, on the one hand, and to idealism, on the other. What then did he see as the alternative?

Materialism

Engels embraced a new or "modern" materialism, which was dialectical, contrasting it with the "old materialism," which was mechanistic; and with idealism as well.

He defined materialism in a very simple way: "The materialistic outlook on nature means nothing more than the simple conception of nature just as it is, without alien addition."[21]** This definition of materialism rested on an equally simple definition of matter:*** "Matter is nothing but the totality of material things from which this concept is abstracted."[22] His assertion of the primacy of

---

* In this, Engels misrepresented neo-Kantianism, in that it tended to eliminate the Kantian notion of the thing-in-itself. Neo-Kantianism is discussed in chapter two.
** Engels's intent in contrasting "nature just as it is" with "alien addition" was to opt for naturalism over any form of supernaturalism. His formulation may seem epistemologically naive, in that we cannot conceive of nature as it is apart from its mediation through the historical evolution of the whole conceptual apparatus of human knowledge.
*** This definition was too simple and begged the question in using a form of the term defined, i.e. "material," in the definition. However, his insight about the assertion of the primacy of matter not being dependent on any notion of prime matter was astute.

matter did not depend on a notion of prime matter. He did not think of matter in terms of a *materia prima*, as a substratum. He made the point quite firmly that the concept of matter was an abstraction and that there was no matter as such apart from definite existing bits of matter. Looking for matter as such was like demanding to see fruit as such, instead of cherries, pears, and apples, or the mammal as such, instead of cats, dogs or sheep.

As to how materialism was to be differentiated from idealism, it was a matter of the relation of spirit to nature. For Engels, this was the paramount question of the whole of the history of philosophy. The answers that philosophers gave to this question split them into two great camps: those in one asserted the primacy of spirit to nature, and those in the other asserted the primacy of nature to spirit. Often the question sharpened into another question: Did God create the world, or has the world been in existence eternally? Engels maintained that idealists, those who asserted the primacy of spirit to nature, always, in the last instance, brought in some force from outside the world to explain the world, some external impulse to set matter in motion. His own contention was that it was the essence of matter to be in motion and that it therefore had no need of a prime mover to begin its motion. He believed that the universe evolved in a natural way by transformations of motion that are by nature inherent in matter and that there was, therefore, no need for recourse to a creator. He argued further that the new advances in the natural sciences, particularly the colossal impact of the Darwinian theory of evolution, had dealt a deathblow to all theological conceptions and had obliterated the last vestiges of plausibility for belief in an extramundane creator.

From the time of the Renaissance, Engels argued, the emancipation of the natural sciences from theology had been under way, though it was still far from completion. This history seems to have captured Engels's imagination and he wrote of it in such phrases as:

> "a time which called for giants and produced giants"
> "only now for the first time was the world really discovered"
> "the dictatorship of the church over men's minds was shattered"
> "the bounds of the old *orbis terrarum* were pierced"[23]

He went on to describe the first episode in the modern struggle of the natural sciences for their right to existence:

> The revolutionary act by which natural science declared its independence was the publication of the immortal work by Copernicus by which he, though timidly and only from his deathbed, threw down the gauntlet to ecclesiastical authority in the affairs of nature.[24]

The struggle, Engels knew, was still far from completion, but he delighted to reflect on how far things had come:

> God is no where treated worse than by the natural scientists who believe in him.... Newton still allowed him the "first impulse," but forbade him any further interference in his solar system.... Secchi only allows him a creative act as regards the primordial nebula.... In biology, his last great Don Quixote, Agassiz, even ascribes positive nonsense to him; he is supposed to have created not only the actual animals, but also abstract animals, the fish as such. And finally Tyndall totally forbids him any entry into nature and relegates him to the world of emotional processes.... What a distance from the old God—the creator of heaven and earth, the maintainer of all things—without whom not a hair can fall from the head.[25]

He was convinced that with the inexorable advance of science this process would advance still further and further and that the materialism implicit in science would become more and more explicit, making idealism increasingly obsolete as the concepts became superfluous.

Dialectics

But others thought this as well. There were, as has been indicated, other varieties of materialism contending for influence in the antiidealist mood prevailing at this time. Engels put the weight of his effort into differentiating between his materialism and theirs, contrasting the new organicist and emergentist form of materialism with the old mechanistic and reductionist form. He did so, however, in a most unfortunate way, setting the issue in terms of a contrast between "dialectics" and "metaphysics," and using the word "metaphysics" in a particularly eccentric way that has been followed by subsequent generations of Marxists* and bringing serious confusion into the development of the tradition through his fixation on the terminology of dialectics. Nevertheless, the point he was intent on making was an important one, for he wished to show the difference between two fundamentally different modes of thinking, even if neither dialectics nor metaphysics were the most appropriate terms for characterizing them.

---

* Not that all have done so uncritically. The British Marxist, Maurice Cornforth, for example, regarded it as unfortunate that Engels used the term in such a unique sense. In its original meaning, it meant simply "after physics" or "beyond physics" and so was the title the editors of Aristotle gave to his treatise on being, which came after his treatise on physics. In his opinion, there was no real basis for the use of the word metaphysics to denote the antithesis of dialectics. Despite his reservations, Cornforth concluded that the use of the term in this way had become so entrenched in the discussion of dialectics that there was no ready alternative (c.f. Maurice Cornforth, *The Open Philosophy and the Open Society*, London 1968, pp. 61–62). Nina Yulina has broken with this convention, using "metaphysics" in the broadest

What Engels called the "metaphysical" mode of thinking, was static and rigid. It operated in terms of hard and fast antitheses, sharp and impassible dividing lines. It saw the world as a series of essentially separate things, only externally related; as a series of essentially static things, only set in motion by an external impulse. It was a mode of thinking that was one-sided and restricted, oblivious to the beginning and the end of things, forgetful of their motion and interrelations, unable to see the wood for the trees. It was a habit of thought discredited by the advance of natural science, for the driving force of the Darwinian idea was that nature had its history in time, that all living things underwent incessant molecular change, that there was no fixity of species. Moreover, in physics, the last true gases had been liquefied, aggregate states had lost the last relics of their absolute character, and the great basic law of motion, the transformation of energy, had indeed put an end to all fixity of categories.

But the "old metaphysics," which accepted things as finished objects, was in line with the natural science of its time, which investigated both dead and living things as finished objects. Up to the end of the eighteenth century, natural science was predominantly a collecting science, a classifying science, a science of discrete and finished things. In the nineteenth century, it had become a systematizing science, a science of processes, a science of the origin and development of things, a science of the interconnections of things, binding all natural processes together into one organic whole. Those surviving habits of thinking connected with the superseded science of the previous century were holding back the advance of science in the new century. They left as their legacy the stubborn tendency to observe natural objects and processes in isolation, apart from their connection with the vast whole, to see them as

sense as a theory of being. Some non-Marxist philosophers have used the term metaphysics in a pejorative sense, e.g., in the modern period both positivists and existentialists have denounced metaphysics in favor of a return to concrete experience. However, recently there has been a reappraisal of metaphysics, with many, even in the Anglo-American tradition, reaffirming continuity with the metaphysics of the ancients, while denying that it can proceed by the deductive method, and affirming that it must be grounded in the concrete situation. The term *metaphysics* is coming to be used as "world hypotheses" based on "root metaphors" (Stephen Pepper) or as a general view of the scope of human knowledge (Stuart Hampshire). For an indication of the terms of this discussion, cf. Kennish and Lazerowitz (ed.), *Metaphysics: Readings and Reappraisals*, Englewood Cliffs, N.J., 1966; D.F. Pears (ed.), *The Nature of Metaphysics*, London, 1962; Ian Ramsey (ed.), *Prospects for Metaphysics*, N.Y., 1961. Only Marxists use the term metaphysical to mean undialectical. There is no sound basis for using the term this way, and it is a hindrance to the interaction between Marxist and non-Marxist philosophers. If Marxists must use the term at all, it should be used as the rest of the world uses it: as synonymous with *Weltanschauung*, as the analysis of the most basic presuppositions about the world underlying all of our other thinking.

constants rather than variables, and to classify them according to fixed lines of demarcation and even polar antagonisms.

What Engels called the dialectical method of thinking, on the other hand, was dynamic and fluid. It put an end to all rigidity and fixity of categories, to all irreconcilable antitheses. It approached the world as a complex of processes, an interconnected totality, with motion immanent in matter and with everything affected by and affecting every other thing. It comprehended things in their essential connection and movement, in their origin and ending, in their wholeness.

But the dialectic was not simply a method of thinking. Indeed Engels's very justification for it as the correct method of thinking rested precisely on the assertion that it was more than this. The reason why it was appropriate to think dialectically, as he explained it, was that reality was dialectical and that a dialectical reality was necessarily distorted by thinking of it any way except dialectically. For Engels, "the dialectics of the mind is only the reflection of the forms of motion of the real world, both of nature and of history."[26] In speaking of dialectical development as prevailing throughout nature, history, and thought, Engels saw the dialectic as the Ariadne's thread binding together spheres that had often been artificially severed from each other by wrong methods of thought. He spoke of the striking parallelism between the laws of thought and the laws of nature, and thought that it must seem strange to those who saw consciousness as something apart from nature or even opposed to nature. But for Engels, it did not seem at all strange, for "the products of the human brain, being in the last analysis also the products of nature, do not contradict the rest of nature's interconnections but are in correspondence with them."[27] He summed up this conception of all the processes of nature systematically interconnected in and through the dialectic in this way:

> In nature, amid the welter of innumerable changes, the same dialectical laws of motion force their way through as those which in history govern the apparent fortuitousness of events; the same laws as those which similarly form the thread running through the history of the development of human thought and gradually rise to consciousness in the mind of man; the laws which Hegel first developed in an all-embracing but mystic form.[28]

This process of the dialectic rising in consciousness in the mind of man was a very complex one. Although Engels believed it was only the stage of development being reached by the natural sciences in his own time that made possible a mature and well-grounded form of dialectical thinking, it nevertheless had a prior history. Engels felt that sometimes philosophy leaped ahead of the actual state of scientific knowledge of its time. It saw the shape of things to come and left the justification in detail to the natural science of the future. It was in this light that Engels seems to have seen the articulation of the dialectical method from Heraclitus to Hegel. They saw "through a glass darkly," so to speak, and were vindicated in Engels's eyes by the revolution in the natural sciences he was experiencing in his own time. The intuitions of the ancient Greeks were being confirmed by modern research.

But in embracing the dialectical method he learned from Hegel, Engels was careful to set out the essential difference between Hegel's idealist interpretation of it and the materialist transposition of it effected by himself and Marx. When writing the preface to a later version of *Anti-Dühring,* Engels reflected: "Marx and I were pretty well the only people to rescue conscious dialectics from German idealist philosophy and apply it in the materialist conception of nature and history."[29]

The main thrust of Engels's argument for dialectics was along essentially materialist lines. The weight of the argument was on the evidence of the natural sciences. He did not simply cite modern natural science in general, but argued the matter by citing particular scientific discoveries. He singled out three great discoveries in his discussions of the dialectics of nature:

1. the discovery of the cell, which demonstrated the unity of the organic world;
2. the discovery of the law of the conservation and transformation of energy, which revealed nature as one continuous process;
3. the discovery of the evolution of the species, which established the natural origins of human history.

Engels believed that the natural sciences were on the point of a crisis in which there were only two choices: either the reign of chaos and incoherence or the achievement of order and coherence through dialectical synthesis. As he put it:

The revolution, which is being forced on the natural sciences by the mere need to set in order the purely empirical discoveries, great masses of which have been piled up, is of such a kind that is must bring the dialectical character of natural processes more and more to the consciousness even of those empiricists who are most opposed to it.[30]

He insisted that dialectics alone offered the analogue, and thereby the method of explaining the evolutionary processes occurring in nature and the inter-connections between one field of investigation and another. It was necessary to understand this in order to overcome the problem that was increasingly presenting itself: the conflict of the results of scientific discoveries with preconceived methods of thinking.

So Engels argued that the discovery of the dialectics of nature was to a degree spontaneous, in the sense that the sheer logic of the development of the natural sciences, which was bursting the bounds of old thought patterns, was simply forcing it. Nevertheless, he thought it best to have the knowledge of the historically evolved forms of dialectical thought:

> It is possible to arrive at this recognition because the accumulting facts of natural science compel us to do so; but one arrives at it more easily if one approaches the dialectical character of these facts equipped with an understanding of the laws of dialectics. In any case, natural science has advanced so far that it can no longer escape dialectical generalization.[31]

Engels believed that, in outlining certain laws of dialectics, Hegel had achieved a conceptual insight of considerable power. Once transposed into materialist terms, Engels felt that this insight provided the key to knowledge in the new era.

The Laws of Dialectics

Engels presented the laws of dialectics as the most general pattern of matter in motion, the underlying assumption being that nature was subject to laws that were more general than those of any one science. These laws were the overall patterns prevailing in nature, in history, and in thought and they governed the path of development by which complex substances, events, or concepts evolved from simple ones. Engels said that the laws of dialectics could be reduced in the main to three:

The first was the law of the transformation of quantity into quality and vice versa. The point was that development was not simply a matter of increase or decrease of what already existed. Development was, on the contrary, a matter of deep transformation, a process in which genuine novelty could emerge in the world. In the process of quantitative change, a threshold was reached and a dramatic leap took place. At certain nodal points new qualities emerged. Engels cited various examples. At a rather mundane level, there was the example of boiling or freezing water, in which there was a gradual rise or decline in temperature, but the point at which the liquid became a gas or a solid was a sudden leap. At a more advanced level, there was the periodic system of the elements discovered by Mendeleyev, in which the chemical properties of elements were shown to be a function of atomic weight, such that an increase or decrease in atomic weight at certain points gave rise to new chemical properties. The idea was that the whole was greater than the sum of its parts. The separate atoms that constitute a molecule had different properties from those of the molecule itself. A large aggregate, taken as a whole, displayed characteristics that were not displayed by its parts. This emphasis on the qualitative distinctness of complex entities evolved from simple ones was the basis of the antireductionism of Engels's philosophy of science. But the emphasis on the quantitative buildup to qualitative change, the evolution that prepared the way for revolution, was important as well, as it was the basis for the rejection of vitalism and mystical interpretations of scientific discoveries.

The second law of dialectics was that of the interpenetration of opposites. This was presented as an explanation of the monumental energy inherent in nature. The basic idea was that, upon analysis, it came to light that matter was not an undifferentiated, inactive, homogeneous mass. There were contradictory aspects or tendencies inherent in everything that existed: action and counteraction, attraction and repulsion, aggregating and segregating processes. It was the interpretation, the struggle of the forces one with another, that gave the internal impulse to development. And it was this process that accounted for the power inherent in matter and dispensed with the need for an external mover to explain motion. It was from this inner tension that the process of development drew its power.

The third law of dialectics was the law of the negation of the negation that described the process by which synthesis occurred. Dialectical movement involved constant regeneration and renewal. Everything carried within itself the conditions of its own annihilation. The old was in the process of dying, while the new was in the process of being born. The new negated the old, replaced the old, while carrying forward certain aspects of it in a new and higher synthesis and in a more vigorous form. Each phase was only a temporary synthesis that contained the seeds of its own supersession and of further development in a new synthesis.*

---

* What Engels was arguing for in the name of the dialectic and the laws of dialectics was something not only plausible, but vitally important: a developmental and integrative way of thinking grounded in a developmental and integrative ontology. However, in tying this project too closely to the notion of dialectic, Engels not only unfortunately tied his insights into an inppropriate Hegelian terminology, but was drawn into a nest of conceptual confusions that have beset the Marxist tradition to this day. The Hegelian terminology of quantity, quality, contradiction, and negation has obscured the explanation of such processes as evolution and revolution, emergence of novelty, tension, reciprocal interaction, internally generated motion, transformation, and synthesis. The formulation of the law of the transition of quantity into quality is particularly quaint and the notion of contradiction is particularly problematic. The argument today associated with Colletti was put to Engels by Dühring: that contradictions properly applied to thoughts, but not to things. Engels insisted otherwise. However, it has never been clear why such blurring of the lines between logical contradiction and ontological opposition of forces should be necessary to guarantee the continuity of that level with other levels. However, even if nature is seen in terms of opposing forces rather than contradictions, the problem remains of whether all development can be explained in terms of opposing forces. (Note continued on page 451.)

The Relation of Philosophy and the Natural Sciences

Engels seemed to believed that two interweaving lines, that is, the mounting empirical evidence and the history of thought, had come to a particularly significant point of convergence. He resolutely opposed any dualism between natural science and philosophy and called attention to the dangers of neglecting one for the other. He attacked the attitude of many scientists reflected in Newton's admonition "Physics, beware of metaphysics." He insisted that it was not possible to get by on the "bare facts"; *thinking* was at all times necessary. After all, atoms, molecules, and such things could not simply be observed under the microscope; they were only discerned by processes of thought. Engels put great stress on the importance of philosophy* and believed

---

* This is the main thrust of Engels's position, though there is a striking inconsistency in Engels's pronouncements on the role of philosophy. In contrast to his spirited defence of the philosophical enterprise and its historical traditions, there are passages in which he spoke of the decline of philosophy made superfluous by the rise of the positive sciences in terms very close to positivism. In *Ludwig Feuerbach*, for example, he spoke of "an end to all philosophy in the hitherto accepted sense of the word ... instead one pursues relative truths along the paths of the positive sciences, and the summation of their results by means of dialectical thinking. At any rate, with Hegel philosophy comes to an end: on the one hand, because in his system he summed up its whole development in the most splendid fashion; and on the other hand, because, even though unconsciously, he showed us the way out of the labyrinth of systems to real positive knowledge of the world" (p. 17). In *Anti-Dühring*, Engels claimed that philosophy had been "sublated," i.e., both overcome and preserved, overcome as regards to its form and preserved as regards to its real content. The new materialism, he said, was "no longer a philosophy at all, but simply a world outlook that has to establish its validity and be applied not in a science of sciences standing apart, but in the positive sciences" (p. 166). In another passage: "As soon as each special science is bound to make clear its position in the great totality of things and our knowledge of things, a special science dealing with this totality is superfluous (or unnecessary). That which still survives, independently of all earlier philosophy is the science of thought and its laws – formal logic and dialectics. Everything else is subsumed in the positive science of nature and history" (p. 36). Worst of all, in *The German Ideology*, co-written with Marx, he gave the analogy: "Philosophy and the study of the actual world have the same relation to one another as onanism and sexual love" (*Collected Works* 5, p. 236). There seems to have been in Engels two lines of reasoning that cannot easily be reconciled, despite the efforts at tortured exegesis on the part of those who treat the classics of Marxism as sacred texts. Nor can Engels's unfortunate formulations on this point be explained away by those exponents of "dialectical" thinking who rejoice in contradictions of whatever sort. In my opinion, the overwhelming weight of Engels's thinking was extremely antipositivist and prophilosophical and these other statements run contrary to the real direction of his thinking. He did in fact see a role for a special science dealing with the totality of results of the individual sciences. Engels was trying to make a vital point: that there must be an end to philosophy unfolding deductively in disconnection from the development of the natural sciences.

that consistency of thought must always be used to help get over defective knowledge. In order to climb the pinnacles of science, it was necessay to assimilate the twenty-five hundred years of the history of philosophy. He felt strongly that the antiphilosophical prejudices of scientists were holding back the advance of science:

> Natural scientists believe that they free themselves from philosophy by ignoring it or abusing it. They cannot, however, make any headway without thought. . . . Hence they are no less in bondage to philosophy, but unfortunately in most cases to the worst philosophy, and those who abuse philosophy most are slaves precisely to the worst vulgarised relics of the worst philosophies. . . . It is only a question whether they want to be dominated by a bad fashionable philosophy or by a form of theoretical thought which rests on acquaintance with the history of thought and its achievements.[32]

Engels warned, however, not only against a one-sided, and therefore distorted, emphasis on natural science over philosophy, but also against the opposite. He sharply criticized philosophers who proceeded down all sorts of speculative paths oblivious of developments in the natural sciences or who arbitrarily forced the new facts into preconceived conceptual moulds. He stressed heavily the importance of proceeding from the given facts. The interconnections were not to be built into the facts, but to be discovered in them, and when discovered, to be verified as far as possible by experiment. In elaborating on this, Engels seems to have anticipated the criticism so often made these days: that the dialectics of nature is a concept arbitrarily foisted upon the world of nature from outside it; that the dialectics of nature is an anthropomorphic projection of human concepts onto nature. Engels's argument was that "nature is the proof of dialectics" and that "modern science has furnished this proof with very rich materials, increasing daily and thus has shown that, in the last resort, nature works dialectically and not metaphysically."[33] Again and again, he emphasized that there could be no question of building the laws of dialectics into nature, but that it was a matter of discovering them in it and evolving them from it. These principles were not to be the starting point of an investigation, but its final result. They were not to be applied to nature and human history, but abstracted from a conscientious study of the real processes at work in them.

In all of this, Engels was proposing a new synthesis of philosophy and natural science: "a comprehensive view of the interconnections in nature by means of the facts provided by empirical natural science itself . . . in order to arrive at a 'system of nature' sufficient for our time."[34] To achieve this, it was necessary for those engaged in the task to locate a balance point in their own thinking between the flow of intellectual history and the new facts emerging from empirical research. It was vital to preserve the continuity of thought, while at the same time coming to terms with the ruptures caused by new discoveries and bringing them to bear on new lines of development. The process was complex and turbulent, but it needed to be a unified process and it needed to be constantly moved forward. Engels understood why there had been an adverse reaction among scientists to the older philosophies of nature with their fact-denying or fact-forcing tendencies, but he hoped that when it had run its course it would "perhaps be possible to pronounce once more the name of Hegel in the presence of natural scientists without provoking that St. Vitus dance."[35]

Evolution and Human Destiny

Engels was a man of his time in sharing the idea of progress that swept so many of his contemporaries up in a wave of excitement about the new possiblities that the natural sciences were bringing to bear on the human future. But in Engels's case he did not hold to it in a naive or simplistic way. There were, first of all, so many unrealized possibilities, and necessarily so, for every advance in organic evolution was at the same time a regression, fixing one-sided evolution in one direction and excluding it along many other directions. There was, moreover, the tragic and painful side of the evolutionary process. Although he believed that a progressive development asserted itself in the end, he knew well that, in the course of evolution, there was also extreme cruelty and colossal waste. Not only that, but he believed that his own kind would be extinguished without trace. In weighing it all, Engels wrote what Robert Cohen has called a "scientific epilogue to Mephistopheles."[36]

> Nevertheless "all that comes into being deserves to perish." Millions of years may elapse, hundreds of thousands of generations be born and die, but inexorably the time will come when the declining warmth of the sun will no longer suffice to melt the ice thrusting itself forward from the poles; when the human race, crowding more and more about the equator, will no longer find even there enough heat for life; when gradually even the last trace of organic life will vanish; and the earth, an extinct frozen globe like the moon, will circle in deepest darkness and in an ever narrower orbit about the equally extinct sun, and at last fall into it.[37]

With tragic irony, he nevertheless held on to an idea of progress, though of a rather subdued sort and in a cosmic dimension:

> However many millions of suns and earths may arise and pass away; however long it may last before, in one solar system and only one planet, the conditions for organic life develop; however innumerable the organic beings, too, that have to arise and to pass away before animals with a brain capable of thought are

developed from their midst, and for a short span of time find conditions suitable for life, only to be exterminated later without mercy—we have the certainty that matter remains eternally the same in all its transformations, that none of its attributes can ever be lost, and therefore also, that with the same iron necessity that it will exterminate on earth its highest creation, the thinking mind, it must somewhere else and in another time again produce it.[38]*

In Engels's whole vision of the evolution and extinction of the human species, there was always the underlying theme of our oneness with nature, a vivid sense of how totally our origin and destiny was bound up with the rhythms of the natural world. He thought it senseless and unnatural to hold to a dichotomy of mind and matter, man and nature, soul and body, for it gave rise to many illusions. As if he had seen the specter of modern technocracy, he warned:

> Let us not, however, flatter ourselves overmuch on account of our human victories over nature.... Thus at every step we are reminded that we by no means rule over nature like a conqueror over a foreign people, like something standing outside nature—but that we, with flesh, blood and brain, belong to nature and exist in its midst, and that all our mastery of it consists in the fact that we have the advantage over all other creatures of being able to learn its laws and apply them correctly.[39]

Dogmatism

Another warning that Engels issued was against dogmatism. Already in the later years of his life, he recognized the development of a "vulgar Marxism" and observed that the ideas developed by Marx and himself were attracting "dangerous friends" who simply labeled things with their terminology as an excuse for not looking into them any further. He wished the ideas elaborated by Marx and himself to be considered a method that could serve as a guide to further study, as a series of pointers to future research. Not that their contribution was a purely methodological one, for, in elucidating their method, they were consciously laying the foundations of a scientific *Weltanschauung* adequate to the demands of their times. In any case, it is a distinction not to be too sharply drawn, for every method has ontological consequences and every ontology presupposes a methodology. The point was that they wanted their positive statements to be tentative ones, open always to constant reassessment in light of new knowledge achieved in the onward advance of science. Fully

---

* This must, of course, be read as an expression of faith, of imaginative vision, and not as a matter of scientific prediction, of deterministic necessity, despite Engels getting carried away with such terms as "certainty," "iron necessity," "must," and "will."

expecting many of his own ideas to be superseded, Engels declared: "With each epoch-making discovery in the sphere of the natural sciences, it [materialism] has to change its form."[40] And indeed it has.

The History of Science

There is one more point worth making about Engels's philosophy of science in light of the subsequent history of the philosophy of science. This is the fact that Engels held views on many matters, such as on the history of science and the logic of scientific discovery, that not only anticipated certain contemporary theories but are in some respects still in advance of them. His account of the history of science, for example, while incorporating some of the best insights now associated with Kuhn, was in some ways far richer than Kuhn's. Engels saw science as a historical process going through periods of evolutionary development (what Kuhn would call "normal science") dominated by received patterns of explanation (Kuhn's "paradigms") interrupted by points of crisis where the received patterns were called into question, giving rise to scientific revolutions in which new patterns came to prevail (Kuhn's "paradigm shifts"). Kuhn tends much more to the internalist end of the internalist-externalist spectrum than his reputation would indicate. Engels most certainly did not. Engels conceived of the history of science within the broader framework of the whole history of nature and human society. For him, science was not a purely self-developing process, but in complex, subtle, reciprocal interaction with a whole range of other processes: social, cultural, political, economic, technological processes. Science was inevitably linked to production, making the bearer of scientific development, not exclusively the scientific community, but the society as a whole. But this emphasis on the wider context for scientific activity did not lead him in the direction of the anti-rationalism that has characterized various trends in contemporary philosophy of science that have looked to the wider context. Engels seems to have felt no conflict between his stress on the historical and contextual nature of science and his affirmation of the rationality of science and the overall progressive character of its development.

Engels's views on the logic of scientific discovery also included many points sometimes thought to be original with Popper, but without the one-sidedness of Popper's account. Engels, in taking exception to the narrowness of his positivist contemporaries, pointed out "how little induction can claim to be the sole or even the predominant form of scientific discovery,"[41] and considered the main form of development in the natural sciences to be the hypothesis. However, he wisely did not throw out induction altogether, and penetrated more deeply than Popper into the complexities of the process, in holding that "induction and deduction belong together as necessarily as synthesis and analysis."[42]

Evaluating Engels: Problems and Pitfalls

These then were the ideas that set the pace for the future development of Marxist philosophy of science and account for its distinctiveness vis-à-vis other trends in the philosophy of science. Engels's work, however, remained largely programmatic. The failure of his successors to understand this has been, to a great degree, responsible for the disrepute into which his work has sometimes fallen. Rough notes, responding to developments in the science of his day, only published posthumously, have been taken as the last word about nature for all times. His flexible and fluid thinking has been rigidified into textbook laws of dialectics, always and everywhere three and unchanging, only to be brought up to date by new examples for new times. Even now, much ill-conceived debate between the critics and defenders of Engels could be avoided if both sides made more discerning judgements about how to characterize Engels's work and about where to locate his essential contribution. What is significant in Engels is his fundamental orientation. His contribution lay in sketching the broad outlines of a processive, antireductionist and humanist materialism; in an integrated way of thinking that bridged the gap between rationalism and empiricism, between organicism and materialism, between humanism and naturalism. Far in advance of his positivist or idealist contemporaries, he set the most advanced science within a new and more appropriate philosophical framework. Far in advance of his more mundane fellow socialists, he gave to the most advanced social theory a cosmological dimension by highlighting its connection with a new philosophy of nature. For this, for grounding Marxism in a holistic vision, binding together politics, science, philosophy, and indeed all else, Engels still needs to be taken seriously; for this, rather than for his specific statements about electricity, heat, magnetism, or tidal friction, rather than for his adherence to the Hegelian terminology of quantity, quality, contradiction, and negation, rather than for his elaboration of laws of dialectics as "the science of the general laws of motion and development of nature, human society and thought."[43] For this, he must be forgiven his positivist statements about the supersession of philosophy, his misconceived attacks on metaphysics, his question-begging definition of matter, and much else besides. Above all, he must be judged by his own intentions and his own thought and not by what his defenders and critics have sometimes unfortunately made of him.

The pioneering nature of Engels's effort and the enormity of his undertaking perhaps made uneven quality and mistaken formulations almost inevitable.

With marvelous daring and exuberance, Engels took on a project of tremendous sweep: to lay the foundations for a synthesis of scientific knowledge that would bring to light the cosmic context within which human history was unfolding. It was not enough for Engels to outline the political economy of capitalism and to organize the proletariat for a revolutionary transformation of the oppressive social order. It was vital that these efforts be grounded in a comprehensive worldview, encompassing all that exists, from the atoms to the stars. In so doing, he charted a new course, setting out the basic concepts of a *Weltanschauung* that would synthesize the best of the history of philosophy with the best of the history of science. This set him apart from the majority of his contemporaries who, in striving to come to terms with one, inevitably seemed to leave the other out of account. He did not shrink back from the great basic questions that have perplexed the philosophers of the ages, but he did insist that any attempts at answers be grounded in the best empirical knowledge of the time. This set his philosophy in advance of its competitors of the time, who were unable to transcend the extremes of positivist reductionism or idealism. In this respect, it still has much to offer to our own contemporaries, still caught up in newer versions of the same trends.

The Philosophy of Marx and the Marx-Engels Relationship

The question remains, however, as to where Marx fits into all this. It still needs to be established what Marx's thinking on these matters was and what relation it bears to Engels's work in this sphere.

It is true that Marx wrote no full-scale works on the philosophical implications of the natural sciences, nor even any extended treatment of the fundamental philosophical assumptions underlying his vision of things. Only Engels engaged in such projects. Marxism, after all, made its entry into the history of ideas primarily as a socioeconomic theory and expounding this was the absorbing lifework of Marx. There is, however, no reason to conclude from this that he was not interested in those questions raised by Engels or that he had no clearly thought-out views on these matters. On the contrary, there is every reason to believe that he was intensely interested in these issues, that he followed Engels's work in this sphere very closely, and discussed the problems with him in great detail. The reason why Marx had considerably less than Engels to say on these subjects* lies in the well-known and well-documented fact that, in the lifelong intellectual collaboration between Marx and Engels, there existed an agreed-upon division of labor between them. In their early years, they embarked upon joint works that are of themselves testimony to a

---

* Over the years, however, it has come to light that Marx did, in fact, have more to say than was heretofore supposed, especially with the discovery of his manuscripts on mathematics and the natural sciences.

unity of fundamental orientation. In their mature years, however, they organized things differently. It was Marx's task to concentrate on political economy and to marshall all his energies to the writing of the major constructive work, *Capital*, whereas Engels was to devote himself to philosophy, to the natural sciences, and to a number of other subjects, and to engage in polemics against their critics. Such sustained and harmonious collaboration is in itself a strong indication of underlying agreement on all essential issues, although they were, of course, two different persons, each with his own distinctive style and distinctive interests and emphases.

There is, moreover, considerable textual evidence for claiming that Marx considered himself to be in accord with Engels in philosophical orientation and shared his explicit materialism and his conception of the dialectics of nature.

The Marx-Engels correspondence testifies with great clarity to the fact that they seemed to have been of one mind on everything that really mattered to them. Marx often spoke of "our conception" and referred to the fact that they were working according to a common plan and doing something purposeful in the world together. He often indicated his high regard for Engels's opinion, as when he sent him a draft of something he had just written, saying "Your satisfaction is more important to me than anything the rest of the world may say of it."[44] Their correspondence also reveals the fact that Marx followed with interest Engels's studies of the natural sciences and undertook similar studies himself. He was fully aware of the philosophical conclusions Engels was drawing from them and approved wholeheartedly. In a letter written to Engels in 1867, for example, he responded specifically to some of Engels's earliest dialectical ideas about the natural sciences, making the point that he thought Engels was right and going on to tell him: "I quote Hegel's discovery regarding the law that merely quantitative changes turn into qualitative changes and state that it holds good alike in history and natural science."[45] He was referring to a statement to this effect that subsequently appeared in the first volume of *Capital*. It is a statement very hard to explain away by those who insist that Marx believed dialectics to be applicable exclusively to the sociohistorical sphere and not to that of the natural sciences, since he said here exactly the opposite. In the following year, he wrote another letter to Engels, making the very explicit statement: "So long as we effectively observe or think, we cannot possibly get away from materialism."[46] Nevertheless, despite the fact that Marx repeatedly stated his explicit adherence to materialism, the critics persist in claiming he was not a materialist. The correspondence, as already indicated, makes many references to Darwin. But the most far-reaching statement about the significance of Darwinism and its implications for the development of the Marxist world view came from Marx himself, who went so far as to say of *Origin of Species*: "This is the book that contains the natural-historical foundations of our outlook."[47] Nevertheless, the critics claim that

Marx left nature, other than human nature, alone and that his theory of history was unconnected with any theory of nature. The texts simply do not bear it out.

In Marx's other writings, from the first to the last,* there are similar affirmations. There is in them all a great stress on the importance of the natural sciences and insistence on unity of method. The *1844 Manuscripts*, in particular, are often given an antiscientific and irrationalist reading that is utterly alien to them. These very manuscripts are in fact infused with a respect for science and reason. In them, he criticized philosophy for remaining alien to the ever-growing mass of material that was being accumulated by the natural sciences.[48] In these manuscripts, too, he spoke of his ideal of a unified science, in which natural science would become the basis of human science, as much as the science of man would come to embrace natural science.[49] He firmly rejected the idea of one basis for life and another for science, believing as he did that natural science was the foundation of all knowledge and that morality needed to be grounded in knowledge. A persistent theme with Marx is that of the essential unity between man and nature. Setting man within nature, human history was seen within the framework of natural history, and continuity of method was grounded in this.

In an often-quoted passage from the *Economic and Philosophical Manuscripts*, Marx remarked, "Nature taken abstractly, for itself, nature fixed in isolation from man, is nothing for man."[50] Nevertheless, almost as if to answer those who assert that for Marx nature in itself is nothing, Marx stated in the same *1844 Manuscripts*: "Only nature is something."[51] Just as forcefully, in *The German Ideology*, he declared: "In all this the priority of external nature is unassailed."[52] Despite what the critics say, he did not shy away from speaking of "prior nature" or nature without man, clearly aware of the irreducibility of nature to a mere object of human praxis. In *The Holy Family*, for example, there is the statement: "Man has not created matter itself. And he cannot even create any productive capacity if the matter does not exist

---

* There is a plethora of literature posing a sharp dichotomy between young Marx and the mature Marx, the neo-Hegelian interpreters coming down on the side of the young Marx and the anti-Hegelian ones renouncing the young Marx and opting for their own interpretation of the mature version. Louis Althusser is particularly prominent in this debate, asserting an "epistomological break," a rupture in Marx's thought, occurring in 1845, in which Marx broke from his Hegelian past and earlier humanistic and historicist leanings and became scientific. The textual evidence against the rigidity of such periodization is overwhelming. The publication of *Grundrisse*, for example, undercuts both of these extreme positions, demonstrating a greater degree of continuity of thought. Nevertheless, while there were no radical breaks, there were distinct stages in the historical development of Marx's thought and Marx's mature work does represent a development beyond his earlier writings, which were closely tied to the Hegelian philosophical mode. Marx moved beyond certain formulations in the process of his development, but never had occasion to renounce any of his writings. There is no evidence he ever changed his mind regarding the sociohistorical character of all human knowledge, nor about the centrality of alienation to the critique of capitalist society.

beforehand."[53] Later, in *Capital*, he took up the point again: "A material substratum is always left, which is furnished by nature without the help of man."[54] Thus, all of man's interactions with nature came up against the sheer existential inexhaustibility of nature, determining the contours and setting the limits. At the same time, nature was only encountered in and through human history and was comprehended only through the socially evolved categories of human thought. In this sense, science came under the category of social labor and always its sociohistorical character was duly stressed. He gave this particularly striking formulation in the *1844 Manuscripts* when he said that the "forming of the five senses is a labour of the entire history of the world down to the present," and that the "senses have therefore become directly in their practice theoreticians."[55]

Marx paid enormous attention to the rapid and dramatic scientific and technological advances of his time that were making possible the reduction of necessary labor time and transforming the very character of necessary labor. He saw even the most "pure" natural science as inextricably linked to technology and industry, as well as to a whole range of human practices. Always science and technology were perceived within their full sociohistorical context. Every social order posed characteristic questions, investigated certain problems, organized the division of labor in its distinctive way, and even thought in different categories. Every social order gave its own characteristic shape to the development of science and technology. Scientific concepts were socially evolved and bore the imprint of technological objectives, class interests, and a host of other forces at work at any given time. Scientific theories were the product, not only of experimentation and thought, but of historical conditions. Certain theories could only arise under certain historical conditions, but not under others.

In the course of his own writings on many subjects—from philosophy to political economy to natural science to technology—Marx was inclined to reflexive considerations on the nature of scientific methodology generally. On this topic, his views evolved somewhat and had reached a point of considerable sophistication by the time he was writing his *Grundrisse*. Whatever the different interpretations of *Grundrisse* (and there are many), it is clear that his conception of scientific method was grounded in an epistemology that stood in contrast to both the classical empiricist theory of knowledge and to the speculative idealist theory of knowledge. Against the empiricist view, he stressed the necessity of abstraction. Against the idealist view, he stressed the foundational role of the concrete reality. In the operation of scientifically correct method, the process started with concrete, complex concepts and proceeded by way of analysis and synthesis to abstract, simple concepts. The end result was the reproduction of the concrete situation. The concrete was not so much the starting point as the result of the process, a process involving both observation and conceptualization. The process was marked by a dialectic of the abstract and the concrete. It was a "method of rising from the abstract to the concrete." Along this path, "the abstract determinations lead towards a reproduction of the concrete by way of thought."[56]

Another theme running through Marx's writings that shows Marx to be far closer to Engels than to the critics is his agreement with Engels's assessment of Hegel. He referred to "Hegel's method" and to his own "critical manner of applying it."[57] In his Afterword to *Capital*, he discussed the importance of Hegel and of discovering the "rational kernel" within the "mystical shell," and spoke of how, when Hegel came into disrepute in Germany, he openly proclaimed himself "the pupil of that mighty thinker."[58] Those who emphasize the paradigmatic importance of *Capital* can find no basis in such statements as this for their counterposition of a Kantian Marx to a Hegelian Engels.

Aside from this textual evidence, there is also the fact that Marx read the whole of the manuscript of *Anti-Dühring* before publication, seemingly had no reservations about it, and even contributed a chapter to it. Engels wrote in the second preface of 1885:

> The mode of outlook expounded in this book was founded and developed in far greater measure by Marx . . . it was self-understood between us that this exposition of mine should not be issued without his knowledge. I read the whole manuscript to him before it was printed, and the tenth chapter . . . was written by Marx.[59]

Neither in their own writings or in that of any of their contemporaries is there any evidence of any major theoretical disagreement between Marx and Engels. There is indeed much evidence to the contrary. This is not to say that they had no disagreements or that they shrank back from expressing any if they had. It is well known that there was bad feeling regarding Marx's reaction to the death of Mary Burns and they did explicitly disagree on one occasion on a question involving natural science, Marx taking to the ideas of Tremaux on evolution and soil properties while Engels dismissed them. But it was only after the death of Engels that attempts to drive a wedge between them on a major theoretical point emerged.

Much is made of the fact that Marx never used the term "dialectical materialism." Neither did Engels for that matter. Indeed it may not be the most appropriate term for designating Marxist philosophy. Nevertheless, the implications that are most often drawn from the repeated assertion of the fact that Marx never used the term are quite unwarranted. Much of the force of the argument is lost when one considers that Marx, like Engels, repeatedly referred to his position as "modern materialism" or the "new materialism" and just as often called attention to the fact that his method was the dialectical method.[60]

Even in spite of formulations in early works that could be cited against this interpretation, the overall evidence would seem nevertheless to be quite clear-cut. Indeed, there is very little except unwarranted speculation, an excessively voluntarist reading of the *Theses on Feuerbach*, and an exaggeratedly anthropocentric reading of the *1844 Manuscripts* to stand against it.

Anti-Engels Literature

Nevertheless, there is a mass of secondary literature standing against this interpretation of Marx and the Marx-Engels relationship, as well as against such a positive evaluation of Engels. Much of it draws the sharpest possible dividing line between Marx and Engels and holds that Engels's concept of the dialectics of nature in particular is utterly alien to the thought of Marx. There has been a flood on the market of books with such titles as *The Tragic Deception: Marx Contra Engels* and *The Betrayal of Marx*.[61]

It is not, however, an altogether new position. Early in the century, such authors as the Italian philosophers Giovanni Gentile and Rudolfo Mondolfo, and the Polish philosopher Stanislaw Brzozowski, drew a distinction between the philosophical positions of Marx and Engels. But it was in the 1920s that the real discussion actually began. An early work was an article by Erwin Ban in *Kommunismus*, the Comintern journal for Southeastern Europe, that held that Engels's role in the development of Marxist theory was a negative one. Ban saw Marx as the heir to classical German philosophy, from which Engels had broken in opting for a positivist and naturalist conception of the dialectic. He also saw Engels's views as having an ontological character foreign to Marx's thought.[62] But it was only with the controversial book published by the Hungarian Marxist György Lukács, a few years later in 1923, that the discussion really got off the ground. Lukács's *History and Class Consciousness*, a very influential book in the history of Marxism, called into question the philosophical accord between Marx and Engels. It was Lukács who introduced the differentiation specifically in terms of the dialectics of nature. Lukács held that Engels was misguided in his attempt to apply dialectics to nature, as the crucial determinants of dialectics, interaction of subject and object, unity of theory and practice, historicity, were absent from our knowledge of nature.[63] In the 1930s, Sidney Hook was again playing Marx off against Engels, rejecting the dialectics of nature and accusing Engels of turning Marx's revolutionary theory into a positivist ideology, of transforming his critical historicism into a vulgar materialism.[64] In the 1950s, it was the rage in the French leftist circles of Jean-Paul Sartre, Maurice Merleau-Ponty, and Henri Lefebvre. Coming into the 1960s, it became the academic orthodoxy in various centers of western Marxology, coinciding with the rise of the New Left. Not that it was entirely a western phenomenon as the pages of the Yugoslav *Praxis* testify. In the 1970s, still new paperback proclamations of this position have appeared, with some new variations of it still appearing on the horizon.

The pattern running through this veritable mountain of literature is to pose the dichotomy along the following overall lines, with various nuances among various authors:

The predominant line of argument is that Engels alone was responsible for laying the basis for the philosophy that has become known as dialectical materialism and that that philosophy is alien to the whole spirit of Marx's thinking. According to this theory, Engels distorted Marx's method by transforming it into a philosophical system. He created a *Weltanschauung* based on a vulgarized form of Darwinism fashioned in accordance with the positivist idea of a unified science, whereas Marx was only interested in a sociohistorical analysis of human alienation under capitalism. Engels wrongly extended the dialectic to nature, unlike Marx, who saw the dialectic as appropriate only to the sociohistorical sphere. For Marx, there could be no dialectic without man; indeed there could be no nature except in and through man. For Engels, on the contrary, there was only opaque, inert matter and all the world was reduced to the same undifferentiated stuff. Marx's method was anthropological, critical, pragmatic, empirical. Engels, in contrast, was objectivist, contemplative, determinist, abstract. There have been a host of variations on this theme.

In George Lichtheim's book *Marxism: A Historical and Critical Study*, Marx's "critical vision of a critical theory" validated by revolutionary action is contrasted to Engels's "dull science of causal evolution." Marx's vision of a "drastic transvaluation of values," a unique historical breakthrough, was transformed by Engels into a "cast iron system of laws." Marx wisely left nature, other than human nature, alone; whereas Engels ventured where Marx feared to tread and thus began the drift towards positivism and scientism. Lichtheim calls it a "travesty."[65]

In Lichtheim's *From Marx to Hegel*, he further elaborates on the contrast. Here he claims that the notion that Marx's anthropological naturalism was anchored in a general theory of the universe finds no support in Marx's own writings. There was no logical link between it and the dialectical materialism of Engels. It was a systematization that has suffered the fate of other systematizations undertaken for nonscientific reasons, as it was required to turn Marxism into a coherent *Weltanschauung* first for the German labor movement and later for the Soviet intelligentsia. Marxism, as interpreted by Engels, eventually did for Central and Eastern Europe what positivism had done for the west: it brought to the public a manner of viewing the world that was "scientific" and it extended to history and society the methods of the natural sciences.[66] The historical parallel drawn between Marxism and positivism in this respect is astute enough, but there is no justification for concluding that this systematization was undertaken only for nonscientific reasons. In characterizing Marx's position as so starkly divergent from that of Engels and in asserting that Marx took no interest in nature and in the problem of working out a theoretical basis for the natural sciences, inconvenient texts (such as Marx's manuscripts on mathematics and the natural sciences) are ignored, and inconvenient historical facts (such as Marx's approval of the text of Engels's *Anti-Dühring*) are lightly skimmed over.

Then there is Z.A. Jordan's *The Evolution of Dialectical Materialism*, which contrasts Engels's metaphysical materialism with Marx's naturalistic activism or anthropological realism. Jordan argues that Marx did not adopt a dialectical but an anthropological conception of nature and that the dialectical and anthropological conceptions of nature are mutually exclusive. Engels supposedly completely misunderstood Marx and indulged in system building with inadequate intellectual resources.[67]

There is a similar condescension in Shlomo Avineri's book *The Social and Political Thought of Karl Marx*, in which he looks to differences in family background and education as the source of philosophical divergence. Marx was from a highly sensitive family of Jewish origins, university educated, mainly interested in philosophy. Engels, on the other hand, was from an industrialist family with pietist leanings and was educated for the commercial world and interested mainly in economics. This supposedly explains why Marx was too subtle to have been a materialist; his epistemology, according to Avineri, occupied a middle position between classical materialism and classical idealism. Engels's crude materialism, based on the mechanistic traditions of the eighteenth century, Avineri claims, only came to light after Marx's death and marked a sharp departure from Marx's thought. Because he was lacking in philosophical sophistication, Engels's use of the Hegelian terminology served only as an external and rather shallow veneer. Whereas Marx tried to find the human meaning of natural science, Engels looked for a natural science methodology to fit the human world. By applying the dialectic to nature, Engels divorced it from the mediation of consciousness, which really meant that his view was not dialectical at all.[68]

The same sort of approach can be found in John Lewis's *The Marxism of Marx*,[69] in which he discovers after so many years that Marx is "The Man Nobody Knows," who must be radically separated from Engels, who was a highly successful Manchester businessman who had no training as a philosopher. Marx's approach was anthropological. His conception of dialectics was concerned entirely with man and history and not with nature at all. Dialectics, for Marx, was not a universal process at work in the stars, in outer space, in the geological origins of the earth, in chemistry, or in biology, for there could be no question of a dialectic of external nature independent of man. Intelligence is the *sine qua non* of dialectical development, and this is peculiar to man. Unlike Engels, Marx did not think of himself as a materialist.

Another such work, Leszek Kolakowski's *Main Currents of Marxism*, draws the lines very sharply indeed. Volume one, *The Founders*, restates the argument that Marx's idea of a philosophy of praxis gave way in Engels to a theory that subjected humanity to the general laws of nature. Whereas Engels believed that man could be explained in terms of natural history, Marx's view was that nature as we know it is an extension of man. He contrasts Engels's naturalistic evolutionism with Marx's anthropocentrism; Engels's technical interpretation of knowledge with Marx's epistemology of praxis; Marx's revolutionary eschatology with Engels's notion of infinite progression. Engels, he claims, alternated between scientistic phenomenalism and materialism, the latter being beyond the reach of scientific rigor and, according to how it is formulated, necessarily either obscure or unprovable. After taking such pains to draw such a sharp contrast between the philosophy of Marx and Engels, Kolakowski at the end of the volume surprisingly asserts that no heuristic role can be ascribed to what suddenly becomes "Marx *and* Engels's philosophy of nature" (my emphasis).[70]

We see one more variation of this trend in an article entitled "Marxism and the Engels' Paradox" that appeared in *The Socialist Register*. The author, Jeff Coulter, contrasts Marx's dialectical rationality with Engels's logic of exteriority, Marx's vision of man becoming an "autonomous subject in a total praxis" with Engels's anthropomorphized naturalism which dehumanizes man by reducing him to the status of a mere predicate in the movement of external events. With Engels, the Marxist analysis of alienation, reification, ideology, commodity fetishism, false consciousness, and praxis counted for nothing. There was only the flat expanse of an arid scientism.[71]

There is obviously a strong element of caricature in it all. The assumption underlying the account of Coulter, as well as that of Kolakowski, Lewis, and Avineri, seems to be that humanism can only establish itself by seeing man stand on his own, totally autonomous and unconnected with anything outside himself. Engels's conception of humanism, however, like that of Marx, did not diminish humanity by placing it within a naturalistic framework. On the contrary, it enhanced it, by giving scope to its full network of possible interactions.

A more sophisticated book belonging to the same school of interpretation is *The Concept of Nature in Marx* by Alfred Schmidt. For Schmidt, Marx and Engels parted ways when Engels engaged in a problematic extension of Marx's original conception into a *Weltanschauung*. Engels's dialectics of nature remained external to its subject matter and was a system built along the lines of the positivist ideal of a unified science. Marx's concept of nature, on the other hand, was of a sociohistorical character—nature only came into consideration through the forms of social labor, only becoming relevant in so far as it was drawn into the web of human and social purposes. From here, Schmidt argues that nature in itself is predialectical and only becomes dialectical by producing men as transforming, consciously acting subjects. It is only the process of knowing nature that can be dialectical, not nature itself, for nature by itself is devoid of negativity. Schmidt argues that a materialist dialectics of nature is impossible and that a purely objective dialectic makes dialectics and materialism incompatible. As a dialectical relation is only possible between man and nature, objectivism is of necessity undialectical. All statements about nature already presuppose social praxis. Marx's position, as it emerges through Schmidt's interpretation, was that of a nonontological materialism. Authentic Marxism, he argues, is not to be seen as a naturalized Hegelianism which simply replaces one ontological substratum, spirit, with another, matter.[72] While our knowledge of nature is always socially mediated and while Schmidt may be right to stress it and to stress that Marx stressed it, it is by no means clear that Engels denied it, nor is it clear why this should rule out ontology, or that it did so for Marx.

The other major line of argumentation in assessing Engels's work comes at it from the opposite direction. Engels is accused of compromising Marx's materialism through a revival of an archaic and essentially idealist *Naturphilosophie*. It is said that Engels's view was in essence antiscientific, because his ideas were derived from Hegel's philosophy of nature and constitute an alien imposition on the natural sciences from outside.

The most prominent exponent of this position is Lucio Colletti, who claims in *Marxism and Hegel* that Engels did not in fact work to emancipate science from any remaining speculative bonds, but did just the opposite. He grafted the old metaphysics onto science, refashioning philosophical Marxism into a cosmological romance. Counterposing an unregenerately Hegelian Engels to an unconsciously Kantian Marx, Colletti argues that Engels distorted Marx's methodology by expanding it into cosmic dimensions. He insists that the Hegelian dialectic is of an intrinsically idealist nature and cannot be transformed in a materialist manner.[73]

Much the same view is taken by Gareth Stedman Jones, writing in *New Left Review*. His argument is that the inconsistency between method and system is to be found, not in Hegel but in Engels. According to him, it is necessary to choose between materialism and dialectics: it is impossible to have both. In borrowing Hegel's method, Engels found himself, despite himself, the unconscious prisoner of his system. This made Engels's work antiscientific in implication, even though it was far from antiscientific in intention.[74]

While Colletti may be on very weak ground in his interpretation of Marx and in drawing such a sharp line between Marx and Engels, it is interesting to note that he has backed down from this somewhat and has conceded the criticism made of himself and others for judging Engels at times too harshly, while creating a kind of sacred zone around Marx.[75] More weight should be given to the larger question raised by Colletti and Jones, the relation of Marxism to its Hegelian origins, which they are right to consider problematic. However, everything depends on what is taken from Hegel and how it is reintegrated within a new philosophical stage. It is a mistake to rule out of court at the start any possibility of appropriating elements of even the most idealist philosophical systems intelligently and integrally within a materialist framework.

The theme of an illicit extension of Marx's method into a philosophical system runs through both schools of interpretation, but they nevertheless reflect two opposite lines of approach, corresponding roughly to the neo-Hegelian and the anti-Hegelian currents within Marxism. Each in its own way also tends to assume the incompatibility of dialectics and materialism, the one side faulting Engels for his attachment to the dialectic and the other side for his adherence to materialism. Undoubtedly, there is some basis for criticism along such lines in Engels's writings, for in striving to achieve a new synthesis of older philosophy and newer science, he did sometimes veer a bit too much in the direction of Hegelianism or positivism. Thus Lichtheim not only criticizes Engels for yielding to the scientism of his time, but also criticizes him for reviving a certain archaic cast of mind associated with the old romantic philosophy of nature.[76] But, in remaining fixed on the excesses in one direction or another, they miss the basic thrust of Engels's thinking, which was to transcend both of these positions in a creative synthesis, rather than to combine the uncombinable in an eclectic jumble. Some authors go even further and employ the two lines of argument in a totally unintegrated way, giving way to the phenomenon of "too many reasons" and making their critique of Engels itself an eclectic jumble. The most extreme example of this tendency is an article by Richard Gunn in *Marxism Today* in which he cites almost all of the above authors: Lukács, Lichtheim, Jordan, Lewis, Schmidt, as well as Colletti, marshalling every possible antimaterialist argument of the one strand as well as all the antiidealist arguments of the other. He also brings into play both the Christian-mystical arguments of Nikolai Berdyaev against Engels, and the positivist-mechanist ones of Jacques Monod, so as to leave no possible stone unturned against Engels.[77]

Of the various philosophical questions raised in this literature, to be taken up many times again in coming to terms with the history of Marxism, the major ones to be assessed at this point are those relating to the validity of the interpretative positions taken vis-à-vis the philosophical views of Marx and Engels. Obviously, none of these interpretations stands on its own and they are inevitably intertwined with various overall philosophical positions, in most cases forming the basis of full-scale schools of thought within Marxism that need to be assessed in their own terms in their own place within the historical development of Marxism. What is at issue here is their status as interpretations of the classical works of the Marxist tradition.

The question remains as to how such interpretations come to terms with the rather formidable historical and textual evidence weighing against them. Sometimes, it is ignored. Sometimes, it is dimly acknowledged and weakly explained away. To exemplify typical lines of reasoning:

Petrović: What is Marx's attitude to the dialectics of nature? Here and there Marx used to remark that dialectical laws hold not only for society but also for nature. But he never became so interested in the dialectics of nature as to try to write more about it.[78]

Jordan: [On the question of what was Marx's opinion of Engels's peculiar combination of science and speculative philosophy] The most plausible answer is that Marx did not trouble to make up his mind about it.[79]

Schmidt: It cannot be established how far Marx was conscious of the difference between his concept of nature and that of Engels.[80]

Lichtheim: The first steps in this fateful interpretation were taken by Engels (in his *Anti-Dühring*) at a time when Marx was still alive, and indeed with his express, though possibly reluctant, sanction.[81]

Lewis: [At the time Engels read the manuscript of *Anti-Dühring* to Marx,] Marx was old and ill, and exercising his remaining powers on the uncompleted volumes of *Capital.*[82]

Colletti: If we leave aside the few and isolated statements where Marx appears to ratify the "dialectics of nature," we must on the other hand take into account that impressive and incontrovertible fact that he left behind him *Capital, Grundrisse, Theories of Surplus Value*—in other words, not a cosmogony but an analysis of capitalism.[83]

The only one to make a full-scale attempt to answer the evidence point by point is Gunn. He argues that the force of such evidence is offset by the following considerations:

Absence of recorded disagreement does not entail absence of disagreement. Marx and Engels may have been unaware of differences or seen reasons not to articulate them. The division of labor suggests a mechanism whereby they could remain unaware of objective disagreement.

As regards *Anti-Dühring*, Marx may have been more interested in the work's political effect in combatting Dühring's influence in the German Social Democratic Party than in its correctness on philosophical topics.

Engels's other main works on this were not known by Marx. He had not read the *Dialectics of Nature* manuscripts and *Ludwig Feuerbach* was written after his death.

Marx's references to a dialectics of natural sciences are isolated and entirely marginal to the argument of *Capital*. When Marx referred to dialectics, his reference was most often to dialectical method.[84] And that is the entire case for maintaining deep-rooted theoretical divergence between Marx and Engels.

It simply leaves any honest and unprejudiced reader unconvinced. In any case, to make such far-reaching assertions on vague inferences about Marx's reluctance, indifference, frailty, or lack of awareness in the face of specific and definite evidence to the contrary is to grasp at straws. During the period in which Engels was working on both *Dialectics of Nature* and *Anti-Dühring*, both lived in London and saw each other every day, discussing their work in the most elaborate detail. Nor is there any basis for implying that Marx was the sort of thinker who could acquiesce to a fundamental philosophical orientation alien to the main thrust of his work or pursue a socioeconomic analysis ungrounded in a well-thought-out set of philosophical assumptions. His thoroughgoing commitment to unity of method would seriously militate against such a way of proceeding with his own work. It is moreover a highly questionable procedure to dismiss so glibly his explicitly philosophical statements in favor of a philosophy hidden in the structure of his work of which he was himself unaware. Not that it cannot happen sometimes that a thinker in fact proceeds on the basis of philosophical assumptions other than those specifically acknowledged. But, in Marx's case, it has not been established, and the argument is based on an untenable forcing of texts. The methodology that emerges from this obscurantist "reading" of Marx is in fact at variance with the methodology he actually employs.

Pro-Engels Literature

There is, of course, another whole body of secondary literature giving a very different interpretation of the work of both Marx and Engels and answering the points made in the other literature. Not all of such writings are of equal value, however, in that some display a tendency to write the biography of Marx and Engels in the style of hagiography, to strain logic to the limits in an effort to explain away unfortunate formulations in the Marxist classics, and to ignore those that cannot be so explained away. These works not only artificially tidy up the loose ends, but misrepresent the nature of the very dynamic, colorful, and daring life and work of Marx and Engels. They were complex men, men of great vision, but men with inconsistencies and moments of myopia like the rest of men.*

The fact that the publishing houses of Eastern Europe have produced a flood of works defending Engels against all possible criticism and asserting complete unanimity between himself and Marx in the context of abusive, heresy-hunting, antirevisionist polemics has not helped matters much. There is a rare stridency in the whole debate that undoubtedly has its roots in divergent political positions. It is likely that much of the motivation underlying the negative evaluation of Engels lies in antipathy to the rigidity of so-called orthodox Marxism and a desire to find every possible reason to be set apart from it. In the process, while much of the critical literature wrestles with very real ambiguities in the Marxist classics, much of it resorts to rather shoddy scholarship, tenuous logic, groundless speculation, and sometimes what seems like outright distortion. In response, the answer all too often comes in the form of a repetitious reassertion of textbook formulations and a glib attribution of unworthy motives to all who come to question them. And the cycle goes on and on.

For instance, Mikhail Iovchuk's *Philosophical Traditions Today*[85] proceeds to write off virtually every thinker as "erroneous" and "revisionist" who has ever even hinted at critical consciousness, whether in regard to the Marxist classics or anything else. John Hoffman's *Marxism and the Theory of Praxis* argues against Lukács, Sartre, Schmidt, Avineri, and Coulter on the question of dialectics of nature, but in doing so takes a completely uncritical attitude to the Marxist classics, glosses over genuine problems of interpretation, and is unnecessarily abusive of any author who has taken a critical attitude or has uncovered problematic areas.[86] S.I. Titarenko also takes on such authors, addressing himself to the views expressed by Avineri, Sartre, Kolakowski, Petrović, Levine, and Gunn, opposing Marx to Engels and asserting that Marx was not interested in the natural sciences, while Engels was not interested in problems of alienation.[87] It is, however, weakly argued, riddled with jargon, and too dismissive of genuine problems of interpretation.

Other authors, while still highly polemical, have carried on the argument on a somewhat more constructive level:

Sebastiano Timpanaro evaluates the whole range of anti-Engels literature, criticizing the line taken by the likes of Fetscher and Schmidt as well as that

---

* It is, for example, a source of embarrassment to some biographers that Engels lived unmarried with the Irish proletarian woman, Mary Burns, and later with her sister, Lizzie. Engels felt no hesitation about defying the moral conventions of the Victorian era. It is unworthy of any serious biography to pass over this. The episode involving Marx's "illegitimate" son, Frederick Demuth, is similarly never mentioned in such works. It is more understandable that Marx's and Engels's views on the Slavs should cause embarrassment, as they did have unfortunate prejudices with regard to Slavic peoples and doubted their ability to form viable national states.

taken by such as Colletti. He tries to explain how and why it is that these trends come into play. His view is that each time a particular intellectual current takes the upper hand in bourgeois culture, certain Marxists rush in to interpret Marx's thought in such a way as to make it as homogeneous as possible with the predominant trend. Timpanaro is quite scornful of the attempt to do this by counterposing a problematic Marx to a dogmatic Engels. In all of these modernizing operations, he says:

> there is a need for somebody on whom everything which Marxists, at that particular moment, are asking to get rid of can be dumped. That somebody is Friedrich Engels. Vulgar materialism? Determinism? Naturalistic metaphysics? Archaic and schematic Hegelianism? Marx turns out to be free of all these vices provided one knows how to "read" him. It was Engels who, in his zeal to simplify and vulgarise Marxism, contaminated it. Thus, whereas Engels is loaded down with materialist ballast, Marx can take on the physiognomy of a profound and subtle (and still uncomprehended) great intellectual who is *de rigueur* in our cultural world.[88]

Although Timpanaro is far from uncritical of Engels, he recognizes that he was a man of genius who, even more than Marx, though not in dissent from him, was conscious of the need to deepen their materialism in response to the radical changes introduced by the Darwinian demonstration of the historicity of nature. He saw the need for a materialism that was not purely socioeconomic but also natural. Timpanaro believes Engels was correct to point out the serious danger of dissolving Marxism into a simple-minded Spencerian evolutionism or into mindless empiricism, but was wrong to react with a hardening of his Hegelianism. Nevertheless, despite his own reservations about the use of the terminology of dialectics, Timpanaro firmly insists that the work of Engels remains a brilliant attempt to fuse historical materialism with the materialism of the natural sciences and must be defended against attempts to downgrade it. In this he is right, as he is also in concluding that those who set out to establish a Marxism without Engels tend to arrive, coherently enough, at a Marxism without Marx. There is, however, in my view, a need for more careful thinking to be given to the relation of Marxism to other intellectual trends, from which it may take much that is of value, without necessarily going over to the side of the bourgeoisie.

Donald Weiss also answers proponents of the variants of the viewpoint defended by Lukács, Lichtheim, Avineri, Dupré, Tucker, and Jordan and argues that those who are unable to consider Marx a materialist, depicting him as a defender of spirit against the crude encroachments of Engelsian matter, are reading Marx with a distinctively ruling-class bias and severely distorting his real views. They have equated materialism with reductionism and have accused Engels of preoccupation with nonconscious factors, thus neglecting the importance of consciousness in history. These critics of Engels have not only fostered the mistaken belief that Engels's philosophical viewpoint was essentially different from that of Marx, but have severely muddled the very notion of materialism in the process. Not only is materialism not to be identified with reductionism, but Engels's materialism is virtually the only philosophical framework within which essentially antireductionist attitudes can be presented in a manner that does not violate the basic presuppositions of modern science. Not only is this materialism a fundamental aspect of the modern scientific world view, but it is the foundation upon which the whole Marxist edifice is built. Weiss insists that the full significance of Marx's theory of history cannot be realized except as based upon Engels's materialism. He shows that there is ample evidence that Marx himself was fully aware of this, there being no lack of passages in which Marx defended the same sort of conjucture between materialism and antireductionism.[89] While Weiss's insinuation of a "ruling class bias" against the critics of Engels is neither accurate nor helpful, his arguments are to the point in so far as the equation of materialism with reductionism has been a recurring theme underlying the critique of Engels. It is also important to set the record straight with regard to the antireductionist character of Engels's materialism and also to assert its importance to both Marxist theory and modern science.

An astute critique of Schmidt's analysis is by G.A. Cohen. Reviewing the English edition of Schmidt's book, Cohen calls into question the validity of his interpretation of Marx in which he inferred that Marx saw nature only in relation to human activity. This inference, Cohen claims, inverts Marx's real meaning, for Marx was not in a quasi-Kantian way, inserting nature within the framework of human activity, but, on the contrary, was inserting human activity within the framework of nature. Cohen also points out that Schmidt underestimates the importance that scientific discoveries had for Marx.[90]

In the same vein, Valentino Gerratana, has shed new light on the issues involved by emphasizing the relationship between Marx and Darwin. On the basis of thorough research, Gerratana shows that even before the publication of *Origin of Species,* Marx was extremely interested in the problem of the relationship between human history and natural history and thus between the science of society and the science of nature. He saw Darwin's work as making it possible for the first time to link human history with the whole rest of natural history into a unitary conception of the world on an entirely scientific basis. Gerratana, however, makes the point that the relationship between Marxism and Darwinism should not be seen in terms of eclectic combination, but as the first historical example of a new conception of the relations between philosophy and science, between the natural sciences and the social sciences. The need to assert the convergence of the rigor of rational thought with empirical evidence was indicated very explicitly by Marx and developed by

Engels. Engels, Gerratana shows, acted as an important intellectual stimulus to Marx and Marx attributed great value to the field that was systematically developed by Engels in *Anti-Dühring* and *Dialectics of Nature*.[91]

The article on Engels in the *Dictionary of Scientific Biography*, written by Robert Cohen, while not altogether clear about the exact extent of agreement between Marx and Engels on the concept of the dialectics of nature, nevertheless outlines a huge area of agreement between Marx and Engels with respect to their thinking on science and on nature. The article is extremely appreciative of Engels, depicting him as a man of remarkable learning and breadth, writing in broad strokes in an exuberant but serious critical style. Engels was sensitive to puzzles, polarities, and contrasts and his analysis was often penetrating and subtle. In contrast to those who claim Engels was negligent in relation to the role of consciousness and to the human dimension, Cohen believes that the role of consciousness was essential for Engels and asserts that Engels as well as Marx understood Protagoras's "Man is the measure of all things," an understanding that manifested itself in the centrality of labor to their thinking, an understanding that apparently did not rule out Engels's project of placing radical social analysis and revolutionary practice within a cosmology of emergent evolution. Cohen emphasizes how much of Engels's work in both politics and in science remained programmatic and interprets him as writing in sweeping but heuristic terms.[92]

But it is not just Marxists of various descriptions who have come to the defence of Engels on these matters. David Joravsky, an American historian of science, has stated that the Marx-Engels correspondence reveals the groundlessness of the notion that Marx did not share Engels's views on the philosophy of natural science.[93] Another American historian of science, Loren Graham, cites other of Marx's writings that were suffused with the realization that an understanding of man must begin with an understanding of nature as evidence of his interest in physical nature and of his basic agreement with Engels. He states that recent writers, Lichtheim and Jordan in particular, who have tried to divest Marxism of all remnants of inquiry into physical nature, have not only misrepresented Marx, but have deprived Marxism of one of its greatest intellectual strengths: its explanation of the organic unity of reality. It is not necessary to restrict Marx's interest to ethics and economics to free him of vulgar materialism. Graham asserts that a reevaluation of Engels by historians of science is overdue.[94] There are indeed signs that the time is coming.

NOTES

1. Georg Hegel, *Philosophie der Geschichte* (Berlin, 1840), p. 535, cited by Engels in *Anti-Dühring*, (London, 1975), p. 26.

2. The full text of the letter appears in the German edition of the Marx and Engels collected works published in Berlin 1972–1974: *Werke* 33, p. 81. The letter appears in the English *Selected Correspondence* (Moscow, 1975) minus the postscript.

3. Friedrich Engels, *Ludwig Feuerbach and the End of Classical German Philosophy* (Moscow, 1946), pp. 42–43.

4. Marx to Engels, 4 July 1864, *Werke* 30, p. 418.

5. J.B.S. Haldane, Preface to *Dialectics of Nature* (London, 1940), pp. xiv–xv.

6. Engels to Marx, 14 July 1858, *Werke* 29, p. 524.

7. Marx to Lassalle, 16 January 1861, *Selected Correspondence*, p. 115.

8. Eugen Dühring, *Kursus der Philosophie als streng wissenschaftlicher Weltanschauung und Lebensgestaltung* (Leipzig 1875), cited by Engels in *Anti-Dühring* (London 1975), p. 43.

9. Marx to Sorge, 19 October 1877, *Selected Correspondence*, p. 290.

10. Engels to Liebknecht, 31 July 1877, *Werke* 34, p. 285.

11. Engels to Marx, 28 May 1876, *Selected Correspondence*, p. 286.

12. Engels, *Anti-Dühring*, p. 389.

13. Ibid., pp. 11, 172.

14. Haldane, Preface to *Dialectics of Nature*, p. xii.

15. Ibid., p. xiv.

16. Engels, *Ludwig Feuerbach*, p. 36.

17. Engels, *Anti-Dühring*, pp. 16–17.

18. Engels, *Anti-Dühring*, p. 394.

19. Ibid.

20. Engels, *Ludwig Feuerbach*, p. 25.

21. Engels, *Dialectics of Nature*, p. 198.

22. Ibid., p. 236.

23. Ibid., p. 21–22.

24. Ibid., p. 22.

25. Ibid., p. 200.

26. Ibid., p. 203.

27. Engels, *Anti-Dühring*, p. 49.

28. Ibid., p. 16.

29. Ibid., p. 15.

30. Ibid., p. 18.

31. Ibid., p. 19–20.

32. Engels, *Dialectics of Nature*, pp. 209–210.

33. Engels, *Anti-Dühring*, p. 33.

34. Engels, *Ludwig Feuerbach*, pp. 46–47.

35. Engels, *Anti-Dühring*, p. 396.

36. Robert S. Cohen, "Friedrich Engels," *Dictionary of Scientific Biography* vol. XV, supplement I(New York, 1978), p. 145.

37. Engels, *Dialectics of Nature*, pp. 35–36.

38. Ibid., p. 39.

39. Ibid., p. 180.

40. Engels, *Ludwig Feuerbach*, p. 27.

41. Engels, *Dialectics of Nature*, p. 229.

42. Ibid., p. 228.

43. Engels, *Anti-Dühring*, p. 169.

44. Marx to Engels, 22 June 1867, *Selected Correspondence*, p. 176.

45. Ibid., p. 177.

46. Marx to Engels, 12 December 1868, *Werke* 32, p. 229.

47. Marx to Engels, 19 December 1860, *Werke* 30, p. 131.

48. Karl Marx "Economic and Philosophical Manuscripts," in *Collected Works* 3 (London, 1975), p. 303.

49. Ibid., p. 304.

50. Ibid., p. 345.

51. Ibid., p. 343.

52. Marx and Engels, "The German Ideology," *Collected Works* 5, p. 40.

53. Marx and Engels, "The Holy Family," *Collected Works* 4, p. 46.

54. Karl Marx, *Capital* 1 (London, 1874), p. 50.

55. Marx *Collected Works* 2, pp. 300, 302.

56. Karl Marx, *Grundrisse* (London, 1973), p. 101.

57. Marx to Kugelmann, 27 June 1870, *Selected Correspondence*, p. 225.

58. Marx, "Afterword to the Second German Edition," *Capital* 1, p. 29.

59. Engels, *Anti-Dühring*, p. 13.

60. There are numerous references that could be given here, e.g., in *Capital* 1, p. 27, Marx spoke of "the materialistic basis of my method," and then on p. 29 of "my dialectic method"; in the tenth of his *Theses on Feuerbach* (Moscow, 1946) he contrasted the "old materialism" with the "new."

61. Norman Levine, *The Tragic Deception: Marx Contra Engels* (London, 1976); Frederick Bender, *The Betrayal of Marx* (New York, 1975).

62. Erwin Ban, "Engels als Theoretiker," *Kommunismus* December 3, 1920.

63. Georg Lukács, *History and Class Consciousness* (London, 1971).

64. Sidney Hook, *Towards the Understanding of Karl Marx: A Revolutionary Interpretation* (New York, 1933), pp. 25–30; *From Hegel to Marx* (New York, 1936), pp. 29–34.

65. George Lichtheim, *Marxism: A Historical and Critical Study* (London, 1961), pp. 235–247.

66. Lichtheim, *From Marx to Hegel* (New York, 1974), pp. 67–73.

67. Z.A. Jordan, *The Evolution of Dialectical Materialism* (London, 1967), pp. 9, 27, 80, 93, 110, 176, 392–393.

68. Shlomo Avineri, *The Social and Political Thought of Karl Marx* (Cambridge, 1970), pp. 3, 6, 65, 69.

69. John Lewis, *The Marxism of Marx* (London, 1972), pp. 18, 62–65, 76, 90.

70. Leszek Kolakowski, *Main Currents of Marxism*, vol. 1, *The Founders* (Oxford, 1978), pp. 181, 337–408.

71. Jeff Coulter, "Marxism and the Engels Paradox," *The Social Register* (1971) pp. 129–149.

72. Alfred Schmidt, *The Concept of Nature in Marx* (London, 1971), pp. 15, 52–60, 166–167, 185, 195–198.

73. Lucio Colletti, *Marxism and Hegel* (London, 1973), pp. 44, 49, 162, 178.

74. Gareth Steadman Jones, "Engels and the End of Classical German Philosophy," *New Left Review* 79(1973):28, 36.

75. Colletti, *New Left Review*, no. 86(July-August, 1974).

76. Lichtheim, *Marxism* pp. 246–247.

77. Richard Gunn, "Is Nature Dialectical?", *Marxism Today* (February 1977), pp. 45–52.

78. Gajo Petrovic, *Marx in the Mid-Twentieth Century* (New York, 1967), pp. 27–28.

79. Jordan, *Evolution of Dialectical Materialism*, p. 10.

80. Schmidt, *Concept of Nature*, p. 207.

81. Lichtheim, *Marxism*, p. 237.

82. Lewis, *Marxism of Marx*, p. 9.

83. Colletti, *Marxism and Hegel*, p. 18.

84. Gunn, "Is Nature Dialectical?", p. 46–47.

85. M. Iovchuk, *Philosophical Traditions Today* (Moscow, 1973).

86. John Hoffman, *Marxism and the Theory of Praxis* (London, 1975).

87. S.I. Titarenko, "Novye Metody Revizionizma Marksizma-Leninizma," *Voprosy Istorii KPSS 6(1977).*

88. Sebastiano Timpanaro, *On Materialism* (London, 1975), p. 74.

89. Donald Weiss, "The Philosophy of Engels Vindicated," *Monthly Review*, January 1977, pp. 15–30.

90. G.A. Cohen, "Marxism: A Philosophy of Nature?", *Radical Philosophy* 2 (Summer 1972):15.

91. Valentino Gerratana "Marx and Darwin" *New Left Review* 82(1973):60–82.

92. Robert Cohen, "Friedrich Engels." pp. 131–145.

93. David Joravsky, *Soviet Marxism and Natural Science* (London, 1961), p. 6.

94. Loren Graham, *Science and Philosophy in the Soviet Union* (London, 1973), pp. 27, 30, 35.

CHAPTER 2

THE NEW GENERATION:
The Marxism of the Second International

The New Era

After the death of Marx and Engels, the task of carrying their theory forward and developing it further fell to the next generation. The leading lights of the day were the leaders of the German Social Democratic Party, who had been acquainted with Marx and Engels and nurtured on Marxist theory during the very process of its formation. But, as has so often been the case in the history of a tradition's development, the immediate successors were somewhat less in stature than the founders. Marx and Engels had at least some notion of what was to come. They were far from uncritical of the rising generation. As Marxism began to become the vogue among the vulgarizers, Marx was once provoked to say, "All I know is that I, at least, am no Marxist."[1] On another occasion he referred quite caustically to Kautsky as "a small-minded mediocrity who busies himself with statistics, without deriving anything intelligent out of them."[2] Engels, for his part, was known to have cast a critical eye at Bernstein, disapproving of his enthusiasm for Fabianism during his period of residence in England.[3] He also expressed a certain general disappointment with the caliber of their successors, commenting that Marx may well have said, in the words of Heine: "I sowed dragons and I reaped fleas."[4]

But there were other factors to be taken into account in tracing the fate of Marxism in the hands of the new generation. The times were changing. Marxism was emerging into a new era.

Engels lived to see in the inauguration of the Second International, being present at its founding congress in Paris in 1889.* At first it existed only in the form of successive congresses, but in 1900 a permanent organ was established. The International Socialist Bureau was not a directive body, but more of an administrative center. Like the First International of Marx and Engels, which fell apart in the complications arising out of the ideological and organizational struggle with the anarchists, the Second International was a loosely organized federation, without clear-cut criteria for membership and without an elaborately defined theoretical basis. Nevertheless, the Second International represented an advance on the First, from the Marxist point of view. Although non-Marxist trends had not disappeared or lost their influence, Marxism had clearly become the dominant ideology of the working-class movement. The Second International was Marxist. Moreover, the movement was no longer simply one contending trend among others and it was a far from marginal phenomenon.

The largest, fastest-growing, and most theoretically consistent party in the socialist movement was the German Social Democratic Party. As such, it held the commanding position within the International and its theoretical journal, *Neue Zeit*, was the most influential forum for the discussion of ideas within the movement. The party had grown into a mass movement. The antisocialist laws had been repealed in 1890 and, in that year, the party won one and a half million votes in the elections and 35 seats in the Reichstag. In the next year, the Erfurt Program, drafted by Bernstein and Kautsky, was adopted, purging party policy of its last Lassallean elements. The SPD** had become officially Marxist. And it went from strength to strength. By 1912, it was able to count on 34.8 percent of the vote and held 110 seats in the Reichstag. By 1914, in the final year before the International crumbled in ruins at the onset of war, the party had over a million members. It grew by leaps and bounds, not only in membership, but in the number and variety of its organizations, in newspaper outlets and other types of publications, and even in its real-estate holdings. There were socialist schools and trade unions, organizations for women, and even for sport and tourism. Some began to feel, however, that within its very success was danger. They worried that this all-encompassing cradle-to-grave presence ran the risk of making the SPD an end in itself, that it blunted the edge of the

---

* Actually, due to the conflict between the possibilists and the impossibilists, i.e., the currents associated with Jaurés and Guesde in France, the congress straightaway split in general confusion. Two congresses were held, but only one mattered, as the other was without any further issue. The International was formed, with representatives of twenty countries present.

** Sozialdemokratische Partei Deutschlands, the German Social Democrat Party.

class struggle for revolutionary transformation. They feared that it gave rise to reformist evolutionism and its theoretical counterparts that were beginning to surface in the complex and multi-leveled debate that soon spread throughout the party, and indeed through the whole International.

Still other factors were setting the scene for the wide-ranging debate that was to sweep over the International. The uncertain development of Germany as a nation created a political instability and a corresponding intellectual vacuum, to which Marx had addressed himself from the left and others from the right. It was, however, part of a larger instability and a larger vacuum. Liberalism was in crisis and the ominous approach of imperialist wars cast a dark shadow over all of European political life. A dramatic shift in the whole intellectual atmosphere was under way. The first wave of excitement about the great advances of nineteenth-century science had spent itself. No longer did the dynamic scientific optimism underlying the liberal faith in progress hold sway. There came a sharp reaction against all forms of materialism, giving way to a new renaissance of idealism comparable to the romantic reaction against the Enlightenment that a century earlier had brought new varieties of idealism in its wake. When on its way up, as a progressive class setting itself against the forces of reaction, against the old order that wished to maintain its privileges on the basis of blood and land, the bourgeoisie had been inclined to stake its lot with science and reason and to put its faith in the progress made possible by science and reason. In its struggle for hegemony against the right, it had been inclined to put its stress on the realism and rationality underlying the new science and to place its confidence in the possibilities of the human future. But once it gained the upper hand against the right, it had to deal simultaneously with the complexities and ambiguities of being in power and with the mounting forces arraying themselves against it on the left. Things began to turn sour and other themes came to the fore, antirealist and irrationalist themes, expressed not only in philosophy and a whole range of other disciplines, but penetrating art forms and popular culture as well. The bourgeois intelligentsia was turning its back on both militant liberalism and militant materialism.

Not that intellectual history follows upon the development of economic and political forces in a simple and uncomplicated way. The connections can be of an extraordinarily subtle nature, with complications arising from overlapping and countervailing tendencies and from others forces coming into play. The antimaterialist reaction of the late nineteenth century took on a variety of forms, some of an extraordinarily sophisticated nature, and stemmed from a number of factors, such as the fact that very real human values were often left out of account by the prevailing positivist ethos of the earlier period and that very serious epistemological problems were beginning to arise within science itself out of the revolution in physics at the turn of the century.

The Neo-Kantian Revival

In Germany, the first and predominant manifestation of this trend was in the form of a neo-Kantian revival.* It had begun in the German universities in the 1870s, but by the 1890s had become an influential force beyond the sphere of the universities, making its impact felt throughout European social democracy in a particularly forceful way. A reaction against the emphasis on science and modernity, the revival was an attempt to ground knowledge and values in a more reliable method, to knit together the social fabric in a way that was less dependent on the uncertain liberal values bound up with scientific rationality and the idea of progress. It represented a rehabilitation of philosophy in opposition to the antiphilosophical sentiments of the positivists, but it was also a rejection of metaphysics in opposition to Hegelianism and indeed to all efforts at all-embracing ontological constructions. Instead, philosophy was to devote itself to the critique of knowledge. The critique of knowledge must logically precede all particular knowledge and opened the way for judgements of value. Science, the neo-Kantians argued, did not provide its own interpretation. In and of itself, it could not validate its own methods and results, nor could it explain value. In their view, science simply provided descriptions of phenomena, i.e., of things such as the structure of our minds permitted us to experience them. It could know nothing for certain about things-in-themselves. Moreover, it could provide no basis for value judgements. Science was therefore not rejected—it had developed too far for that and the neo-Kantians were too sophisticated for that—but it was severely restricted in scope. The tendency was to draw it away into a truncated realm of its own, shutting it off from its role in providing an all-pervasive illumination of the whole of the natural and human world. Back again were the sharp and impassable dividing lines between fact and value, science and ethics, nature and history, causality and teleology, natural science and social science, determinism and freedom. Back again were the most fundamental forms of dualism.

---

* Immanuel Kant had been a towering figure in the revolt against the age of reason. Taking up the post-Enlightenment problematic, but without giving all ground over to a sweeping irrationalism, Kant advanced a new synthesis. Reason and science were valid within their limits, but beyond such limits they had to give way to other methods. What he denied to knowledge was given over to faith. Though he pursued this road himself with a certain respect for reason, others latched on to Kant's "critical philosophy" as a license to believe virtually whatever they were disposed to believe. Kant's philosophy later receded into the background, overshadowed by Hegelianism and positivism. After a few decades, it was Hegel's turn, as Marx observed, to become a "dead dog," with Kant once more ascendant.

The Neo-Kantian Critique of Marxism

Neo-Kantianism brought a new level of sophistication to the criticism of Marxism. Its epistemological orientation not only brought rejection of the fundamental realism underlying the Marxist interpretation of the natural sciences, but called into question the very possibility of a causal social science that could account for the development of human history, thus undermining the Marxist thesis of the inevitability of socialism. Moreover, even if its inevitability could be established, this left unanswered the question of its desirability. Thus the argument was that Marxism had not provided socialism with proper epistemological or ethical foundations.

The Neo-Kantian Interpretation of Marxism

While some marshaled these arguments in the hope of undermining Marxism in a way that had not been possible with the previous crude and scornful attacks on it, others used them to justify efforts to supplement Marxism with neo-Kantianism, claiming thus to put socialism on a firmer epistemological and ethical basis. It was a response to the whole mood of the times and a reaction to very real problems. New knowledge was bringing new complexities. Historical developments were posing new perplexities. The turn of the century brought the decline of the idea of progress and an atmosphere of crisis—epistemological crisis and moral crisis. There was a sense of old foundations having collapsed and a need for new foundations to take their place.

Despite the fact that neo-Kantianism was associated with the decline of liberalism, many neo-Kantians were liberals. They were liberals anxious to move with the times and therefore moved to fuse the ascendant philosophy thrown up by the bourgeoisie's move to the right with something of the drive embodied by the powerful proletariat building up on the left. Some, very early on, looked with interest towards socialism. By the 1890s, some began to advocate a full-scale synthesis of socialism with neo-Kantianism. The notion of "ethical socialism" was first introduced by F.A. Lange and Hermann Cohen. Cohen claimed that the Kantian ethic represented the true philosophical expression of the human aims of socialism, and that Kant's categorical imperative made him the actual founder of German socialism. Such a socialism, morally founded, could, he insisted, have nothing to do with philosophical materialism.[5]

In spite of these overtures towards socialism on the part of the university neo-Kantians, most active social democrats at first paid little attention. However, when certain influential socialists like Conrad Schmidt, Karl

Vorländer, Ludwig Woltmann, Franz Staudinger, and Kurt Eisner, also began proposing a fusion of Marx and Kant, the scope of the discussion broadened dramatically. And there was far more at stake in it. Not that they all agreed on just exactly how Marx and Kant were to be fused. Schmidt's acceptance of neo-Kantianism was highly qualified. He felt that it was necessary to turn away from the metaphysical speculations of Hegel to the true science that was Kant's critical philosophy. However, he insisted, turning to Kant on questions of epistemology did not mean accepting the Kantian system as a whole. He did not accept Kantian ethics, believing that moral duty was defined to some extent by social necessities. Other, like Woltmann and Vorländer, were even more critical of Hegel and far more thoroughgoing in their acceptance of Kant. As they saw it, Marx's theory of society should be completed by the whole of Kant's social philosophy, maintaining that Kant anticipated many of Marx's theories. In general, the neo-Kantian Marxists embarked on a critique of Darwinian evolutionism and on all forms of materialist determinism that saw the human sphere as an extension of the natural world. History was inexplicable in naturalistic terms, for nature knew nothing of freedom: history and nature belonged to different spheres altogether and therefore required different methods of understanding and explanation. No scientific analysis could provide answers to moral questions. It could disclose what is, but not what ought to be. In epistemological terms, Kantianism made it clear that socialism could not be justified scientifically but only morally. In ethical terms, it established the categorical imperative as the moral foundation of socialism.[6]

Bernstein

By far the most influential adherent of the neo-Kantian interpretation of Marxism, though certainly not the most philosophically accomplished, was Eduard Bernstein. Bernstein, an active social democrat since 1872 and editor of *Sozialdemokrat*, the SPD's organ in exile, was an established leader of the party. Engels's *Anti-Dühring* converted him to Marxism after he had come under the spell of Dühring. He lived for many years abroad and, as there were charges against him in Germany, he remained in exile even after the repeal of the antisocialist laws. His association with Engels in London culminated in his being named as literary executor in Engels's will. But his English sojourn also brought a close association with Fabianism, the latter influence coming to dominate in the end. From the beginning of his painful departure from the mainstream, and even before his return to Germany in 1901, when his life changed rapidly from one of an outlaw to that of a Reichstag deputy, his enunciation of unorthodox views carried enormous weight and won a large number of adherents.

The Revisionist Controversy

By this time, the "revisionist controversy" was in full swing, with the battle lines drawn on an exceedingly wide number of issues. In the foreground of the debate were the economic and political issues. It began with a dispute on agricultural policy, but rapidly broadened into a wide-ranging debate on the nature of capitalist development and the road to socialism. As the 1890s wore on, Bernstein came increasingly to question the analysis that he and Kautsky had worked out in 1891 and that had been enshrined in the Erfurt Program. This program had envisaged a future of increasing centralization of wealth, progressive impoverishment of the proletariat, and mounting class tension. But, as Bernstein began to see it, the opposite was happening and the future promised to be instead one of more equitable distribution of wealth and of increasing prosperity, security, order, and tranquility. This meant a different political strategy, a more gradualist one. Bernstein argued that the party had in fact already adopted a reformist strategy, at least as far as its day-to-day practice was concerned, no matter how tenaciously it clung to the revolutionary rhetoric that had marked the explosive entrance of Marxism into the political arena. In one of his more extreme statements on the matter, Bernstein declared: "To me that which is generally called the ultimate aim of socialism is nothing, but the movement is everything."[7] No longer did he look to a once-and-for-all collapse of capitalism and consequent triumph of socialism. Indeed, the socialist movement was steadily bringing improvements in the everyday life of the workers as it achieved basic social reforms. This struck a note in those for whom the party had become an end in itself; for the natural extension of this syndrome was the conviction that one could build the new society by degrees within the existing framework without a violent revolution to bring in a new social order from the ashes of the old.

Eventually, the controversy became, in fact, a full-scale reappraisal of many of the most fundamental premises of Marxism. Bernstein did not shy away from this. Although at first he claimed to represent the truest line of continuity with the founders, he soon abandoned this claim and frankly admitted his enterprise was a "revisionist" one. He simply argued for the validity of his own premises and for the necessity of criticizing what he considered to be the negative aspects of Marxism, without however ceasing to consider himself Marxist. Ironically, in light of the later casting of Engels into the role of dogmatic distorter of the complex and sensitive Marx, Bernstein considered Marx to be the more dogmatic one and Engels a more flexible thinker.

The philosophical dimension of revisionism was logically connected with its political dimension. Bernstein set the tone for the struggle to establish this with his high-flown declaration: "Under this banner—Kant, not Hegel—the working class fights for its emancipation today."[8] Bernstein thought that the Hegelian heritage of Marxism was to blame for its exaggerated revolutionism. Marxism had been led astray by the "treacherous" Hegelian dialectic that set a "snare" on Marx's thought.[9] It constituted an arbitrary imposition of a structure of explosive development through contradictions on a reality that actually developed in a much more gradual, less explosive way. It resulted in a tendency to develop revolutionary expectations on the basis of a priori schemata, instead of acknowledging the evolutionary development of reality on the basis of an empirical study of the actual facts. Bernstein spoke of the "great danger to science of Hegel's logic."[10] He thought that, whatever Marx and Engels had accomplished, they had not achieved it by means of the dialectical method, but in spite of it. The dialectical form of Marx's work he considered to be a sort of "scaffolding," from which his followers had been unable to detach themselves. This had led to "reasoning based on what the dialectical scheme prescribes, so as not to let a post fall out of the scaffolding."[11] For dialectical revolutionism, Bernstein substituted an "organic evolutionism."[12] Development through leaps and contradictions was replaced by a smooth and unilinear type of progress. Despite such a drastic revision, he still felt the Marxist theory of history stood intact. Somehow there was some core left that could be separated from its pattern of development. As Bernstein put it, looking back after a number of years:

> I had not turned away from the fundamental concepts of the Marxian view of history, but had always merely opposed certain interpretations and applications. Further I had sought to explain hasty conclusions of Marx and Engels as the consequence of their being seduced by the Hegelian dialectic, which after all is not integrally connected with the theory.[13]

Not only did Bernstein not consider the dialectic to be integral to Marxism, he did not consider materialism to be either. The neo-Kantian movement was a source of inspiration for his rejection of both. As early as 1892, he wrote in *Neue Zeit* that the movement represented "a reaction against the shallow naturalistic materialism of the middle of this century on the one hand, and the excesses of speculative philosophy on the other."[14] By the end of the decade, Bernstein allied himself with the neo-Kantian philosophy in its critique of both dialectics and materialism. In the conclusion to his important pace-setting book of 1899, *Die Voraussetzungen des Sozialismus und die Aufgabe der Sozialdemokratie,* which he entitled "Kant Against Cant," he explained why:

> And in this mind, I, at the time, resorted to the spirit of the great Königsberg philosopher, the critic of pure reason, against the cant which sought to get a hold on the working class movement and to which the Hegelian dialectic offers a comfortable refuge. I did this in the conviction that social democracy required a

Kant, who should judge received opinion and examine it critically with a deep acuteness, who should show where its apparent materialism is the highest—and is therefore the most easily misleading—ideology and warn it that the contempt of the ideal, the magnifying of material factors until they became omnipotent forces of evolution, is a self-deception.[15]

Bernstein's arguments against materialism actually took the form of a voluntaristic rejection of determinism, based on a dualism of nature and history, necessity and freedom, fact and value, science and ethics. For him, to be a materialist was to trace back all phenomena to the necessary movements of matter, which was "a mechanical process." To be a materialist was to be "a Calvinist without God."[16] He seemed to have no objection to determinism in the natural sciences, but he did object to its extension to history. As he put it: "Philosophical materialism, or the materialism of the natural sciences, is in a mechanical sense deterministic. The Marxist conception of history is not."[17] He proceeded to argue for a multiplicity of factors to be considered in understanding the historical process, for the conditional autonomy of political and ideological forces, and in particular for greater space to be allowed for the independent activity of ethical factors. But in philosophical terms, it was the Kantian assertion that the mode of thought appropriate to the realm of natural necessity was not appropriate to the sphere of human freedom, that science could not provide a basis for ethics.

It is, however, open to question just how deep Bernstein's Kantianism actually went. Neither friend nor foe rated him very highly in philosophical matters at the time. The secondary literature that has come since has tended to concur with this judgement. Not only did his major opponent Karl Kautsky think him to be "theoretically a cypher,"[18] but others who stood on the same side of the debate in both political and philosophical terms, such as Kurt Eisner, cast doubt on the depth of his understanding of Kant.[19] Peter Gay's work on Bernstein argues that Bernstein's philosophical case against Marxism was something of an afterthought. Gay argues that, highly empiricist in temperament, Bernstein simply appended it to his attempt to refute Marxist conclusions on empirical grounds. He distrusted philosophical speculation as prone to lead to utopian constructions. Gay claims that Bernstein had not actually become a neo-Kantian, that his echoing of the call "Back to Kant" as valid for social democracy did not mean an acceptance of Kant's theory of knowledge or his ethics. It simply meant that he felt German social democracy to be in need of objective and keen criticism that rejected Hegelian dialectical apriorism and accepted the validity of ethical judgements.[20]

While there is something in this analysis, it goes too far. Though it is evident that Bernstein neither fully understood nor fully adhered to Kantian epistemology or ethics, he did accept certain basic Kantian philosophical assumptions and these were logically connected with his political theses. Of particular importance was his hostility to monism and his underlying acceptance of a dualistic methodology that involved one set of rules for the sphere of natural sciences and another for the sphere of human history. It was, in other words, the approach Marx had set himself against from his earliest years: one basis for science and another for life. Such a way of thinking was necessary to provide a justification for suddenly severing socialism, and indeed the whole process of historical development, from the sphere of science,* and bringing it into a quite separate sphere of ethics, thus removing the force of necessity and opening the door to voluntarism. There is also the factor that his rejection of Hegelianism was more than a rejection of apriorism and a distaste for philosophical system building. It was also an aversion to the dialectical form of development giving rise to what he often called the "revolution-mongering" of the left wing party. Its replacement with a more evolutionary pattern of development was the logical counterpart of his justification of the political strategy of reformist gradualism.

There can be no doubt, however, that Bernstein's excursions into the philosophical realm were marked by considerable confusion and inconsistency. At one point, he criticized the Fabians for reducing socialism to "a series of socio-political measures, without any connecting element that could express the unity of their fundamental thought and action."[21] At other times, he seemed to care little for such a connecting element or such fundamental unity. He was actually known to defend eclecticism as a "natural reaction to the doctrinaire desire to deduce everything from one thing and to treat everything with one method." He upheld it as "the rebellion of sober reason against the tendency inherent in every doctrine to fetter thought."[22] There were times when he seemed to believe he could leave behind him all doctrines and all theoretical constructions and move into an ontologically neutral realm of presupposition-less empirical investigation. He never really seemed clear about how to characterize himself philosophically. Once he flatly asserted: "My way of thinking would make me a member of the school of positivist philosophy and sociology."[23] Yet he could also be quite dismissive of positivism, extolling instead the virtues of idealism, especially in his disputations on ethics.

Liebknecht

Bernstein was not the only one whose ideas often seemed inconsistent. Others displayed even stranger combinations of ideas. Although there was a certain historical logic to the connection between political revisionism and philosophical idealism and to the corresponding connection between political radicalism and philosophical materialism, and although most disputants did

---

* Later he introduced various qualifications to this, but this was the basic direction of his thinking.

conform to this pattern, there were nevertheless some who crossed these lines. Karl Liebknecht, for example, one of the younger leaders of the party's left wing, was in fact a neo-Kantian in philosophy. He too denied that there was any necessary connection between the Marxist theory of history and philosophical materialism. His rather confused statement on the matter was to the effect that the materialist conception of history was not really materialist, for it "has no thread of materialism running through it, at least not in the actual philosophical sense. At the most, it has an undertone of materialism in the vulgar moralizing sense."[24] And there were others who crisscrossed these boundaries in even more complicated ways, particularly the very complex Austro-Marxists.

The Defence of Orthodoxy

The debate rapidly spread throughout European social democracy. The German SPD remained at the center of it, but all parties were drawn into it in one degree or another, particularly those in Austria and Russia. Bernstein's declarations brought an unprecedented flood of attack from many quarters. He drew upon himself, not only the disapproval of the veteran party leaders, Wilhelm Liebknecht and August Bebel, but also the fiery wrath of Rosa Luxemburg and bitter charges of treachery from Georgi Plekhanov. With varying degrees of passion and conviction, most of the leading figures of the International took him to task. At successive party congresses, there were lively and protracted debates on revisionism lasting days on end and finally a strongly worded resolution against revisionism was adopted by the Dresden Congress in 1903.[25] A similar resolution was passed by the 1904 Amsterdam Congress of the International. But everything really was kept at bay. The radicals' demand for the expulsion of revisionists was not successful, although Kurt Eisner and several of his colleagues did lose their posts on the party daily.

The critics who were grouped together under the category of "orthodoxy" were in fact of the most diverse sort. On certain issues they managed to put up a common front, at least for a time, but on other issues they held widely divergent views. Some were actually highly unorthodox figures, however broad an interpretation is put on the concept of orthodoxy.

Kautsky

The prime contender against Bernstein in the first phase of the controversy was Karl Kautsky. Kautsky too had long been an active social democrat and was a prominent party leader. He also lived for a long period in exile, and had associated during that time with Marx and Engels. He was founder and editor of *Neue Zeit*, the leading Marxist journal. In 1890, after having been won over to both Marxism and Darwinism while still a student in the 1870s, he returned to Germany and thereafter was an important figure at all congresses both of the SPD and of the International. He was the chief opponent of the center point of view, standing between the revisionists on the right and the radicals who had begun to emerge on the left. At the turn of the century, he was at the peak of his influence and his authority as the major spokesman of "orthodoxy" was virtually unchallenged.

Kautsky took on Bernstein on all points. In 1899, the very same year as Bernstein's major book appeared, Kautsky published a major book entitled *Bernstein und das Sozialdemokratische Programm: Eine Anti-Kritik* answering Bernstein's political and economic arguments and insisting that the perspective for the future was for crises, wars, and catastrophes, which were endemic to capitalism. It was not the absolute impoverishment of the working class that was the spur to the proletariat's struggle for socialism, but the intensification of class tensions, which was not necessarily incompatible with an increase in the living standards of the working class. Capitalism would necessarily collapse because private ownership of the means of production ruled out the most rational use of the technology it developed. But socialism could not simply evolve through reforms and concessions within the capitalist framework. Revolution was necessary in order to bring about the seizure of political power by the proletariat, a condition for the socialization of the means of production. Belief in the natural necessity of social processes, and in particular in the inevitability of socialist revolution, was to Kautsky the very cornerstone of Marxism. Determinism was absolutely essential to the Marxist scheme of things.

Kautsky is no easier to characterize philosophically than Bernstein is. He was by no means so consistent as commentators most often imply. The general line is that he carried forward the philosophical tradition of Engels, albeit in a scholastic spirit and in a relatively uncreative way, along the same lines as did Plekhanov and Lenin. This analysis is put forward by writers as diverse as Lichtheim, Kolakowski, Colletti, and Timpanaro. Although there is much textual and historial support for it, it is an inadequate analysis, in the light of his writings taken as a whole.

Kautsky did see himself in the same philosophical tradition as Engels, Plekhanov, and Lenin. Against Bernstein's attacks, he defended both dialectics and materialism. He asked polemically, "What remains of Marxism if it is deprived of dialectics, its best working tool and its sharpest weapon?"[26] About materialism, he seemed just as firm. A Darwinist even before becoming a Marxist, he had very early on embraced a materialist world view. He spoke of human activity always within its natural setting and within the framework of biological evolution. While criticizing the simple extrapolation of biological laws to human laws and insisting on the specificity of laws governing each level, he nevertheless saw human history as the organic extension of natural history. He looked on science with great respect and argued for unity of method, the bringing together of all spheres into a coherent theoretical system embracing the whole of nature and history.[27]

But there is reason to call into question just how deep was Kautsky's commitment to unity of method and just how serious was his defense of a dialectical and materialist philosophy. When Plekhanov first pressed him to take a stand against the neo-Kantian philosophy of the revisionists, he replied:

> In any case I must openly declare that neo-Kantianism disturbs me least of all. I have never been strong in philosophy, and, although I stand entirely on the point of view of dialectical materialism, still I think that the economic and historical viewpoint of Marx and Engels is in the last resort compatible with neo-Kantianism. . . . If Bernstein was moulting only in this respect, it would not disturb me in the least.[28]

He did proceed to take a stand against neo-Kantianism and to criticize Bernstein on philosophical grounds, but it was not as if he had changed his mind about the issue of the connection between the Marxist theory of history and dialectical materialism. By 1909, there were other philosophies besides neo-Kantianism being proposed to take the place of dialectical materialism, most prominently Machism,* which was gaining ground among Russian and Austrian Marxists in particular. When Kautsky was asked for his opinion of it by the Austrian Marxists, he wrote in the pages of their journal *Der Kampf* that he was himself a dialectical materialist and that he therefore opposed it. He nonetheless accepted the legitimacy of the endeavor, conceding that Marxist social and economic theory could be combined, not only with neo-Kantianism, but now with Machism or other philosophical variants as well. To justify this, he employed two contradictory lines of reasoning. (Ironically, both of these contradicted the line of reasoning he had employed against Bernstein). The first was to assert that "Marx proclaimed no philosophy, but the end of all philosophy,"[29] implying that Marxists should have transcended all the current philosophies. The second was to claim that Marxists should have a certain philosophical position, although it was a private matter that should not enter into the political arena. What really mattered politically was the Marxist conception of history and

---

* Machism refers to the philosophy of Ernst Mach, an experimental physicist and professor of inductive sciences in Vienna, who was the main representative of a trend known as empirio-criticism, that at this time was reviving the traditions of Hume. As its impact on Marxism came primarily in the form of controversy within the Russian party, it will be treated more extensively in the next chapter. In other parties, including the Austrian one where it had some influence, neo-Kantianism was in the center of the philosophical debate.

whether the conception is based on eighteenth-century materialism or on Machism or on Dietzgen's dialectical materialism, is not all the same for the clarity and unity of our *thought*, but it is a question that is entirely inconsequential for the clarity and unity of our *party*. Individual comrades may study this as private people, as they may the question of electrons or Weissmann's law of heredity; the *party* should be spared this.[30]

As the years went on, Kautsky continued to affirm this belief that the Marxist theory of history was independent of any particular philosophical position:

Philosophy concerns us . . . only in so far as it touches the materialist conception of history. And the latter seems to be compatible both with Mach and Avenarius and with many other philosophies.[31]

This took Kautsky a long way from Marx and Engels, as well as from Lenin and Plekhanov. Though he seemed to think this a minor matter, it was a matter of the most central significance to others. The unintegrated character of Kautsky's Marxism explains the contradictoriness of his thinking whenever he did venture close to the philosophical sphere. Despite a superficial tendency to schematization, Kautsky was far from a monist in his method of thinking. The overall direction of his thought was determined by a deeply entrenched dualism. This not only took him out of the mainstream development of Marxism, despite his reputation as the great defender of orthodoxy, but put him far closer to the revisionist trend than he cared to admit. There were times when he seemed to accept the methodological principle so anathema to Marx: one basis for science and another for life.

This came out most clearly in his case against the concept of ethical socialism. He asserted that the ethics was actually the weakest point of Kantian philosophy and was alien to the revolutionary process, for its sociohistorical tendency was one of toning down contradictions, of reconciling antagonisms, not of overcoming them through struggle. He argued too against its shifting of the theoretical basis for socialism from science to ethics, from inevitability to desirability, from determinism to voluntarism. But his arguments lost all their force when he conceded all essential ground to them by accepting an underlying dualism of fact and value, science and ethics, determinism and freedom. The argument of his major work on this subject, *Ethik und materialistische Geschichtsauffassung*, was extremely contradictory. On the one hand, he seemed to be going for a monistic solution, subsuming the sociohistorical world into the framework of biological evolution, and subsuming ethics within the system of causal determinism, going so far in this as to effectively reduce moral choice to natural instinct.[32] On the other hand, as the pages went on, he spoke of the inadequacy of scientific explanation and invoked the "moral ideal" as vital to the struggle for socialism. Here ethics was set in a realm quite outside the sphere of science, assuming human freedom beyond the reach of natural necessity, and endorsed as having its essential place in the life of social democracy. As he put it:

> Even social democracy, as the organisation of the proletariat in its *class struggle,* cannot do without the ethical ideal, without ethical indignation against exploitation and class rule. But this ideal has nothing to do with *scientific* socialism, which is the scientific study of the laws of evolution and the motion of the social organism. . . . It is, of course, true that in socialism the investigator is always both a militant and a man, who cannot artificially be cut into two parts having nothing to do with each other. Even in a Marx the influence of a moral ideal sometimes breaks through in his scientific research. But he rightly sought to avoid this as far as possible. For in science the moral ideal is a source of errors. Science is always only concerned with the knowledge of the necessary.[33]

It becomes obvious just how thoroughly Kautsky misunderstood Marx and how fundamental was his divergence from the path staked out by the founders of Marxism. Monism was of the very essence of Marxism. Despite the fact that Bernstein and Kautsky both still adhered to certain Marxist tenets, when it came to underlying patterns of reasoning, the dispute was conducted at times on basically non-Marxist terrain. Both failed to grasp the significance of the Marxist commitment to unity of method, to understand the degree to which Marx and Engels strove to break through all forms of dualism: between nature and history, between fact and value, between science and ethics, between determinism and freedom.

Mehring

Bernstein and Kautsky were not alone in this failure of understanding. Some of Bernstein's other critics were inclined to concede the same ground to him as did Kautsky. Others also brought back the sharp, impassable dividing lines between nature and history reflected in the tendency to separate the philosophy of nature from social theory. Another pillar of orthodoxy, biographer of Marx and party leader, Franz Mehring, was another who took this path. Mehring articulated the dichotomy in no uncertain terms:

> Historical materialism is a self-contained theory, designed for the cognition of the historical development of human society, a theory that draws its power from itself alone and allows itself to be amalgamated with the methods of natural science just as little as it raises natural scientific claims for itself.[34]

He seems to have grasped not at all the most basic presuppositions with which Marx and Engels proceeded in the philosophical sphere, for he projected his own views back onto them, flying in the face of all they had to say of themselves in philosophical terms:

> Marx and Engels always retained Feuerbach's philosophical standpoint, although they broadened and deepened materialism by applying it to the sphere of history. They were frankly just as much mechanical materialists in the field of natural science as they were historical materialists in that of social science.[35]

It is hard to know how this could be seriously maintained after the publication of Marx's *Theses on Feuerbach* and Engels's *Ludwig Feuerbach and the End of Classical German Philosophy* and so many other writings that contradicted this so very obviously. The irony of it was that these assertions came from a man who described himself as, more than anyone else, having concerned himself with the philosophical origins of Marxism. Mehring seemed to have rigidly dichotomized Marxism into two parts having nothing to do with one another and to have taken a somewhat flippant attitude to the philosophical part. At times he even bordered on repudiating it, making random anti-philosophical remarks of a somewhat philistine nature. In *Neue Zeit* he wrote that the "rejection of all philosophical fantasies" was "the precondition of the masters' immortal achievements."[36]

Austro-Marxism

Mehring's way of looking at things was not confined to the German party. There was the Austrian party as well. By 1907, the Austrian Social Democratic Party had become the largest party in parliament. Like the German SPD, it had become a mass party with a complex network of organizational forms. One could join a socialist glee club or take part in socialist Friends of Nature excursions. There was work for socialist singers and actors and folk dancers. James Joll points to those who, like Joseph Buttinger, could reminisce that it was from the workers' co-operative that he bought his first toothbrush and from the workers' library that he borrowed his first book.[37]

Viktor Adler, the founder and acknowledged leader of the party, ironically took the occasion of an article marking the death of Engels to suggest that, although Marxism was not only an economic doctrine but a world outlook, it was possible to replace its materialist foundations with Kantian philosophy.[38] This justifiably met with a query from Plekhanov who asked "What is one to think of a universal doctrine whose philosophical foundation is in no way connected with its entire structure?"[39]

Other Austrian Marxists explored further the possibilities of neo-Kantianism for Marxism. The Austro-Marxists, as the younger generation of party intellectuals came to be called,* came into the picture around 1904 and constituted a distinctive trend within the Second International. Vienna became at this time an influential center of Marxist thought, and the Austro-Marxist's journal *Der Kampf* soon came to occupy a leading place beside *Neue Zeit*. Stimulated by the revisionist controversy, they took up an intermediate position, setting themselves against revisionism on some issues, against the center on others and against the left wing on yet others. They were arrayed with the critics of Bernstein in that they wanted to preserve the revolutionary character of Marxism and to defend the scientific justification for socialism. They stood by the labor theory of value and the traditional concepts of class struggle, the historical inevitability of socialism and dictatorship of the proletariat. However, they differentiated themselves from Kautsky, Mehring, and other leaders of the SPD by their interest in questions of epistemology and philosophy of science and their concern with examining the broad theoretical foundations of Marxism in a serious way. In pursuing this, they differed too from those others who shared this interest within the International, such as Plekhanov and Lenin, by looking to neo-Kantianism and Machism for their orientation in this endeavor. They rejected Engels's materialism in matters of philosophy of science, subjecting it to a Kantian critique and declaring it to be full of holes. They proposed to reconstitute historical materialism on the foundations of a critical epistemology.

Max Adler

The most prominent among these young intellectuals when it came to philosophical problems was Max Adler. Adler was an active party member, a lawyer by profession, and founder of Marx-Studien and of Zukunft, a workers' school in Vienna. Adler set himself the task of formulating, in a critical and rigorous manner, the methodological principles of Marxism as a social science. In this, he found inspiration in Kant, although he insisted he did not see the relation between Kant and Marx in the same way as other neo-Kantians who wanted to supplement Marx with Kant. He then proceeded to do so himself in his own way. He criticized revisionists for eclectically linking Marx and Kant and then went on to claim that the Kantian critique actually provided the best method of defending the Marxist science of society against revisionist dilution. He rejected the claims of ethical socialism that socialism needed the ethical justification of the goals that could be provided by Kant's practical philosophy. What he felt should be taken from neo-Kantianism was

---

* The term was coined, during this period, by the American socialist Louis Boudin, who maintained an association with this group of thinkers over many years.

not an ethical theory but the idea of a critique of knowledge to be applied to the theoretical foundations of a science of society. Just as Kant's questions about nature as an object of thought arose from the emergence of Newtonian physics, so similar questions about society as an object of thought followed from Marx's construction of a scientific sociology. Adler saw the convergence of Marx and Kant primarily in epistemological terms, based on a reading of Marx that ascribed Kantian transcendentalism to him and that interpreted Marx as extending the notion of experience as a logical construct into the sociological domain.

For Adler, Marxism was defined as a social science, an investigation of the causal regularities of social life. On the question of the relation between social science and natural science, it must be said that he displayed a baffling inconsistency. One aspect of his thinking on this came through in his opposition to the Baden and Heidelberg neo-Kantians in the great methodological debate, the *Methodenstreit*, that was dominating German academia. This debate had to do with the neo-Kantian insistence on a special methodology for the humanities based on the view that the natural sciences reduced all phenomena to universal laws, whereas the human sciences were concerned with phenomena that were unique and unrepeatable. The method of the natural sciences was causal, whereas that of the human sciences was teleological. Adler argued, on the contrary, for the unity of science, a unity based on the causal methodology of the natural sciences. He believed there needed to be a rapprochement between the natural sciences and the social sciences that would allow them to be brought together within one framework. It was, he said, the entry of Marxism into the history of thought that made such a prospect viable. It was the Marxist concept of society that "prepared the way for that new attitude of scientific thought according to which human processes can be conceived in the same manner as natural processes."[40] Expanding on this theme, Adler continued,

Hence it is Marx's concept of sociation with its more precise content which makes it possible, for the first time, to overcome the division between nature and society which has been developed in modern thought, and to bring them together in a single scientific conceptual scheme. Nature and society now comprise the causal regularity of events as a whole.[41]

So, with Marx, it became possible for the first time, to extend the realm of science and to put social science "logically on the same footing as natural science."[42] However, on another occasion he contradicted this, asserting that "by its logical character, natural science is stationary, but social science, by its character, is revolutionary."[43] Going even further, he announced that "social science cannot be tackled with natural science methods" and that Marxism "rejected the natural scientific view of social phenomena and substituted it for a historical conception."[44]

Adler did add, however, that its rejection of the methodology of natural science did not mean a rejection of causality. He did adhere to an emphasis on causality with some consistency, but not even an equivocal use of the term natural science is enough to explain away the overall inconsistency of his views on the relation between natural science and social science.

When it came down to it, Adler was a dualist. In the last analysis, nature and history were separate and discrete spheres that could not be bound together within a single system of causal determinism, whatever gestures he made in this direction. The same pattern forced its way through in his discussions of the relation of science to ethics that was so topical then. In his critique of ethical socialism, he sought to integrate ethics into his concept of social science, to show its place in the system of causal determinism. Morality and valuation were explained as immanent causal factors of human history. Only when objective historical conditions were ripe for it could moral will become historically causal on any significant scale. In socialism, historical inevitability coincided with moral value. Adler opposed the reduction of *Sein* to *Sollen* on the part of Bernstein and the reduction of *Sollen* to *Sein* by Kautsky. But the integration did not really work and he constantly reverted to speaking of ethics in a typically neo-Kantian fashion, bringing to the surface the assertion that the moral dimension could have nothing to do with science, with any system of causal explanation. No scientific analysis could yield values. Only the critique of practical reason could provide the ground for moral will.[45] Both science and ethics were to be kept in their separate spheres, lest the one distort the other. For Adler, "the person who has fully understood this will never again allow the sphere of science to be disturbed by the concept of value."[46] The Kantian antinomies won out in the end.

Adler's dualism asserted itself fully in his overall characterization of Marxism as a social science that was ontologically neutral. It was a scientific explanation of social phenomena that did not imply either logically or historically, any particular ontological foundation. Like most other major theorists of the Second International, he felt that historical materialism could stand on its own, without any broader philosophical materialism underlying it. Marxist social theory needed no comprehensive world view to sustain it, least of all a materialist one, a closed world view reducing everything to matter in motion. He criticized philosophical materialism at length, arguing that philosophy could not take as its starting point the question of the primacy of mind or nature. Here the Kantian trancendentalist theory of knowledge obviously asserted itself over the Marxist realist one, which Adler inaccurately characterized as a passive and reductionist theory of knowledge, the supposedly inevitable result of materialist ontology. Philosophical materialism logically yielded a reduction of mind to matter and this logically yielded a passive epiphenomenal mind. His dualism complicated matters here, because at times he accepted realism in the realm of social theory, claiming that the materialist premise about the determination of thought by being was a useful maxim of investigation that facilitated causal explanation of social phenomena. But at the same time he insisted that any epistemological extension of this premise brought a recurrence of the most archaic forms of epistemological naiveté.[47]

Because of this epistemological orientation, Adler tended to waver in his philosophy of science between empiricism and transcendentalism. The empirical cast of his thinking came out in his arguments against other neo-Kantian tendencies which, in his opinion, were too formalistic, and concentrated too exclusively on the construction of abstract models divorced from concrete empirical investigation. It showed through also in his defence of causality over various forms of teleology in the social sciences. In contrast, the transcendentalist emphasis came to the fore in his rejection of philosophical materialism. Here at times it seemed as if there was no real need for the empirical world to actually exist as a grounding for the Kantian categories, existence being only a predicate of an existential judgement.

Adler argued that both the social sciences and the natural sciences were ontologically neutral. But evidently the natural sciences were not so neutral as not to exclude materialism. Materialism, he claimed, was utterly without foundation in the natural sciences, for the natural sciences had demolished the unintelligible abstraction—*matter*. Here Adler was influenced by the Machist interpretation of the new discoveries in physics and he read Mach back into Marx, writing that he saw in Marx "only a form of natural science positivism, more or less in the manner of Ernst Mach."[48] Later he wrote that Marx's so-called materialism was only "the positivism of modern science" and not an application of philosophical materialism in the traditional meaning of the term.[49] He saw Marxism as the completion of critical idealism, its extension into the domain of scientific knowledge and revolutionary action. As to the natural sciences, they must look to critical idealism not in order to extrapolate its results into a philosophical world view but to bring the critique of knowledge into play in the clarification of its methodology.

Adler did, however, defend dialectics and displayed a much more positive attitude towards Hegel than was current among the neo-Kantians. From Hegel came the idea of law-governed development, unfolding itself through opposition and resolution of contradictions. This Adler considered a step beyond Kant, that revealed the actual processes going beyond formal logic underlying the real flow of thought. It was a mode of thought based on the idea that the world constantly developed and obeyed a force immanent and inherent in itself. Calling upon the distinction between method and system, Adler took dialectics to be a method, not to be extended into a system, either in the idealist manner of Hegel or in the materialist manner of Engels.

This idea brought Adler directly into conflict with Engels's dialectics of nature. For Adler, the dialectic was the law of the mind and the law of the mind was not the law of things. Only thought was dialectical and nature was dialectical only in so far as it was an object of thought. The dialectic, as method, applied equally to social science and natural science. There was, however, for him, no justification for holding that nature itself was dialectical, for that would be to impose the forms of thought on being.[50] Adler never left aside his critique of a materialist *Weltanschauung*[51] and in fact went very far in the direction of providing a philosophical justification of religion, seeing it in time as the only way of bridging the yawning gap between the natural order and the human spirit that so bedevilled his thought.[52]

The Austro-Marxists were not entirely homogeneous in their thinking, though the main lines of Adler's philosophical approach were broadly shared by the group as a whole.

Friedrich Adler

However, Friedrich Adler had a distinctive twist to his thought in that he was more fully influenced by Machism than the rest. Adler was son of the party leader, lecturer in physics at Zürich, later party secretary, editor of *Der Kampf* and assassin of the prime minister during the war. He proposed a thoroughgoing synthesis of Marx and Mach[53] and, in this respect, was closer to the Russian Machists than to anyone else in the International in philosophical terms. He excused this most obvious departure from the classical sources with the comment, "Engels did not yet know Mach," which met with a rather biting reply from Lenin in his extended critique of the Machist interpretation of Marxism.[54] Adler's influence was felt primarily in his native Austria, especially among those who became known as the Vienna Circle. He helped to create the atmosphere in which the epistemological problems of the empirical sciences were discussed quite seriously and was particularly involved in examining the theoretical foundations of physics. His analysis of all these problems was consistently in the tradition of Mach. He saw Einstein's theory of relativity as confirmation of the Machist epistemology.

Bauer

Otto Bauer, a founder and editor of *Der Kampf*, at one time a government minister and best known for his work on nationalities, agreed that historical materialism did not imply any particular ontology and spoke of the importance of the Kantian critique of pure reason as the one solid barrier that would protect the working class against the sort of scepticism that would transform knowledge into deceptive appearance and moral action into the play of blind chance. But he opposed those who argued that an ethical justification for socialism was to be found in the critique of practical reason, recommending that those who wished to know what moral ideas could provide the driving force of class struggle should study, not Kant, but the proletariat. But even on this, he was more Kantian than he realized, and proceeded to discuss ethics in a typically neo-Kantian fashion, holding that no moral clues could be derived from historical necessity. This opinion, expressed in the midst of an ongoing debate with Kautsky in the pages of *Neue Zeit*, [55] met with a counter statement by Kautsky that neither could any moral clues be derived from the categorical imperative.

As the years went on, however, Bauer changed his mind about neo-Kantianism. Inclined to interpret philosophical movements in relation to social ones, he explained that just as Calvinism had responded to the needs of the bourgeoisie in the early stages of capitalist accumulation and Darwinism reflected the laws of capitalist competition, so neo-Kantianism was a manifestation of new needs on the part of the bourgeoisie. It was the expression of philosophical reaction following upon the defeat of liberalism and the bourgeoisie's abandonment of the militant materialism of its youth. Its fusion with Marxism corresponded to their need to win over the working class by eliminating the revolutionary character of its socialist ideology. Bauer criticized Max Adler, in retrospect, for never cutting the umbilical cord that tied his thought to the ideology of the bourgeoisie and for failing to recognize the social roots of Kant's philosophy in the bourgeois individualist eighteenth century. [56]

Hilferding

Like so many of the other major theorists of the Second International, the Austro-Marxists, despite all of their constructive efforts, seemed to be reverting to pre-Marxist modes of thought, rather than superseding Marxism with a more progressive theory. The Kantian dualism came into all their work, even when not dealing specifically with philosophical questions. Thus Rudolf Hilferding, who devoted himself to economics and the analysis of imperialism, wrote in his preface to his study of finance capital:

> It has been said that politics is a normative doctrine ultimately determined by value judgements. Since value judgements do not belong within the sphere of science, the discussion of politics falls outside the limits of scientific treatment. Clearly, it is not possible here to go into the epistemological debate about the relation between a science of norms and the science of laws, between teleology and causality . . . a politics is scientific when it describes causal connec-

tions. . . . Marxist politics is exempt from value judgements. It is therefore incorrect . . . to identify Marxism and socialism. Considered logically, as a scientific system alone, apart, that is, from the viewpoint of its historical affectivity, Marxism is only a theory of the laws of motion of society. . . . To recognise the validity of Marxism . . . is by no means a task for value judgements, let alone a pointer to a practical line of conduct.[57]

And so on all questions, the dichotomies set the tone and paved the way for the conclusions, making their position a very disjointed and unintegrated one, however praiseworthy their efforts to make it otherwise. The gap between science and ethics, between nature and history, between knowledge of the world and the struggle to transform it, was always there, unable to be bridged. It was not enough to want to bind the diverse spheres together. It was necessary to adhere to a viable and consistent methodology to make it possible, and in this they failed, despite the sincerity of their intentions and the seriousness and sophistication of their efforts.

The same may be said of other major theorists of the Second International, outside of Germany and Austria, who participated in these great debates. Various others added their own distinctive twists and turns to the search for the best direction in which the tradition should forge ahead into the new century.

Jaurès

In France, the two streams of thinking associated with Guesde and Jaurès entered into an uneasy alliance in 1905. Jean Jaurès, the leader of the socialists in parliament, assassinated on the eve of the war, was most certainly not of the main line of development of Marxist thought. Despite his professions of orthodoxy, described by Hook as "Pickwickian,"[58] Jaurès was openly critical of Marx and Engels. Nevertheless, he declared that Marxism had within itself the means to supplement and revise itself. Politically, Jaurès was a conciliator, accused by the orthodox of being a backslider, of being all things to all men, of glossing over differences, of blunting the edge of class struggle through generalized moralizing. Hook, more kindly, interprets him as "a socialist humanist more anxious to find a basis for common faith and action in a political opponent than a disguised class enemy in a critical party comrade."[59] There was in fact more at stake. He was in any case far more inclined to appeal to universal human values than to proletarian class interests. Even Jaurès, however, queried Bernstein's opinion that class divisions were becoming more fluid. There were still the haves and the have-nots. In the debate between Bernstein and Kautsky, he stood in the middle. He saw socialism primarily in ethical terms, putting ideas of justice at the very center of the picture. To him, socialism was the realization of all human values to which humanity had aspired over the ages. Nothing that the human spirit had ever created would be lost. The working class was building elements of socialism within the capitalist framework and from this effort would come the unity of politics and morality, revolution and evolution, materialism and idealism.

Although a professional philosopher, it was not to Marxism that Jaurès looked for his philosophy. He resisted any effort to interpret Marxism as implying an answer to the basic philosophical questions. To these questions he had his own answers, and they were very different from those of Marx and Engels. He opposed any attempts to commit the party to any particular ontology and wrote to Briand about a proposed resolution: "Renaudel wants the party to declare itself materialist and anticlerical. It is absurd. We are not metaphysicians. I will never abandon my freedom of thought."[60] His own thought in this area was influenced by neo-Kantianism, though he parted ways with the neo-Kantians by engaging in metaphysical system building over and above the Kantian critique. Jaurès's system was a kind of evolutionary pantheism, described by Kolakowski as "an attempt to reconcile almost every conflicting metaphysical standpoint and to show that all of them are basically right, but each is incomplete in the light of his universal theory of being."[61] In the general trend of the universe towards an absolute and final unity, all antitheses were overcome, all being turned towards perfect Being. All history moved towards final harmony. In this scheme, there could be no conflict between what is and what ought to be, for the course of nature and the rules of morality were derived from the same source and were heading towards the same goal. It was an idealist monism in the manner of Hegel, dealing with all that was out of bounds for the Kantians. Jaurès spoke favorably of the dialectic as a theory of natural evolution through which one social formation engenders another through the process of internal contradiction and he criticized Bernstein for putting it aside. Nevertheless, he often seemed closer to Bernstein here in assuming an evolutionary and unilinear progress in which development took place in a cumulative and continuous way. His rejection of materialism was quite definite; he explained that he could not accept the idea that the ultimate explanation of the universe could be found in matter. For Jaurès, it could be nothing less than the Absolute.[62] Although the influence of Jaurè's political ideas may have been considerable, in his philosophical efforts he was not much influenced by Marxism, nor did he much influence it.

Lafargue

The French thinker closest to the orthodox wing of the SPD was Paul Lafargue, son-in-law of Marx, a doctor by profession, participant in the Paris Commune, and cofounder along with Guesde of the French Socialist Party. He set himself firmly against all forms of political and philosophical revisionism and devoted himself to answering whatever manifestations of these made their way into the French scene. He interpreted the philosophical developments in a political context:

> At the beginning of the nineteenth century our bourgeoisie, having completed its task of revolutionary destruction, began to repudiate its Voltairean and free-thinking philosophy. Catholicism, which the master decorator Chateaubriand painted in romantic colours, was restored to fashion, and Sebastian Mercier imported the idealism of Kant in order to give the *coup de grâce* to the materialism of the Encyclopaedists, the propagandists of which had been guillotined by Robespierre.
>
> At the end of the nineteenth century, which will go down in history as the bourgeois century, the intellectuals attempted to crush the materialism of Marx and Engels beneath the philosophy of Kant. The reactionary movement started in Germany. . . . Bernstein, and the other disciples of Dühring, were reforming Marxism in Zurich. It is to be expected that Jaurès, Fournière and our intellectuals will also treat us to Kant as soon as they have mastered his terminology.[63]

Lafargue wished to carry forward the tradition of Marx and Engels amidst all this, but it must be said that he did not really understand fully the nature of the new stage in philosophical development represented by their thought. He was far more interested in the natural sciences and more serious about the philosophy of science than were most of his contemporaries, but in pursuing this interest he tended to revert to the old materialism of the French Enlightenment rather than develop the new materialism of Marx and Engels. His scheme of things was highly simplified and amounted really to an elementary and popularized sensationalism. Joravsky characterizes his views as "a startling union of mechanistic materialism with Cartesian rationalism."[64] This indeed came through in his discussion of the relation between science and ethics, motivated by his desire to answer Jaurès's contention that ideas of justice, rather than scientific ideas provided the theoretical basis for socialism. Lafargue wished to overcome the radical disjunction between scientific and ethical justifications for socialism and sought to do this by means of a sweeping reduction of all ideas, scientific and ethical, to sensations. Both were ultimately empirical in character, based on the physiological functioning of the human brain that transformed sensations into ideas as a dynamo transformed motion into electricity. In the course of this reasoning, Lafargue engaged in a rather idiosyncratic defence of innate ideas in empiricist terms; according to this, ideas became hereditary through certain molecular arrangements of the human brain transmitted genetically—a rather extreme form of the then wide-spread belief in the inheritance of acquired characteristics.[65]

But the synthesis did not really work, for it was based on a reductionism that left too much out of account. In the end, it brought him to see idealism as a necessary complement to materialism, for it supplied the ideals that were

missing in his view of nature as lacking any inherent purpose in historical
evolution, all thought having been reduced to sensation, all will to natural
necessity. His materialism was a rather crude sort—in his view, the brain was
an organ for thinking as the stomach was for digesting—which was why his
materialism was not self-sufficient and could not hold its own against idealism.
A reductionist materialism can never maintain its materialism, because what it
leaves out is always still there, demanding to be explained in one way or
another, if not in a materialist way, then in some other way.

Sorel

Another French thinker who considered himself a Marxist and took part in
the great debates of this period, though without any participation in the
organizational forms of the socialist movement, was Georges Sorel. To him,
Marxism was most emphatically not a scientific theory, but the historical
instrument of the revolutionary proletariat. This set him in firm opposition to
both the revisionists and the orthodox. While denying the scientific status of
Marxism put him at odds with such as Kautsky, his intransigent rejection of
class alliances and of any species of political compromise or participation in
bourgeois institutions, in fact, his total intolerance of anything less than total
revolution, put him far outside the orbit of Bernstein or Jaurès.

Sorel's ideas amounted to a fusion of those of Marx and Nietzsche. The need
of the hour, as he saw it, was for the rebirth of greatness, for the recovery of
primal values, for some monumental revivifying myth. Socialism was the
answer to this need and the militant proletariat was the instrument that would
bring it into being. Socialism was seen in moral, activist terms as a radical
revaluation of all values, and he was implacably hostile to all attempts to put it
in a scientific determinist context. In the spirit of the neo-romantic movement,
represented in France by Bergson, Sorel opposed all forms of both positivism
and rationalism. The notion of the "inexpressibility of the concrete" protected
all other notions against demands for rational or empirical justification. For
analytical reason, regarded by him as a source of decadence, he substituted the
primal mythopoetic act. All issues were set in apocalyptic terms. All
revolutionary activity was guided by one indivisible, unanalyzable idea,
pressing towards the final catastrophic and purifying destruction, making way
for a monumental outrush of creative energy that would make all things new.[66]

Addressing himself to questions of the philosophy of science, Sorel took
cognizance of the atmosphere of epistemological crisis and articulated an
extreme position against all forms of epistemological realism. His theory of
knowledge was an activist, intuitionist, pragmatist one. According to his
interpretation of scientific method, scientists were not discovering laws of
nature, but simply manufacturing models. Indeed, there were no laws of nature
for them to discover. Sorel believed that there was no sign of determinism of

any sort operating in the world. Hazard, blind chance, waste, and entropy prevailed in the world of nature. Science was a sort of rule-of-thumb intervention in this world, seeking to establish regions of determinism in a radically indeterministic universe. In this he was influenced to see mankind as struggling hopelessly to create islands of disentropy in a malevolent entropic universe by the most pessimistic conclusions that were being drawn at this time from the second law of thermodynamics.[67]

Obviously there was no idea of progress here. Nevertheless, it was not for the proletariat to be concerned about such things. Sorel said of himself that he "never tired of urging the workers to avoid being drawn into the rut of bourgeois science and philosophy."[68] Theirs was to destroy and to create. It was the colossal effort that mattered rather than what could be achieved. It was the heroic energy of the struggle that was more important than anything that could be won.

These ideas obviously took Sorel far from any effective influence on the main line of development of Marxist thought, but they did strike a chord with the adherents of Italian fascism in the 1920s and with certain sections of the New Left in the 1960s.

Labriola

A critic of Sorel who was much more influential within the International was Antonio Labriola. Although he stood aside at the formation of the Italian Socialist Party and was not a political activist, he nevertheless played a preeminent and crucial role in bringing Marxism into Italy. He was generally categorized as orthodox, although he was an original thinker and an unorthodox one in some respects. He did join in the criticism of revisionism, accusing Bernstein of having thrown in his lot with the liberal bourgeoisie and of abandoning socialism because the expected changes did not immediately come. There were, however, in his approach, hints of something akin to Bernstein's evolutionary socialism, in that he believed that new social forms could be grafted by degrees onto existing institutions. His critique of revisionism also took issue with those who he thought were depriving historical materialism of its philosophical foundations and leaving it suspended in midair.

Labriola was a professor of philosophy and was one of the most prominent theoreticians in the International when it came to discussing the philosophical foundations of Marxism. He disapproved of the prevailing neo-Kantianism as a regression, Kant having been superseded by Hegel. However, he displayed an extreme distrust of metaphysical system building, as well as of the positivist reaction it bred. By way of an alternative, he steered a historicist path, a line of development to be continued among later generations of Italian Marxists. He was the first major proponent of the trend interpreting Marxism as the "philosophy of praxis." His starting point was human historical practice, pursued in such a way as to imply rejection of epistemological realism. The cognitive process was seen as a historical and pragmatic one, rooted in the labor process, to be tested according to its functional success in human situations, and not as a discovery of truth already there to be discovered. Knowledge was not a matter of penetrating the secrets of nature, but of articulating human behavioral situations. All ideas, whether in science or philosophy or in any other field, were to be evaluated in functional terms, as human historical activity that could not be judged by its correspondence to anything outside itself. With praxis as his point of departure, he treated science itself as labor. Such a historicist approach, grounded in social labor as constitutive of the human historical world, he proposed as undercutting all dualisms. It exhibited, he said, a tendency to monism, but it was a critical tendency, keeping it from falling back into a simple empiricism or from ascending to a transcendental hyper-system.

As to the relation between philosophy and the natural sciences, Labriola firmly rejected the positivist predictions of the demise of philosophy. He did not believe philosophy had been made redundant by the positive sciences, as it still had its own role to play in formulating the general concepts necessary to unify experience and to summarize and scrutinize the results of scientific discovery. Science and philosophy were interwoven in a historical process that was always developing and leading to their mutual reconstitution. Philosophy was vital, due to the necessity of epistemological reflection on the sciences, of the integration of the sciences within an overall world view, of the anticipation of what science could not yet resolve. But philosophy, as critical thought, never completely coincided with the material of knowledge. Labriola at times seemed to see the fusion of science and philosophy as an ideal to be approximated but never fully achieved. At other times, however, he seemed doubtful of it, even as an ideal, casting a somewhat scornful eye on efforts in this direction, castigating the "mania of many to bring within the scope of socialism all the rest of science which is at their disposal."[69]

In his discussion of dialectics, Labriola warned that dialectics was neither an instrument of research nor a method of proof. It was "a comprehensive formula, valid, indeed, but *post factum*."[70] It was a rhythmic movement of understanding that strove to capture the general outline of reality in the making. It was vital for the mental faculty to remain fresh and alive and not to fall into a priori diagrammatic patterns.

Labriola not only criticized other trends in the international movement, but addressed himself to other trends in Italian Marxism. One the one side, he took exception to the views of his student Bendetto Croce, who at one time considered himself a Marxist, although even then he was a sort of crypto-Marxist. Most theoreticians of the International who referred to Croce classified him as a straightforward critic of Marxism, as they tended to do with Sorel as well. Croce represented an extreme neo-idealist reaction against positivism that went so far as to deny all cognitive value to the natural sciences. While Labriola had no time for such a regressive rejection of science, he also distanced himself from those he felt had been swept off their feet by the science of the time. He particularly criticized Enrico Ferri for subsuming the materialist understanding of history in the general theory of evolution and for attempting to place historical materialism under the protection of the philosophy of Comte or Spencer. Labriola believed that Ferri had lost himself in Darwinism and failed to emphasize the distinctiveness of Marxism in relation to it. Labriola stressed the continuity of human history in relation to natural history, but insisted that the social could not be reduced to the natural. If anything, Labriola was inclined to veer in the opposite direction, being, in Timpanaro's view, "a too impatient adversary of 'vulgar materialism.'"[71]

The political debate over revisionism in Italy took the form of opposition between the "integralists," a centrist trend in the Italian party led by Ferri, who was editor of *Avanti*, and the "reformists" who took up positions similar to Bernstein's.

Polish Marxism

Poland was another center of debate. The major theoreticians there were highly unorthodox and somewhat on the periphery of the international debate. Partly owing to the linguistic barrier, their ideas were not very well known in other sections of the International and their main influence has been exerted on subsequent generations of Polish Marxists. Even today, their major works are not in translation and they are barely known outside of Poland. This is true of many Polish works, which is unfortunate, because of the particularly high level of Polish logic and philosophy of science over decades.

Krzywicki

Ludwik Krzywicki was a founder of the first Marxist group in Poland. This group was in touch with the first socialist party in Poland, an underground organization known as the Proletariat that soon broke up under the weight of the political persecution of Russian Poland. Later Krzywicki was associated with the Union of Polish Workers, but when the socialist movement split into two camps, the PPS and the SDKPiL,* Krzywicki joined neither. He did, however, contribute to the journals of the PPS from time to time. As the years went on, his revolutionism was toned down considerably, and he began to take a more evolutionary approach to socialism, emphasizing the possibilities for the rationalization and democratization of the capitalist system.

---

* The PPS was the Polish Socialist Party, which could only be considered Marxist in the loosest sense. The SDKPiL, the Social Democratic Party of the Kingdom of Poland and Lithuania, cofounded by Rosa Luxemburg, professed its adherence to Marxist orthodoxy.

Philosophically, Krzywicki believed that historical materialism could stand on its own with no need for any particular ontological foundation. His own orientation showed the influence of neo-Kantianism and Machism. In epistemology, he was essentially a phenomenalist, believing that man apprehended the world in his own human fashion, making distinctions and categories that were instruments of prediction, but not reflections of objective reality. For him there were no laws of nature independent of human perception. Men imposed laws on nature and the whole scheme of evolution was simply a construction of the human mind.

Krzywicki was in fact hostile to Darwinism and particularly critical of the strain of Polish rationalism built on the theory of evolution. His own scheme was essentially a back-to-nature idealization of primitive communism that involved an extreme hostility to industrialization. There was in his thinking a great nostalgia for lost innocence and a belief that with the growth of urban civilization human creativity inevitably became drowned in a sea of mediocrity. His version of historical materialism was a much watered-down one, amounting really to a multifactoral analysis, with nothing left of Marxism but a vague reminder to take production relations into account when analyzing historical processes.[72]

Evaluating his role, Kolakowski takes the view that Krzywicki's eclectic approach was the reason why Polish Marxism failed to take on orthodox forms and tended to dissolve into a general rationalist or historicist trend. From the Marxist point of view, he was "not so much a battering ram as a Trojan horse."[73]

Brzozowski

An even more unorthodox Polish thinker of this period was Stanislaw Brzozowski, a provocative and explosive writer. Inclined to react violently and take up extreme positions, Brzozowski was at odds with all the political forces of his day. He died of consumption at the age of thirty-three and spent his last years under the shadow of suspicion arising out of the "Brzozowski Affair," his name having appeared on a list of Okhrana agents (tsarist secret police), under circumstances that are still unclear. He was one of the first to draw a sharp dividing line between Marx and Engels and to interpret Marxism in a historicist fashion as a philosophy of praxis. He thought more highly of Labriola and Sorel than of any of his other contemporaries. As for the rest, they were distracting attention from the original impulse of Marx's thought and had been led astray by Engels's diametrically opposite point of view.

What Brzozowski objected to in Engels was his linking of human history to biological evolution and the idea that history was subject to natural laws. In his radically anthropocentrist view of things, there was no nature already there to be subject to any laws. There was no nature except in and through human praxis. The world was coextensive with labor. It was a new twist to Kantian epistemology, the concept of labor standing in for the transcendental conditions of experience. He accepted the empirio-critical rejection of the idea of objective truth and interpreted science as the human organization of experience, rather than the discovery of any preexisting laws of nature. He repudiated all objective criteria of evaluation, whether in the sphere of science or philosophy or art or morality, putting forward the idea of creativity as a challenge to all forms of naturalistic determinism. He was implacably opposed to Darwinism and the idea of progress.

Brzozowski presented his "philosophy of labour" as an alternative to both the evolutionist faith in progress and the romantic cult of the self-sufficient ego. He was not really so far from the latter as he believed, the major difference being that he substituted the proletariat for the individual. Kolakowski points out that, for Brzozowski, the proletariat had all the traits of a Nietzschean hero. His notion of socialism was vague and highly adventurous, as was Sorel's. Moreover, he defined the proletariat by the fact that it performed physical labor and not by its place in the process of production. He was criticized for dissolving Marxism into an undifferentiated cult of labor.

Brzozowski rejected materialism as a philosophy that accepted the results while trying to forget the process. In any case, he believed his philosophy transcended both materialism and idealism. If there was nothing not immanent in human history, he claimed, the dispute between materialism and idealism fell to the ground.[74]

In the end, Brzozowski was drawn back to Catholicism. Kolakowski very perceptively comments that his conversion threw doubt on the possibility of maintaining a strictly anthropocentric viewpoint, radical anthropocentrism being impossible and contradictory because it implies that human existence is at the same time both contingent and absolute. Kolakowski describes Brzozowski's development as a road "from activistic narcissism via collective solipsism . . . to the church."[75]

Kelles-Krauz

The Polish thinker closest to SPD orthodoxy in this period was Kazimierz Kelles-Krauz, the chief theoretician of the PPS, who joined in the attack on revisionism within the International and defended Marxism against its critics from without. Like Kautsky, Mehring, and so many others at this time, Kelles-Krauz believed that Marxist social theory implied no particular solution to epistemological or ontological questions. His own epistemological position was a vaguely phenomenalist one. He looked upon knowledge in a typically historicist fashion, akin to Labriola in seeing the cognitive process as a sociohistorical phenomenon and not as a reflection of things in themselves. According to Kolakowski, during the period of the Second International, the idea of Marxism as a form of philosophical materialism scarcely existed in Poland.[76]

Plekhanov

In fact, the idea of Marxism as a form of philosophical materialism scarcely existed anywhere in the International, except among the Russians, and they were divided on this as well. The only major theoreticians of the period to argue for it as crucial and central to Marxism were Plekhanov and Lenin.* To them it was a very serious matter and they argued with extraordinary vigor and passion against those who called it into question.

In Plekhanov's polemics against revisionism, there was not a trace of moderation. To him, Bernstein was not a comrade of mistaken views, as he was to most German social democrats; he was an apostate, a traitor, an out-and-out enemy. He considered all those like Bernstein and Schmidt who sought to undermine the philosophical foundations of Marxism to be "stalking horses for the class enemy." They represented a "reapproachment with the advanced sections of the bourgeoisie" and they were "befogging the class consciousness of the workers."[77] "The bourgeoisie are now rejoicing," he said.[78] Against them all, he proclaimed all-out war. To those who tried to prevail upon him to temper his attacks, he responded sharply that he did not see why he should, argued violently against any trace of a conciliatory attitude in polemics, and urged on all the need to wage a struggle to the death. He called for the expulsion of revisionists from the party and was dissatisfied with the resolutions against revisionism passed at congresses of the SPD and of the International. Following the Stuttgart Congress of the SPD at which Plekhanov felt Kautsky had been too temperate in his arguments against Bernstein, Plekhanov addressed a fiery open letter to Kautsky, warning him that there was much at stake. The question was "who is to bury whom, whether Bernstein will bury social democracy or social democracy will bury Bernstein."[79]

On the philosophical front, Plekhanov was the first to join battle with the revisionists. He persuaded Kautsky to publish a series of articles against them in *Neue Zeit* in 1898. Even after Kautsky's plea for moderation and his pruning of the articles, they were still violently abusive in tone and Kautsky finally decided to curtail the discussion before the last of them had appeared.

---

* Among all the trends of the Second International, this strand in Russian Marxism was most influential in giving rise to a new line of development. As such, its story belongs to the next chapter. A fuller exposition of the views of Plekhanov and Lenin will come in chapter 3, along with the account of the controversies within Russian Marxism.

Plekhanov took Kautsky to task for his underestimation of the role of philosophy, insisting with regard to *Neue Zeit*: "It is essential to *force* the readers to interest themselves in philosophy . . . *it is the science of sciences*."[80] He was extremely disturbed by Kautsky's reluctance to cross swords over philosophy and insisted that this struggle was not only not superfluous but was obligatory.

Plekhanov analyzed the neo-Kantian revival as a reaction against materialism rooted in the psychology of the bourgeoisie of that epoch. They were inclined to resurrect Kant's philosophy as a bulwark against materialism, because they were alarmed by the appearance of the revolutionary proletariat on the scene of world history. The critical philosophy, for Plekhanov, was closely linked up with the class struggle. Its purpose was "to lull the proletariat into quietude."[81] Everything about it made it a perfectly suited instrument for this task:

> Kantianism is not a philosophy of struggle, or a philosophy of action. It is a philosophy of half-hearted people, a philosophy of compromise.[82]

The neo-Kantian anti-Hegelianism represented the philistine displeasure at the idea of dialectics, its fear of major upheavals. Its ethics supported the individualistic bias of the bourgeoisie against the proletarian realization of the social basis of moral values. Its inherent dualism tore assunder the living tie between thinking and being and thus inclined its adherents to lose their bearings. Its agnosticism opened the door again to religion and held science at bay.

Plekhanov believed that neo-Kantianism represented, not only a deep decline in philosophical thinking, but a threat to science. Much of it was directed to undermining confidence in scientific knowledge and to finding a place again for God and it was becoming an obstacle to further scientific development. As Plekhanov put it: "The philosophy of Kant, for me, signifies nothing else but an armistice between the discoveries of natural science and the ancient religious tradition."[83]

Against this, Plekhanov came to the defence of the scientific materialist world outlook and condemned the indifference towards philosophical materialism prevailing in the International. He put particular stress on the monist character of Marxist thought and argued forcefully against all forms of dualism, eclecticism, multi-factoral analysis, against anything which severed the wholeness of Marxism. He emphasized the inextricable unity of dialectical materialism and historical materialism. Marxism was an integral world outlook and dialectical materialism was the necessary theoretical foundation for socialism. With no equivocation, he parted ways with those he considered to have been only "temporary Marxists," stating very definitely: "One cannot be a Marxist yet reject the philosophical basis of Marxism."[84]

Lenin

Lenin agreed with Plekhanov's views and he praised Plekhanov as "the only Marxist to criticize the incredible platitudes of the revisionists from the standpoint of consistent dialectical materialism."[85]

To shed further light on the historical roots of revisionism, Lenin analyzed it in terms of the theoretical victory of Marxism against all opposing points of view within the labor movement. The Second International

> from the outset, and almost without a struggle, adopted the Marxist viewpoint in all essentials. But after Marxism had ousted all the more or less integral doctrines hostile to it, the tendencies that expressed themselves in those doctrines began to seek other channels. The forms and causes of this struggle changed, but the struggle continued. And the second half-century of the existence of Marxism began (in the nineties) with the struggle of a trend hostile to Marxism within Marxism itself.[86]

The historical triumph of Marxism had been so thoroughgoing that it had compelled its contenders to struggle against it, no longer on their own independent ground, but on the ground of Marxism itself. Indeed, in passing over into revisionism, those embodying essentially hostile trends came to represent themselves as Marxists. As Lenin saw it, non-Marxist trends were gaining hold in the labor movement because small producers were being forced into the ranks of the proletariat without having lost the attitudes typical of the petty bourgeoisie. The transitory, unstable, and ambiguous status of such social groups made them inclined to accept halfhearted and eclectic views and to tolerate contradictory principles and attitudes. Lenin noted that the division in international social democracy was everywhere proceeding along basically the same lines, but at least this was an advance over the situation in the First International. The German Bernsteinians, the English Fabians, the French Ministerialists, the Italian Reformists, and the Russian Criticists all belonged to the same family.

Bernstein's catchphrase about the movement being everything and the goal nothing expressed, in Lenin's opinion, the substance of revisionism better than many long disquisitions:

> To determine its conduct from case to case, to adapt itself to the events of the day and to the chopping and changing of petty politics, to forget the primary interests of the proletariat and the basic features of the capitalist system, of all capitalist evolution, to sacrifice these primary interests for the real or assumed advantages of the moment—such is the policy of revisionism.[87]

In the sphere of philosophy, the revisionists followed the bourgeois philosophy professors "back to Kant" and indeed, as Lenin saw it, back to a repetition of all the platitudes that priests had uttered thousands of times against materialism. In their rejection of dialectics, they were floundering "into the swamp of philosophical vulgarisation of science" and replacing revolutionary dialectics with a "simple and tranquil evolution."[88] One reason for the faint-heartedness of the heroes of the Second International, Lenin believed, was their complete failure to understand what he considered most decisive in Marxism, namely its revolutionary dialectics.

Lenin reacted sharply against any suggestion that philosophy (or religion) be treated as a private affair in relation to the party of the working class. Philosophical idealism was "nothing but a disguised and embellished ghost story."[89] Religion was reactionary; materialism was revolutionary: that left little scope for choice among the leadership of the revolutionary proletariat. He vigorously defended the principle of partisanship in philosophy and condemned in no uncertain terms those who employed high-sounding phrases about "freedom of criticism" to conceal unconcern, eclecticism, helplessness, or lack of principle in theoretical matters. He felt no attraction to the "freedom of the empty barrel." Those who had brought forth revolutionary advances in science were imbued with a sense of progress and truth and not inclined to demand freedom for their new discoveries simply to coexist with the old ideas. He reminded his opponents as well that, although freedom was a grand word, it was in the name of freedom for industry that the most predatory wars were waged; it was under the banner of freedom of labor that the working people were robbed.

Unlike those who argued calmly and dispassionately, Lenin as well as Plekhanov believed there was too much at stake for them to be in any way detached about the outcome. As Lenin saw it:

> We are marching in a compact group, along a precipitous and difficult path, firmly holding each other by the hand. We are surrounded on all sides by enemies and we have to advance almost constantly under fire. We have combined, by a freely adopted decision for the purpose of fighting the enemy, and not retreating into the neighbouring marsh, the inhabitants of which, from the very outset, have reproached us with having separated ourselves into an exclusive group and with having chosen the path of struggle instead of the path of conciliation. And now some among us begin to cry out: "Let us go into the marsh." And when we begin to shame them, they retort: "What backward people you are. Are you not ashamed to deny us the liberty to invite you to take a better road." Oh yes, gentlemen. You are free not only to invite us, but to go yourselves wherever you will, even into the marsh . . . Only let go of our hands, don't clutch at us and don't besmirch the grand word freedom, for we too are free to go where we please.[90]

The Russian Marxists had to oppose what they felt were revisionist ideas, not only among other sections of the International, but within their own party. There were very few anywhere who stood with them in their defence of the integrality of Marxism, fewer in fact than they seem to have realized themselves. Their references to Kautsky, Mehring, and Lafargue in philosophical matters reveal that they were under considerable misapprehension regarding where these others stood in philosophical terms.

Boudin

But here and there there were others, though not many, who insisted on the organic unity of Marxism, others who believed it must be accepted or rejected as a structural whole. A particularly vigorous rejection of revisionism and a particularly eloquent defence of the integrality of Marxism came from a not very well-known author of this period, Louis Boudin of the American Socialist Party. He believed that the fact that opponents of Marxism had taken up the more modern tactics of the revisionists showed that there was no longer any theory capable of competing with it outright. He protested against attempts to separate the philosophy of Marxism from its sociology and economics. In his words:

> Like the stones under the head of Jacob, so have the different elements which go to make up the Marxian system been welded by a superior power into one whole. . . . From the explanation of the hoary past, through the appreciation of the contending forces of the present, to the vision of the future—from the preface to *Zur Kritik*, declaring the laws of the historical march of civilization, through the intricacies and subtleties of the laws of value governing the capitalist system, to the sounding of the bells ringing out the old and decrepit capitalist system and ringing in the new and vigorous socialist society—the whole of the grandiose structure reared by Marx is hewn from one stone. Its foundations lie in the past, its framework embraces the present and its lofty tower pierces the future.[91]

Boudin, like Lenin and Plekhanov, seems not to have realized the extent to which even most of the so-called orthodox have deviated from the philosophical roots of the tradition and how few of his contemporaries at this time believed in the essential wholeness of the theory. After the war and the collapse of the American Socialist Party, he was no longer politically active, although he did maintain his association with the Austro-Marxists, with whom he had not so much in common as he seemed to believe.

The Marxist Classics in the Period of the Second International

The theoreticians of the Second International were almost all activists, whose way of life did not permit extensive and thorough scholarship. Moreover, publication and research facilities were far from what they are today. Many of the Marxist classics were still unpublished. Some that were published, along with various theoretical works published during the period of the Second International, were not widely known. As a result of this, many theoreticians of this period were unfamiliar with the views of Marx and Engels on many themes and were under various misconceptions about the views of their own contemporaries. A full picture of the range of theoretical positions at this time could only become clear in light of later scholarship.

The Influence of Dietzgen

Because of these widespread misconceptions, those who believed Marxism should be more than historical materialism, and even those whose philosophical inclinations were in the direction of those of Marx and Engels, looked to writers other than Marx and Engels for an epistemological and ontological supplement to historical materialism. For example, leading Dutch Marxists, Hermann Gorter, Anton Pannekoek, and Henriette Roland-Holst looked to the philosophy of Joseph Dietzgen, a philosophical autodidact and contemporary of Marx and Engels, who arrived independently at philosophical views strikingly similar to those of Marx and Engels and who did use the term "dialectical materialism." Pannekoek in 1906 recommended Dietzgen's philosophical works as constituting an "indispensable auxiliary" to the works of Marx and Engels.[92] In similar terms, Ernst Untermann, like Dietzgen a German social democrat who emigrated to America, saw Dietzgen as supplying to Marxism the epistemology it lacked.[93]

Writing in 1910 of the philosophy of Joseph Dietzgen and its relevance for the proletariat, Henriette Roland-Holst extolled Dietzgen's theory of unity in diversity, an infinity of interrelated objects and concepts, ever-changing, always connected. It was a philosophy, she explained, that opened up the possibility of a unitary materialist approach not only to social struggles, but also to science, art, and morality. Such a philosophy added a cosmic aspect to historical materialism. It brought together mind and nature, thought and action, ethics and aesthetics.[94]

Through the English translations of Dietzgen's philosophical works published by Charles Kerr in Chicago in 1906, Dietzgen achieved enormous popularity among working-class autodidacts in Britain and America. His books became the basic texts for courses in philosophy given within the network of labor colleges in Britain. There were even special Dietzgen Study Circles in Glasgow. Indeed for a whole generation of working-class militants, the name of Joseph Dietzgen was synonymous with the discipline of philosophy. The philosopher Dorothy Emmet reported on how, when taking classes among Welsh miners and announcing philosophy as the subject matter of the class, the response was "That's Ditchkin, isn't it?"[95] Marx and Engels were the authorities in political and economic matters and Dietzgen was the authority in philosophical matters. This state of affairs was due largely to lack of availability of the philosophical writings of Marx and Engels.

Dietzgen's philosophical writings, although irritatingly rambling and repetitious and full of points that seem obvious, even banal, to the modern reader, were pioneering efforts in their time, efforts of which Marx and Engels were duly appreciative. His ontology was one that saw reality as one, changing and interconnected. He was a forthright critic of both idealism and mechanistic materialism. His ethics was one in which the end justified the means. His epistemology was one which, although it put heavy stress on mental constructivism, was nevertheless firmly realist. He was very explicit in his affirmation of the independent existence of the external world.[96]

This, however, did not stop his son Eugene Dietzgen from interpreting his father in a phenomenalist direction and from himself adopting a radically phenomenalist position, that moved him from seeing knowledge as the organization and classification of experience to denying the existence for man of a world independent of human consciousness.[97]

Dietzgen put great stress on the inductive method in fact. His writings were permeated with a great respect for science, its methods, its achievements, and its goals. He strongly identified the mission of the proletariat with that of science. The proletariat was the appropriate bearer, not only of science, but of monism. It was in the interest of the enemies of socialism, advocates of a class-divided society, to adhere to dualism.

A Golden Age?

Louis Boudin, years later, in his correspondence with Friedrich Adler, looked back on the first decade of the century as a "Golden Age."[98] If he was thinking of the growing strength of the movement during these years, especially in light of the dissolution that came in the next decade, one can see why he looked back at the earlier period in this way. But certainly, as far as the development of Marxist theory was concerned, it is hard to see it as a golden age.

The generation that immediately followed, in the context of the Third International, did not think it was, no matter which side they took in the philosophical debates of the Comintern period. Lukács insisted there was a hidden connection between the two sides of the debate represented by Bernstein and Kautsky, for economic fatalism entailed ethical socialism in that such a severely limited perspective did not provide an adequate justification for revolution. Nothing, he asserted, was further from the minds of Bernstein's philosophical opponents than a defence of the Marxist philosophical tradition with its roots in the Hegelian dialectic.[99] Korsch also took a "plague on both your houses" approach. Kautsky's orthodoxy was the theoretical obverse and symmetrical complement of Bernstein's revisionism. Korsch objected vigorously to the then prevailing atmosphere in which it was not regarded as impossible for a leading Marxist theoretician to be a follower of Arthur Schopenhauer in his private philosophical life. As he summed up the relation of Marxism to philosophy in this period:

> Bourgeois professors of philosophy reassured each other that Marxism had no philosophical content of its own—and thought they were saying something important *against* it. Orthodox Marxists also reassured each other that their Marxism by its very nature had nothing to do with philosophy—and thought they were saying something important *in favour* of it. There was yet a third trend that started from the same basic position; and throughout this period it was the only one to concern itself somewhat more thoroughly with the philosophical side of socialism. It consisted of those "philosophizing socialists" of various kinds who saw their task as that of "supplementing" the Marxist system with ideas from *Kulturphilosophie* or with notions from Kant, Dietzgen or Mach, or other philosophies. Yet precisely because they thought that the Marxist system needed philosophical supplements, they made it quite clear that in their eyes too Marxism in itself lacked philosophical content.[100]

Another to look back on this period from the standpoint of the Hegelian revival that succeeded it was Siegfried Marck who characterized it as an era of doubt of the kind that usually followed on the heels of a time of dogmatic creativity.[101] The Soviet historian David Riazanov looked back in the 1920s on what he considered to be the deplorable state into which Marxist theory had fallen under German social democracy. The pages of one party's central organ were, he thought, filled with the most grotesque mixture of elements of various social systems, manifesting an utter lack of comprehension of the theory of dialectical materialism, the materialist theory of history, class struggle, or many of the foundational concepts of Marxism. Even Wilhelm Liebknecht, he said, had such a tenuous grasp of Marxist philosophy that he constantly confused the materialism of Marx and Engels with that of Büchner and Moleschott.[102] In any case, the subsequent generation were far from thinking of it as a golden age.

Kolakowski, however, is of the contrary opinion. In his *Main Currents of Marxism*, he has entitled the second volume, which deals with this period, *The Golden Age*. His justification for this is that during the years from 1889 to 1914, the doctrine had been clearly enough defined to constitute a recognizable school of thought, but was not so rigidly codified or subjected to dogmatic orthodoxy as to rule out discussion or advocacy of rival solutions.[103] In this respect, he is right to see it as a more attractive period than the one that came after. But what Kolakowski's analysis misses is the fact that many of the contending parties had failed in the first place to grasp the real historical distinctiveness, the organic wholeness of the theory. This is not to say that the debates of the period did not address themselves to serious and important problems, problems left far from resolved by the writings of the founders of Marxism. Still, in theoretical terms, it is hard to see it as a "golden age" when one considers that the tenuous hold the leading figures of the day had on Marxism from the beginning put them in a dubious position as far as developing it further was concerned. Many of the rival solutions were, in consequence, not rooted in a clear definition of what constituted the core of the theory that made it a recognizable school of thought. As Marxist scholarship was still in a somewhat underdeveloped state, so, sometimes, as a result, was the level of theoretical discussion.

Very few commentators today have shed much light on the Second International in philosophical terms. There is little written about it and much of what has been written is quite wide of the mark.

Althusser calls Stalinism the posthumous revenge of the Second International, for it took up the essential legacy of the earlier period that he defines as "the ideological pair, economism/humanism " that necessarily eliminated class struggle.[104] But it is exceedingly eccentric and against all historical logic to assert an organic and consubstantial link between economism and humanism, not to mention the fact that neither economism nor humanism is necessarily antithetical to class struggle. In any case, this does not take us to the essential philosophical point to be made about the Second International.

Colletti's analysis concentrates on what he considers to be the link between vulgar materialism, identified as the legacy of Engels, and social democratic revisionism. Supposedly, the main problem causing the distortion of Marxism during the Second International was its contamination by Darwinism. Asserting the very opposite of what in fact was the case, Colletti maintains that Engels's emphasis on philosophical-cosmological developments, his dialectics of nature, had a determinant weight for the succeeding generation.[105] Engels's philosophy of science, for better or worse, did not have such a determinant weight for the succeeding generation, nor is it fair in any case to identify Engels's philosophy with vulgar materialism.

A further point against Colletti's analysis is made by Timpanaro, who argues against his putting the weight of his emphasis on underscoring a connection between vulgar evolutionism and revisionism. He admits a connection, though he refuses to associate Engels and Kautsky with it, but insists that the accounts do not balance out so easily. What escapes Colletti is the significance of the shift in bourgeois culture around the turn of the century. Precisely because it reflected the hegemony of the bourgeoisie, in Timpanaro's analysis, revisionism followed the bourgeoisie in its passage from positivism to idealism. Colletti does not take account of the fact that revisionism actually leaned much more heavily on neo-Kantianism, empirio-criticism and semiidealism of various kinds than on any kind of materialism.[106]

The lines, of course, cannot be too sharply drawn, at least as far as the historical lineup of forces is concerned. Timpanaro also calls attention, for example, to the fact that within Marxism there were those who believed that the idealist renaissance could serve as a tonic against the gradualism and parliamentarism of the Second International. Although he seems to have in mind those of the following generation such as Lukács, Korsch and Gramsci, who came into prominence after the collapse of the International and who applied this sort of critique to their immediate predecessors, it is also true of the Austro-Marxists within the period of the International itself.

Ironically, it is Alfred Schmidt who puts his finger on the problem in maintaining that the inadequate understanding of Marx throughout the whole period of the Second International was essentially a consequence of the failure to grasp the connection between philosophical and historical materialism. He properly calls attention to the negation of this connection by Kautsky and Mehring, as well as in Karl Liebknecht and Max Adler. It is ironic, because in the very book where he makes this point, he implicitly negates this connection himself with his own rejection of ontological materialism.[107]

Nevertheless, it is the essential point. It was only a few who carried forward the mainstream line of development in insisting that Marxism implied a specific philosophy of science that was integrally connected with its philosophy of history and with its political program as well. The rest experimented with bits and pieces of other philosophies that could not always be satisfactorily integrated into a coherent framework for analysis. They failed to realize the nature of the revolutionary breakthrough achieved by Marx and Engels in transcending all forms of dualism at a higher level than previous monistic philosophies. Many fell back on the old dualisms. A few reverted to monisms superseded by the new stage of the development of science. The inability to bridge the gap between facts and values, between science and ethics, between the world being revealed in the development of the natural and social sciences and the revolutionary activity necessary to transform it, may in fact have limited their vision, sapped their energy, and paralyzed their will to act.

The Fate of the Second International

The philosophical debate in the International, such as it was, ended in deadlock. Political factors intervened and, at the onset of war, the International was in ruins. Despite all the reaffirmations of proletarian internationalism, the nationalism of the bourgeoisie, intent on pursuing imperialist wars, was the more powerful force. The various sections of the working class went their separate ways, allying themselves with those they had previously stood against in slaughtering those with whom they had vowed to stand always shoulder to shoulder. Not that all gave in. There was socialist opposition to the war and the brave efforts of those who dared stand against the tide were met with assassination or imprisonment. During the period of the International, many socialists had gone from jail to ministerial posts and then back again.

The socialist movement was rent down the middle and on different lines than in any of the previous divisions. Within the parties themselves, many who had heretofore stood together found themselves on opposite sides. The lines were drawn in unexpected ways. One would expect the "orthodox" to have held out against the war with the revisionists joining in, as this would seem to be the logical way of things and many historically did follow this logic. But there were many surprises. The war set Plekhanov against Lenin. Among the Austro-Marxists, it set father against son; Victor Adler for and Friedrich Adler against. In France, Guesde and Jaurès were still on different sides; only it was Guesde who was for the war and Jaurès who was against it. In Germany, the orthodox Cunow supported the war, while Bernstein was against it, although he did fail to vote against war credits from the floor of the Reichstag, unlike Karl Liebknecht. The SPD was so disrupted by the war that it could no longer maintain itself as a single party. In 1917, a new party, the Independent Social Democratic Party of Germany (USPD) was formed by those opposing the war, encompassing not only both Bernstein and Kautsky but also the left wing of Rosa Luxemburg and Karl Liebknecht. It won over a million members by the end of the first year.

With the victory of the October Revolution in Russia and with the end of the war, followed by widespread militancy among the working class, there came the really bitter and decisive parting of ways. For or against the new Soviet state, for or against the new militancy, for or against the Comintern, the movement split finally and irrevocably between communists and social democrats.

In postwar Germany, the socialists held the reins of power during the first period of the Weimar Republic. The SPD and USPD governed jointly, but the majority were so obsessed by fear of bolshevism on the one hand and so intimidated by the old bureaucracy and the old army on the other, that they could not govern decisively. The USPD left the government in December 1918 and at the same time the left wing of the USPD, Spartakusbund, broke away and founded the new Communist Party of Germany (KPD). With the Spartakus uprising in 1919, the SPD turned to the old army against the communists, crushed the rebellion and cruelly murdered its leaders Karl Liebknecht and Rosa Luxemburg. Kurt Eisner, at the head of government in Bavaria, was assassinated. In 1920, the USPD itself split between communists and social democrats, the majority joining the KPD. The wave of militancy eventually subsided, with the left divided into warring camps too strong to be suppressed by reaction and too weak to effectively take power. Eventually, their time had passed and in 1933 the Brownshirts marched through the Brandenburger Tor.

Even the more militant Italian Socialist Party, which had stood firm against the imperialist war, found itself unprepared for the tumultuous wave of postwar insurgency that spread through Italy and paralyzed by the effective counter-measures taken against it. Ironically, it was the former socialist, Benito Mussolini, who led the Italian fascist movement to power.

In March 1919, came the inauguration of the Third International, the Communist International. Nineteen-twenty saw the formation of communist parties in one country after another, whether in the form of the unification of hitherto disunited groups, as was the case with the Communist Party of Great Britain, or in the form of a split within the socialist parties, in France the majority forming the Communist Party, in Italy the minority.

Meanwhile, there were attempts to resurrect the Second International. After failing to do so in both Bern and Luceren in 1919, it formally reconstituted itself in Geneva in July 1920, the German SPD and the British Labour Party managing to rally others round. But it was really only a ghost of the past.

Through all this, the Austro-Marxists tried to steer a middle course between the increasingly reformist social democratic parties and the newly formed and fiercely militant communist parties. They were instrumental in the formation of the International Working Union of Socialist Parties which came to be known as the Second and a Half International, founded in Vienna in February 1921. For a time there were attempts at forming a united front of all three Internationals, but these in the end came to nothing. The short-lived Second and a Half International dissolved itself in May 1923 merging with the Second and forming the Labour and Socialist International. From 1923 to 1939, Friedrich Adler was secretary of this merged group.

In their theoretical work too, the Austro-Marxists sought to fashion a position that would transcend the communism/social democracy split. With the collapse of the Austro-Hungarian empire after the war, they were in government and in "Red Vienna" they worked to chart a road to socialism on the middle ground between bolshevism and reformism. But in 1934 came civil war in which the forces of reaction did not share their reluctance to take up arms. That for the time was the end of the middle ground and so the debate with bolshevism was settled for the moment in the bolsheviks' favor, if not in words, then in deeds.

It was the bolsheviks who were to open up the next path of dramatic development for Marxism, both in theory and in practice. In politics, in the philosophy of science, and in most other matters as well, the center of gravity had shifted eastwards.

NOTES

1. Karl Marx, cited by Engels, to P. Lafargue, 27 August 1890, *Correspondence* 2 (Moscow, 1960), p. 386.

2. Marx to J. Longuet, 11 April 1881, *Werke* 35, p. 178.

3. Engels to Bebel 20 August 1892, *Werke* 38, p. 433.

4. Engels to Lafargue, *Correspondence*, p. 386.

5. The classical study of the neo-Kantian revival and the various shades of opinion involved in the philosophical dimension of the revisionist controversy is Karl Vorländer's *Kant und Marx* (Tubingen, 1926). Vorländer himself was a prominent participant in the events under study.

6. Ibid.

7. Eduard Bernstein, *Evolutionary Socialism* (New York, 1961), p. 202. This is the English edition of his classical book expounding the revisionist position, which appeared in 1899: *Die Voraussetzungen des Sozialismus und die Aufgaben der Sozialdemokratie* (Stuttgart, 1899).

8. Eduard Bernstein, *Dokumente des Sozialismus*, 5 (1905), p. 421.

9. Bernstein, *Die Voraussetzungen des Sozialismus und die Aufgaben der Sozialdemokratie*, p. 26, 51. Certain passages in the German original do not appear in the English translation.

10. Ibid., p. 53.

11. Bernstein, *Evolutionary Socialism*, p. 211.

12. Ibid., p. 137.

13. Bernstein to Kautsky, 16 December 1927, Bernstein Archives, cited by Peter Gay, *The Dilemma of Democratic Socialism* (New York, 1952), p. 136.

14. Bernstein, "Zur Wurdigung Friedrich Albert Langes," *Neue Zeit* 10 no. 2 (1892), p. 102.

15. Bernstein, *Evolutionary Socialism*, pp. 222–223.

16. Ibid., p. 7.

17. Ibid., p. 18.

18. Karl Kautsky to V. Adler, 5 June 1901, *Viktor Adler, Briefwechsel mit August Bebel und Karl Kautsky* (Vienna, 1954), p. 355.

19. Vorländer, *Kant und Marx*.

20. Gay, *Dilemma of Democratic Socialism*, pp. 145–146, 152.

21. Bernstein, *Zur Geschichte und Theorie des Sozialismus* (Berlin, 1901), p. 177.

22. Bernstein, *Evolutionary Socialism*, p. 14.

23. Bernstein, "Entwicklungsgang eines Sozialisten," in *Die Volkswirtschaftslehre der Gegenwart in Selbstdarstellungen* 1 (Leipzig, 1924), p. 40.

24. Karl Liebknecht, *Studien über die Bewegungsgesetze der gesellschaftlichen Entwicklung* (Munich, 1922), p. 107.

25. *Protokoll über die Verhandlungen des Parteitages der Sozialdemokratischen Partei Deutschlands*, 1903, p. 103. The resolution was adopted 288 to 11.

26. Karl Kautsky, *Bernstein und das sozialdemokratische Programm: Eine Antikritik* (Stuttgart, 1899), p. 22.

27. Ibid.; as well as: *Ethik und materialistische Geschichtsauffassung: Eine Versuch* (Stuttgart, 1906), *Vermehrung und Entwicklung in Natur und Gesellschaft* (Stuttgart, 1910); *Die materialistische Geschichtsauffassung* (Berlin, 1927).

28. Kautsky to Georg Plekhanov, 22 May 1898, *Gruppa 'Osvobozhdeniia truda'* ed. Lev Deutsch, 6 vols. (Moscow, 1924–1928), 5:227.

29. Kautsky, "Ein Brief über Marx and Mach," *Der Kampf*, 10 (1909), p. 452.

30. Ibid.

31. Kautsky, *Die materialistische Geschichtsauffassung* 1, p. 28.

32. Kautsky, *Ethik und materialistische Geschichtsauffassung*, pp. 63–67.

33. Ibid., p. 141.

34. Franz Mehring, *"Zur Geschichte der Philosophie,"* *Gesammelte Schriften und Aufsatze* 6 (Berlin, 1931), p. 239.

35. Ibid., p. 337.

36. Mehring, *Neue Zeit* 1, 28, p. 686.

37. James Joll, *The Second International* (New York, 1966), p. 191.

38. Viktor Adler, in an introduction to Engels, *Contribution to the Critique of Political Economy*, the first Italian edition published in Milan on August 5, 1895, cited by Georg Plekhanov in *Fundamental Problems of Marxism* (London, 1969), pp. 21–22. (Originally published in 1908.)

39. Plekhanov, *Fundamental Problems of Marxism*, p. 22.

40. Max Adler, *Der sozialogische Sinn der Lehre von Karl Marx*, (Leipzig, 1914); pages 11 to 18 of this book are translated into English in a selection of texts of the Austro-Marxists edited by Tom Bottomore and Patrick Goode, *Austro-Marxism* (Oxford, 1978), p. 57.

41. Ibid., p. 60.

42. Ibid.

43. M. Adler, *Kant und der Marxismus* (Berlin, 1925); English translation in *Austro-Marxism*, p. 68.

44. M. Adler, "Zur Kritik der Soziologie Othmar Spanns," *Der Kampf* 20, pp. 265–270; English translation in *Austro-Marxism*, p. 71.

45. M. Adler, *Marxistische Probleme* (Stuttgart, 1913), pp. 108–109, 136–137.

46. M. Adler, *Kausalität und Teleogie im Streite um die Wissenschaft* (Vienna, 1904), p. 431.

47. M. Adler, *Marxistische Probleme*, pp. 66–67, 82–83.

48. Ibid., p. 62.

49. M. Adler, *Engels als Denker* (Berlin, 1920), p. 50.

50. M. Adler, *Marxistische Probleme*, pp. 20–21, 36–39.

51. M. Adler, *Lehrbuch der materialistischen Geschichts auffassung* 1 (Berlin, 1930); 2 (Vienna, 1969).

52. M. Adler, *Das Soziologische in Kants Erkenntniskritik* (Vienna, 1924).

53. Friedrich Adler, *Ernst Machs Uberwindung des mechanischen Materialismus* (Vienna, 1918).

54. F. Adler, cited by Lenin in *Materialism and Empirio-Criticism* (Moscow, 1970), pp. 50–51.

55. Otto Bauer, "Marxismus und Ethik," *Neue Zeit* 35 no. 2, (1905–1906), pp. 485–499; Karl Kautsky, "Leben, Wissenschaft und Ethik," pp. 516–529.

56. Bauer, "Max Adler: Ein Beitrag zur Geschichte des Austro-Marxismus," *Der Kampf* 4, August 1937, pp. 297–302.

57. Rudolf Hilferding, Preface to *Das Finanzkapital: Eine Studie uber die jungste Entwicklung des Kapitalismus* (Vienna, 1910).

58. Sidney Hook, *Marx and the Marxists* (New York, 1955), p. 74.

59. Ibid., p. 71.

60. Jean Jaurès, cited by Joll, p. 98.

61. Leszek Kolakowski, *Main Currents of Marxism* 2 (Oxford, 1978), p. 120.

62. Jean Jaurès, *Oeuvres* (Paris, 1931–1939).

63. Paul Lafargue, "Le matérialisme de Marx et l'idéalisme de Kant," *Le Socialiste*, February 25, 1900.

64. David Joravsky, *Soviet Marxism and Natural Science* (London, 1961, p. 15.

65. Paul Lafargue, *Social and Philosophical Studies* (Chicago, 1906), pp. 56, 74–76, 161, 354.

66. Georges Sorel, *Réflexions sur la violence* (Paris, 1908); *La Décomposition du marxisme* (Paris, 1908); *Matériaux d'une théorie du prolétariat* (Paris, 1919).

67. Sorel, *Les préoccupations métaphysiques des physiciens modernes* (Paris, 1907).

68. Sorel, *Les Illusions du progrès (Paris, 1947), p. 135.*

69. *Antonio Labriola, Socialism and Philosophy* (Chicago, 1934), p. 100.

70. Ibid., p. 209; Cf. also Labriola, *Essays on the Materialist Conception of History* (New York, 1966).

71. Sebastiano Timpanaro, *On Materialism* (London, 1975), p. 49.

72. Kolakowski, *Main Currents of Marxism* 2, chapter 9. Z.A. Jordan, *Philosophy and Ideology: The Development of Marxism-Leninism in Poland since the Second World War* (Dordrecht, 1963), pp. 56–57.

73. Kolakowski, *Main Currents of Marxism* 2, p. 207.

74. Ibid., chapter 11.

75. Ibid., p. 238–239.

76. Ibid., chapter 10.

77. Georg Plekhanov, "Cant against Kant, or Herr Bernstein's Will and Testament," *Selected Philosophical Works* 2 (Moscow, 1976), p. 365, 371.

78. Plekhanov, "On the Alleged Crisis in Marxism," *Selected Philosophical Works* 2, p. 316.

79. Plekhanov, "What Should We Thank Him For," *Selected Philosophical Works* 2, p. 351.

80. Plekhanov to Kautsky, in Lev Deutsch, ed., *Gruppa "Osvobozhdenia Truda"* 6, p. 257.

81. Plekhanov, "Materialism or Kantianism," *Selected Philosophical Works* 2, p. 413.

82. Plekhanov, *Fundamental Problems of Marxism*, p. 97.

83. Plekhanov to Kautsky, undated, probably 1895, at the International Institute of Social History, Amsterdam, cited by S.H. Baron in *Plekhanov* (London, 1963), p. 179.

84. Plekhanov, *Fundamental Problems of Marxism*, p. 95; *Materialismus Militans* (Moscow, 1973), p. 60.

85. Lenin, "Marxism and Revisionism," *Selected Works* 1, (Moscow, 1975), p. 51.

86. Ibid., p. 50.

87. Ibid., p. 54.

88. Ibid., p. 51.

89. Lenin, *Materialism and Empirio-Criticism* (Moscow, 1970), p. 170.

90. Lenin, "What Is To Be Done?" *Selected Works* 1, pp. 97–98.

91. Louis Boudin, *The Theoretical System of Karl Marx: In Light of Recent Criticism* (New York 1967), p. 255–256. (First published in Chicago, 1907.)

92. Anton Pannekoek, Introduction to *The Positive Outcome of Philosophy* by Joseph Dietzgen (Chicago, 1928), p. 63. (Originally published 1906.)

93. Ernst Untermann, *Science and Revolution* (Chicago, 1905), p. 161; and "Antonio Labriola and Joseph Dietzgen," in *Socialism and Philosophy*, op. cit., pp. 222–260.

94. Henriette Roland Holst, *De Philosophie van Dietzgen en Hare Beteekenis Voor Het Proletariaat* (Rotterdam, 1910). I am indebted to Dirk Struik for bringing this work to my attention and for a summary of its contents.

95. Dorothy Emmet, "Joseph Dietzgen, the Philosopher of Proletarian Logic," *Journal of Adult Education* 3 (1928):26. I am grateful to Jonathan Rée for bringing my attention to the extent of Dietzgen's influence in Britain. Cf. also the recently published book by Stuart Macintyre, *A Proletarian Science* (Cambridge, a1980), for an outline of this phenomenon.

96. Joseph Dietzgen *Philosophical Essays* (Chicago, 1906); *The Positive Outcome of Philosophy*.

97. Eugene Dietzgen, "The Proletarian Method," included in Joseph Dietzgen, *Philosophical Essays*, p. 61.

98. Boudin, cited by Tom Bottomore in his introduction to *Austro-Marxism*, p. 1.

99. György Lukács, *History and Class Consciousness* (London, 1971).

100. Karl Korsch, *Marxism and Philosophy* (London, 1970), p. 32.

101. Siegfried Marck, "Die Philosophie des Revisionismus," *Grundsätzliches zum Tageskampfe* (Breslau, 1925), pp. 23–30.

102. David Riazanov, *Karl Marx and Friedrich Engels* (New York, 1973), p. 207.

103. Kolakowski, *Main Currents of Marxism* 2, p. 1.

104. Louis Althusser, *Essays in Self-Criticism* (London, 1976), pp. 85–89.

105. Lucio Colletti, *From Rousseau to Lenin* (London, 1972), pp. 21–75.

106. Timpanaro, *On Materialism*, p. 119.

107. Alfred Schmidt, *Concept of Nature in Marx* (London, 1971), pp. 197–198.

CHAPTER 3

THE SHIFT EASTWARD:
Russian Marxism and the
Prerevolutionary Debates

Prerevolutionary Russia

It is one of history's ironies that the world's first proletarian revolution took place in a country in which 80 percent of the population were peasants. Tsarist Russia was extremely backward. The vast geographical expanse, the harsh climate, the poor distribution of mineral resources, the low level of development of science and technology, the slow pace of urbanization, the degree of control by foreign capital would seem to favor neither the development of advanced forms of production and social relations, nor the development of advanced ideas. What development did take place was extremely uneven, resulting in the simultaneous existence of tribal, feudal, capitalist and socialist elements side by side. It was a land of sharp contrasts in which could be found the extremes of the most radical ideas and the most primitive superstitions. There was the enlightened intelligentsia, under the influence of western ideas and reasonably affluent, and there was the dark mass of the Russian peasantry, plunged in grinding poverty and dark illiteracy. Above it all was the autocracy, its repressive apparatus so severe and far-reaching as to bring, not only violent uprisings among the peasantry, but the massive alienation of the intelligentsia. Its rule was such as to leave no middle ground between servility and rebellion.

The character of the Russian intelligentsia was markedly different from its western counterparts. Russia had been untouched by scholasticism, by the

113

Renaissance, by the Reformation. Its Enlightenment came late and took its own distinctive form. It had never been rent by the dualisms that came with the decline of the medieval synthesis and the ascendancy of capitalism, protestantism and modern science. It has not been torn by the wave of scepticism and relativism that had so tempered Western European intellectual life. The Russians were inclined to passionate monism, but they were not notorious for logical rigor.

Throughout the nineteenth century, all intellectual debate centered around the question: for or against the west. As the twentieth century approached, there emerged a new question, one of class. The question became: for the old aristocracy and the new bourgeoisie, or for the old peasantry and the new proletariat. The clash over the new question reopened and intensified debate over the older question in a new setting. This debate brought some of the narodniks, the more radical Slavophils, advocates of a kind of peasant anarchism, to reconsider their position. Leading the way, the young narodnik Georgi Plekhanov became a Marxist. As the first Russian to have embraced Marxism as an integral and all-embracing world view, Plekhanov is called the father of Russian Marxism. Those who came after all acknowledged themselves to be pupils of Plekhanov.

The Character of Russian Marxism

Russian Marxism merged within itself the radicalism of the Russian revolutionary tradition with the respect for reason and science characteristic of the westernizing intelligentsia. In so doing, it did have impressive predecessors— Belinsky, Herzen, and Chernyshevsky were among the great nineteenth-century Russian revolutionary thinkers charting such a path.

In 1883, Plekhanov and others formed the Society for the Emancipation of Labour, devoted to bringing Marxist ideas to bear on the Russian situation. Among those who came to join it was Vladimir Ilyich Ulyanov, who became known to the world as Lenin. For the most part, their activity was confined to writing and to the organization of study circles. Lenin's group, however, was involved in some degree of active work among the working class. The society's members were faced with isolation, subterranean existence, harrowing poverty, arrest, and exile, but they did introduce Marxism to Russia.

In the 1890s, Marxism began to achieve a considerable degree of influence among the working class and intelligentsia. It became, in fact, as Plekhanov later said, "the vogue at every St. Petersburg chancery."[1] Because tsarist Russia was a society of such extremes, Marxism came to take the place of liberalism for the bourgeois intelligentsia. In this, as both Plekhanov and Lenin saw the situation, there were both advantages and dangers for the working class. On the one hand, they were valuable allies and in fact the bearers of socialist ideology to the ranks of the proletariat, which could only generate of itself a trade unionist and economist consciousness at this time. Lenin pointed out that Marxism was brought to the proletariat from the outside by the most progressive elements of the bourgeois intelligentsia, including such people as Marx and Engels and, of course, Lenin himself. On the other hand, because of the absence of liberalism, Marxism appealed to people who, in terms of social status, mentality, or moral qualities, did not in fact identify with the proletariat and its liberation and who were inclined to adapt Marxism to the tastes and requirements of the bourgeoisie.[2] According to this analysis, these were only "temporary Marxists" and were seeking the liberal alternative within Marxism itself. This was the social basis of the Russian variants of German revisionism.

Because leading theoreticians of Russian Marxism were in exile in Western and Central Europe and because the Russian intelligentsia always turned an acute ear to western trends anyway, Russian Marxists were very much involved in the wide-ranging debate that was spreading through the International. It was, however, not only foreign trends to which Russian Marxists were addressing themselves, but trends within their own ranks. The Russians were extremely serious about philosophy, and coming from a tradition not shaped by protestantism and Kantianism, they were disinclined to believe it could be set apart in a separate sphere of its own or to think it was not politically important. Even those who came forward with epistemological or ontological alternatives to dialectical materialism were not so prone to argue that Marxism was philosophically neutral or that its politics could be combined with numerous alternative epistemologies. Some, directly under the influence of German neo-Kantianism and revisionism, did so, but the dominant tendency was to claim that a different philosophical standpoint was more in harmony with Marxist political strategy.

At any rate, Russian Marxism, from the very moment of its inception, became thickly embroiled in philosophical debate. The dominant stream, the dialectical materialists, as they began to call themselves,* had various opponents to be taken on.

---

* There is some dispute over whether it was Plekhanov or Dietzgen who was first to introduce the term "dialectical materialism." It has been customary to credit Plekhanov with it, though Adam Buick (*Radical Philosophy*, 10, Spring 1975) has claimed that it was Dietzgen. Jonathan Ree, in an unpublished manuscript entitled *Proletarian Philosophers*, seems to have established that it was Plekhanov after all. Whatever the facts are, it was Plekhanov who brought the term into general use in the movement as the designation of the Marxist position in philosophy.

The "Legal Marxists"

The first round of the battle was against the criticists: P.B. Struve, N.A. Berdyaev, M.I. Tugan-Baranovsky, S.N. Bulgakov, and S.L. Frank. They are most often referred to as "legal Marxists," as they lived and wrote under their own names and did not participate in underground activity, unlike those who lived hunted lives under pseudonyms or were sent into exile. The "legal Marxists" in some respects anticipated Bernsteinian revisionism and in others reflected it. Politically, they were in favor of evolutionary socialism, the gradual achievement of reform within the existing institutions. They went so far in their insistence on the necessity of the capitalist stage of development as to regard it as an end in itself, with the socialist movement as the means to achieve it, rather than the other way round. Struve, like Bernstein, influenced by the Fabians, enjoined Russian social democrats not to concern themselves with unrealistic projects of "heaven-storming" but instead to "learn in the school of capitalism."[3]

Like their foreign counterparts, the "legal Marxists" sought to disengage Marxist social theory (such as remained of it after their revisions in that sphere) from its philosophical foundations. They too argued that historical materialism could be reconciled with neo-Kantianism or positivism or other possible alternatives as well. For themselves, they tended to neo-Kantianism, believing that Marxism as a scientific determinist account of history did not encompass the sphere of ethical principles and that these must be derived from other sources. Both Struve and Berdyaev put heavy emphasis on the need for the sort of ontological foundation for values that led them eventually to the reality of the Absolute as a grounding for absolute values. Even during their period of active involvement in the social democratic movement, they were Marxists with so many reservations as to make their self-identification as Marxists highly questionable. They soon followed their line of thinking to its logical conclusion and abandoned the field, ceasing to consider themselves Marxists at all.

This coincided with the rise of a genuine liberal movement in Russia and the falling away of this group from Marxism reflected the emergence of a liberal alternative to Marxism within the bourgeois intelligentsia. Years later, Lenin summed up their place in the movement's history: "They were bourgeois democrats for whom the breach with narodism meant a transition from petty bourgeois (or peasant) socialism, not to proletarian socialism, as in our case, but to bourgeois liberalism."[4]

As they were essentially liberals and liberals everywhere were abandoning materialism, this group too believed that scientific optimism, faith in progress, and all forms of naturalism, had had their day and that idealism was on the march everywhere. Being Russian, they took the process to a more extreme stage than their opposite numbers in the west. Struve, Berdyaev, Bulgakov, and Frank all found their way back to Christianity. This transition was marked in 1903 by the publication of *Problems of Idealism*, in which these authors and others explained the reasoning behind their evolution from Marxism to idealism. It was a concerted attack on atheism, materialism, determinism, evolutionism, rationalism, and collectivism, in short, on all the hitherto accepted liberal values, as symptoms of intellectual impoverishment. They condemned Marxism for moral nihilism and contempt of personality, for sacrifice of the individual to the collective, for sacrifice of the present to the future. All the essays were pervaded by an intense personalism and longing for transcendence. The intrinsic value of human personality and the absolute validity of moral norms required, for them, the postulation of the existence of God and the immortality of the soul. They insisted that materialism was incompatible with morality, that laws of nature and history were incompatible with the freedom of the individual. They even invoked Nietzsche in associating socialism with mediocrity and the values of the herd.[5] Berdyaev, who became the best known in later years, continued for many years to engage in criticism of Marxism, both of its philosophy and of its politics and constantly reiterated the contention that dialectical materialism was an inherently contradictory philosophy based on absolutely irreconcilable elements: "Dialectics, which stands for complexity, and materialism, which results in a narrow one-sidedness of view, are as mutually repellent as water and oil."[6] Berdyaev was a powerful and brilliant writer and his later works often were remarkable for powerful insights into the Russian context of Soviet Marxism.

Marxists criticized this group for having turned their back on the oppressed in their glorification of egocentrism and in their adherence to religion, which they saw as an instrument of oppression. The foremost critic against this group was Lyubov Axelrod, who wrote under the pseudonym "Orthodox." A pupil of Plekhanov's while in exile, she held a Swiss doctorate in philosophy, was an active social democrat, and returned to Russia in 1906. She stressed the continuity between the natural sciences and human history and defended materialism against this reversion to idealism on the part of those who had previously stood with them. She defended determinism, both natural and historical, against the neo-Kantians, particularly Stammler who had objected that it was inconsistent to believe in both historical inevitability and revolutionary will. For Axelrod, this contention only reflected the fear of the future characteristic of a class that history had doomed to destruction.[7]

The Economists

Another trend the Marxists needed to take on was economism, which began gaining ground with the spread of a labor movement in Russia. This was the belief that the working class should confine itself to economic struggle and leave political (and of course scientific and philosophical) matters to the liberals. The economists believed that revolution would come spontaneously and inevitably (though in an ever more receding future) and they disputed the need for the formation of an independent working-class party.

The Bolshevik-Menshevik Split

Despite the controversy, an independent working-class party was formed. In 1899, a congress was held in Minsk for the purpose of launching a Russian Social Democratic Party, but it was unsuccessful and the small number of delegates were arrested soon after. However, in 1900, the weekly *Iskra* began publication, along with the theoretical journal *Zarya*, under Plekhanov, Lenin, and others. These organs were used to lay the foundations for a proletarian party. The first years of the century saw a massive strike wave, led mostly by Marxists, but there was as yet no organized party to provide overall coordination of this militancy. In 1903, a party congress began in a rat-infested flour mill in Brussels surrounded by Russian and Belgian detectives. After moving the congress to London, a bitter debate broke out over the nature of the party, with Martov advocating a loose mass organization and Lenin fighting for a tightly knit, disciplined, centralist party. When it came to the vote for the central committee and editorial board of *Iskra*, Lenin and his supporters were in the majority, becoming known as bolsheviks. the minority were called mensheviks. The mensheviks foresaw a long period of capitalist development and a bourgeois democratic revolution led by the bourgeoisie; their immediate goal was to form a legal party sharing power in a liberal government. The bolsheviks, however, believed that it would be possible to proceed directly from a bourgeois to a proletarian revolution and that even the democratic stage must be led by the working class in alliance with the peasantry. Their goal was revolutionary insurrection and to achieve it, both legal and illegal forms of work were deemed necessary. The bolsheviks were unable to maintain the edge they had secured at the congress, as the mensheviks succeeded in winning over Plekhanov and a few others and gained control of the central committee and *Iskra*. The bolsheviks split away and formed their own party, with its own central committee and paper. The mensheviks were supported by most parties of the Second International. Even Karl Kautsky and Rosa Luxemburg came out against Lenin. This was the situation on the eve of the revolution of 1905: in effect, two distinct social democratic parties with two distinct policies, each claiming to be the leadership of the working class.

The Revolution of 1905

Nobody planned the revolution of 1905. Out of the massive discontent following the defeat in the Russo-Japanese War, in a period of severe economic depression, and coinciding with a new wave of strikes, came the spontaneous revolt. On "Blood Sunday" a vast procession of workers, who had marched to the Winter Palace to present a petition to the tsar, was fired on by the troops. The massacre was followed by riots, street fighting and further strikes in city after city, town after town. The peasants also arose burning the houses and seizing the lands of their masters. The sailors of the battleship Potemkin mutinied. During the October general strike in St. Petersburg, the soviets sprang up and these councils of workers' representatives acted as the effective authorities in those towns where they became the focus of revolutionary activity. Most members of the soviets were arrested, but certain concessions were won from the tsar. The *October Manifesto* provided for the convening of a constituent assembly, the Duma. In December, there was a final armed uprising from Finland to the Ukraine. The Moscow Soviet controlled the city for nine days, but the tsarist government was still strong enough to crush these revolts and it did. Thousands upon thousands were killed in fighting, executed, imprisoned, or exiled. Punitive expeditions scourged the country. Troops burned workers' quarters wholesale, watched babies go up in flames, dragged the wounded from the ambulances. The tide had turned. The autocracy survived.

The years following were years of reaction. It was a time of the most severe repression, a veritable reign of terror. Even members of the Duma were arrested and imprisoned. It was a period of extreme depression, despair, and intellectual confusion, even within the ranks of social democracy. Plekhanov commented: "We are passing through a period of unprecedented intellectual decline."[8] Lenin remarked on how the disintegration, the disunity, and the vacillation reflecting the abrupt change in the conditions of social life had brought about a serious and profound internal crisis in Marxism, bringing those who previously had assimilated Marxism in "an extremely one-sided and mutilated fashion to revise the very theoretical foundations of Marxism."[9] The bolsheviks and mensheviks held conflicting interpretations of the 1905 revolution and of the political strategy flowing from it as a consequence.

Russian Marxism and the Philosophy of Science

But this was not the only area of debate. There emerged at this time a full-scale controversy in the area of philosophy of science, crossing the bolshevik-menshevik divide. There were a number of factors at work in bringing conflicting philosophical tendencies into play. One was the fact that the trend sweeping through the European intelligentsia early in the century was one of the antimaterialist reaction. The mood was against scientific rationalism and historical optimism. There was a resurgence of idealism and many varieties of mysticism. This had already touched the Russian intelligentsia, as was demonstrated in the defection of the legal Marxists from the ranks of Marxism and their conversion to out-and-out idealism, which struck an intensely anti-scientific and anticollectivist note. The disillusionment following the defeat of the 1905 revolution did much to strengthen this mood and bring it to bear in sectors that had heretofore not felt the force of it. It is undoubtedly a pattern running through the history of ideas that periods of historical optimism give the upper hand to materialist currents of thought, whereas periods of historical pessimism give the edge to idealism. It was this mood that began to penetrate more pervasively into Russian social democracy reeling under defeat.

The Crisis in Physics

Another major factor coinciding with this to bring about a major debate in philosophy of science was the effect of new developments within the natural sciences themselves. The turn of the century saw a crisis in physics in which the fundamental concepts of classical physics were being subjected to the most searching reexamination. The new discoveries of radioactivity, of the electron, of the structural complexity of the atom, of the electromagnetic field, of transformations in mass effected by transformations in velocity, called into question established notions of time, space, motion, matter and energy.

As scientists penetrated ever more deeply into the level of the microcosm and discovered there new properties of matter, some were inclined to believe the very concept of matter discredited and to interpret the new revelations as a refutation of materialism. Such scientists as the French physicist Louis Houllevique declared "The atom dematerialises, matter disappears."[10] Ostwald's energeticism regarded pure energy, motion without matter, as the basis of all change. Far-reaching epistemological conclusions were being drawn. For some, it was not simply the concepts of classical physics that were thrown into doubt, but the very cognitive validity of the scientific enterprise. Scientific concepts began to be seen as merely subjective means of systematizing and coordinating experience, from which no conclusions could be drawn regarding nature itself. In fact, for some, the very notion of nature itself was held to be meaningless. Necessity and causality were denied and it was being declared that there was no logical connection between facts and events, but only simple succession. The old confidence of scientists that they were unveiling the truth about the world was shattered. The old certainty that modern science was bringing to mankind reliable knowledge of nature was gone. Now, an all-pervasive perplexity prevailed among those who discussed such questions.

There was a great flurry in writing in the field of philosophy of science. Various schools of thought emerged that reflected both the antimaterialist/antipositivist mood of the time and the epistemological crisis within the natural sciences: in Germany and Austria, the empirio-criticism of Mach and Avenarius, in England, the phenomenalism of Pearson, in France, the conventionalism of Duehin and Poincaré. These trends represented a new critical form of positivism that brought the critique of positivism within positivism itself. They were attempts to defend the cognitive validity of the scientific enterprise in light of the new challenges to it by dissociating it from materialism and realism, believing that science could stand on its own with no need of such ontological or epistemological claims.*

Most Marxists responded differently, seeing no need to compromise so much ground. Rather than seeing holders of these views as allies in the struggle to defend science against obscurantism, Marxists tended to view them with implacable hostility. Indeed, they seemed to perceive them as a greater danger than the explicitly idealist and anti-scientific philosophers, accusing them of bringing idealism into science itself. They could not be forgiven for contesting the claim of materialism to be the philosophy in the closest alliance with contemporary natural science. The new trends presented materialism as the product of prescientific common sense or of science in a lesss mature stage of its development and so the new scientific discoveries and the new examination of the epistemological foundations of the sciences were said to mean that materialism had been superseded by science itself. Timpanaro even today analyzes these early twentieth-century antiobjectivist tendencies as corresponding to the need to accept and promote scientific progress, while stripping science of its charge of combative secularism. These schools rejected the idea that science had an irreparably materialist nature and that the only refuge from materialism lay in rejection of science. Instead, they sought to show that the new science was actually providing the best possible refutation of materialism. They represented an upgrading of science within the context of idealism. Materialism had no longer to contend only with an idealism denying science, but also with an idealism within science.[11]

The crisis in physics undoubtedly did mean a crisis for the old materialism. The notion of the qualitative homogeneity of material substance had formed the basis of the mechanistic picture of the world, and the foundations of this picture were decisively undermined by new theories such as radioactivity and the electromagnetic field. The question, then, was the status of the new materialism in all this. Marxists could argue with justification that it was mechanism that was in crisis and not materialism, for the dialectical form of materialism had decisively broken from the mechanistic presuppositions that

---

* Positivism was characterized by the assertion that science represented the only valid claim to knowledge. Its view of science was inductivist, progressivist, and reductionist. It represented a combative secularism, hostile to theology and idealist philosophy. Where the earlier positivists—Vogt, Comte, and others—had argued on a materialist and realist basis, later positivists shifted to a phenomenalist basis.

were being undermined. Indeed, some Marxists, proceeding on the basis of the structural heterogeneity and inexhaustibility of matter, did argue this way. They believed that the antireductionist philosophy developed by Marx and Engels had within it ample resources to develop further and to come to terms with the new forms of matter and new principles of motion being brought to light. Engels had said that with every new development in the natural sciences, materialism had to change its form. This was its chance to do so. Marxists at this time were faced with a very different situation from the one in which the founders had first enunciated the new philosophy of science. For Marx and Engels, materialism was in the air. Their task was to develop it to a higher stage and to stress what they called its dialectical character. But in the new situation, certain aspects of dialectics, such as the relativity of knowledge, had become accepted, but materialism was under attack.

This was the new situation in which Marxist philosophy found itself early in the century. Russian Marxists responded to it quite differently and so found themselves engaged in philosophical polemics dealing with very complex issues and calling for considerable sophistication. It was not simply a matter of contrasting a Marxist interpretation of the new discoveries with a Machist one, for in one group within Russian social democracy, Marxism merged with Machism and so there was a Machist-Marxist interpretation to contend with as well. Nevertheless, it is to the great credit of the Russians on both sides that, of all the Marxists of this period, it was they who faced the new situation most forthrightly and most explicitly. Ironically, it was the Russians who were on the cutting edge of the most advanced philosophical debate of the time and not the Germans, who, although from a far more advanced intellectual culture, were half-heartedly engaged in a struggle with a neo-Kantianism already superseded by Mach.

Machism

Ernst Mach was an experimental physicist, who sought to explore the epistemological roots of science in order to distinguish true knowledge from false or illusory claims to the title. In so doing, he came to the conclusion that it was necessary to purge physics of metaphysics. His philosophical starting point was Kant, but as time went on he began to realize "what a superfluous role was played by the thing-in-itself," and he returned to views like those of Berkeley and Hume "as far more consistent thinkers than Kant." He told how, on strolling in the open air on a fine summer day, the whole world all at once seemed to be "one complex of interconnected sensations."

In outlining the role of science in the light of this paradigmatic intuition, he took the view that science was an attempt to organize sensations in the most concise possible way, without any ontological significance. Its purpose was to select, classify, and record the results of experience in order to facilitate manipulation and prediction, not to discover the truth about the world. Explanation was a matter of condensed description and there could be no meaningful way of raising questions of causality or ontological status. A developed science simply expressed itself in terms of functional relationships of sensations to one another, according to the principle of economy of thought.

In pursuing the ideal of the unity of science, Mach's unifying concept was experience rather than matter. Experience in and through sensations was the only meaningful epistemological category, excluding any notion of source, cause, or reference. Physical objects were only relatively constant groupings of sensations. It was out of order to raise the question of the relation of experience to a world above and beyond itself. He rejected the charge of solipsism or subjectivism, maintaining that he was not treating material qualities as mental states but simply declaring meaningless the question of the relation of mind and matter. A resolute monist, Mach declared that, in the notion of experience, the dualism of subject and object was vanished. A further argument against the charge of solipsism or subjectivisim lay in the public character of science, in which evidence was a matter of shared sense data.[12]

Mach had pushed a phenomenalist account of science as far as it would go. It was a philosophy that obviously left a lot out of account, as its critics were quick to point out, but it was a serious philosophy that had its roots in the desire to give a new account of science that would meet the very real epistemological problems that were arising within science at this time. The characterization of Machism as a reactionary and fideist trend on the part of its Marxist critics was wide of the mark. Mach was himself a socialist and an atheist and his inspiration in formulating his philosophy of science was a thoroughly progressive and rationalist one. He wished to protect science from all forces that would intrude on it or threaten it. As one commentator has so deftly put it: Mach wished to keep "the door between laboratory and church firmly shut" and so he "barred the door of the laboratory from within."[13] Mach himself knew that he had left much out, but chose to endure an incomplete world view rather than find satisfaction in a seemingly complete, but inadequate, system.

Similar conclusions had been arrived at by others, such as Richard Avenarius, who also wished to free philosophy from the dualism of mind and matter by reducing all being to experience. Any question of the existence of a world outside human perception was wrongly framed, according to Avenarius, as even any notion of a universe with no one perceiving it inevitably posited an imaginary observer. Science, as the experience upon which it was based, was ontologically neutral.[14] Phenomenalists did not consider themselves idealists, but claimed to have transcended the distinction between materialism and idealism.

The Russian Machists and Their Critics

Among Russian Marxists who saw in empirio-criticism the basis for a monistic philosophy embracing the whole of experience in accord with both modern science and their revolutionary aspirations were Bogdanov, Bazarov, Lunacharsky, Gorky, Yushkevich, and Valentinov. These thinkers varied somewhat in their approach to the fusion of Marxism and Machism, some such as Valentinov being strict Machists, whereas others devised their own philosophical variants of empirio-criticism. Thus, there was Bogdanov's empirio-monism and Yushkevich's empirio-symbolism,* not to mention the more exotic offshoot of this movement called "God building" associated with Lunacharsky and Gorky.

The principal critics of this group, Plekhanov, Axelrod and Lenin, refused to be sidetracked by the various differences within what they called empirio-criticism. It didn't matter so much, they insisted, about how the Machists differed with each other or with Mach, and they concentrated on what united them: their rejection of the realist theory of knowledge, that, in they eyes, was essential to both modern science and to the revolutionary movement. The Russian Machists believed, on the contrary, that both science and revolution demanded an abandonment of epistemological realism, which they regarded as naive and precritical. Their critics responded that the opposite was the case. It was necessary, they said, to choose between Marx and Mach, for Machism undermined confidence in the cognitive power of science and sapped the will to act.

Bogdanov

The most important of the Russian Machists in terms of philosophical output was Alexander Alexandrovich Malinovsky, who wrote under the name of Bogdanov. A doctor by profession and an active bolshevik, he had very early on been influenced by Ostwald's energeticism. In Machism, however, he found a scientific monism that would overcome all dualism and fetishism and provide the foundation for the unification of all of his efforts both in the intellectual sphere and in the political sphere. He saw Machism as a form of social adaptation, aiming at the purest description of experience within the maximum economy of thought. In his three-volume work, *Empirio-monism,*

---

* Bogdanov's empirio-monism took as its point of departure Mach's empirio-criticism. Bogdanov felt, however, that Mach's theory had not overcome the dualism of mind and matter. He put forward a monism that considered the psychical and physical as two different modes of organizing the same experience. Yushkevich's empirio-symbolism was in the tradition of Berkeley's substitution of the notion of mark or sign for that of corporeal causality.

published from 1904 to 1906, Bogdanov put forward a full-blown philosophy of science shaped under the influence of Mach. He took as his starting point the Machist derivation of all knowledge from experience and interpretation of scientific concepts as the forms of coordination of perceptions. The main task, as he saw it, was to eliminate the chasm between mind and matter: "to find the way by which it would be possible to reduce all interruptions in our experience to the principle of continuity."[15] Empirio-monism was the solution to the problem of the dualism of the physical and the psychic. Experience was the unifying factor. Both the physical and the psychic were to be regarded as differently organized elements in one and the same experience: the psychic was individually organized experience, while the physical was socially organized experience. The criterion of truth was collectivity: what was true was what came to be socially accepted. This did not mean majority rule. In his time, Copernicus alone embraced the accumulated astronomical experience of the species. Others possessed only fragments of it and so it remained unorganized in its fullness before Copernicus. Knowledge was founded upon the collective labor process. The question of the conformity of experience to anything outside itself was meaningless, as experience encompassed all there was, there being nothing outside it. Bogdanov rejected the "thing-in-itself" as a useless multiplication of entities. He regarded it as an attempt to explain the known by the unknown, to explain what was accessible and experienced by what is inaccessible and unexperienceable.

Philosophy, as Bogdanov saw it, was never neutral; it always served a given class. The stress on the active, even constitutive, power of the human mind, as well as the emphasis on collectivity, were seen as appropriate, even necessary, to the revolutionary temperament. An understanding of causality as grounded in social labor constituted the truly proletarian perspective, according to Bogdanov. His interpretation of causality broke from the deterministic model. Human labor, involving forethought, decision, and skill, introduced a new factor into the process. This equation of instrumentalism with activism and of realism with passivity and quiescence was an influential factor in bringing Bogdanov's ideas a significant following among Marxists.

As to the relation of his collectivist subjectivism to the philosophy of Marx and Engels, Bogdanov argued that there should be less attention paid to the letter than to the spirit. He explicitly located himself within the Marxist tradition, especially as, "in the teaching of Marx, philosophy for the first time found itself, its place within nature and society, instead of above and beyond them." On the issue of materialism, he asserted that materialism was an inappropriate term for describing Marx's views, because "although Marx called his doctrine 'materialism,' its central concept is not 'matter,' but practical activity, live labour."[16] By his own complex logic, Bogdanov described his own system as, "although not materialist in the narrow sense," belonging "to the same category as materialist systems."[17] Materialism was a more progressive philosophy than idealism, which failed to grasp the fundamental role of human labor. On the question of dialectics, he was a bit clearer about the area of disagreement. He criticized the tendency to leave the limits of dialectics undetermined. Defining dialectics as "an organizing process going on by way of struggle between opposite tendencies,"[18] he admitted that dialectical processes existed, but denied that they were universal. He took Engels to task for attempting to apply laws governing ideal processes to real phenomena. For Bogdanov, the emphasis was on equilibrium, as a more fundamental state than struggle. The concept of organization was the keynote of his whole system. Empirio-monism in philosophy laid the foundations for the harmonization of all intellectual experience. Socialist revolution in the sphere of political activity would lead to the harmonization of all social experience.

If Mach was not to be treated kindly by his Russian critics, even less so was Bogdanov.

Plekhanov

The first to take up the challenge of the empirio-criticists on behalf of a more orthodox interpretation of Marxism was Georgi Plekhanov, who had been directly attacked by them. He brought to his criticism of the Russian Machists the same passion and outrage that he had brought to bear on his criticism of the German revisionists. Again he responded bitterly to pressures to tone himself down, asking, "Have you ever seen a cat with a mouse in its mouth? Try advising it to shorten or postpone its prey. . . . The same with me . . . Bogdanov must die now."[19] Again, too, he saw the matter, not only in terms of truth and error, but in terms of loyalty and treachery. He answered Bogdanov's complaint that he had been insulted at Plekhanov's addressing him as "Mr." rather than "Comrade" by sharply asserting "You are no comrade of mine . . . because you and I represent two directly opposed world-outlooks. . . . I call comrades only those who hold the same views as myself and serve the same cause."[20] Numerous personal insults were interspersed throughout his philosophical polemic.

Materialismus Militans, published in 1908, was an extended reply to Bogdanov, though Plekhanov dealt with the trend represented by Bogdanov in various other works as well. Disputing Bogdanov's claim that Machism was "the philosophy of twentieth century natural science," he appealed to the evidence of the natural sciences as providing the decisive refutation of Machism. The empirio-critical reduction of the external world to sensations coincided with the traditional idealist thesis "there is no object without subject," a thesis that could not be sustained in light of the fact that the growing knowledge of the history of the earth showed that the object existed long before the subject appeared. It could not be that reason dictated its laws to nature, as the discovery of evolution had made it clear that reason appeared in the organic world only at a very high rung in the ladder of development.

Machism, in Plekhanov's view, despite its pose of being the last word in modern natural science, was really a backward step, a return to the views of Berkeley and Hume. As Plekhanov described it: "Machism is only Berkleyism refashioned a little and repainted in the colours of twentieth century natural science."[21] He grounded his argument regarding the retrogressive character of Machism in his overall interpretation of the history of modern philosophy. Following Marx and Engels, he extolled both the eighteenth-century materialists and Hegel and regretted the tendency, which came with the neo-Kantian revival, to distort materialism and to pass over Hegel in silence. He thought that Kant should be given his due, but believed that the revival of his thought was called forth not by his strong side but by his weak one. It was his dualism that was most attractive to the ideologists of the bourgeoisie. Dualism, however, was the Achilles heel of Kantianism, and, as it could not be maintained, with consistency, one had to declare for one or the other of the two irreconcilable elements in Kantianism. Criticism of Kant could proceed either from the right or from the left. It could develop either towards idealism or towards materialism. Mach and Bogdanov had chosen the former, whereas Marxists chose the latter. The touchstone was one's attitude to the unknowable thing-in-itself. Both trends rejected it, the one in the direction of subjectivism, the other in the direction of objectivism. The choice was either solipsism and "the blind alley of the absurd" or the recognition of the objective existence of the external world, the decisive step of thought that cut the Gordian knot of Humian scepticism.[22]

Plekhanov believed the epistemology of empirio-criticism was riddled with contradictions and absurdities. No matter what contortions of logic Mach and Avenarius might resort to in order to safeguard their system against solipsism, he insisted that it couldn't be done. But even the occasional gestures towards materialism that emerged from time to time in Mach were absent in Bogdanov. According to Plekhanov, Bogdanov had made this philosophy "idealist from A to Z."[23] He sarcastically pointed out that Bogdanov's "higher criterion of objectivity," i.e., intersubjectivity, implied a time when the hobgoblin had objective existence.

In the same tradition as Marx and Engels and his contemporary, Lenin, Plekhanov was impatient with epistemological agnosticism and identified it with sophistry and cowardice. He agreed that it was merely "shamefaced materialism," a cowardly materialism that tries to perserve an air of decency. The English, he contented, were particularly prone to hold back and be afraid to go through to the end. He repeatedly referred to Engels's assertion that the best refutation of Kantian agnosticism was provided by experiment, by industry, by daily practical activity: "The proof of the pudding's in the eating." He believed that agnosticism contradicted the facts accomplished in time and space.

The fact remains that he did at times veer towards Locke in his own epistemological orientation. His theory of knowlege was a representational one, based on the notion of hieroglyphics: "representations of the forms and relations of things are no more than hieroglyphics—enough for us to be able to study how these things-in-themselves affect us, and in our turn, to exert an influence on them."[24] Lenin took exception to this as an unwarranted concession to agnosticism.

Lenin also criticized Plekhanov's critique of empirio-criticism as failing to set this new philosophical trend in relation to the new situation existing in science. While he had much to say that was of value regarding the relationship of science to philosophy and its relevance in the development of Marxism, he did fail to come to terms fully with the significance of the newer discoveries and the challenge they entailed for the further development of materialism. Even as regards older discoveries, he addressed them only in a very general way. When he discussed the relationship between Marxism and Darwinism, he stressed the continuity between the study of nature and the study of history and the significance of natural necessity and dialectical development, revealed by Darwin in the sphere of biology and by Marx in the sphere of history. In reply to the charge of the Czech philosopher Thomás Masaryk that Darwinism was not consonant with Marxism, Plekhanov took Darwin to task for not sufficiently separating biological categories from social ones and proceeded to assert that Marx and Engels adhered to the viewpoint of Darwinism in the sphere of biology, but that in sociology their viewpoint was that of historical materialism.

Plekhanov fought hard for his conception of the integrity of Marxism. Its organic unity was to be defended at all costs. The materialist explanation of nature was the foundation upon which the materialist theory of history rested. These could not be torn asunder without doing the most severe damage to the revolutionary potential of Marxism. The Machists were a threat to this. Although they did not hold that the Marxist theory of history was philosophically neutral, they did propose to substitute a different philosophical position, which they claimed was more consistent with it. But for Plekhanov, this was impossible. It would undermine the foundation upon which the whole edifice had been constructed and which could not be replaced without the whole structure falling to the ground.

Plekhanov was implacably hostile to the sort of eclecticism that he believed formed the basis of hasty syntheses of Marx with any and every thinker who came into fashion. He wondered how long it would take for someone to come along and supplement Marx with the thinkers of the Middle Ages. Speaking of the insoluble connection between the socioeconomic and philosophical aspects of Marxism, he complained:

> And since these two aspects cannot but hang in mid air when arbitrarily they are torn out of the general context of cognate views constituting their theoretical foundation, those who perform the tearing out operation naturally feel an urge to "substantiate Marxism" anew by joining it again quite arbitrarily and most frequently under the influence of philosophical moods prevalent at the same time among ideologists of the bourgeoisie—with some philosopher or other: with Kant, Mach, Avenarius or Ostwald, and of late with Joseph Dietzgen. . . . No attempts have yet been made to "supplement Marx" with Thomas Aquinas.[25]

When Plekhanov spoke in such terms in relation to Russian Machism, he tended to ignore Bogdanov's monism, to forget that Bogdanov too maintained an indissoluble connection between the socioeconomic and philosophical aspects of his interpretation of Marxism. Whatever else about it, Bogdanov's synthesis of Marxism and Machism was anything but eclectic and arbitrary.

Plekhanov paid a great deal of attention to the role of philosophical tradition in the formation of Marxism. He agreed with Engels that the great basic question in the history of philosophy was between materialism and idealism and he was intolerant of all who tried to steer a path midway between the two. The Russian Machists, according to him, were in this category. His deeply felt monism made him far more sympathetic to consistent idealism than to what he believed to be a philistine dualism that smugly stood halfway in between. As he saw the history of philosophy, the great philosophical systems had always been monist, that is, they had regarded spirit and matter merely as two categories of phenomena whose cause was inseparably one and the same. Profound minds were always inclined to monism, unable to feel satisfaction with many-sidedness. The great merit of Hegelianism was that it contained not a trace of eclecticism. Whoever went through this stern school learned consistency of thought and acquired forever a salutary repugnance for "eclectic mish mash." This was perhaps fair enough as regarded certain varieties of neo-Kantianism, but hardly fair in relation to Machism, which so much stressed consistency of thought and so ardently pursued the unity of science.

Nevertheless, for Plekhanov, the monism of the twentieth century needed to be one "thoroughly imbued with materialism."[26] In insisting that the material was the basis of the psychic, his monism was to be distinguished from that of the Machists. Materialism, he contended, was the philosophy most in accord with modern science; since a time was coming when what had happened in the natural sciences would be repeated in the social sciences, all phenomena would be given a materialist explanation. Materialism Plekhanov defined as "a doctrine that wishes to explain nature by its own forces."[27] It took nature as its point of departure and asserted the primacy of matter to spirit. Matter was "nothing but the totality of things-in-themselves, in as much as the latter are the sources of our sensations,"[28] a far from satisfactory definition that actually conceded far more ground to sensationalism than he seemed to have realized. In any case, Plekhanov believed that materialism was confirmed by the natural sciences and was the necessary theoretical basis of socialism. As such, he defended it against the prevailing antimaterialist "isms" and against all the popular misconceptions of it as drab, gloomy, and deadening, as immoral, as reducing man to "merely as a wavelet in the ocean of the eternal movement of matter."[29] He made the point that, after Feuerbach and Marx, materialism coincided with humanism.

Essential to this new stage of materialist philosophy, as Plekhanov saw it, was its convergence with dialectics: "The dialectical method is the most characteristic feature of present day materialism."[30] Darwinism had shed once and for all the notion of immutability of species and in all fields dynamic theories were replacing static ones. The idea of evolution had transformed everything, moving things beyond the stage at which earlier materialists could only resort to what they called reason. Dialectics, of course, was not identical with evolution; it implied leaps and abrupt transformations. It was a veritable "algebra of revolution," which to Plekhanov explained the psychological substratum of all the attacks on Hegelian dialectics: it directly confronted the philistine fear of revolutionary upheaval. In Hegel, however, there was a contradiction between method and system that only came to a resolution with the materialist transformation of dialectics. He conceived the relation between method and system differently in different contexts. However, the conception most in harmony with Plekhanov's monist inclinations was his assertion that method was the soul of any system. He regarded dialectics as the soul of Marxism. In his exposition of the laws of dialectics, he often left logical gaps between the illustrative instances and the principles they were supposed to illustrate, bringing criticism from Lenin, who accused him of reducing dialectics to an aggregate of examples.*

Plekhanov's most explicit discussion of the dialectics of nature came in his critique of Masaryk,** who had contended that dialectics and materialism were incompatible and there could therefore be no question of dialectics in nature. According to Masaryk, materialist dialectics was a *contradictio in adjecto*. Hegel's dialectics was "simple hocus pocus" and a "metaphysical cobweb."[31]

---

* Despite Lenin's criticism, this was a mode of presentation to be taken up by subsequent generations of Marxists.

** Thomas Masaryk was a Czech philosopher and politician who wrote an extended critique of Marxism. He was later, from 1918 to 1935, President of the Republic of Czechoslovakia. He was a serious thinker, who deserved better than Plekhanov's flippant exposition and critique of his views.

Marx and Engels were eclectic; they did not understand that dialectics was not for them, if they wanted to be materialists. The many contradictions in the details of Marxism sprang from the contradiction in the theoretical foundation of the entire system. Masaryk asserted that objective dialectics simply did not exist. In nature itself, there was no dialectical contradiction; there were no leaps. The examples cited by Engels as evidence of dialectics of nature were, according to Masaryk, only examples of nondialectical development, as they didn't conform to the Hegelian triad. Plekhanov came to the defence of Engels's conception and pressed the point that dialectics could by no means be reduced to the Hegelian triad. Masaryk, Plekhanov insisted, was incapable of grasping the distinction between a materialist and an idealist interpretation of the dialectic. It was one of many such debates on the nature and status of the dialectic that left the issues far from clear and quite unresolved.

So Plekhanov was determined to stand by the philosophy of science sketched out by Marx and Engels and to set himself against all comers. As far as he was concerned, the theory must remain intact, "an impregnable fortress against which all hostile forces hurl themselves in vain."[32]

Axelrod

Another who immediately came to the defence of orthodoxy against the empirio-criticists, Lyubov Axelrod, was "Orthodox" herself. She contended that a subjective idealism that treated objects as collections of sensations, thus making consciousness the creator of the world, could in no way be reconciled with Marxism. On the contrary, such an orientation was logically linked to social conservativism. Moreover, it made science impossible. Science both presupposed and substantiated the existence of the external world. Echoing Plekhanov, she held it was impossible to maintain that there could be no object without subject, as science had demonstrated that the earth existed long before man. Therefore, mind was the product of nature and not a condition of it. Human knowledge had its source in matter. Axelrod was not, however, so facile as Plekhanov in defining matter. She held, on the contrary, that matter could not be defined, for it was the primal fact, the original substance, the beginning, the essence and cause of all things. Even here she fell into the trap she warned against, for the history of science was developing towards the disclosure of forms of matter that could not be considered as substance. In epistemology, Axelrod was of the same mind as Plekhanov and held to a representational theory of knowledge. Human sensations were in a relation of correspondence to the external world. Against the Machists, she argued that this must be the case, otherwise it would be impossible to distinguish between true and false perceptions. There would be no point in it, if reality and sensation were one and the same. Against Lenin's critique of Plekhanov, she argued that sensations were not reflections of the world in the sense of copies or mirror images, but in the sense of their content depending on the objects producing them. She accused Lenin of holding, not to philosophical materialism, but to naive realism.[33]

Lenin

Axelrod's assessment of Lenin was not altogether fair. It was true that he sometimes went on for pages and pages seeming to reduce materialism to realism, and a very naive realism at that. But if his philosophical works were looked at as a whole, it would be clear that there was more to the picture. There were indeed passages lacking in philosophical sophistication, in which he skimmed lightly over matters to which professional philosophers gave much attention, or in which he was unnecessarily abusive of opponents.* But anyone put off completely by these things would be badly mistaken. For in terms of Lenin's grasp of certain philosophical issues at this historical turn, he left many in the shade. Although he was not a professional philosopher and although he turned his mind to many things besides philosophy, Lenin took philosophy extremely seriously and he displayed an interesting insight into the state of philosophy at this time and into the political and scientific context for its development at this historical juncture. In 1908, in response to the writings of the Russian Machists, and because he was not entirely satisfied with the efforts of Plekhanov and Axelrod, Lenin threw himself into an intensive study of philosophy and current scientific theory that culminated in the publication, in 1909, of a major full-scale philosophical work, *Materialism and Empirio-Criticism*.

The difference between Lenin and others was in his orientation to science. In his critique of the Russian Machists, he echoed many of the arguments already

---

* Lenin referred to his opponents as: "wretched eclectics," "pettifoggers," "fleacrackers" and "buffoons of bourgeois science." As to their ideas: "quasi-scientific tomfoolery," "conciliatory quackery," "pitiful sophistries," "ancient trash" (cf. *Materialism and Empirio-Criticism*). Lenin rejected the fundamental premises common to these various thinkers. Empirio-criticism, empirio-monism, empirio-symbolism, energeticism, etc., were among the "thousands of shades of varieties of philosophical idealism," always with another on the way. There was some justification for this; Lenin preferred to keep a sense of perspective and to focus on the underlying presuppositions. The scope of his research, shown in all of his philosophical works, especially *Materialism and Empirio-Criticism*, was extensive and impressive, enough to show that he was aware of most of the differences between the varieties of philosophical thought.

His standards of scholarship often left something to be desired. His assertion that "Marx and Engels scores of times termed their philosophy dialectical materialism" (ibid. p. 7), was untrue. His references to Kautsky, Mehring, and Lafargue reflected misapprehensions as to their views. But he was well acquainted with the literature relating to the philosophical discussion of the natural sciences at this time.

put over by Plekhanov and Axelrod, but, unlike them, he placed Machism within the context of the newest developments in the natural sciences and the whole complex of new philosophical trends sparked off by them. He addressed himself, not only to the empirio-critical revision of Marxism within Russian social democracy, but to the whole range of philosophical speculation on the new science. His analysis embraced all of the current trends, not only Machism, but conventionalism, instrumentalism, immanentism, energeticism, and the rest. He looked at the theories of Mach, Avenarius, and their Russian counterparts, along with those of Duhem, Poincaré, Le Roy, Pearson, Ostwald, and a host of others. Lenin realized, in a way that the others did not, the full meaning of the radical change in philosophical mood and the really far-reaching character of the epistemological crisis in the natural sciences. He knew that a crucial turning point had come. Science had erupted into a sharp and formative controversy, a controversy that was profoundly philosophical in character, and philosophy could not be allowed to develop in disconnection with it, Marxist philosophy least of all. For attempting to do so, he openly criticized, not only such compatriots as Plekhanov and Deborin, but also others, such as Joseph Diner-Dénes, who wrote an article in *Neue Zeit* entitled "Marxism and the Recent Revolution in the Natural Sciences," making the point that new discoveries, such as X-rays and radium, confirmed various conclusions of Engels, but ignoring the epistemological conclusions being drawn from the new physics.[34] To criticize Machism while ignoring its connection with the new physics, Lenin said, was to fall short of what was demanded of Marxists in this new period.

Lenin was acutely aware of the fact that the new period demanded of him something very different from what the previous period had demanded of Marx and Engels.* They had entered the philosophical arena at a time when materialism reigned and they therefore had devoted their attention to the theoretical development of materialism, to bringing it to a higher stage. They had not been so much concerned with the defence of materialism as with the vulgarization of it. They had warned against inconsistent, badly thought-out materialism, "materialism below, idealism above," and also against forgetfulness of the valuable fruits of idealist systems. They had not concerned themselves much with epistemology, as the objectivity of science was for the most part taken for granted. As Lenin explained it:

This is why Marx and Engels laid the emphasis in their works rather on *dialectical* materialism than on dialectical *materialism*, and insisted on

---

* Lenin saw Marx and Engels as of one mind in philosophical matters and challenged the assertions of the Russian Machist Viktor Chernov to the contrary. Chernov was one of the earliest to counterpose Marx to Engels and to accuse Engels of a crude, naive, and dogmatic materialism, while exonerating Marx.

historical materialism rather than on historical *materialism*. Our would-be Marxist Machists approached Marxism in an entirely different historical period, at a time when bourgeois philosophy was particularly specializing in epistemology, and, having assimilated in a one-sided and mutilated form certain of the component parts of dialectics (relativism, for instance), was directing its attention chiefly to a defense or restoration of idealism below and not of idealism above.[35]

Lenin, therefore, concentrated on the defence of materialism and epistemological realism, against the most challenging criticisms that had yet been made on this philosophy. The main thrust of his argument was that the crisis in physics was a crisis of growth. It was engendered by the very progress of science, though, to those in the middle of it, it mightn't seem so. The breakdown of old principles had brought a breakdown of confidence in the whole scientific enterprise and a rash of theories to the effect that science was not a matter of description, but simply convenient fiction, pure artifice, utilitarian technique, a method of notation, a process of symbolization. These ideas reflected real and serious problems in formulating for a new era the epistemological foundations of science. Lenin felt that the new physics was to be welcomed and due regard taken of the new level of complexity that had been introduced into the sphere of the philosophy of science. Nevertheless, as Lenin saw it, the essential point was to make a distinction between the new and progressive scientific discoveries and the reactionary philosophical implications that were being drawn from them. The new physics most emphatically did not necessitate the abandonment of materialism, as he saw it, but furthered the movement within science itself from a mechanistic stage to a dialectical one, even if scientists were not able to rise directly and at once from the old mechanistic materialism to the new dialectical materialism. Lenin felt that the logic of the new physics made this transition inevitable, however difficult and complex the process would be:

> This step is being made, and will be made, by modern physics: but it is advancing towards the only true method and the only true philosophy of natural science, not directly, but by zigzags, not consciously, but instinctively, not clearly perceiving the "final goal," but drawing closer to it gropingly, unsteadily, and sometimes even with its back turned to it. Modern physics is in travail; it is giving birth to dialectical materialism. The process of childbirth is painful.[36]

Within this context, Lenin made his defence of materialism. Against such as Valentinov who contended "the statement that the scientific explanation of the world can find firm foundation *only* in materialism is nothing but a fiction, and what is more so, an absurd fiction,"[37] he stated that the exact opposite was the case. More than ever did science require the solid grounding that only materialism could give it, but it needed to be a far more sophisticated and flexible materialism than heretofore. The concept of matter was as essential to physics as ever it was, but it needed to be extended beyond its previous boundaries. To Lenin, the idea that matter was disappearing was an indication of the fact that the limits within which matter had hitherto been known were disappearing, that certain properties of matter which were once thought to be absolute, immutable, and primary—such as impenetrability, mass, inertia—were revealed to be relative, mutable, and characteristic only of certain specific states of matter. What was happening was that human knowledge was penetrating more deeply into the structure of matter and discovering new forms of matter in the process, and would continue to do so. The electron would prove as inexhaustible as the atom. Older definitions of matter were becoming obsolete, but the concept of matter was not; it was simply expanding. Older forms of materialism had been superseded, but materialism had not; it was developing to a higher stage. It was pointless to throw out the baby with the bath water. According to Lenin, the sole property of matter with which philosophical materialism was bound up was its objectivity. Although he claimed that matter could not be defined, as a definition entailed bringing a concept within a more comprehensive concept, and this could not be done in the case of the ultimate and most comprehensive concepts, he nevertheless stated: "Matter is a philosophical category denoting the objective reality which is given to man by his sensations."[38]*

The new materialism, incomparably richer than the old, recognized the temporary, approximate, and relative character of every scientific theory of the structure and properties of matter. This was where dialectics came into it for Lenin.

Dialectics highlighted the inadequacy of all polar opposites. It brought to bear a healthy relativism, scepticism, and negation, without being reducible to these, always moving on to higher affirmations. It was the failure of scientists to grasp the dialectical nature of development that accounted for the tendency

---

* This is an unsatisfactory formulation, but highlights the problem of giving a definition of the primary category in a monistic philosophy, as well as the necessity of openness to new research in the natural sciences. It should be possible to acknowledge the comprehensiveness and flexibility of the concept, while accounting for its specificity vis-à-vis other possible primary categories giving rise to alternative monistic philosophies. For instance, materiality, as the basis of philosophical materialism, is intrinsically bound up with spatiality, temporality, motion, etc., in contrast to spirituality in so far as it forms the basis of philosophical idealism. In practice, Lenin operated with a more complex and specific concept of matter than that expressed in his stated definition. This is clearest in passages where he contrasted materialism with idealism and in his arguments against Ostwald's energeticism. Lenin, in contrast to Ostwald, who held there could be motion without matter, held that there could be no motion without matter and no matter without motion. Engels had earlier argued that it was in the essence of matter to be in motion and hence in Engels's thought, too, there was a more complex implicit definition of matter than the one explicitly

of a minority of scientists, who previously held mechanist views, to slip by way of relativism into idealism. The new physics, as Lenin saw it, was wavering between dialectical materialism and phenomenalism.

The majority of scientists, however, still adhered to a spontaneous materialism, though it needed to be philosophically developed. Lenin analyzed the thinking of such scientists of the day as Huxley, Hertz, Helmholtz, Volkmann, Boltzmann, and Rey, who could not suppress their instinctive materialism, but were unable to carry it through to an integral and consistent philosophy of science. It was the natural way of things that, even at this time of crisis, most scientists would persist in learning more in the direction of materialism than idealism. For every scientist who had not been led astray by professorial philosophy, or indeed for any healthy person who had not been an inmate of a lunatic asylum or the pupil of an idealist philosopher, Lenin insisted, sensation was the direct connection between consciousness and the external world. For the idealist philosopher, on the contrary, sensation was a wall separating consciousness from the external world.* Materialism, based on epistemological realism, took the objectivity of science as the starting point, whereas idealism, including the new varieties of it represented by empirio-criticism, conventionalism, and the like, had to construct elaborate detours to deduce the objectivity of science from the collectivity of consciousness.

The difference between materialism and idealism for Lenin lay in their respective answers to the question of the source of knowledge and its relation to the physical world. Materialism, in accord with the evidence of the natural sciences that had decisively established the existence of the earth prior to man, asserted the primacy of matter and regarded sensation, consciousness, and thought as secondary, as products of a long evolution, as the historical outcome of a very high stage in the development of matter. Its fundamental premise was the existence of the external world, outside the human mind, of objects independent of sensations. Idealism, on the contrary, rested on the primacy of consciousness over matter, of mind over the external world, of sensations over objects. Materialism also adhered to the objectivity of causality, time, and space. Against all the various idealist interpretations of these concepts as *a priori* forms of sensibility or as organizing categories of human experience, they were real properties of the physical world. They did not exist in themselves, apart from matter, but they were real qualities of

---

stated. Another problem with Lenin's definition of matter, as given, was that it left no way of distinguishing ontological materialism and epistemological realism. These do need to be distinguished.

* Lenin made much of the fact that the "naïve realism" of everyday life formed the basis of the materialist theory of knowledge. There is a point in this line of argument, but it is dangerous to carry it too far. An exaggerated emphasis on common sense can become reactionary in a period when science is leading further and further away from the world of common sense. Engels warned against this even before science had begun to move so dramatically in this direction. Lenin was aware of this danger as well, although there are passages where it certainly did not show.

matter. There was true necessity in the world as well as objective succession in time and arrangement in space.

Lenin was anxious to answer the educated philistine's criticism of materialism: that it considered personality to be a *quantité négligeable* and converted man into an automaton; that it made the world dead, devoid of all sound and color; that it did not grasp the epistemological critique of concepts; that it simply imposed preconceived solutions. Lenin showed that the new materialism did indeed give scope to human personality and human freedom, within the context of natural and historical necessity; that it was disclosing the world to be even richer, livelier, and more varied than it seemed; that it had already moved beyond the Kantian critique to a higher stage; that it was open to each new step in the development of science, always discovering ever new aspects in nature.

Between the two fundamental trends in the history of philosophy, Lenin declared, there could be no third way. Lenin was as contemptuous as Plekhanov of all the then current attempts to bridge the gap, to transcend both materialism and idealism through the concept of experience. He insisted that it simply couldn't be done; the middle ground could not be maintained with consistency and always tended to lean one way or the other. Machism, he maintained, led inevitably to either solipsism or supernaturalism. The difference between materialism and idealism was fundamental and irreconcilable and it was necessary to choose. In the case of the Machists, they too had chosen. As Lenin saw it, despite all their proclamations to the contrary, they were subjective idealists; they opened again the door to fideism that had been closed by materialism and, whatever their intentions, they were acting in the interests of the clergy. Mach, although personally an atheist, by proclaiming the neutrality of philosophy on this question and holding that religious opinion was a private affair, had shown too much servility to the theologians. If Bogdanov were right in believing truth to be only an organizing form of human experience, then clearly the way was open for religion as another organizing form of human experience. In undermining confidence in science, they encouraged others to seek for other paths to the truth, to look for alternatives to science, most often in the direction of religious mysticism. Like Marx and Engels, Lenin was an atheist of the most militant kind and fought very passionately against all justifications of religious belief. His response to Lunacharsky's flirtation with religion, known as "God-building," was particularly harsh. Reflecting both the influence of Machism and the increase in religious interest after 1905, Lunacharsky and Gorky proposed a reconstructed religion, an anthropocentric and purely immanent religion for socialism. The idea was to embrace all that was positive in traditional religion, that is, the sense of community and man's yearning for transcendence, without adhering to belief in God, the supernatural world, or the immortality of the soul.[39]

The position of the Russian Machists, Lenin concluded, was a logically untenable one: hostility to dialectical materialism, and, at the same time, the claim to be Marxists in philosophy. It was, he contended, impossible to maintain. The very distinctiveness of Marxism in philosophical terms lay in the refusal to recognize such hybrid projects for reconciling materialism and idealism as Machism, and the determination to move forward along a sharply defined philosophical road. Marxism was an integral philosophy, "cast from a single piece of steel," from which not one basic premise or one essential part could be taken away without its being transformed into something quite different. This did not mean that it could never change or develop. Lenin distinguished here between revision, which was an essential requirement of Marxism, and revisionism, which entailed betraying the essence under the guise of criticizing the form. So highly did he value integrality that he quoted with favor Dietzgen's expression of preference for the honesty and consistency of religious believers to the halfheartedness of agnostics, dualists, and all halfway elements like the Machists, for "there a system prevails, there we find integral people, people who do not separate theory from practice."[40] There was a point worth making in standing on the side of integrality and wholeheartedness, but it was not fair to categorize the Russian Machists as among the halfway elements.[41] In Bogdanov, too, a system prevailed. Nor was he one to separate theory and practice.

As time went on, Lenin became more discerning in relation to philosophical positions with which he disagreed. This showed itself in his *Philosophical Notebooks*, writings dating from 1914–1915, published posthumously in 1929–1930. In this period, when he was concentrating on coming to grips with Hegel, as well as continuing his studies of the history of philosophy and current scientific theory, the emphasis was somewhat different than in *Materialism and Empirio-Criticism*. In the earlier work, he had directed all his fire against attacks on materialism, whereas in the later notes he turned his attention to the positive aspects of idealism and the danger to materialism in neglecting such insights. This attitude was summed up in his exceedingly wise statement: "Intelligent idealism is closer to intelligent materialism than stupid materialism."[42] There was a new level of philosophical sophistication in that he no longer tended to depict philosophical idealism in terms of stupidity or trickery on the part of fools and charlatans, but instead saw it as a one-sided, myopic development of human knowledge. Philosophical idealism was not groundless. It was a "sterile flower" growing on the living tree of human knowledge. Its roots were both social and epistemological. It was not so much a matter of blindness or deceit, but of myopia:

> Philosophical idealism is only nonsense from the standpoint of crude, simple, metaphysical materialism. From the standpoint of dialectical materialism, on

the other hand, philosophical idealism is a one-sided, exaggerated . . . development . . . of one of the features, aspects, facets of knowledge into an absolute, divorced from matter, from nature, apotheosized. Idealism is clerical obscurantism. True. But philosophical idealism is (more correctly *and* in addition) a road to clerical obscurantism through one of the shades of the infinitely complex *knowledge* (dialectical) of man.

Human knowledge is not (or does not follow) a straight line, but a curve, which endlessly approximates a series of circles, a spiral. Any fragment, segment, section of this curve can be transformed (transformed one-sidedly) into an independent, complete straight line, which then (if one does not see the wood for the trees) leads into the quagmire, into clerical obscurantism (where it is *anchored* by the class interests of the ruling classes). Rectilinearity and one-sidedness, woodenness and petrification . . . voilà the epistemological roots of idealism.[43]

Most of what Lenin had to say about dialectics dates from this period, although he also discussed dialectics in the earliest of his philosophical works "What the Friends of the People Are and How They Fight the Social Democrats," written in 1894. Here he differentiated the Hegelian dialectic from the Marxist one and tried to dispel the popular notion that Marxists tried to prove things by triads. He claimed that the dialectical method coincided with the scientific method in sociology, regarding society "as a living organism in a state of constant development and not as something mechanically concatenated and therefore permitting all sorts of arbitrary combinations of separate social elements."[44] In "The Three Sources and the Three Component Parts of Marxism" written in 1913, Lenin defined dialectics as "the doctrine of development in its fullest, deepest and most comprehensive form, the doctrine of the relativity of human knowledge that provides us with a reflection of eternally developing matter."[45] In "Karl Marx," written in the same year, he saw dialectics as:

a development that proceeds in spirals, not in a straight line; a development by leaps, catastrophes and revolutions; breaks in continuity; the transition of quantity into quality; inner impulses towards development, imparted by contradiction and conflict of the various forces and tendencies acting on a given body . . . ; the interdependence and the closest and indissoluble connection between all aspects, a connection that provides a uniform and universal process of motion, one that follows definite laws."[46]

But Lenin's most elaborate unfolding of the dialectical method was in the *Philosophical Notebooks* where his attitude to Hegel changed somewhat from the time in which he wrote "What the Friends of the People Are" and he became more inclined to emphasize continuity with the Hegelian dialectic, though he was still at pains to point out the differences. Here he saw the essence of dialectics to be the "all-sided flexibility of concepts," "living, many-sided knowledge with an infinite number of shades," "the internal source of all activity." The dialectical concept of development was far richer than the simple notion of evolution. What distinguished it was the leap, the contradictoriness, the interruption of gradualness. He put great emphasis on unity of opposites as the kernel of the whole process. He went into very great detail about the characteristics of the dialectical process, at one point listing sixteen elements of dialectics, although these could be summed up in three: (1) the thing itself considered in the totality of its relations and in its development, (2) the internally contradictory tendencies in every phenomenon, and (3) the union of analysis and synthesis.[47] Lenin's thought during this period was far more subtle than his earlier work, though it was unfortunate that he did not explore the "living many-sideness of things" in a way less tied to the texts of Hegel.

Another of the features of his new concentration on dialectics was his announcement of the identity of logic, dialectics, and the theory of knowledge. Despite the possible obscurity of this formulation, epistemology was an area in which Lenin's thought underwent an obvious development, tied to this new attention to the living many-sidedness of knowledge. There is a noticeable difference in emphasis and degree of sophistication in Lenin's treatment of the theory of reflection in *Materialism and Empirio-Criticism* and *Philosophical Notebooks*. In the earlier work, he spoke of sensations as "copies, photographs, images, mirror-reflections of things."[48] However, in the later work, he dealt with epistemology more subtly, moving beyond his earlier, essentially passivist formulations, to a new emphasis on the active side of the cognitive process. He put it quite sharply: "Man's consciousness not only reflects the world but creates it."[49] He also saw the path of knowledge as less of a straight line but more full of twists and turns. Traversing it was a complex, halting, indirect process. The rhythm of the process of knowing was the rhythm of the dialectic, moving through the power of internal contradictions, each resolution always approximate and incomplete, with ever-new aspects yet to be discovered. Science was inextricably bound up with fantasy. Induction was connected with analogy and surmise.

Lenin, however, continued to believe that human ideas were not simply created by thought out of itself, but reflected the reality of the external world in however complex and indirect a way. He still held to a realist theory of knowledge, though certainly to a more subtle and highly developed one. Commentators who see some kind of radical "epistemological break" here are quite wrong, whether their preference is for the later Lenin interpreted in the radically voluntarist manner of such as Petrović or Avineri, or for the earlier Lenin as interpreted by anti-Hegelian authors like Colletti. It is equally wrong, however, to gloss over the differences in different stages of Lenin's thought as Hoffman does. In doing so, he engages in an extremely strained line of argument to defend Lenin's earlier formulations and imply that his later insights were already there in his earlier thinking though unexpressed. He does state the problem quite pointedly: "Does not the imagery of the 'photograph' or the 'copy' imply a measure of passivity in the process of thought and sensation, so that the later statement in the *Notebooks* plays an important role in *correcting* an earlier contemplative bias?" He thinks not, because Lenin's polemical target in *Materialism and Empirio-Criticism* was subjective idealism and he did in any case mention the criterion of practice in cognition. Hoffman then goes on to deal with the question of whether the formulations of *Materialism and Empirio-Criticism* are misleading, whether the imagery of photographing and copying are liable to passivist misconstruction. His answer is: "Unless we follow Plekhanov's position and question the very premise that ideas do resemble reality in some intelligible way, how can we possibly avoid imagery which is liable to be misconstrued by those who cannot understand the practical nature of the reflection process?"[50] Aside from this serious misrepresentation of Plekhanov's position, this is a thoroughly untenable line of reasoning. Not only is it possible to find a more activist imagery for the knowing process, but many have and Lenin himself did.

As far as the secondary literature giving an overall evaluation of Lenin as a philosopher goes, there are great extremes of opinion. Not surprisingly, most of the authors hostile to Engels are equally hostile to Lenin and speak of him in the very same terms. In the same way, those who take an idolatrous attitude to Marx and Engels do so to Lenin also, only even more so. There have, however, been some distinctive features in the history of commentary on Lenin's philosophical work.

Lenin's critics

The earliest to attack Lenin's philosophical work were his immediate opponents. The Russian Machists, against whom so much of his philosophical argument was directed, did not hesitate to answer back. Bazarov, Bogdanov, and Yushkevich published their reply in 1910 in *Pillars of Philosophical Orthodoxy*. Yushkevich was particularly abusive, claiming that Plekhanov and Lenin exemplified the decadence and dogmatism of Russian Marxism and accusing Lenin of "bringing the habits of the Black Hundred into Marxism."[51]*

---

* The Black Hundreds were monarchist gangs formed by the tsarist police to combat the revolutionary movement. They murdered revolutionaries, assaulted progressive intellectuals and organized pogroms against the Jews.

By the next generation of Russian Marxists, of course, Lenin's philosophical works were highly esteemed, and in fact elevated to the position of classics of Marxism, as they were by the overwhelming majority of Comintern activists. Notable exceptions, however, were Anton Pannekoek and Karl Korsch. Pannekoek, under the pseudonym of J. Harper, first published *Lenin als Philosoph* in Amsterdam in 1938, the period during which Lenin's philosophical works were being published abroad in foreign language translations. Pannekoek, who had been one of the Dutch leftists attacked by Lenin in *Left Wing Communism: An Infantile Disorder*, called *Materialism and Empirio-Criticism* a "confused tirade," in which Lenin distorted the views of Mach and Avenarius and failed to criticize them from a genuinely Marxist point of view. Lenin, according to Pannekoek, based his criticism of Machism on "middle class materialism," which was the philosophical counterpart to the half-bourgeois, half-proletarian nature of bolshevism.[52] It was true that Lenin was unfair to Mach and Avenarius, but it was not true that his was not a legitimately Marxist point of view. Just what Pannekoek's genuinely proletarian alternative was to Lenin's supposedly bourgeois or half-bourgeois position remained altogether unclear.

Even harsher was Korsch's assessment of Lenin as philosopher. He approved Pannekoek's criticism and carried it further. Also a leftist, Korsch too accused Lenin of failing to transcend the boundaries of bourgeois materialism. According to Korsch: "Lenin attacked the later attempts of bourgeois naturalistic materialism, not from the viewpoint of the historical materialism of the fully developed proletarian class, but from a preceding and scientifically less developed phase of bourgeois materialism."[53] He also asserted in an altogether unsubstantiated way that Lenin was completely unaware of the real achievements made by science in the days since Marx and Engels. Another charge against Lenin was that he preferred immediate practical utility to theoretical truth. That was altogether unfair, for Lenin was far closer to science than Korsch ever was and was far from indifferent to considerations of theoretical truth.

This has been the tone of many commentaries since, most recently echoed by Kolakowski, who has characterized Lenin's philosophy as "crude and amateurish," based on vulgar commonsense arguments and unbridled abuse of his opponents and adding nothing new to Engels and Plekhanov. Lenin, according to Kolakowski, failed to answer satisfactorily the philosophical questions with which he dealt, because of his "indolent and superficial approach and his contempt for all problems that could not be put to direct use in the struggle for power." Moreover, it "had a deplorable effect in furnishing pretexts for the stifling of all independent philosophical thought and in establishing the party's dictatorship over science."[54] Kolakowski is wide of the mark, for Lenin was far more serious about philosophy and had far more worth saying about it than he is willing to concede. However, certain texts of Lenin and the use to which they have been put by Leninists do lend a certain support to his criticisms.

Lenin's defenders

Predictably, this type of commentary has been answered by others. Timpanaro, himself a leftist,* has specifically answered Pannekoek, Korsch, and more contemporary leftists as well, by pointing out that voluntarism, subjectivism, and the refusal of science may constitute a momentary revolutionary stimulus, but cannot in the long run provide the basis of a solid revolutionary doctrine. Timpanaro argues that Lenin's defence of materialism was far more revolutionary than the semiidealism of Korsch and Pannekoek. Convinced that he was far in advance of Lenin, Korsch was actually very far behind him in thinking that the dominant orientation in bourgeois philosophy was materialism colored by the natural sciences and in waging an anachronistic battle against it in the name of his own idealism represented as authentic Marxism. Lenin had already discerned the reactionary character of the idealist renaissance that was under way, even when it presented itself as a revival of revolutionary activism against social Darwinist quietism. Lenin foresaw that empirio-criticism was only the first phase in an involution towards forms of fideist spiritualism that would lead from a methodological critique within science to a negation of science itself or to a falsification of it in a teleological or providential direction. This Korsch did not see, as his basic background had been alien and hostile to the natural sciences, unlike Lenin who was acutely aware of the most recent developments in the sciences and their philosophical significance. There was not, on Lenin's part, the slightest repudiation of the new physics, which he understood well, despite a number of inaccuracies with regard to particular physical theories. He was basically right in his belief that the philosophy of Mach and Avenarius led logically to idealism, although they were not prone to the open idealism of their successors. In distinguishing between the new science and the antimaterialist use to which it was being put, Lenin managed to redeem the historicity of science as well as its objective truth. This concern for objective truth, Timpanaro insists, was the animating force of *Materialism and Empirio-Criticism*, despite Korsch's extremely prejudicial reading of it, which led him to assert otherwise. Lenin believed idealism reactionary because it was untrue and not vice versa.[55]

---

* This term is being used in the historical descriptive sense for those who have defined themselves as being to the left of the communist parties.

Timpanaro is not himself uncritical of Lenin's philosophy, but his criticisms are of a far more constructive nature. He believes that there was in Lenin's work a certain wavering between the need for an integral materialism and a tendency to fall back on objectivism or realism. On the other hand, he does note that it is clear that Lenin was aware of the fact that materialism was much more than an epistemology. Timpanaro also thinks that Lenin's enthusiasm for certain Hegelian propositions in *Philosophical Notebooks* was excessive. He does not believe that the profound transformations in the traditional concept of matter can be so easily accommodated through the framework of the Hegelian dialectic. He prefers the formulations in "Karl Marx" where Lenin dealt with contradictions more in terms of conflicts between forces than in terms of identity of opposites. Also, he holds that in "What the Friends of the People Are," Lenin displayed a more perceptive awareness of the intrinsically idealist nature of the Hegelian dialectic, that put it beyond any redemption through any "standing on its head" or "breaking its shell."[56]

Another striking contrast to Kolakowski's assessment of Lenin is that by the Irish scientist, J.D. Bernal. To him, what distinguished Lenin was his sense of perspective, the range of vision that brought understanding of the grand movements of nature and history. Lenin saw that the early twentieth century was a great period of transition in the scientific and philosophical world and that the old controversy between materialism and idealism was reappearing in new guises in the form of disputes between atomism and energeticism, between evolutionism and vitalism. Lenin entered into all these disputes and advanced materialism to the level of the science of his day. He also saw that the tendencies common to Mach, Ostwald, Poincaré, and Bogdanov marked a drift towards a recurrence of fideism, which had its political counterpart in a tendency to blur more and more the definiteness of class struggle as part of the general trend of political compromise after the defeat of the 1905 revolution. The political aspect of Lenin's own philosophy was shown in the tone he set for the development of a new kind of science that developed after the revolution, when he was able to see the historical effects in a completely different way from Kolakowski.[57]* While Bernal's analysis is perhaps overly eulogistic and does not venture into the problem areas, it is nevertheless significant that one who has known science so well has regarded Lenin's philosophy of science so highly.

Praise for Lenin's philosophical work has sometimes come from surprising quarters. The Sovietologist and priest, J.M. Bochenski, has commented that Lenin worked out an epistemology that was original in many respects, combining realism with rationalism and paying due regard to the difficulties for

---

* An assessment of the subsequent development of Soviet science, under the impact of Lenin's philosophy of science, is, however, a matter for later chapters.

realism engendered by the latest results of research in physics.[58] Another remarkable comment has been made by Feyerabend: "There are not many writers in the field today who are as well acquainted with contemporary science as was Lenin with the science of his time, and no one can match the philosophical intuition of that astounding author."[59]

But others can damn by their praise. Such is the case with the supposedly pro-Leninist essay by Althusser entitled "Lenin and Philosophy." By Althusser's eccentric logic, it is not possible to raise the question of whether or not Marxist philosophy has a history. Here a "symptomatic difficulty" exists, Althusser declares, because "philosophy has no history"; it is "that strange theoretical site where nothing really happens, nothing but this repetition of nothing . . . philosophy leads nowhere because it is going nowhere."[60] According to Althusser, Lenin simply explained that Mach merely repeated Berkeley and counterposed to this his own repetition of Diderot. Nothing happened but a repetition of nothingness. Althusser explains that what is called Marxist philosophy presents a rather curious spectacle. After Marx's "epistemological break" there was a long interval of philosophical emptiness. In this philosophical silence, only the "new science" spoke. After thirty years, it was broken only by an "unforeseen accident," Engels's "precipitate intervention" in which he replied to Dühring for political reasons and was unfortunately constrained to follow Dühring onto his own territory. The ultimate reason for this philosophical silence is that "the times were not ripe, that dusk had not yet fallen, and that neither Marx himself, nor Engels, nor Lenin could yet write the great work of philosophy which Marxism-Leninism lacks."[61] Althusser announces that Lenin's philosophical work must be perfected, that is, completed and corrected, because Lenin was "unfortunate enough to be born too early for philosophy." But by the bottom of the same page, he reverses this verdict and says that Lenin was not born too soon for philosophy because: "If philosophy lags behind, if this lag is what makes it philosophy, how is it ever possible to lag behind a lag which has no history?"[62] As Althusser interprets the situation in which Lenin found himself, Lenin responded to the "scientific pseudo-crisis of physics" by drawing the crucial distinction between science and philosophy. Lenin supposedly proclaimed that philosophy was not a science, that philosophical categories were distinct from scientific concepts, that the meaning of the philosophical category of matter did not change for it did not apply to any object of science, but was absolute.[63]

Nothing could be further from the spirit of Lenin, nor from the letter for that matter. Lenin explicitly affirmed the historicity and scientificity of philosophy, the relativity of the concept of matter, its basis in concrete experience and its development in response to new scientific research. Althusser's analysis of Lenin is sheer nonsense and it is disedifying nonsense at that. The spectacle of the professional philosopher denigrating philosophy presents a very unworthy picture, as does such flagrant violation of logic and such overt distortion of another philosopher's position.

Another aspect of Althusser's interpretation concerns Lenin's attitude to philosophy in the political situation that existed in 1908. He suggests that Lenin's reticence to engage in philosophical debate with the bolsheviks who were also Machists in order to safeguard the political unity among bolsheviks was more than a tactical attitude, that it was a new "practice of philosophy," for dialectical materialism was "not a new philosophy, but a new practice of philosophy."[64] Such was not the case. Lenin's attitude was a tactical one and his pursuit was a philosophical one, as a closer look at historical events will show.

The Politics of Philosophy: Bolsheviks/Mensheviks/Machists/Materialists

The philosophical debate among Russian social democrats cut across different lines than the political debate. Among the dialectical materialists in philosophy, Plekhanov, Axelrod, and Deborin were mensheviks, whereas Lenin was a bolshevik. Among the empirio-criticists, Yushkevich and Valentinov were mensheviks, while Bodganov, Bazarov, Lunacharsky, and Gorky were bolsheviks. Philosophical discord was always on the verge of breaking the fragile political unity of the bolsheviks. Several times it actually did. These tensions forced them into certain tactical concessions regarding publications. Three of the seven editors of the bolshevik paper were Machists, and there was an agreement in 1904 that publications be considered philosophically neutral. In 1908, the arrangement was spelled out more specifically. In *The Proletarian*, the illegal official newspaper, no philosophical articles were to be published. In the legal publications, however, philosophical articles could be published on the condition that both sides be given equal space.

Lenin was at first reticent about participating in the philosophical debate, but only for fear of contributing to a political split. Lenin was of an intensely philosophical frame of mind and believed philosophical matters to be of crucial importance. As time went on, the tension mounted, and, by 1908, Lenin had been provoked to boiling point and he could no longer refrain from intervening. Various developments had convinced him that it had become necessary.

One was the fact that mensheviks such as Plekhanov, Axelrod, and Deborin insisted on drawing attention to the fact that they saw a logical correlation between Machism and bolshevism. They claimed that the bolsheviks were deviating towards Blanquism in politics and philosophical subjectivism was naturally linked to political voluntarism. They did everything to instigate philosophical polemics among the bolsheviks. Finally, in a bolshevik publication, Bogdanov began to reply to Plekhanov. Lenin worried all the more about the image being projected of the mensheviks as orthodox and the bolsheviks as revisionist in the philosophical domain and when he wrote *Materialism and Empirio-Criticism*, he accused Plekhanov of being less concerned with refuting Mach than with dealing a factional blow at bolshevism. For this exploitation of such differences, Lenin remarked, Plekhanov was deservedly punished by the subsequent publication of two books by mensheviks advocating Machism.

Another development that convinced Lenin to enter the debate was the publication by the Machists of their most audacious works to date. Their collective work *Essays in the Philosophy of Marxism* caused him to "rage with indignation"; it was, he felt, becoming intolerable to have as allies people who believed such things. *Neue Zeit*, the most important theoretical organ of Marxism, had just published Bogdanov's article "Ernst Mach and Revolution." To add fuel to the fire, Gorky had written an article in which he seemed to be confirming, from the other side, the correlation the mensheviks were making between the philosophical and political aspects, and depicting the debate as a clash between a philistine materialism linked to historical fatalism and a philosophy of activism. When the article arrived at the bolshevik editorial office, Lenin vigorously protested against its publication at a meeting of the editorial board.

Meanwhile, there were new political debates among both the bolsheviks and the mensheviks that had in any case destroyed the tactical political unity. The bolsheviks were divided on the question of *otzovism*, a leftist move to recall delegates from the Duma, opposed by Lenin and supported by Bogdanov and Lunacharsky, who formed a faction known as the vperëdists. The mensheviks, at the same time, were split on the question of liquidationism, an exaggerated parliamentarism that called for the abolition of the structure of illegal party work and that was challenged by the "party mensheviks" such as Plekhanov. Plekhanov withdrew from the menshevik paper and established one of his own. For a time, Lenin thought there might be a possibility of an alliance with the party mensheviks who were dialectical materialists and no longer worried as much about a split with the vperëdists-Machists; the latter were busy propagandizing their views at a school they organized on the island of Capri. In 1909, Lenin violated the 1908 agreement and, overriding Bogdanov's protest, published in the illegal press a severe critique of Lunacharsky. At a 1909 meeting, the bolsheviks ended the arrangement giving equal space to opposing philosophical persuasions and passed a resolution declaring the commitment of their editorial board to dialectical materialism without however mentioning Machism or publishing the resolution. A split ensued. In 1912, however, the bolsheviks readmitted the vperëdists with the stipulation that, although they could continue to publish Machist views, they could not do so in the bolshevik press. This arrangement broke down again in 1914.

Throughout all of this, Lenin never attempted to demonstrate any necessary connection between Machism and menshevism, but he did very firmly insist that the defence of dialectical materialism in philosophy was integrally bound up with the defence of Marxist theory in every other sphere. Marxism was "cast from a single piece of steel" and was the essential weapon of the proletariat in the revolutionary struggle.

The Turn of the Tide

Meanwhile other events were under way that would be decisive in determining the fortunes of the contending philosophies.

In 1912, the tide in Russia began to turn again. In the Lena goldfields of Siberia, in a land of a six-month night in winter and plagues of mosquitoes in summer, where workers labored for long hours in degrading conditions for hopelessly low wages, and their wives were obliged to wait on and sleep with the owners,wages were cut by 25 percent and horses' genitals were found in the food. When they went on strike, five hundred were shot. This touched off a new wave of strikes and demonstrations. All over Russia, there was renewed unrest among workers, soldiers, and peasants. In the same year, at a congress in Prague, the bolsheviks broke finally and irrevocably with the mensheviks and declared themselves to be the Russian Social Democratic Labour Party.

In 1914 came the war and, as one year passed into another, military defeat followed upon military defeat. Discontent was rampant among the troops and the country was thoroughly exhausted. People were hungry, disillusioned, and angry. By 1917, the tsarist régime was unable to suppress discontent. Troops called in to put down a mass strike in Petrograd refused to obey orders and within two days the struggle was over in Petrograd. It then spread throughout the country and to the armies at the front. In February, the tsar abdicated and a Provisional Government was formed. The members of the government did not make the revolution. They merely stepped into empty positions of power. The government, dominated by business interests, continued the war and at home alternated between reform and repression. However, real power shifted to the soviets. A Central Congress of Soviets was formed. At first, mensheviks and social revolutionaries were in the majority, but the more other parties supported the government's war efforts, the more did the bolsheviks isolate them from the hungry and war-weary people with their simple call for land, bread, and peace. Meanwhile, revolutionaries were returning from exile. In April, Lenin arrived at the Finland Station, startling even his own supporters with his militancy against the Provisional Government.

By October, the bolsheviks were very strong indeed. They were still vastly outnumbered by other parties, but they were the clearest, the best organized and the most disciplined. It was Lenin who had the most accurate reading of the pulse of the times.

Soon the political debate about party organization and methods of achieving power was over. It was settled on the streets of Petrograd. The philosophical debate, however, was far from resolved, but the setting in which it would be pursued was changed forever.

NOTES

1. Georgi Plekhanov, "Karl Marx," *Selected Philosophical Works* 2 (Moscow, 1976), p. 676.

2. Lenin, "What Is To Be Done?" *Selected Works* 1 (Moscow, 1975), p. 114; Plekhanov, "Karl Marx," p. 676.

3. Pyotr Struve, *Critical Notes on the Question of Economic Development in Russia* (1894), cited by E.H. Carr in *The Bolshevik Revolution* 1 (London, 1966), p. 21.

4. Lenin, "Preface to the Collection *Twelve Years*," in *Against Revisionism* (Moscow, 1966), p. 94.

5. *Problemy idealizma*, (St. Petersburg, 1903).

6. Nikolai Berdyaev, *Wahrheit und Lüge des Kommunismus* (Lucerne, 1934), p. 84.

7. Lyubov Axelrod, *O problemakh idealizma* (Odessa, 1905); *Filosofskie ocherki* (St. Petersburg, 1906).

8. Plekhanov, *Materialismus Militans* (Moscow, 1973), p. 118.

9. Lenin, "Certain Features of the Historical Development of Marxism," in *Against Revisionism*, pp. 113–134.

10. L. Houllevique, *L'évolution des sciences* (Paris, 1908), pp. 87–88.

11. Sebastiano Timpanaro, *On Materialism* (London, 1975), pp. 127–128.

12. Ernst Mach, *The Analysis of Sensations and the Relation of the Physical to the Psychical* (New York, 1959).

13. Otto Blüh, in *Ernst Mach: Physicist and Philosopher*, Robert Cohen & Raymond Seeger (Dordrecht, 1970), p. 18.

14. Norman Smith, "Avenarius' Philosophy of Pure Experience," *Mind* (1906).

15. Bogdanov, *Empiriomonizm* 1 (Moscow, 1905), p. 184.

16. Bogdanov, *Izpsikhologii obshchestva* (St. Petersburg, 1906), pp. 274–275.

17. Bogdanov, *Empiriomonizm* 3, pp. 148–149.

18. Bogdanov, *Filosofiya zhivogo opyta* (Petrograd-Moscow, 1923), pp. 180, 242–252.

19. Plekhanov, *Materialismus Militans*, p. 6.

20. Ibid., p. 8.

21. Ibid., p. 72.

22. Plekhanov, "For the Sixtieth Anniversary of Hegel's Death," *Selected Philosophical Works* 1, p. 467.

23. Plekhanov, *Materialismus Militans*, p. 113.

24. Plekhanov, "Materialism Yet Again," *Selected Philosophical Works* 2, p. 419.

25. Plekhanov, *Fundamental Problems of Marxism* (London, 1969), p. 22.

26. Ibid., p. 81.

27. Plekhanov, "Conrad Schmidt versus Karl Marx and Friedrich Engels," *Selected Philosophical Works* 2, p. 395.

28. Plekhanov, "Materialism Yet Again," p. 418.

29. Plekhanov, *Fundamental Problems of Marxism*, p. 27.

30. Plekhanov, "Essays on the History of Materialism," *Selected Philosophical Works* 2, p. 125.

31. Thomas Masaryk, *Otázka sociální* 1 (Prague, 1946), p. 63, cited by M. Silin in *A Critique of Masarykism* (Moscow, 1975), p. 94.

32. Plekhanov, "On the Alleged Crisis in Marxism," *Selected Philosophical Works* 2, p. 317.

33. L.I. Axelrod, *Filosofskie ocherki*.

34. Lenin, *Materialism and Empirio-Criticism* (Moscow, 1920), p. 238.

35. Ibid., p. 319.

36. Ibid., p. 301–302.

37. Ibid., p. 246.

38. Ibid., p. 116.

39. Anatoly Lunacharsky, *Sotsializm i religiya* (St. Petersburg, 1908).

40. Lenin, *Materialism*, p. 315.

41. Ibid., p. 330.

42. Lenin, "Conspectus of Hegel's Science of Logic," in *Philosophical Notebooks* in *Collected Works* 38 (Moscow, 1972), p. 276.

43. Lenin, "On the Question of Dialectics," *Philosophical Notebooks*, p. 363.

44. Lenin, "What the Friends of the People Are and How They Fight the Social Democrats," *Collected Works* 1 (Moscow, 1972), p. 165.

45. Lenin, "The Three Sources and the Three Component Parts of Marxism," *Selected Works* 1, p. 45.

46. Lenin, "Karl Marx," *Selected Works* 1, p. 23.

47. Lenin, "Conspectus of Hegel's Science of Logic," pp. 87–238.

48. Lenin, *Materialism and Empirio-Criticism*, p. 220.

49. Lenin, "Conspectus of Hegel's Science of Logic," p. 212.

50. John Hoffman, *Marxism and the Theory of Praxis* (London, 1975), pp. 79–81.

51. P.S. Yushkevich, *Stolpy filosofskoy ortodoksii* (St. Petersburg, 1910).

52. Anton Pannekoek, *Lenin as Philosopher* (London, 1975), pp. 68–91.

53. Karl Korsch, "Lenin's Philosophy," appended to Pannekoek's *Lenin as Philosopher*. This edition attributes Korsch's essay to the former German Spartacist, Paul Mattick. Both were living in the U.S. in the 1930s and Mattick was editor of *Living Marxism*, in which Korsch's review first appeared. After checking with other sources, I am satisfied that this essay was written by Korsch and not Mattick.

54. Kolakowski, *Main Currents of Marxism* 2 (Oxford, 1978), pp. 457–466.

55. Timpanaro, *On Materialism*, pp. 225–250.

56. Ibid.

57. J.D. Bernal, "Lenin and Science," *Lenin and Modern Natural Science*, ed. M.E. Omelyanovsky (Moscow, 1978), pp. 40–47.

58. J.M. Bochenski, *Soviet Russian Dialectical Materialism* (Dordrecht, 1963).

59. Paul Feyerabend, "Dialectical Materialism and Quantum Theory," *Slavic Review*, September, 1966, p. 414.

60. Louis Althusser, *Lenin and Philosophy* (London, 1977), p. 56.

61. Ibid., p. 46.

62. Ibid., pp. 48–49.

63. Ibid., pp. 50–51.

64. Ibid., pp. 31, 67.

CHAPTER 4

THE OCTOBER REVOLUTION:
Marxism in Power

The Early Days of the Revolution

In October 1917 the revolution came, transforming utterly the whole context for every debate and giving birth to unanticipated new ones.

Under the leadership of the bolsheviks, the Soviets took state power, nearly without opposition; resistance came only from a handful of cadets and a women's death battalion. The Provisional Government was dissolved and a new government was formed with Lenin as chairman of the Council of People's Commissars at its head. The new Soviet government immediately issued a flurry of decrees, giving land to the peasants, nationalizing key industries, proclaiming the equality of women, and announcing an imminent end to the war. The bolsheviks went from town to town and from village to village, proclaiming the Soviet government and calling on peasants to confiscate land, on soldiers to arrest counterrevolutionary generals, and on workers to assume state functions. And so they did.

It was an extraordinary time. There had never been a time like it before, nor could there ever be a time like it again. It was the world's first socialist revolution and as such it had no precedents. They were doing something that had never been done before, opening up a new line of historical development. There seemed to be unlimited scope for human creativity in this monumental enterprise of building from scratch a whole new social order. The air was alive with possibilities and everything was up for grabs. It was challenging and exhilarating. Lenin exhorted:

Comrades, working people! Remember that now *you yourselves* are at the helm of state. No one will help you if you yourselves do not unite and take into your hands all affairs of the state. . . . Get on with the job yourselves; begin right at the bottom, do not wait for anyone.[1]

And many took him up on it.

But this heady experimentation with previously unknown and untested forms of social organization quickly came up against the sobering fact that the revolution had not yet been won. People were still hungry. Industry was at a standstill. Indeed the whole economy had virtually collapsed. There was widespread chaos and all the while the German troops were advancing. The army was so beaten down and demoralized that it could not even continue a fighting retreat. The treaty of Brest-Litovsk gave the regime a breathing space, although the terms, involving annexation of huge territories under Soviet rule, were exceedingly harsh. At first, the bolsheviks had stalled, appealing to the workers of the world to unite for the last fight they had so often sung about in the *Internationale* and hoping that the German regime would soon go under beneath a great revolutionary upsurge in Europe. But as month followed month and no such world revolution came, their hopes receded and Lenin realized it was necessary to yield space in order to gain time.

Then came the "white terror" of civil war, aided by foreign intervention. The new Red Army fought against tremendous odds on seventeen fronts simultaneously. Not only that, but there was no food; people were fainting in the streets; workers were dropping over onto their machines; babies were whimpering in their cots; the old and sick were simply dying. And the kulaks were hoarding grain. In this period of war communism, every aspect of life was subordinated to the central authority. There was enforcement of compulsory purchase of food and there was swift treatment of suspected saboteurs. Amidst all this, there was mounting evidence of treachery on the part of the other leftist parties, and, after Lenin was shot in an assassination attempt by one of their members, the bolsheviks moved against them, raiding their offices and suspending their newspapers.

The civil war lasted for three years. Although finally it had been won, the country was nevertheless torn apart from one end to another and, on top of everything else, stricken by famine. Production had come to a halt. The working class had been decimated: killed fighting at the front, massacred in towns taken by the whites, or driven back to the land by massive unemployment. In the one country on earth ruled in the name of the proletariat, the proletariat was hardly there.

Proposals to accomplish these tasks of reconstruction through the continuation and intensification of the measures associated with war communism were debated, but Trotsky's call for the militarization of labor was defeated and a very different course was taken. Lenin's New Economic Policy carried the day, giving the régime another breathing space by making economic concessions. Restrictions on small business and private agriculture were relaxed in order to consolidate power and to increase productivity before embarking on a full-scale program for the socialization of industrial production and the collectivization of agriculture.

At that time the question of opposition parties asserted itself again in a situation in which other left-wing parties, associated in people's minds with the defeat of the autocracy and the ideals of social progress, would be immensely popular, yet they did not have the liability of having been responsible for the crippling hardships of the past few years. In 1921, discontent was rampant, symbolized by the Kronstadt mutiny, which the bolsheviks forcibly suppressed, and the opposition parties made the most of it. Martov had already emigrated to Germany where he had given a searing anti-bolshevik speech at the Halle congress of the USPD. Lenin issued a stern warning to the remaining mensheviks that they could either join Martov or go to prison, if they persisted in opposition to the Soviet régime. Menshevik leaders departed for Berlin, with no obstacles placed in their way, and the rank and file either renounced their views or abandoned political activity altogether. The social revolutionaries, however, proved more recalcitrant and thirty four of its leaders were brought to trial on charges of conspiring against Soviet power. The existence of these parties was not yet illegal, but from this time they effectively disappeared from Soviet political life.

This was not, however, the end of political dissent, but from this time on it became concentrated within the party. A debate centering around the "Workers' Opposition" emerged. This group demanded that the workers be given direct power and protested against the employment of "bourgeois specialists," the appointment of managers in industry, and the organization of the army on the basis of traditional military discipline. One of its chief spokesmen, the fiery feminist Alexandra Kollontai, a central committee member at the time of the revolution, castigated the party for not immediately bringing in such measures as free medical service, free education, and subsidized rent. The position of the majority on the central committee was that these demands made an impossible claim on scarce resources and that the immediate task was to begin the economic and social reconstruction that would provide a firm foundation for such measures in the future. Factions within the party were banned at the Tenth Party Congress, a measure which was to have serious consequences for the future of Soviet political life.

A Cultural Revolution

It was against the background of all these events that Soviet intellectual life began to take on its own distinctive form. From the first days of the revolution, there was immense ferment in virtually every discipline. Marxism, for the first time, was in power. It was no longer an opposition current, but the official ideology of the new Soviet state. The time had come to bring this vision to fruition and to extend its influence into all areas of life. This meant interpreting the relevance of the theory to the new situation and developing it further in light of new experience. But when it came to discussing just how to do so, there was great divergence of opinion, opening the way to raging controversies in every field of thought. There were debates about the strategy for industrialization and its relation to the strategy for the collectivization of agriculture; debates about nationalities policy; debates about the nature of the state and the status of law under socialism; debates about the nature of proletarian morality; debates about the liberation of women, "free love," and the future of the family; debates about avant-garde educational theories; debates about different schools of literary criticism and the forms of proletarian culture. And, of course, there were debates about philosophy, about the natural sciences, and about the philosophy of science. There were debates about nearly everything. There were so many ideas to be sifted through and there was so very much at stake.

In their struggle to come to terms with these issues, especially when it came to necessary policy decisions, the bolsheviks were up against enormous odds. For to arrive at such decisions and to implement them, they needed expertise: they needed economists, scientists, technicians, managers, lawyers, university professors, school teachers, writers, and artists. The problem was that the overwhelming number of supporters of the régime were uneducated and uncultured, whereas most of the experts, the highly educated and the highly cultured, tended to be hostile to the régime. Lenin felt the problem deeply and he declared that the lack of culture was the major obstacle to the building of socialism.

The bolsheviks—or the communists as they soon came to be known— therefore embarked on a two-pronged strategy, a struggle on two fronts. On the one hand, they sought to win over the existing intelligentsia, to win them to cooperation with the Soviet régime and, if possible, to intellectual commitment to Marxism. On the other hand, it was necessary to begin to create a new intelligentsia drawn from the proletariat and peasantry and firmly communist in its outlook. The task was to carry out a "cultural revolution." To even begin work on the second front, however, it was necessary to achieve some success on the first, for the "bourgeois specialists" were necessary to train the new generation of "red specialists." M.N. Pokrovsky, historian and leading figure in the transformation of higher education, outlined the dangers of being overly conciliatory towards the intelligentsia of the old order on the one hand or of being too dismissive of what they had to offer on the other. Looking back after the first ten years, he explained:

We were faced with two dangers: on the one hand, there was the danger of remaining in the old rut, since we had a certain fear of too abrupt and decisive a demolition, and, as a result, we could be held prisoner by bourgeois specialists . . . on the other hand, the danger consisted in the fact that there were some comrades who said: "All bourgeois education is worth absolutely nothing. It is necessary to throw this out and to begin anew."[2]

The problem of achieving a correct balance between the old and the new was a theme that recurred in a multitude of forms during these first years. Lenin felt the need to address himself to it on a number of occasions. He vigorously criticized those who took a nihilistic approach to the cultural and intellectual heritage of the past. It was the task of communists, he believed, to critically assimilate, to make fully their own, the best knowledge, the best technique, the best art that had been achieved by human effort over the centuries and to carry it forward in a creative and revolutionary way. However, Lenin did not approve of those who took an uncritical or uncreative attitude towards the past and he admonished them that they were revolutionaries and not archivists. The stock of human knowledge was not only to be drawn on, but to be radically remolded in building the new society. Within this scheme, a preeminent position was accorded to science. Marxism had from the first staked its lot with science and in this tradition Lenin and the bolsheviks believed firmly that science was essential to socialism.

Lenin, in these first years of the revolution, was at his best. Even when the very existence of the Soviet state was in jeopardy and sheer survival was an achievement, he displayed an impressive breadth of vision. He was the architect of the revolution, and, as such, it was his lot to have to apply his mind to an extraordinarily wide number of issues, from problems involving military tactics, food supply, and sabotage, to problems of religious belief, women's rights, homeless children, communist morality, and the philosophy of science. Amidst the welter of details and the monumental tasks, he never lost a sense of perspective and he brought to every issue reflection, balance and sanity. He often had to restrain the overzealous and shortsighted. There was, for instance, the question of the status of religion in a socialist society. Although he was an atheist of the most militant sort, he spoke out against a certain wildness, a certain hooliganism, in the antireligious campaign and forbade Komsomol members on one occasion from holding an antiEaster demonstration. His view was that religion must be a private matter vis-à-vis the state, although certainly not vis-à-vis the party. It was vital to win people over to atheism and to a scientific view of the world by intelligent argumentation and he criticized the crudity of the propaganda that was often employed. On other questions also he took his stand against those who went to extremes and became myopic, forgetting the many-sidedness of things. Thus his opposition to Proletkult,* to the Workers Opposition, indeed to the views of all those who wished to sweep aside the existing intelligentsia, the "bourgeois specialist," claiming that the proletariat stood alone and had to build socialism entirely from its own resources. Lenin insisted that it was necessary to embrace bourgeois culture, its science, its knowledge, its art, and to critically reshape it. In doing so, it was vital to make the fullest use of the "bourgeois specialists" who were the bearers of that culture. Thus too his very wise advice to Komsomol members advocating a "glass of water" approach to sexuality: "Be neither monk nor Don Juan, but not anything in between either, like a German philistine."[3] It was an integrated way of thinking that he brought to bear on every subject that came up for discussion.

The Bolsheviks, the Scientists and the Philosophers

Because of his concern for winning over the existing intelligentsia and because of his special interest in the philosophy of science, Lenin was particularly interested in gaining the support of natural scientists. When the distinguished biologist K.A. Timiriazev announced his fervent loyalty to the new régime, Lenin was overjoyed. The bolsheviks received him with open arms and named a research center after him: the State Timiriazev Scientific Research Institute for the Study and Propaganda of Natural Science from the Point of View of Dialectical Materialism.** Another success was the geneticist N.I. Vavilov, who was put in charge of a whole network of biological research institutions. The physicist A.F. Joffe had actually joined the anti-bolshevik exodus of scholars to the Crimea in 1917, but returned to Petrograd and resolved to connect his fate with that of the land of the soviets long before it was by any means clear that the civil war would be won. He became a member of the Leningrad Soviet and doyen of Soviet physicists, although it was not until 1942 that he joined the party.

There was a concerted campaign to win working natural scientists over to a materialist position in the philosophy of science. Through such agencies as the Union of Scientific Workers, the All-Union Association of Workers of Science and Technology for Assistance to the Construction of Socialism

---

* Proletkult was a movement for "proletarian culture." A section on its ideas and its fate follows.

** Mercifully reduced to Timiriazev Institute in all but the most formal references to it.

(VARNITSO), and the Central Commission for Improving the Condition of Scholars, various societies for materialist natural scientists corresponding to the various scientific disciplines,* the bolsheviks fought to win their hearts and minds.

Lenin took the occasion of the second issue of *Pod znamenem marksizma* (Under the Banner of Marxism), the new journal of Soviet Marxism launched in 1922, to call for an alliance between communists and natural scientists inclined to materialism. The first issue had not mentioned the philosophy of science in its editorial declaration of its aims and purposes, nor had Trotsky in his letter welcoming the new publication. Lenin felt it was necessary to draw their attention to the importance of this field. In his article entitled "On the Significance of Militant Materialism," he urged the journal to attend to the philosophical problems raised by the sharp upheaval in the natural sciences and to enlist natural scientists in the work of the philosophy journal in order to tackle these problems most effectively. He called attention to the latest episode in the ongoing attempt to undermine the foundations of materialism by appeals to modern science by those who were seizing on Einstein's theory of relativity for such purposes. Unlike A.K. Timiriazev, the physicist and son of the prominent biologist, who wrote a review of a Russian translation of Einstein in the first issue, Lenin suggested that they should take care to distinguish between Einstein's physical theories and the philosophical speculation sparked by them. Lenin believed that the natural sciences could not hold their own against the "onslaught of bourgeois ideas" unless they stood on solid philosophical ground. As to where they should turn for such a grounding:

Modern natural scientists (if they know how to seek, and if we learn how to help them) will find in Hegelian dialectics, materialistically interpreted, a series of answers to the philosophical problems which are being raised by the revolution in natural science. . . . Without this, eminent natural scientists will as often as hitherto be helpless in making their philosophical deductions and generalizations. For natural science is progressing so fast and is undergoing such a profound revolutionary upheaval in all spheres that it cannot possibly dispense with philosophical deductions.[4]

In his opinion, the editors and contributors to *Pod znamenem marksizma* should constitute a sort of "Society of Materialist Friends of Hegelian Dialectics."

---

* There were, for example, the Society of Materialist biologists, the Society of Materialist Physicians, and the Society of Materialist Physicists and Mathematicians, organized under the Section for the Natural and Exact Sciences of the Communist Academy.

158

MARXISM AND THE PHILOSOPHY OF SCIENCE

It was far from the only occasion in which Lenin turned his mind to questions of philosophy. In "Once Again On The Trade Unions," in the context of a criticism of the current political thinking of Trotsky and Bukharin, he put great emphasis on the difference between dialectical thinking and eclectic thinking. He called upon such comrades to put aside one-track thinking, compulsiveness, exaggeration, obstinacy, and rigidity and to learn to examine all facets of a thing, all its connections and mediacies. Dialectics, he insisted, required an all-round consideration of relationships in their concrete development, but not a patchwork of bits and pieces. He asked them to look beyond formal logic that dealt with formal definitions of the sort that drew on what was most common or glaring about a thing and stopped there. The point was that a really full definition of anything "must include the whole of human experience."[5] In the same article, Lenin exhorted communists to engage in a serious study of Plekhanov's philosophical works and insisted that the workers' state should demand of professors of philosophy in particular that they have a thorough knowledge of Plekhanov's exposition of Marxism. By this time, Plekhanov was gone, having died in 1918, an unregenerate opponent of Soviet power. Nevertheless, Lenin was magnanimous enough to ensure that his contribution to the development of Russian Marxism be duly recognized.

By 1924, Lenin was gone as well and thenceforth Soviet philosophy had to develop without him. Indeed, many fields were deprived by the loss of his fine mind and the Soviet state suffered greatly by the premature death of the great "dreamer in the Kremlin," as H.G. Wells called him.

In Lenin's day, it was not required that a person be "100 percent pure" and in agreement with all bolshevik policies in order to make a contribution to the new social order.* Although every effort was made to persuade natural scientists to look to Marxism in matters of philosophy of science, it was not demanded of them. Willingness to pursue their research within the new order was enough. They were assured by the highest government officials, confident that communism was being built by everyone working honorably and conscientiously in his own field, that it was their legitimate right not to be Marxists. Indeed, Kalinin, the Soviet president, declared that communism was being built even by the man who proclaimed himself against communism, for if he were a doctor and raised the people's health, he was doing communist work. They were convinced in any case that most natural scientists were spontaneous materialists and nearly Marxists without realizing it. Soviet

---

* Pogodin's famous play *Kremlin Chimes*, still performed in Moscow, shows the authorities coming for a famous engineer. Instead of being under arrest, as he expected, he is brought to Lenin, who enthusiastically explains his plans for the electrification of Russia and invites him to put his skills to work in achieving this great task.

leaders did not feel themselves threatened by those who obviously were not Marxists, allowing them considerable latitude as long as they showed themselves to be in good faith. When Khvolson, who had an international reputation both as a physicist and as an exponent of a fideist interpretation of the new trends in physics, hesitantly approached the new government with queries concerning the fate of various research projects that had been given governmental support before the revolution, he found them ready to devote far greater resources to education and research than were their predecessors. He continued to propagate his views on the reconciliation of the new physics with religious belief, although he was persuaded to remove from the later edition of a very well-received popularization of physics published in 1922 a passage explicitly rejecting the rather Promethean notion of science held by the bolsheviks. Nevertheless, he was awarded in 1926 the title of "Hero of Labor." There was also the case of V.I. Vernadsky, a distinguished geochemist with a lively interest in philosophy of science, who had also joined the anti-bolshevik flight to the Crimea. When this last stronghold of the whites fell, he was arrested by the Cheka and brought to Moscow, where Lenin insisted on his release. He played a leading role in the Academy of Sciences and its organization of the scientific work of the country, although his opposition to certain aspects of bolshevik rule continued. He respected the bolshevik desire for a flowering of science, but warned them against administrative intervention in the creative process of scientific work and steadfastly refused to consider Marxism as a philosophy of science. And there was the mathematician V.A. Steklov, vice-president of the Academy of Sciences, who also put great stress on the importance of academic freedom for the progress of science and, just to reinforce his point, published a popular biography of Galileo in 1923.[6]

Academic freedom was, in fact, enjoyed by the Academy of Sciences, although occasionally a strident voice was raised against it, such as that of the anonymous author "Materialist," who, in the pages of *Pod znamenem marksizma*, castigated the Academy of Sciences as a "reactionary nest."[7] The Soviet government's policy was to finance it generously without infringing on its autonomy. Its publishing house was the only one exempt from state censorship. Not that state censorship, where it did exist, was very strict. The scientific section of the state publishing house, Gosizdat, regularly published non-Marxist works on the philosophy of science, even works clearly antithetical to the Marxist position. In his communication to *Pod znamenem marksizma*, Lenin had expressed his view that books embodying positions with which he disagreed most should be translated and published.

It was not just in the area of publications that such tolerance was extended even to those hostile to Marxism. Outright proponents of idealist and even mystical positions in philosophy were at first allowed to stay at their posts, if they had chosen to remain and fight for their views, rather than emigrate or join the exodus of scholars to White Crimea. Such philosophers as Berdyaev, Frank, Lossky, Ilyin, and Florensky in fact went on the offensive against Marxist philosophy. They established a Free Philosophical Academy in Moscow and a journal *Mysl* in which they boldly proclaimed their mission to struggle against Soviet power in order to protect religion against the brash atheism of the new order. Tension heightened as communist students returned from the front and felt resentful at having to accept as teachers the philosophical counterparts of the enemy they had just defeated. They had sacrificed much to establish the new order and were offended at having to submit to elements hostile to it. In 1922, the government did ask those teachers to leave and join their like abroad. Thenceforth, they continued their campaign in the capitals of Europe, becoming more and more reactionary. Berdyaev devoted himself to propagating his "philosophy of inequality," a "new medievalism," eulogising aristocracy and the "arbitrariness of divine caprice." Others in the same spirit called for the epoch of science to give way to a new epoch of faith.

The New Institutions

Meanwhile, it was becoming clearer than ever that new professors were needed. Thus it was that the Institute of Red Professors was set up in 1921.* Indeed, there was a whole network of communist institutions, paralleling the traditional academic institutions, pursuing bolshevik policy on the second front, the recruitment and training of a new socialist intelligentsia. Workers' faculties, attached to the universities, prepared the sons and daughters of the proletariat for higher education. There were even special universities such as the Sverdlov Communist University in Moscow. The Socialist Academy of Social Sciences, organized in 1918 as a directing center for Marxist research, became the Communist Academy in 1924 and branched out to include the natural sciences as well as the social sciences. Under the academy were such bodies as the Institute of Scientific Philosophy, the Marx-Engels Institute, and the Section of the Natural and Exact Sciences.

There was another network of institutions midway between the other two, with Marxist and non-Marxist scholars collaborating, often with the non-Marxists in the ascendancy. RANION, the Russian Association of Scientific Research Institutes of the Social Sciences, consisted of some fifteen institutes engaged in research and postgraduate training. Work in the philosophy of

---

* Stephen Cohen has described the Institute of Red Professors as "a milieu combining aspects of a university, a political salon, and a monastery." In its early years it reflected the personality of Bukharin, "an aura of Bohemia come to power" (cf. *Bukharin and the Bolshevik Revolution*, New York, 1975, p. 219).

science within the network was concentrated in the Institute of Scientific Philosophy, which, as Marxism came into the ascendancy in the institute, was transferred to the Communist Academy.

Even within the communist institutions, where the hegemony of Marxism was established, there was scope for non-party Marxists to hold prominent positions and indeed there was scope for considerable diversity of opinion. Even at the Institute of Red Professors, half the teachers were not party members and the same was true of the membership of the Communist Academy. Sverdlov Communist University referred to the party organizational bureau (orgbureau) the cases of Deborin and Axelrod, who both had menshevik backgrounds, before giving them teaching posts. The decision was for Deborin and against Axelrod, most likely because Deborin had belonged to the wing of the party that opposed proletarian support for the imperialist war during the debate within the Second International, whereas Axelrod, like Plekhanov, came in behind Russia's war effort. However, being so pressed for qualified professors, the university appealed to Lenin, who replied unhesitatingly in favor of both. Deborin became far and away the most influential Soviet philosopher of the 1920s, although he did not join the party until 1928. He was the leading philosopher, not only at Sverdlov Communist University, but in the Institute of Red Professors and in the Communist Academy. From 1926, he was editor-in-chief of *Pod znamenem marksizma*.

Contending Positions in the Philosophy of Science

It was within the network of communist institutions that there was the greatest intellectual vitality and the liveliest clash of contending positions. In philosophy of science, as in all other fields during the 1920s, there was struggle not only between Marxist and non-Marxist views, but among an extraordinary number of conflicting positions battling for recognition as the correct Marxist position in the field.

Machism

In the first instance, there was the continuation of the pre-revolutionary debate. The vperëdist-Machists were still there after the revolution and were given extraordinarily high positions in ideologically sensitive fields, without renunciation of Machism being required as a condition of their appointment. Lunacharsky was made Commissar of Education, Pokrovsky was the first director of the Institute of Red Professors, Liadov was appointed rector of Sverdlov Communist University; Bogdanov and Bazarov were prominent members of the Communist Academy, and along with Yushkevich, one of the former menshevik Machists, continued to argue their empiriocritical interpretations of Marxism. Like their counterparts elsewhere, the Russian Machists were bringing their claim to be the "philosophy of twentieth-century natural science" to a new level by citing Einstein's theory of relativity as a new confirmation of their theories. In this they were aided in no small measure by Einstein's own acknowledgement of his debt to Mach. Throughout the 1920s, they vigorously argued for their distinctive philosophy of science in academic gatherings, in a series of books, and in *Vestnik kommunisticheskoi akademii* (*Herald of the Communist Academy*) putting particular emphasis on their counterposition of relativity to materialism.

Bogdanov's philosophy had meanwhile evolved to yet another new phase. He put forward the notion of tectology,* the basis of a universal organizational science comprising physics, technology, sociology, philosophy—indeed every possible discipline. According to his idea, the world was conceived as a system of interacting systems. All systems were to be studied in their internal relations and in their external relations, that is, in relation to all other systems as a whole. All systems were organizing or disorganizing. The greater the difference between the system as a whole and the sum of its parts, the better organized it was.

Tectology was an extension of the approach Bogdanov had always advocated, a scientific monism based on the universal law of equilibrium, drawing inspiration from the theories of thermodynamics formulated by Gibbs and Le Châtelier. Within the framework of this new science, he envisaged the future as progress to an all-embracing collectivism.

A constant theme in Bogdanov's participation in the philosophical discussions of the 1920s was the need for philosophy to develop in harmony with the methods and results of the natural sciences. He constantly exhorted his colleagues "to go to nature, to go to life, to prove every chain of reasoning in nature, in life" and to stop arguing by authority, by classical citation, by "grave-robbing."[8]

Proletkult and Proletarian Science

Bogdanov's primary preoccupation in the early days of the revolution was "proletarian culture," an idea he had been propagating in the workers' schools organized by the vperëdists in Capri and Bologna, although it was only with the revolution that it won a mass audience. In 1918, Bogdanov organized the First All-Union Congress of Proletarian Cultural Organizations. This group launched a movement known as Proletkult that quickly became influential on a

---

* Tectology has been seen as a forerunner of later systems of thought, such as cybernetics and general systems theory.

grand scale. The resolution adopted at this congress stated that the proletariat must create a new culture, of a strictly class character, based on social labor. To head off the charge of nihilism, of destroying all the values of the past, of advocating a clean sweep, the resolution ordered that everything in preexisting culture that bore the imprint of common humanity be assimilated, but that this material be recast by the proletariat in the crucible of its own class consciousness. In this task, the proletariat was to act alone, independently of other classes. While asserting that the cultural movement of the proletariat should be autonomous, even though going on at the same time as the political and economic movements, the resolution nevertheless called upon the institutions of government to bring their power to bear in vanquishing the bourgeoisie not only materially but spiritually. In short, Proletkult saw the cultural sphere as relatively autonomous, as demanding a special form of class struggle; culture was saturated with class ideology* and so bourgeois culture could not serve the interests of the proletariat. Any attempt to make it do so was backtracking from the revolution and jeopardizing its gains. Nor could bourgeois specialists serve the interests of the proletariat. The proletariat was the bearer of values specific to itself, a collectivist spirit alien to other classes, which could not therefore be integrated into the new society or have anything to contribute to it. The proletariat would form its own intelligentsia. In all fields, the proletariat would create its own specialists and do things in its own way.

This mode of approach applied along the entire spectrum of cultural forms— from art and literature to science and technology. Just as there would be a proletarian art, there would be a proletarian science. Science was connected to the mode of production and it could not be taken as it stood, as it had been shaped by the capitalist mode of production. It needed to be revolutionized from within. The key to this process was collectivization. This would put an end to the fragmentation of scientific knowledge linked to the competitive drive of capitalist production. It would integrate the diverse sciences within a monist philosophy of science. Distinguishing between a democratization of science that simply disseminated bourgeois science among the proletariat without fundamentally changing its character and the collectivization of science that fundamentally transformed science in a proletarian way, the Proletkult position was:

> Democratization enlarges for the masses the area of mastery of bourgeois science, both in breadth and in depth, but this science as such remains

---

* Bogdanov saw ideologies as forms of social bonding, organizing a certain content and determined by modes of production. Psycho-physical dualism, for example, was the reflection of authoritarian labor relations in the realm of abstract thought. The active, organizing and authoritarian element of production was transformed into soul, while the passive, submissive, and executing element was body.

unaffected. The collectivization of science affects also its essence, method, form and scope. Our task consists in bringing the content and methods of science into line with the requirements of socialist production.

Emphasizing the importance of science and proletarianization for socialism, they concluded:

Without science socialism is impossible, and it is also impossible with *bourgeois* science.[9]

Like all bolsheviks, the Proletkult partisans stood wholeheartedly with science and defended its vital role in the new society, even if they differed from others in understanding the nature of science and its relation to socialism. In fact, in this thorny nest of questions, they not only differed with others, but they differed among themselves. They were by no means homogeneous on the issue of proletarian science. Whereas some implied that science was a superstructural phenomenon determined by its economic base and that any distinction between science and ideology had been obliterated, others were more reticent. Bogdanov himself certainly seemed to believe that bourgeois scientists, such as Einstein, Gibbs, and Le Châtelier, were arriving at conclusions valid for the proletariat and he showed an enormous respect for nonsocialist science.

Lenin, at all events, was unhappy with Proletkult and its proposals. At first he had sent his best wishes to Proletkult, but its unpredictable and dramatic success alarmed him, as he had strong feelings against what he saw as the separatism and nihilism of Proletkult's views on culture. He believed firmly in the value of bourgeois culture, bourgeois science, and bourgeois specialists for the cause of the proletariat. He did not think communism could be built on a firm foundation without them and he would not allow it to be put at risk for the sake of a sectarian "hothouse" culture. A new proletarian culture could not be instantly called into existence. Because of its role under capitalism, the proletariat was still divided, degraded, and corrupted. It needed time, experience, and knowledge in order to mature. Genuine proletarian culture would be developed in an organic, protracted process and would come as the logical development of the whole store of knowledge mankind had accumulated throughout its history, even under the yoke of capitalist, feudal, and primitive production. For Lenin, there was clearly a distinction between knowledge and ideology and there was always the assumption that there was something in science transcending the bourgeois world view. In any case, be began to see such claims of proletarian purity as potentially dangerous and he declared his "merciless hostility" to Proletkult. He decided to clip the wings of the organization that threatened to exclude elements he thought necessary to the construction of socialism and threatened to promote Machism as the "proletarian philosophy" among the millions. There was a definite connection between the phenomenalist epistemology, the notion of proletarian science and the political program of Proletkult, tightly knit together as these were within the logic of a collectivist relativism. At all events, at Lenin's initiative, the Central Committee deputed Lunacharsky to go to the second congress and bring the organization into line under the Commissariat of Education. There was a struggle against this loss of independence and, having lost it, the organization dissipated within a few years.

Another measure that Lenin took to curb the influence of Bogdanov's ideas was to authorize a second edition of *Materialism and Empirio-Criticism* in order to reassert his arguments against Bogdanov's philosophy of science. He explained in the new preface that his duties as head of state had not allowed him the time to update it. However, as an appendix, there was V.I. Nevsky's essay "Dialectical Materialism and the Philosophy of Dead Reaction," a reply to the new edition of Bogdanov's *Philosophy of Living Experience* that had just appeared.[10]

After the decline of Proletkult, Bogdanov's activities were confined to the academic world and, even there, his influence steadily decreased. As a doctor, he became director of the Institute for Research in Blood Transfusion where he died in 1928 in the course of an unsuccessful experiment on himself. Thereafter, his views were, in Schaff's words, "doomed to the biting criticism of oblivion,"* although Schaff and other recent commentators have noted him as a predecessor of the general systems theory and of structuralism.[11]

However, the prerevolutionary debates, from the early 1920s on, became interlaced with, and to some extent overshadowed by, new postrevolutionary debates. There were a succession of new positions in the philosophy of science vying for support as the most appropriate to Marxists in the new social order.

Enchmenism-Mininism

The first, which emerged in the civil war years, was a stridently antiphilosophical shallow materialism. One variant was put forward by Emmanuel Enchmen, an exponent of an "evolutionary theory of historical physiology." An admirer of the Russian scientist Pavlov, Enchmen took as his point of departure a caricature of the neurophysiological school of psychology, which proclaimed the withering away of all thought and all speech to be replaced by a single system of organic reflexes. It was a crude sensationalism that reduced all other levels to that of the biological and proclaimed philosophy to be an

---

* On this perhaps Schaff has spoken too soon, as there are signs these days of a revival of certain ideas associated with Proletkult in certain sections of the radical science movement.

invention of the exploiters to be swept aside with the arrival of the epoch of proletarian dictatorship. Dialectical materialism would be reduced to ashes with the rest.

This "theory of new biology" soon began to spread among the communist youth as they returned from the civil war and became students. It struck a note in the mood that prevailed among certain strata with the ending of the civil war and the launching of the New Economic Policy. Its appeal for the revolutionary youth seems to have been the boldness with which it attacked all that had been previously held sacred. Its extreme sort of nihilism towards the culture of the past and its desire to start from scratch was akin in some respects to Proletkult. It also, in the view of Bukharin, corresponded to the attitude of the "nepman," the new type of Soviet bourgeois that sprang up as a result of the NEP. N.I. Bukharin, editor of *Pravda* and a politbureau member, called attention to this connection and tried to persuade the young communists not to settle for anything so vulgar and so simplistic as Enchmen's views:

> First of all, we have here elements of the new *trader*. The new trader is an individualist. He "accepts the revolution" (of course in parenthesis). This new trader is first of all a vulgar materialist; in every day common affairs he holds nothing "sacred and elevated"; he is used to looking at things "soberly"; he is not bound by any traditions of the past, not burdened by folios of wisdom and heaps of old relics—these the revolution has thrown overboard. He himself is not a descendant of the "spiritual aristocracy," no, he has risen from the bottom; he—the unkempt—has quickly climbed to the top, he is the . . . new bourgeois without any intellectual scruples. He wants everything to smell, to feel and to lick at. He trusts only his own eyes . . . he is physical. From whence his *vulgar materialist superficiality*. Finally, the new trader is crudely practical and vulgar; he is a great simplifier.[12]

Bukharin, along with their teachers in the communist institutions, did their best to convince these students that these views were in the interests of a class with which they had nothing in common.

However, among the teachers there was one who held views akin to Enchmen's, S. Minin, a lecturer at Sverdlov Communist University who, in 1922, published a programmatic article in *Pod znamenem marksizma*, the message of which was "overboard with philosophy." Minin took the French positivist Comte as his point of departure and his critics charged that he was simply accepting Comte's scheme of history and dressing it up in revolutionary phraseology. His idea was that religion was the tool of the landed aristocracy, who needed no science; philosophy was the tool of the bourgeoisie who needed science for production, but needed also to confuse the masses. The proletariat needed neither religion nor philosophy, but only science. According to Minin, the postrevolutionary proletariat had finally outgrown even Marx and Engels who, in pursuing philosophy, were children of their times. Minin's message was a call to arms against philosophy as inseparable from exploitation. The need was for "science, only science, simply science."[13] Minin's first article was accompanied by an extended rebuttal, followed by polemics against it in subsequent issues and in other journals.

Joravsky draws attention to the connection of the various positivistic tendencies surfacing at this time with the whole mood prevailing in the early 1920s. Staggering from one crisis to another, many bolsheviks developed an impatience with abstract theory, if not an outright contempt for it, a revolutionary pragmatism or "Soviet Americanism," as the phrase of the time went. The obverse was a hunger for panaceas, a search for simple, all-embracing formulas that would, on application, quickly dissolve complex difficulties and realize the communist paradise. The theories of Enchmen and Minin captured both sides of this.[14]

Mechanism

This strident ultra-revolutionary positivism, however, was soon superseded by another trend, a less strident and less simplistic positivism. Mechanism as a philosophical current crystallized only gradually and those who became caught up in it never constituted an entirely homogeneous grouping. The mechanists were adherents, for the most part, of more or less unacknowledged positivism. They sometimes paid lip service to the dialectic, but proceeded nonetheless on the basis of a highly mechanistic and reductionistic form of materialism that excluded the dialectical notion of irreducible qualities. In a spirit of pronounced anti-Hegelianism, they treated the dialectic as either superfluous or actually contrary to scientific method. Their conception of science involved the reduction of all qualitative changes to quantitative terms, the reduction of all phenomena, including thought, to physicochemical terms. The Hegelian dialectic, in their view, ran contrary to this, imposing on reality categories unknown to science, like qualitative leaps and internal contradictions, not derived from empirical data.

The mechanists called for the liquidation of philosophy as a specific discipline, believing essentially that science was its own philosophy. Philosophy, in so far as some of them did speak of it in a positive way, was nothing more than the sum total of the natural and social sciences. This was an extremely empiricist, antitheoretical temper of thought, a tendency to stand on the "bare facts" and to believe that they needed no further elaboration. Because of its emphasis on the self-sufficiency of science, it was an extremely popular school of thought in the early 1920s, especially among scientists and students of natural science. Its popularity lay in its sweeping simplicity, and its confident materialism suited the need of those in the grip of fervent revolutionary optimism.

Bukharin

A very powerful figure in setting the tone for Soviet intellectual life was the prominent party leader, Nikolai Bukharin, who interested himself in the philosophy of science. In relation to the earlier philosophical debate, Bukharin admitted "a certain heretical inclination towards the empirio-critics."[15] What Bukharin seems to have found attractive in Bogdanov was his insistence that Marxism was a growing body of thought, his openness to non-Marxist trends, his orientation more towards science than towards Hegel and his emphasis on the concept of equilibrium. In relation to the issues raised by Proletkult, Bukharin took a position somewhere between Bogdanov and Lenin. Regarding proletarian culture, he said:

> I think we can and must try to bring it about, in every sphere of ideological and scientific power, even in mathematics, that we eventually have a certain approach which is specific to us. From this a new spirit will develop in cultural relations.[16]

He nevertheless stood firmly opposed to all forms of coercion in this sphere, to the implicit "reign of terror" in the Proletkult demands, to all forms of official favoritism in such matters, to sectarian appeals to working class origins and proletarian virtue. He believed the proletariat needed time to ripen and needed the bourgeois specialists. Bukharin was indeed the protector of important sections of the old intelligentsia.

Bukharin's influential manual *Historical Materialism*, published in 1921 and used as a basic text in the higher party schools, had a direct bearing on the field of philosophy of science. Like Bogdanov, Bukharin sought to purge Marxism of its lingering Hegelianism and proposed to reinterpret the meaning of dialectics in the process. He thought that dialectics needed to be cleansed of all idealist elements, of the "theological flavor inevitably connected with the Hegelian formulation." This was to be done by giving a theoretical-systematic exposition of dialectics in mechanist terms, that is, by placing it on the foundation of the theory of equilibrium. The dialectical point of view, in Bukharin's exposition, was that all that existed, both natural and historical, was in motion and that motion had its source in the conflict of forces internal to a system. Every system tended to a state of equilibrium. With the conflict of forces came disturbance of equilibrium, followed by a restoration of equilibrium on a new basis, with a new combination of forces. No state of equilibrium was to be seen as absolute or static. Every state of equilibrium was relative and dynamic. Moreover, every system was simultaneously involved in two types of equilibrium: internal and external. The internal equilibrium, that is, the structure of a system, was a function of the external equilibrium—the interaction between a system and its environment.

Bukharin was a mechanist and he did not shy away from the use of mechanist terminology:

It is quite possible to transcribe the "mystical" . . . language of Hegelian dialectics into the language of modern mechanics. Not so long ago, almost all Marxists objected to mechanical terminology, owing to the persistence of the ancient conception of the atom as a detached isolated particle. But now that we have the electron theory, which represents atoms as complete solar systems, we have no reason to shun this mechanical terminology. The most advanced tendencies of scientific thought in all fields accept this view.[17]

The argument thus went that, because in modern times mechanics had itself become dialectical, there was no longer any point in counterposing mechanistic terminology to dialectical terminology. Thus, dialectical materialism had now become indistinguishable from mechanistic materialism. This raised the rather intriguing question of how the Marxist assertion of the universality of dialectics could be reconciled with making "mechanistic" the polar opposite of "dialectical," assuming that the realm of mechanics was undialectical. It was not clear, however, what was entailed in defining mechanics as dialectical, that is, whether this made possible a claim to the universality of dialectics or whether the scope of dialectics was to be restricted after the fashion of Bogdanov, such that some processes were dialectical and others were not. It seemed in some formulations as if dialectics were identified with internal motion. In this case, the status of external motion was unclear. In any event, with his assertion of the priority of external motion to internal motion and the dependence of internal motion upon external motion, Bukharin seemed to be adopting a position contrary to what had been to both Engels and Lenin as essential characteristic of a dialectical point of view, that is, the stress on internal motion. This perhaps is why Lenin in his testament commented that Bukharin "never understood dialectics."

While Bukharin, like Bogdanov, sought after the ideal of the unity of science, he did so in a radically different way. Bukharin differed substantially from Bogdanov in his firm objectivism. He never went the way of empirio-criticism, which he thought of as "psychologizing Marxism." He argued moreover against all the neo-Kantian and Machist denials of causality and assertions of indeterminism and held to an all-pervasive system of causal determinism. There was for him no difference in methodology between the natural sciences and the social sciences.

Most of Bukharin's critics were to be far harsher than Lenin, especially after his political fall from power. There was an *ex post facto* scouring for proof of his long-standing and thorough-going revisionism, reaching the nadir of branding petty thefts in factories as "Bukharinism." The most frequent criticism made was that his emphasis on equilibrium made balance more basic than conflict and presupposed the normality of social harmony as opposed to the traditional Marxist thesis on social struggle. Others centered their critique on his emphasis on determinism, as encouraging social passivity and as undercutting the revolutionary will to act. Bukharin did not believe his philosophical model of change implied political conservatism, asserting that the transformation of quantity into quality was a fundamental law at every step along the way. In both nature, and in human society, there was both evolution and revolution, both gradual adaptation and violent upheaval. The critics, however, were to have their way, for his philosophy did, after his political demise, fall into disrepute, never to be resurrected. However, some have come to reconsider his emphasis on the organizational structure of systems and to consider Bukharin, like Bogdanov, a forerunner of modern systems of theory.

Trotsky

Another politbureau member who turned to philosophy of science and who has been associated with the mechanistic trend, was Leon Trotsky, born Lev Davidovich Bronstein. In his major statement on philosophy of science, his address to the Mendeleyev Congress in 1925, Trotsky unfolded before the assembled chemists a sweeping reductionism:

> Psychology is for us in the final analysis reducible to physiology, and the latter to chemistry, mechanics and physics. . . . Chemistry . . . reduces chemical processes to the mechanical and physical properties of its components. Biology and physiology stand in a similar relationship to chemistry. . . . Psychology is similarly related to physiology.[18]

Extending this scheme under the influence of Pavlov, Trotsky held that the "so-called soul" was nothing more than a complex system of conditioned reflexes, completely rooted in elementary physiological reflexes that, in turn, through the potent stratum of chemistry, found their root in the subsoil of mechanics and physics. Society was simply a more complex system of conditioned reflexes, in the final analysis, a combination of chemical processes. It was "a product of the development of primary matter, like the earth's crust or the amoeba."

Within this framework, the role of philosophy was to systematize the generalized conclusions of all of the positive sciences, linking all phenomena into a single system. To Trotsky's way of thinking, reductionist assumptions were necessary to the achievement of this enterprise. The essence of Marxism, according to Trotsky, consisted in the fact that it examined human history "as one would a colossal laboratory record." There should be no gap between the methods of the natural sciences and those of the social sciences.

To be sure, Trotsky did introduce a cautionary note against transplanting the methods and achievements of one science into another. The methods of chemistry or physiology could not simply be applied to human society. Public life was neither a chemical nor a physiological process, but a social process, shaped according to its own, sociological, laws. Each field of complex combinations of elementary phenomena required a special method. Yet it was not always clear whether these special methods were required by the special character of each level of phenomena or by the fact that science had not yet developed far enough to reduce the analysis of all phenomena to a single method. Sometimes it seemed that Trotsky believed the obstacle to a total reductionism to be only one of time. When it came down to it, Trotsky had occasion to say, the final goal of science was the reduction of all phenomena, whether chemical, biological, social, or intellectual, to the ultimate reality of elementary material particles:

> Scientific thought with its methods cuts like a diamond drill through the complex phenomena of social ideology to the bedrock of matter, its component elements, its atoms with their physical and mechanical properties.[19]

Although every phenomenon of chemistry could not be reduced directly to mechanics, even less could every social phenomenon be reduced directly to physiology and then to chemistry and mechanics, Trotsky said it was the uppermost aim of science to do so.

Yet at other times it seemed that Trotsky held to the dialectical notion of irreducible qualities. This ambivalence was particularly apparent in his discussion of psychology, which he felt was the link between the natural and social sciences. At one moment, Pavlov's reflexology explained all, from the saliva of dogs to the poetry of genius. Then qualifications would be introduced as, for example, when he took issue with Pavlov over his placement of the achievement of human happiness entirely on the shoulders of psychophysiology. No, Trotsky resisted, natural science was powerful, but by no means all-powerful, when it came to the realm of human nature and social relationships. Society was also objectively conditioned and subject to laws, but it did not follow the same laws as the individual human organism. To this methodological problem, Trotsky's answer was that Marxism stood in the same relation to social phenomena as reflexology to psychology.

Trotsky was inclined to a sweeping reductionism, whatever ambiguous qualifications he introduced, that, along with his adherence to a rather thoroughgoing determinism, has been responsible for his being classified as a mechanist. There were, however, other aspects of his thought that made the classification problematic, most particularly his admiration for Hegel and the concept of the dialectic. He saw science as embracing the dialectic of Hegel and Marx, with Mendeleyev's periodic law of the elements as the most outstanding victory of dialectics. So too was the dialectic to be found in Darwin's theory of evolution and in the chemistry of radioactivity. He had no reservations about the language of quantity and quality or the triadic pattern of thesis, antithesis, and synthesis. He did warn against imposing dialectical patterns upon facts rather than finding them there, and he remarked that only painstaking work on a vast amount of material enabled Marx to advance the dialectical system of economics to the conception of labor as social value.

This raised the question of the role of Marxism in relation to the sciences and other disciplines. On this, Trotsky warned against seeing Marxism as an ever-ready master key that could be applied to any sphere of knowledge:

> Whenever any Marxist attempted to transmute the theory of Marx into a universal master key and ignore all other spheres of learning, Vladimir Ilyich [Lenin] would rebuke him with the expressive phrase "communist swagger". Communism is not a substitute for chemistry. But the converse theorem is also true. An attempt to dismiss Marxism with the supposition that chemistry (or the natural sciences in general) is able to decide all questions is no less erroneous and in point of fact no less pretentious than communist swagger.[20]

As to how science, philosophy and politics came together within Marxism, Trotsky addressed himself to the nest of issues involved on a number of levels. He wished to put to rest any lingering fears that Soviet scientists might have akin to the fear expressed by Mendeleyev that science and indeed all enlightenment and morality were threatened by state socialism. On the contrary, Trotsky emphasized that socialism was an organized struggle for civilization, for culture, for science, for morality. Science was to be given maximum scope in the workers' state. Science and technology were essential and decisive in carrying through the process of cultural revolution. They were the way out of the overwhelming backwardness that was weighing upon the Soviet people so heavily and the road to a higher morality. Trotsky was at this time far more optimistic about the future of science and culture under socialism than he would be later, when at times it seemed that the worst fears of Mendeleyev were justified. In these early days, however, he opposed the liberal belief that progress was continuous and accumulative and argued that the history of science, like the whole history of culture, was curved, broken, and zigzagging, with interruptions and failures along the way. With the achievement of socialism, science would advance without such zigzags, breaks, and failures. It was a judgment he would revise in less sanguine days.

But science was not simply taken as given, for class interests introduced false tendencies even into natural science. The wider the field of generalization, the nearer natural science approached to questions of philosophy, the more it was potentially subjected to the influence of class interests. However, in sorting out such matters, it was necessary to proceed with caution. Amidst the many paths by which the proletariat could advance to mastery of science, it was necessary to "learn to learn." The problems of science could not be solved by decree. Marxism was to be brought to bear on new spheres of knowledge by mastering them from within. Taking issue with proponents of a purely proletarian science, Trotsky proposed that the purging of bourgeois science demanded a mastery of it that called for learning, application, and critical reworking, not for sweeping criticism and bald comment. Marxist criticism in science had to be not only vigilant but prudent, lest it degenerate into mere sycophancy. In this respect, Trotsky criticized those who simplistically dismissed Freudianism, declaring psychoanalysis to be incompatible with dialectical maerialism. He had, to be sure, his own criticisms of Freudianism, which he contrasted to Pavlov's reflexology. For Pavlov, generalizations were won step by step, from the saliva of dogs to the poetry of man. Freud, on the other hand, attempted to take all these intermediate steps in one jump and proceeded from above downwards. Pavlov's method was experiment, whereas Freud's was conjecture. Nevertheless, all of the related issues, including the role of the sexual factor, were to Trotsky disputes within the frontiers of materialism. Unlike idealists, who believed the soul to be an independent entity, a bottomless well, both Pavlov and Freud thought the bottom of the soul was physiology. The difference was that Pavlov was like a diver who descended to the bottom and laboriously investigated the well from there upwards, whereas Freud stood above the well and with a penetrating gaze tried to pierce its ever-shifting and troubled waters and to guess at the shape of things below.

Trotsky, in the course of the various zesty speeches in which he outlined his views on these matters through the 1920s obviously had much that was worthwhile to say and, indeed, had issued many a salutary warning. One of these, which had more to do with the hazards for science and for all culture under socialism than he foresaw at the time, was his insistence on the importance of political truthfulness, which he saw as a fundamental requirement of revolutionary policy. It was true, he admitted, that the existence of powerful enemies made necessary a certain cunning, but this could never be an excuse for delusion and deceit with regard to one's own people. Another theme he stressed was the necessity of combining professional specialization with all-embracing syntheses.

However, there were certain problematic areas in Trotsky's analysis of science in the new society. Trotsky often seemed to imply that no special effort was needed to elaborate a Marxist philosophy of science, Marxism in the field of nature science being what natural scientists were already doing. When he more specifically addressed himself to the question of the relation between the new discoveries and dialectical materialism, he floundered somewhat, unable to specify where and when and by whom such questions could be answered. He thought that there were domains in which the party's role was to lead, directly and imperatively, whereas there were other domains in which its role was only to cooperate. For Trotsky, art and science belonged to the latter category. His position was that "Marxist methods" were not the same as the artistic or scientific. Although Trotsky opposed the "revisionist" belief that Marxism was not a universal philosophy, but only a social theory, he often spoke of Marxism as if it were not a universal philosophy applicable to natural science and social science alike, but as if it were only a social science, as if it were a specialized science alongside the other sciences. "Marxist methods" did not, in the context of many of his statements on these matters, encompass the realms of art and science, and so the implication was that the sociopolitical realm was the proper domain of Marxism, whereas the other realms were outside it. Therefore, he seemed to be giving the signal to the artists and scientists to leave politics to the communists, promising that the communists would leave art and science to them. It was not a very integrated position. His insistence on mastery of every field of knowledge from within, his repudiation of solving problems of knowledge by decree was healthy enough, but his haziness about what exactly constituted the domain of Marxism in it all confused matters somewhat.

Trotsky was, however, very definite about where he stood in the debate over Machism. He stood with Lenin. The more science learned about matter, the more unexpected properties of matter it found, the more zealously did "the decadent philosophical thought of the bourgeoisie" try to use these new properties in their struggle against materialism, by implication linking party members under the influence of Machism to alien class interests. Despite his expression of openness to bodies of thought outside Marxism and his appreciation of many of the complexities of philosophy of science, Trotsky could be as ungenerous to opponents as other Russians. In his letter to the first issue of *Pod znamenem markizma*, he wrote that all the variety of idealist, Kantian, empirio-critical philosophies were "in essence a translation of religious dogmas into the language of sham philosophy." He argued, however, not only against Machism, asserting that space and time were not merely categories of the human mind, but also against a vulgar materialism that recognized as matter only what could be felt with the bare hands. This, he said, was sensualism, not materialism. The new discoveries of radioactivity and electrons in no way constituted a threat to materialism. Rather they represented the "execution of the last remnants of metaphysics in materialism" and the "supreme triumph of dialectical thought."[21]

In the course of his address to the chemists at the 1925 Mendeleyev Congress, Trotsky made several other significant points. One concerned the

Soviet attitude to science as part of the cultural heritage of the past. The October Revolution was taking possession triumphantly of the whole of the history of knowledge. The scientific contributions of the past were now theirs, an inheritance of very high value. But it was not to be seen as a matter of simple concretion. There were periods of organic growth as well as periods of rigorous criticism, sifting, and selection. The new social order, in its respect for the future as well as in its respect for the past, could appropriate the cultural heritage of the past, not in its totality, but only in accordance with its own structure.

Another interesting point bore on the logic of scientific discovery and was derived from Mendeleyev. The process of formulating scientific hypotheses was compared to the projection of a bridge across a ravine, it being only necessary to erect a foundation on one side and project an arc in such a way as to find support on the opposite side, and not necessary to descend into the ravine to fix supports at the bottom. So it was with scientific thought. It could base itself only on the granite foundation of experience, but its generalizations, like the arc of the bridge, could rise above the world of facts in order later to touch it again. Trotsky then showed how Mendeleyev's own research methods in discovering the periodic law of elements fit Mendeleyev's own analysis of scientific creativity.

The Formation of Factions

Bukharin and Trotsky were, however, somewhat in the background in the philosophical debates of the 1920s. They were for the most part busy with the affairs of state and, although their writings exercised considerable influence on the debates, they were not among the most direct and active contenders.

The most prominent proponents of the mechanist trend were I.I. Skvortsov-Stepanov, A.K. Timiriazev and L.I. Axelrod. One of the issues around which this trend shaped itself was that of the status of philosophy as an academic discipline. Enchmen and Minin were not the only ones to call into question the place of philosophy in the new order. Certain communists in the field of higher education also began to waver on the question. M.N. Liadov, the rector of Sverdlov Communist University, responded to the susceptibility of the students to the ideas of Enchman by calling, in *Pravda*, for the elimination of philosophy as a separate discipline. His somewhat incongruous line of reasoning was that the surest way of complying with Comrade Lenin's call for solid philosophical training was to stop giving courses in Marxist philosophy. The existing curriculum, he claimed, inevitably led to academicism, to making students' training in Marxist theory too abstract and too isolated, into separate subjects, and ran the risk of turning the revolutionary students into "menshevizing intellectuals" who would become a source of opposition to bolshevik policies.[22]

This provoked the young philosopher S.L. Gonikman to leap to the defence of his profession in *Pravda* and to follow this up with a more extended reply in *Pod znamenem marksizma* in which he was not content to pose philosophy courses as the most effective antidote to the corruption of proletarian students by bourgeois ideology, but went so far as to declare the preeminent importance of Hegel in such courses. To express his position in the sharpest possible terms he designated Hegel as "the only springboard from which the leap into Marxism is possible."[23]

This early exchange contributed to the formation of philosophical factions and was later to erupt into full-blown controversy. But for the moment, mechanism held sway. It was the predominant philosophy held by natural scientists and students of natural science. It could claim as adherents many figures prominent in Soviet public life, particularly in the antireligious campaign.*

A new twist to the argument on the role of philosophy was introduced by V.V. Adoratsky, editor of the first edition of the Marx-Engels correspondence, who wrote a long preface emphasizing the more positivistic utterances of Marx and Engels on the role of philosophy.** His preface served to orient the discussion in some quarters in terms of the definition of ideology*** and of the question of whether or not philosophy was ideology. For his part, Adoratsky defined ideology as a partial and therefore unrealistic scheme of thought and considered philosophy, as necessarily a form of ideology, something to be overcome. Beyond the methodology of dialectics and concrete material of the positive sciences, there was no need for anything further, no need to weave a web of ideology around it. No new philosophy of Marxism was needed.[24]

Another to take up this position was the philosopher I.P. Razumovsky, who asserted his agreement with Adoratsky's pejorative definition of ideology and designation of philosophy as a form of ideology. In the communist society of the future, there could be no ideology but only positive science.[25]

---

* At this time, the campaign was in full swing. The League of Militant Atheists was set up in 1925 under the leadership of Yaroslavsky. Its journal, *The Anti-Religionist,* was decidedly mechanist in tendency.

** See chapter one for a discussion of these statements of Marx and Engels and an evaluation of their relative weight from the point of view of their works taken as a whole.

*** The debate among Marxists over the definition of ideology is still in progress. Many Marxists today still counterpose science to ideology as did Adoratsky and many of his contemporaries, as did Marx and Engels. Others use the word ideology in the broader and nonpejorative sense that there could be scientific or nonscientific ideologies, bourgeois or proletarian ideologies. Lenin employed the term in this sense. This latter approach is, in my opinion, the most sensible.

Stepanov

An influential publication that appeared in 1924 also raised the question of the role of philosophy and opened up the controversy over it on a wider front. I.I. Skvortsov-Stepanov, editor of *Izvestiya*, central committee member, and a leading theorist of the antireligious campaign, published *Historical Materialism and Contemporary Natural Science*, intended to sketch out an atheist *Weltanschauung* and to serve as a manual for antireligious activists. In it he explicitly denied the need for philosophy as a separate discipline, philosophy being synonymous with the latest and most general conclusions of the natural sciences. The position Stepanov unfolded in this manual and in his other writings followed the positivist program for the unity of science, according to which all phenomena, whether physical, biological, or intellectual, were reducible to combination and separation of electrons and protons. The basic law underlying all was the law of the conservation and transformation of matter and energy. The entire universe consisted of material particles, differing only in charge and mass but otherwise identical.[26]

Stepanov was very forthright in his espousal of mechanism. As if to reply to those of his critics who reproached him for not mentioning dialectics in his manual, he boldly entitled an article in *Pod znamenem marksizma* "The Dialectical Understanding of Nature is the Mechanistic Understanding."[27] In the party journal *Bolshevik*, he wrote:

> For the present time the dialectical understanding of nature takes concrete form precisely as the mechanical understanding, i.e., as the reduction of all nature's processes exclusively to the action and transformation of those forms of energy that are studied by physics and chemistry.[28]

Timiriazev

A scientist with similar views on philosophy of science was A.K. Timiriazev. Professor of physics at Moscow University before the revolution, Timiriazev joined the party in 1921 and immersed himself in the feverish ideological activity of postrevolutionary Russia. He kept his post at Moscow University and also became head of the Department of Natural Science at Sverdlov Communist University, a leading member of the Communist Academy, and an editor of *Pod znamenem marksizma*. Timiriazev embodied a sort of *laissez-faire* attitude towards Marxism in the field of the philosophy of science, reassuring communists who had scruples about the influence exercised by non-Marxist natural scientists in the universities that they had no reason to fear, for natural scientists were already unwitting dialectical materialists and it was only necessary to demonstrate this to them. His notion of dialectics too was decidedly mechanistic, dialectical contradiction boiling down to disturbance and restoration of equilibrium on the model of classical mechanics. He did not accept the notion of irreducible qualities characteristic of the dialectical approach, as he thought it an infringement on the scientific method, the essence of which consisted in the reduction of complex phenomena to simple ones. Science proceeded by an ever more extensive reductionism and Timiriazev saw it as his mission to defend it against the encroachments of an alien metaphysics.[29]

Such themes recurred over and over again in the writings of numerous other authors, particularly in those of natural scientists who were stimulated to philosophical pronouncements about science by the prevailing atmosphere. For other physicists, such as Z.A. Tseitlin and I.E. Orlov, quantitative analysis was the *sine qua non* of scientific method. It was the foundation of the drive towards the unity of science and it was not to be obstructed by the Hegelian notion of irreducible qualities at higher levels of integration. I.A. Borichevsky of the Timiriazev Institute flatly stated that science must be emancipated from such philosophical terms as quality and declared that qualities were nothing but certain changes of quantities. Among biologists, there was a certain attraction to mechanism as an alternative to vitalism and as a form of materialism committed to physicochemical explanation of biological phenomena and opposed to the introduction of categories superadded to the scientific process from outside it.

Axelrod

A surprising proponent of mechanism was L.I. Axelrod, no longer the "Orthodox" she once was. A devoted disciple of Plekhanov, she was no longer inclined to put so much emphasis on dialectics and the Hegelian origins of Marxism once her mentor was gone. To be sure, she had unabashedly used the term "the mechanistic outlook" to describe her position in her prerevolutionary days, but with the main battle being the defence of materialism against empirio-criticism, this difference between her thinking and that of Plekhanov and Deborin, her colleagues at that time, was not in the foreground. But whatever mechanistic tendencies she had became more pronounced in her postrevolutionary phase. She began to suggest that the Hegelian terminology that Marxists were using was empty, binding them to nothing, suggesting no questions, and devoid of any serious significance. A constant theme of hers was the "empty and meaningless scholasticism" involved in the elaboration of dialectics from Hegel and the need to look instead to positive data. In stressing this and in criticizing a dogmatic attachment to the classics, she crossed the lines that had been drawn in the old battles and found herself closer to Bogdanov than to Deborin.

Axelrod's attitude became increasingly positivistic. Once, when asked her views on the thesis that dialectics was the most general synthesis of the results of the sciences, she boldly asserted:

> I dare to reply to this question: the synthesis is in the aeroplane, in the radio receiver, and in general in all the great practical results of contemporary natural science. I dare to reply in this way, because I do not forget Marx's great thought that one may explain the world one way or another, but the most important thing is to change it.[30]

The "soul of materialism" was mechanistic causation and, as a consequence, the rejection of teleology and religion. This, the real concrete base, was being set aside, in her view, and all thought was being drowned in barren and meaningless metaphysical abstractions.

Although she never joined the party or exhibited great fervor toward Soviet power, Axelrod, along with Deborin, was far and away the most honored and respected philosopher of the 1920s. They worked together not only at Sverdlov Communist University, but at the Institute of Red Profssors and at the Communist Academy. As time went on and the philosophical factions took shape, they came to symbolize the two opposing trends. All of her carping about "old abstract scholasticism clothed in Marxist terminology"[31] was aimed at Deborin and the students who had come under his influence. In 1926, controversy, in the form of weekly public debates between Deborin and Axelrod, erupted in the Institute of Scientific Philosophy. These did a great deal to define the two main trends and to clarify the issues at stake.

Deborinism

By the mid-1920s the mechanistic tendency was coming increasingly under criticism. At the center of the opposition was Abram Moiseivich Deborin (born Joffe), a professional revolutionary from an early age who, like Axelrod, was a disciple of Plekhanov and gained formal training in philosophy while in exile in Switzerland. Around Deborin gathered a group of younger philosophers, mostly students of his at the Institute of Red Professors, who devoted themselves to studies in the history of philosophy with particular emphasis on Hegel scholarship. They were militantly pro-Hegelian and resentful at attempts to downgrade philosophy in general and that of Hegel in particular.

At first, the Deborinites were not heavily involved in the sphere of the philosophy of science. Their main interest was in the continuity between Marxism and pre-Marxist philosophy. Deborin saw dialectical materialism as the synthesis of the Hegelian dialectic with Feuerbachian materialism. It was a universal philosophy encompassing both the dialectic of nature and the dialectic of history, and the inevitable outcome of the whole of the history of philosophy. The emphasis in the philosophy course at the Institute of Red Professors was on Spinoza, Kant, and Hegel as the sources of dialectical and historical materialism. Philosophy of science came in only as a kind of footnote to Hegel, that is, as an illustration of the universality of the Hegelian dialectic.

However, as the main challenge to Deborin's position came from those whose primary interest was in the philosophy of science, he was drawn onto their terrain for polemical purposes. In 1925, he published *Dialectics and Natural Science* and went on the offensive, declaring boldly the supremacy of philosophy in the field of the philosophy of science. Science, left to itself, was no more than a "collection of facts" and only philosophy could weave these facts together into a coherent whole. Science on its own could not rise above the level of formal logic, whereas the Marxist dialectic introduced a higher logic that made it possible to study phenomena concretely rather than abstractly, in motion rather than simply at rest, in their mutual relations rather than in isolation. Philosophy was the search for the "universal connection of everything with everything." It did not merely *summarize* the results of the positive sciences, as the mechanists claimed, but *synthesized* these into a unitary world view and, moreover, suggested new paths of development. The role of philosophy was to "lead," not to merely "hobble after" the brilliant successes of the natural sciences. It was needed as well to elucidate the method of cognition and to analyze scientific concepts. The dialectics of nature was the "algebra of the natural sciences."[32]

Deborin did affirm the need for a firm union between philosophy and the natural sciences, but it was based on a conception of science that was extremely condescending and at the same time overly deferential. On the one hand, he somewhat brazenly advised natural scientists to abandon their "crawling empiricism" and to study Hegel. Through knowledge of the dialectic, with Hegel as its source, they could overcome their various crises and make progress. He infuriated natural scientists with his Olympian style, his disdain for empirical methods and his tendency to an irritating apriorism, prone as he was to such pronouncements as:

> The question of the possibility of reducing chemistry and biology to mechanical laws is a question of principle. Its methodological formulation and solution cannot be dependent on whether such a reduction has or has not been achieved already in practice.[33]

As if to wave a red flag before a fall, Deborin indicated the sort of methodology that should be the norm in radical contradistinction to that proposed by the mechanists and to that instinctively followed by most natural scientists:

We demand the re-working of the new data in each field of knowledge from the point of view of materialist dialectics, while various "critics", often without being aware of it, are inclined towards the "re-working" of dialectical materialism from the point of view of *particular* facts, of a *particular* science. . . . The method of dialectical materialism is the result of the entire accumulation of human knowledge. Therefore, it cannot be overthrown by *particular, contingent* facts, which are themselves subject to critical examination from the point of view of the general methodology.[34]

On the other hand, because of his own remoteness from the particular facts of the particular sciences, Deborin was inclined to stand aside in debates between rival scientific theories with far-reaching philosophical implications and to leave the natural scientists to get on with it. Only at a fairly advanced stage of the controversy on the theory of relativity in physics or on Morganism versus Lamarckism in biology did Deborin comment on their philosophical import. On some occasions, he echoed the mechanists in asserting that natural scientists were spontaneously coming to dialectical materialism in and through their own research in their own sciences. At times, he even went so far as to say that dialectics could be enriched through the data of the positive sciences conceived of as "practical dialectics." In fact, he said, philosophy would be empty without the "material" provided to it by the natural and social sciences, although, of course, he was quick to add that science was blind without philosophical methodology to guide it.

But, on the whole, Deborin was anxious to keep natural scientists in their place, which, in his view of things, was subordinate to the philosophers. He reproached them for their ontological neutralism that came out in the tendency to define matter exclusively in terms of current physical theory, and he offered his own definition of matter as "the whole, infinite, concrete aggregate of mediations, i.e., relations and connections"[35] as an example of the sort of broader perspective that was needed. He also rebuked them for their reductionism, putting great stress on irreducible qualitative leaps in the higher levels of the organization of matter. The higher forms arose out of the lower forms, but could not be reduced to them.

In these views Deborin was vigorously supported by his students, who were, in time, joined by a growing number of adherents to this fairly homogeneous school of thought. Among the most prominent was I.K. Luppol, N.A. Karev, and Y.E. Sten, who revered Deborin as a great philosopher and began to talk of his works as classics of Marxism.* And like Deborin, they revered Hegel

---

* M.B. Mitin, a student of Deborin's at the Institute of Red Professors and soon to become famous as Deborin's main opponenet, told me that at this time they all thought Deborin to be "Engels's heir on earth" (interview with M.B. Mitin, Moscow, 18 April 1978).

and esteemed his philosophy far above concrete research in the natural sciences. Luppol commented on the relation between two spheres in characteristically stark terms: "The results change, but the dialectics remains the same."[36] Karev, responding to the charge often made against them of having gone over to Hegelianism, once brashly exclaimed: "Yes . . . we are Hegelians. Everything great in modern history has been in one way or another connected with Hegel's name."[37]

Although the Deborinites, or the "dialecticians" as they were sometimes called, were for the most part professional philosophers, while the mechanists were drawn heavily from the ranks of natural scientists, there were as a matter of fact both philosophers and natural scientists on both sides. As the controversy proceeded, several natural scientists threw in their lot with Deborin's "social movement about dialectics." A.A. Maksimov and B.M. Hessen, both of Moscow University's Department of the History and Philosophy of Natural Science, worked in the borderland between philosophy and physics and were instrumental in getting the Deborinites to involve themselves in the debate over relativity theory, just as I.I. Agol, a geneticist, did in the debate over the mechanism of heredity. Maksimov criticized the mechanists for stressing the method of analysis to the neglect of synthesis and for paving the way for the vulgarization of science.[38]

The Deborinites, as time went on, also won a number of supporters, such as Adoratsky and Razumovsky, from the other side. An interesting case was that of S.I. Semkovsky, a menshevik and Machist before the revolution and the most prominent Marxist philosopher of the Soviet Ukraine after the revolution, who exhibited marked mechanist tendencies early in the 1920s. Semkovsky attached great importance to the spontaneous materialism of the natural sciences and showed a certain contempt for Hegelian dialectics, criticizing the "fruitless" philosophizing of the "pure" philosophers. There would be a great flowering in the field of Marxist philosophy, Semkovsky asserted, if it kept to the real living problems being raised by the natural and social sciences and did not draw back into sterile "school philosophy." As the forces began to line up on opposite sides, he called for a "third force" between "vulgar materialism" and "Hegelian scholasticism."[39] In the end, he came over to the Deborinite position on the active and leading role of philosophy in the "dialectical interaction between philosophy and science," although he did still warn against an inordinate attachment to the Hegelian form of dialectics.[40]

By the late 1920s, the Deborinites were in effective control of all philosophical institutions and most organs of philosophical publication; all, that is, except the Timiriazev Institute and its organ *Dialectics in Nature*. In 1928, the Institute of Scientific Philosophy was fused with the Philosophical

Section of the Communist Academy to form the Institute of Philosophy of the Communist Academy with Deborin as director. The Deborinites also had the support of the large organizational network associated with the Society of Militant Materialists, formed in 1924 and amalgamated with the Friends of Hegelian Dialectics in 1928 to become the Society of Militant Materialist Dialecticians.

Neither Trotsky nor Bukharin, the party leaders inclined to mechanism, came to its defence when it came under attack. Presumably they both had troubles enough in the political arena by this time.

Polarization of Positions

When mechanism came under attack, however, its adherents did not allow themselves to be overwhelmed and they fought back with determination; as Stepanov formulated it, the nature of the debate was "a struggle between two irreconcilable points of view":

> From one of these, dialectics is a method which should be used for knowledge of nature and society, in as much as the use of it leads to fruitful results. From the other point of view, all the fundamental relationships of the real world are given in advance in the ready-made conditions laid down in the dialectical philosophy of Hegel. The study of actually existing things can at the very most afford only additional verification of what are essentially a priori assertions.[41]

The contrast was drawn in the sharpest possible terms: a priori philosophy or experimental science. The mechanists chose experimental science.

In further defining the differences, the mechanists were far more active in the main tasks posed by the party to the philosophers of the new era, that is, in aiding the antireligious campaign and in winning over the natural scientists, and they charged the dialecticians with hindering their efforts in both of these tasks. Stepanov accused them of aiding the cause of religion by denying the adequacy of physicochemical explanations and of alienating the natural scientists with their Hegelian scholasticism.

Timiriazev posed the issue in similar terms, setting the "vague and diffuse formulations" of the Deborinites against their own attention to the actual results of research and experiment:

> Now we come to the very root of our disagreements. Those who are called mechanists propose that the study of the concrete facts and phenomena of nature and society should be brought to such a level that the dialectics of these processes would emerge from the processes themselves. Our opponents believe that in the field of natural science one needs to formulate once and for all general propositions like the following: "There is no positive electricity without negative;

there is no dispersion of energy without its concentration; there is no action without a reaction, etc."

This was the theme the mechanists hammered home again and again: it was not possible to write on any subject simply on the basis of a study of Hegel's logic. It was necessary to engage in persevering and specialized research in the particular sciences and to constantly test conclusions by experiment.

Naturally, the dialecticians did not let this way of posing the issues go unchallenged. Luppol responded that their conception of dialectics was not in opposition to natural science. On the contrary, it explained and enriched the conclusions reached by the natural sciences. He insisted that the mechanists' strictures against Hegelianism being imposed on the natural sciences from above did not apply to them, for they were proceeding not on the basis of Hegel's dialectic but on the basis of Marx's and Engels's materialist reconstruction of it.[42] Others added that the mechanists failed to realize the significance of the fact that Marx and Engels had consciously built on Hegel.

And so it went, back and forth. Charges of "crawling empiricism" were met with countercharges of "Hegelian scholasticism." Attacks against "positivism" brought counterattacks against "idealism." Timiriazev called Deborin the "liquidator of the natural sciences." Deborin accused Timiriazev of "khvostism" (tailism). To the mechanists, the dialecticians were hindering their efforts to win natural scientists over to Marxism. To the dialecticians, the mechanists were pandering to the prejudices of bourgeois scientists and obstructing the penetration of dialectical materialism into the realm of the natural sciences. According to the mechanists, the dialecticians were dictating laws to nature instead of inducing them from it. According to the dialecticians, the mechanists were unable to understand the reciprocal interconnection between theory and practice.

The crux of the issue was the relationship between philosophy and science within Marxism, more specifically the relative weight given to the modes of philosophical thought emerging from the development of the history of philosophy vis-à-vis the concrete data derived from the natural sciences in working out a Marxist philosophy of science. No doubt the two groups took up rather extreme positions, swinging too far in the one direction or the other, but the issues were real enough and by no means settled to this day. Still debated are issues relating to the relationship of philosophy to science and the status of the Hegelian heritage within Marxism and great differences have continued to exist among Marxists along these lines.

Two historical factors came into play during this debate which served to discredit the position of the mechanists and give the edge to the dialecticians.

The first was the timely publication of two heretofore unpublished classics of Marxism. Both sides claimed Engels's *Dialectics of Nature* as support for their own positions. The mechanists pointed to the more positivistic utterances of Engels about philosophy being superseded by the positive sciences. The dialecticians were, however, on surer ground, insofar as the overall thrust of the book was a critique of reductionism and a vindication of the role of philosophical thought vis-à-vis the natural sciences. Stepanov published a commentary in which he maintained that the text showed Engels's development evolving through two phases, that is, from a Hegelian metaphysical phase in which Engels was still in the grip of bourgeois philosophy to a phase in which he attempted to eliminate philosophy and dialectics from the realm of the natural sciences.[43]

Both sides also claimed Lenin's *Philosophical Notebooks*. The mechanists called attention to Lenin's statements about the importance of the natural sciences, whereas the dialecticians pointed to the significance that Lenin attached to the study of Hegel. However, there could be no getting away from the fact that the overall direction of thinking in both of these new works did serve to undermine the credibility of the mechanist position. The Deborinites were quick to exploit this advantage and to label their opponents as "revisionists." There was a somewhat reverential attitude to the classics becoming established and Luppol dogmatically stated that the publication of Engels's book should put an end to the controversy.

The second factor coming into play was that developments in the natural sciences themselves were undercutting the mechanist position. Their position in the philosophy of science was tied to the science of an earlier age that was in the process of being overtaken by the new science. These new developments had already been taken into account to some extent and assimilated within the tradition of Marxist philosophy of science by Engels and by Lenin. The mechanist position was to some extent a reversion to a philosophy of science already superseded by Marxism.

Mechanism did, however, reflect the Promethean attitude towards science characteristic of a class that was on its way up. There was a brave defiance of the old modes of thought and a sweeping revolutionism that in the early days was inclined to underestimate the complexities and to charge ahead in the grip of an overly simplistic optimism. There was also a healthy respect for science and experiment and the empirical world and a legitimate wariness of a priori theoretical constructions blocking the way.

The Victory of Deborin

Things came to a head in 1929. In April of that year, the Second All-Union Conference of Marxist-Leninist Scientific Institutions took place. All the leading figures in the philosophical controversy were present and took part in the proceedings and both points of view were expressed in very sharp terms.

The main speeches were given by Deborin as a philosopher and by Otto Shmidt as a natural scientist. Shmidt, a prominent mathematician, headed the Communist Academy's Section of the Natural and Exact Sciences.

Deborin's report summarized the points at issue as he saw them. The difference between the two groups consisted in their "attitude to the facts":

> The mechanists stand on the basis of bare facts, revealing an incomprehensible fear before theoretical thought, but we dialecticians "tie" facts with theory, understanding these as a unity and as a relation.[44]

He reproached the mechanists for revisionism, for failure to see the role of the dialectic in the natural sciences, for lack of understanding of the methodological crisis in science.

Shmidt's speech was somewhat more conciliatory. Although he spoke as a dialectician, he did note the tendency of some philosophers "to underrate positive knowledge, which portended scholasticism." The emphasis, however, was on his rejection of the attempts of the mechanists "to build their own philosophy on the basis of contemporary natural science in isolation from the development of dialectical philosophy and social theory in general." He concluded with what he thought the attitude of natural scientists towards philosophy should be:

> The ridiculous dispute between philosophy and natural science does not exist, for, as dialectical materialists, we recognise the leadership of philosophy in the sense of a general knowledge and a general viewpoint. On the other hand, the philosophers, through the mouths of Deborin and his students who have spoken here, have refuted the absurd accusation sometimes made against them, that they scorn concrete natural science and prescribe their own laws to nature. While continuing their own work in the development of dialectical materialism, the philosophers are in need of facts and concrete applications of the method from us. We, in our turn, need the leadership of dialectical materialism for the solution of the tremendous task before us, that of mastering all natural science as a whole, and we shall utilize that leadership.[45]

The mechanists fought back. They accused the Deborinites of being more intolerant of mechanistic materialism than of idealism and of diverting attention from the tasks given to philosophers by the party, especially in the antireligious campaign. Tseitlin restated their grievances against the dialecticians:

> Natural scientists will not accept Deborin's formalism. It is inapplicable to exact scientific research. . . . Here formalism leads absolutely nowhere, and those materialist natural scientists who want to carry on a genuine, dialectical materialist investigation of nature, and not a game of jackstraws, not a game of "dialectics", inevitably clash with the philosophers, who are proceeding along

the wrong path. All this is explained by the fact that our philosophers do not study the natural sciences . . . they learn Hegel and other philosophers by rote from Kuno Fischer.[46]

However, the consensus was for Deborin. The conference adopted two resolutions, one in accordance with Deborin's report and one in accordance with Shmidt's.* The resolution, based on Deborin's report, acted as a kind of official condemnation of mechanism:

The most active revisionist philosophical tendency during latter years has been that of the mechanists (L. Axelrod, A.K. Timiriazev, A. Varjas and others). Carrying on what was in essence a struggle against the foundations of materialist dialectics, substituting for revolutionary materialist dialectics a vulgar evolutionism, and for materialism, positivism, preventing, in point of fact, the penetration of the methodology of dialectical materialism into the realm of natural science, this tendency represents a clear departure from Marxist-Leninist philosophical positions. The conference considers it necessary to continue the systematic criticism and exposure of the mistakes of the mechanist school from the point of view of consistent Marxism-Leninism. The most important problems confronting the philosophy of Marxism-Leninism are the further development of the theory of dialectics, and the thorough application of the method of dialectical materialism both in the field of social science . . . and natural science. The crisis through which the contemporary theory of natural science is passing is a continuation of that crisis, which has already been analyzed by Lenin. The present successes of natural science do not fit into the pattern of the old, mechanistic, formal logical theories. Here, bourgeois philosophy paralyzes itself attempting to use the crisis in natural science for its own ends. However, a genuine solution of the fundamental difficulties of natural scientists can be attained only by applying the method of materialist dialectics.[47]

The mechanists protested against closing the discussion in this way, some complaining that the issues had not been fully aired, some querying the propriety of settling philosophical issues by majority vote. Some, under pressure of the mounting consensus against them expressed a change of heart. Others, like Axelrod, were quite recalcitrant. Tseitlin declared to the assembled delegates that the passing of a hundred resolutions would not alter the facts that had made him a mechanist.

From this time on, mechanism, like Machism, became something of a dead letter among Soviet Marxists. Those who continued to hold such views were free to persist in arguing for them, but they did so in an atmosphere in which their views were considered discredited by their colleagues, by the authorities, and by the general public. The only administrative consequence of the

---

* The Deborinite drafting committee, however, eliminated from the latter the implied criticism of themselves in his thesis concerning philosophers underrating positive knowledge.

conference was the fusion of the Timiriazev Institute with the Section for the Natural and Exact Sciences of the Communist Academy, although it remained a distinct entity within the other, with a certain relative autonomy.

The Deborinite victory, however, did not last for long. No sooner had they triumphed over the mechanists than they found themselves with a new opposition to face, an opposition of a very different sort. Indeed, in every question, the battles of the 1920s were drawing to a close and giving way to new ones.

The New Turn

Nineteen twenty-nine was the year of "the great turn on all fronts of socialist construction." It was the year of "shattering transformations." The day of the NEP was over. The hopes of immanent revolution throughout Europe had long since faded away. The "left deviation" associated with Trotsky, Zinoviev, and Kamenev, emphasizing world revolution and rapid industrialization, had been defeated by 1927 and its exponents, once politbureau members, were expelled from the party. Trotsky, once thought to be a successor to Lenin, had been sent into exile. Stalin and his "socialism in one country" had come to prevail. Henceforth, it was accepted that the Soviet Union would build socialism from its own resources and not be awaiting revolution in the west to facilitate its movement into the next stage. Nor, it was said, would industrialization proceed at the expense of the peasantry. But in 1929 there was another turn. The right deviation, based on the theories of Bukharin, who advocated a moderate approach to the peasantry, even the kulak section, in the hope of achieving such economic stability as would allow steady industrialization, was defeated. Now, with Stalin reversing himself, after standing with Bukharin against Trotsky, Zinoviev, and Kamenev, it was declared that the peasantry could no longer be placated at the expense of the proletariat. Nothing could any longer be allowed to hold back the great push for rapid industrialization that was now deemed necessary to begin the new era of socialist construction.

The years from 1928 to 1932 were characterized by the frenzy of the first Five Year Plan—the titanic and turbulent struggle to build heavy industry and to collectivize agriculture, to transform a backward land into an advanced industrial nation. The architect of the new policy was Joseph Stalin, as Yosif Vissarionovich Dzhugashvili was known to the world, who by this time had emerged supreme from the complex power struggle that had followed upon the death of Lenin. Audaciously he exhorted the Soviet people to achieve what seemed to all the world impossible. He declared with a breathtaking wilfullness that the Soviet people must reach for dizzying heights and that it was a matter of life or death:

> If there is a passionate desire to do so, every goal can be reached, every obstacle
> overcome. . . . We lag behind the advanced countries by fifty to a hundred years.
> We cover that distance in ten years. Either we do it, or they will crush us.[48]

He declared that there were no fortresses that bolsheviks could not storm,
indeed that they were bound by no laws. Tempo of development was purely
and simply a matter of human decision.

In the intellectual sphere, the new policy signaled the time to effect the
scientific changeover from "bourgeois specialists" to the new "red specialists."
The traditional intelligentsia were served notice that the time for ideological
neutrality was at an end. They were asked to declare themselves for or against
Marxism or, as Karl Radek chose to put it, "on one side of the barricades or the
other."[49]

University professors were required to undergo reelection based on their
political views as well as on their professional competence. In 1927, new
statutes were decreed for the Academy of Sciences that brought the academy
under the jurisdiction of the Council of Peoples' Commissars, and provided for
a new nomination procedure in the election of academicians* with a view to flooding the academy with
communist members. The existence of the communist network of institutions
parallel to the noncommunist ones was to be abolished. The central
committee decreed in March 1931 that the distinction was to be erased and
that the Communist Academy was to be the directing center of all. The
academy was to establish methodological control over all academic institutions
by supervising the drawing up of their plans. Also, communists were no longer
to confine themselves to the special scientific societies organized under the
Communist Academy but were to penetrate the traditional scientific societies.
Each of the annual conventions of the various scientific societies that met in
1930 was pressed to give declarations of loyalty, to pass resolutions on the
role of science in the construction of socialism, and to turn its attention to the
problems involved in the dialectical materialist reconstruction of their science.

Most acquiesced, if only outwardly. There were, however, several instances
of rebellion. The Moscow Mathematical Society, in December 1930, refused
to expel the eminent mathematician D.F. Egorov, who had declared that the
binding of a uniform *Weltanschauung* on scientists amounted to wrecking.
The society ceased functioning for a time, but emerged reconstructed in 1932.

---

* In 1927, the number was raised from 45 to 70. In 1928, it was raised again from
70 to 85. When the elections took place in 1929, there were 42 new academicians
to be elected. Among the new academicians elected at this time were Bukharin,
Deborin, Riazanov, Mitkevich, Pokrovsky, Priashnikov and N.I. Vavilov. For a
detailed account of events relating to the Academy of Sciences at this time, see
Loren Graham, *The Soviet Academy of Sciences and the Communist Party*
(Princeton, 1967).

Another example was the failure of the renovation of the Academy of Sciences to proceed precisely according to plan. A determined opposition was led by Vernadsky and Pavlov. Vernadsky based his attack on Deborin, at this time at the summit of his career as the leading Soviet Marxist in the field of philosophy, on an attack on Marxist philosophy in general. Dialectical materialism was only a survival of Hegelianism that was out of step with science, he claimed. He further argued that science was above all class and party conflicts and that the elections should proceed on the basis of the internationalism of science and not on the basis of specific Soviet conditions. Pavlov based his attack on Bukharin on the claim that he was a person "up to his knees in blood". In any case, when the elections came, Deborin, along with Fricke and Lukin, although sponsored by the party, failed to be elected. There were angry protests from the institutions that had nominated them, with the Institute of Red Professors furiously demanding a purge. The academicians soon capitulated and in extraordinary session soon after, they elected the three further members. Even so, the party set up a special governmental commission to investigate the academy and later issued the Figatner report designating the Academy of Sciences "an asylum for alien elements hostile to Soviet power." The employees of the academy were called to numerous meetings to subject one another to "socialist criticism". A sweeping purge ensued and in December 1929 alone, 128 of its employees were arrested and another 520 were dismissed. In March 1930, a new presidium enacted sweeping organizational changes. Another new charter in 1930 brought in a political requirement for membership of the academy that disqualified scientists with a "hostile attitude". Eventually, at least in principle, the traditional intelligentsia "disarmed itself before the central committee", as the phrase of the time went. In practice, however, most continued through this period undisturbed in the pursuit of their specialized research.

During this period as well, there was a drastic expansion of higher education; between 1928 and 1932, the teaching staff doubled and the student body trebled. This resulted in the traditional intelligentsia's becoming suddenly swamped by the new forces. The "pushed-up ones," who now greatly outnumbered them, had a completely different style of work and introduced many innovations. Ideological requirements were introduced, even into graduate studies. "Shock brigade" methods of teaching and grading were adopted. The emphasis was on harnessing every discipline in the service of the Five Year Plan.

In December 1929, Stalin appeared before the Conference of Agrarian Economists to express his dissatisfaction at the way things were going in the academic world. He spoke of the great drive to collectivize, which was reaching its peak and was at a point of crisis, demanding the utmost in dedication from its practical workers, even while some theorists were still doubting and questioning. Stalin made it clear that he was addressing himself not only to agricultural economists but to intellectuals in all fields, for he felt that Marxist theory was lagging behind the actual stage reached in the practical process of socialist construction.

> Theoretical thought is not keeping pace with our practical progress. There is a gap between our practical progress and the development of theoretical thought, which should not merely keep pace with the practical, but indeed should move in advance of it, arming our practitioners in their struggle for the victory of socialism.[50]

The New Turn on the Philosophical Front

Intensified pressure to "bolshevize" every institution and every discipline followed. At the Institute of Red Professors, philosophy's bolshevizers emerged to make their case. A group of younger philosophers, former students of Deborin at the institute, who had taken Deborin's side against the mechanists, began to express dissatisfaction with the philosophical leadership of Deborin and to declare that neither of the two contending factions in the 1920s had worked out a philosophical position worthy of Soviet communists. This group, led by Mark Borisovich Mitin, acknowledged the contribution of Deborin in exposing the errors of the mechanists, but claimed that he had gone to the opposite extreme in the direction of idealism. They accused Deborin of adopting a formalist attitude, of a habit of scholastic theorizing on the categories of dialectics, that did not take full account of the actual concrete results of the natural sciences or of the practical tasks of socialist construction. Deborin, in their view, had nearly gone over to Hegelianism and had left too wide a gap between theory and practice. They also challenged his perspective on the history of philosophy, not only disputing the emphasis on Hegel in the task of the elaboration of dialectics, but disputing the priority of Plekhanov over Lenin in the philosophical sphere. It was not right, they insisted, that Plekhanov be considered the great philosophical leader of Russian Marxism and Lenin the great political leader. Lenin should be recognized as the greater in the philosophical domain as well as in the political.

The new group called for "a new turn on the philosophical front," one that would be more attentive to the duties of philosophers in serving the needs of the party in its great historical tasks, one for which Lenin would be the supreme philosophical and political authority, one that transcended both the positions of mechanism and of semiidealism that had held sway in the 1920s.

The Deborinites, to be sure, had also taken the cue from Stalin's speech and on several occasions had called for "a turn to the actual problems of socialist construction." In a spirit of self-criticism, Deborin took note of the negligence of philosophers in not criticizing the methodological basis of Trotskyism. This did not satisfy the young bolshevizers, who thought this approach insufficiently self-critical. Deborin accused them of manifesting "unhealthy symptoms" and a "thirst for power."[51]

A turning point in the new controversy was reached on June 7, 1930, when an article sharply critical of the philosophical leadership written by Mitin, Yudin, and Raltsevich was published in *Pravda* with editorial endorsement. They deplored the fact that the situation on the theoretical front was far from satisfactory, with theory lagging behind socialism's actual achievements. They criticized the failure of philosophers to do their part in illuminating the laws of the transition period and accused the existing leadership of a lack of party-mindedness and a formalistic separation of philosophy from politics. The new policy of socialist construction, they declared, called for a reconstruction of science. Communists must initiate a philosophical offensive to intensify the class struggle in science. They called for "a struggle on two fronts"—in philosophy as in politics.[52]

After this, the Deborinites were on the defensive. *Pod znamenem marksizma* was now split three ways, Timiriazev remaining a mechanist, Maksimov supporting the bolshevizers, but with the Deborinites still holding the majority. Deborin was still editor-in-chief. The Deborinites therefore used the journal to reply to the *Pravda* article and expressed extreme resentment at the accusations. Deborin and his colleagues insisted that everything that was correct in the article was simply a restatement of their own views. They too had called for philosophy to turn itself to the service of the party's political needs. They had already acknowledged their shortcomings in not exposing the theoretical foundations of Trotskyism. They denounced the young philosophers as "militant eclectics" and alleged that their denunciation of Hegelian studies and their assimilation of theory to practice amounted to a revival of mechanism. The "new turn" should be built on their previous work, which had been faithful to the philosophical will of Lenin in developing the theory of materialist dialectics as the knot which tied together theory and practice. They took strong exception to the "shocking falsehood added to astounding confusion and gross political errors" in the *Pravda* article.[53]

The next episode came when a resolution giving a detailed survey of philosophical work in the preceding years was adopted by the party bureau of the Institute of Red Professors. The resolution repeated the accusations against the leadership and reasserted the need to take up the philosophical legacy of Lenin. It also pointed to the necessity of making maximum use of the materials gathered by the natural sciences and to the importance of not ignoring the antimechanist contributions of modern bourgeois science in developing the theory of materialist dialectics.[54]

In October, the bureau adopted another resolution in which they again outlined their criticisms of the existing philosophical leadership: (1) giving preference to Plekhanov over Lenin, (2) considering materialist dialectics the same as idealist dialectics, (3) drifting into Kantian formalism, and (4) flirtation with Trotskyism. It then put forward twelve recommendations, which included struggle against philosophical idealism, as well as persistence in the struggle against mechanism, and analysis of the achievements of contemporary natural science and its penetration by dialectical materialism.[55]

The decisive battle then opened with the philosophical conference beginning on October 18, an extended meeting of the presidium of the Communist Academy. For the occasion, all the leading figures were present to argue their positions. The main report was given by V.P. Milutin, chairman of the presidium of the Communist Academy, who proclaimed the central problem to be that of the place of theory in practice and criticized the philosophical leadership for their formalist approach that ignored the vital problems of the day. Deborin's coreport followed. He conceded certain of the criticisms, but pointed to the achievements of himself and his associates and defended their work as fundamentally sound, despite certain acknowledged shortcomings. The militant bolshevizers then came in on the attack, reiterating their criticisms. Mitin elaborated on what the charge of formalism, specifically denied by Deborin, meant. It was, he said, empty scholastic theorizing on the categories of dialectics, transforming dialectics from a vital method of cognition into the coordination of abstract formulas. It was the break between form and content, between the logical and the historical, between philosophical theory and the sociohistorical situation. The Deborinites fought back with great spirit, bringing a wide assortment of labels into play in expressing their contempt for the new trend. Sten accused the new group of "theoretical liquidationism." Also present was Timiriazev, who claimed that the new trend amounted to a vindication of mechanism. Yaroslavsky had by this time gone over to the side of the bolshevizers.[56]

All in all, the consensus was clearly in favor of the new trend. By the end of the conference, Deborin was more subdued. He backed down somewhat from the sentiments expressed in his coreport and expressed his distress at the distance that had come between himself and the "young cadres" with whom he had worked so closely for so long, but he stopped short of denouncing his philosophy as a formalist deviation. The conference concluded with a resolution to the effect that philosophical work should henceforth proceed along the lines indicated by Mitin and his associates.[57]

There were no significant organizational changes as a result of the conference. The Deborinites stayed at their posts and were as powerful as

ever. The younger philosophers were still very dissatisfied with the situation.*
Presumably because of these tensions, *Pod znamenem marksizma* ceased to
appear. Mitin and his colleagues decided to appeal directly to Stalin, but were
extremely unsure of what sort of response to expect. As it turned out, he
summoned them immediately and received them warmly. Having listened to
their case, he replied that they were right but not well oriented politically. They
needed to specify what sort of formalist deviation Deborin and his group were
advocating. It was, Stalin told them, "menshevizing idealism." He then asked
them what they felt should be done and they expressed their desire to see
changes in the editorial board of *Pod znamenem marksizma*. When Stalin
drew them out as to whom they felt should be on the editorial board, they
suggested Mitin and Yudin. Stalin then added, "but also Deborin and
Timiriazev."**

On December 29, there was yet another resolution from the party bureau of
the Institute of Red Professors. In it, the group incorporated the new term
"menshevizing idealism" and characterized it as an outlook reflecting the
pressure of hostile class forces surrounding the proletariat. They also asserted
that the Deborinites had in many ways made common cause with the
mechanists, even while criticizing them. This resolution was followed by a
report given by Mitin on January 1, 1931, at the Communist Academy in
which he gave great emphasis to the correlation of mechanism with Bukharin
and the right deviation and the correlation of "menshevizing idealism" with
Trotskyism and the left deviation.[58]

The Meaning of the New Turn and the Politics of Philosophy

This became in the 1930s the standard formula: the party had been fighting
a war on two fronts. The philosophical debate was linked with the political
debate in that mechanism served as the theoretical justification for the right
deviation, just as Deborinism provided the theoretical justification for the left
deviation.

Explaining the alleged parallels for a foreign audience, the Russian Dmitri
Mirsky, resident in Britain,*** said that, in the 1920s, class enemies had been
allowed to dominate cultural life, just as they had been allowed to carry on in
economic life. However, with the new revolutionary advance of 1929–1930,
these deviations, both philosophical and political, became ideological sabotage

---

* In a long interview with M.B. Mitin devoted to a discussion of the events of this
time, Mitin claimed that Deborin and the others were obstructing the younger
philosophers in their work and they were finding their situation intolerable (interview with M.B. Mitin, Academy of Sciences, Moscow, 18 April 1978).
** My account of this meeting of 9 December 1930, is based on the very detailed
description given to me by Academician Mitin in the above cited interview.
*** Mirsky was Reader in Russian Literature at the University of London.

akin to the industrial sabotage uncovered in "the trials of the wreckers" of 1930–1931. Both Bukharinism in politics and mechanism in philosophy represented the outlook and interests of the survivals of capitalist economy. The one represented the surrender of economic autonomy to bourgeois entrepreneurs and the other amounted to the surrender of ideological autonomy to bourgeois scientists.[59]

Looking back at the historical evidence in relation to the parallel debates of the 1920s, there can be little doubt that Mitin's formula was a *post factum* and highly oversimplified statement of the situation. During the 1920s, there were few indications of any consciousness of connections between the factional struggle within the party and the debates in philosophy of science. What odd references to it there were as often as not linked Trotskyism with mechanism, not with Deborinism. At the time of the political defeat of Trotsky and his expulsion from the Communist Academy, Deborin's prestige was higher than ever and yet to be higher still. Moreover, most Deborinites were anti-Trotskyist. As to the philosophical mechanists, there is no evidence that they constituted a hotbed of support for Bukharin's political proposals.

This is not, however, to deny any lines of connection whatsoever. Commentators such as Kolakowski who assert that there was "no logical connection" and that the formula was "a fabrication of the most arbitrary kind" are quite wrong. This, too, is an oversimplification. The argument linking mechanism to a tendency to determinist evolutionism, to overemphasis on continuity in denying qualitative leaps, and linking Deborinism to a tendency to voluntarism, to an overemphasis on leaps that could lead to revolutionary adventurism, is not really so ridiculous as Kolakowski makes it out to be.[60] Joravsky's approach, arguing against any simple and direct connection, but for a certain ideological affinity that became apparent in retrospect, is more plausible. At the center of the politicians' arguments was the problem of carrying peasants along in the drive to industrialize. At the center of the philosophers' arguments was the problem of winning scientists to support the bolshevik ideology that inspired that drive. Both the mechanist philosopher and the rightist politician, each in his own area, urged caution and gradualism—the coaxing of the refractory elements. On the other hand, both the Deborinite philosopher and the Trotskyist politician argued boldness and haste—the compelling of the refractory element.[61] While such was roughly the case, it was not quite so simple. While the mechanists were inclined to patience vis-à-vis natural scientists, they were inclined to militancy where religious believers were concerned. The Deborinites were inclined to a certain conservatism in relation to the history of philosophy and they lacked the sharp combative edge of those who emphasized the materialist implications of the natural sciences in sweeping away all forms of philosophical idealism. Still there was a subtle and indirect connection between the political and philosophical debates. Both reflected the extremely complex tensions involved in trying to achieve a synthesis of seemingly diverse elements in order to bring them into a new harmony for the new revolutionary social order.

The Soviet attitude to this question has changed somewhat over the years. Adoratsky's *Dialectical Materialism*, published in 1934, was a typical work of that period, in claiming that both tendencies converged in acting as channels of bourgeois influence over the proletariat.[62] An influential collective publication, the *Philosophical Encyclopedia*, as late as 1962 still described the struggle in philosophy as connected in a simple and straightforward way with the struggle against the right and left deviations and described the discredited trends as "opportunist views which tried to undermine the significance of Leninism."[63] However, few Soviet philosophers today would assert such a direct connection. The term "menshevizing idealism" is no longer used to describe the trend associated with Deborin.

But in 1931 Mitin's formula held sway. On January 25, 1931, a central committee decree was issued "Concerning the journal *Pod znamenem marksizma*," criticizing the editorial policy of the journal for turning it into an organ of "menshevizing idealism" and ordering a reorganization of the editorial board.[64] Deborin became an ordinary member of the editorial board, no longer the editor-in-chief. Karev and Sten were dropped and Mitin and Yudin were brought on. The "new turn on the philosophical front" had thus decisively triumphed and consolidated itself.

Until recent years, there has been very little in the way of secondary literature, even in the Soviet Union, evaluating the development of Soviet philosophy during these years. Not that there is very much even today, but several significant works have emerged.

To the foreign audience, one of the earliest accounts came in 1933 in Julius Hecker's *Moscow Dialogues*, an extraordinarily engaging account of the events of the 1920s as they looked from the perspective of the early 1930s. Hecker, a Russian émigré, who returned to Russia after the revolution, presented his interpretation of these years in the form of a dialogue between a mythical Soviet philosopher "Socratov" and a delegation of visitors from abroad. The story was told from a point of view enthusiastically in favor of the "new turn."[65]

In the 1940s, the American philosopher John Somerville spent two years in the Soviet Union, and in 1946, he published the book *Soviet Philosophy*. Again, the events were recounted in a light favorable to the "new turn." Somerville assessed the net effects of the two successive major controversies in terms of scope, method, and temper. As to scope, the first controversy brought about a widening and broadening of philosophical effort, such that the whole range of traditional problems of the history of philosophy were to come under its aegis; the second controversy emphasized that there should be no diminution of scientific spirit or social commitment. In terms of method, the first controversy resulted in a greater receptivity to generalization, theory, and synthesis, to which the second added the necessity of seeing these methodological instrumentalities in the focus of practice. Regarding temper, the first controversy meant an effort to substitute a certain integrative urbanity for insular specialism, while the second meant the effort to harden the integrative quality, giving it the sharp edge of social realism.[66]

In the foreign language literature (that is, non-Russian), the next to take an interest in the early history of Soviet philosophy were a network of "Sovietologist-philosophers," many of them Catholic priests, centered in Western Europe and America and assessing these ideas from a neo-Thomist orientation. Thus came Gustav Wetter's *Dialectical Materialism: A Historical and Systematic Survey of Philosophy in the Soviet Union* in 1958, J.M. Bohenski's *Soviet Russian Dialectical Materialism* in 1963, T.J. Blakeley's *Soviet Philosophy* in 1964, and Richard de George's *Patterns of Soviet Thought* in 1966. Through all of these runs a common pattern of interpretation. One theme is that the controversies highlighted the irreconcilability of dialectics and materialism, the ambiguous relationship between philosophy and science, the tension between the philosophical and the positivistic strands in Marxist thought. Another often-repeated assertion in these works is that the autonomy of Soviet philosophy came to an end by decree in 1931, becoming thereafter totally subservient to the voice of Stalin.[67]

A particularly interesting analysis of this period can be found in Loren Graham's *The Soviet Academy of Sciences and the Communist Party*. Graham argues that writers who have characterized the bolsheviks as "wreckers of science" have missed much of the drama of the situation, for commitment to science was more fundamental to bolshevism than was political control over scientists. Thus the irony of the very serious damage done to science by the bolsheviks during this period by the particular interpretation of the idea of unity of theory and practice that became accepted in the early 1930s, by the growing tendency for terror to stifle the intellectual dimensions of socialism and for apparatchiks to replace theoreticians under Stalin. Also, seeing an aspect of the philosophical debates of the 1920s missed by most commentators, Graham analyzes their meaning in terms of the notion of "proletarian science." The mechanist-Deborinite controversy, he contends, was connected indirectly with the discussion of the place of science within the Marxist schema of base and superstructure. Mechanists such as Stepanov believed science to be nearly homogeneous in different social systems, whereas such Deborinites as Egorshin insisted on the class nature of science itself and argued that modern natural science was bourgeois in its theoretical foundations. The victory of the Deborinites marked an increasing willingness to put science within the superstructure and to uphold the active role played by Marxist philosophy of science. This tendency survived their supersession by the Stalinists. The 1931 group accepted a somewhat more involved analysis of the relationship between base and superstructure than either the mechanists or Deborinites had granted, but they still included science in the superstructure. Shmidt, for example, argued in his 1932 article on "Natural Science" in the *Large Soviet Encyclopedia* that in each epoch the level of natural science was determined in the final analysis by the level of development of the forces of production. However, natural science, as part of the superstructure, was also influenced by other elements of the superstructure, while at the same time, it influenced in its turn the forces of production.[68*]

The really definitive work on Soviet philosophy of science during this period has been that of David Joravsky's *Soviet Marxism and Natural Science 1917–1932*. Although the book is historically accurate and extremely well-researched, Joravsky's judgement on these events is sometimes exceedingly harsh, not that harsh judgement is not sometimes well deserved. He is highly critical of the method of conciliarism, that is, the establishment of a philosophical line by resolutions at conferences. According to his interpretation, 1931 marked the "great break" for Marxist philosophers of science. The events of that year eliminated the issues that had divided philosophers in the past, restricted the field, changed the rules of argument, and brought to the fore "a new generation of court philosophers, ready to continue the restricted argument under the new rules, in the presence of the close-mouthed, unblinking chief of the revolution."[69]

A more recent and even less sympathetic treatment of these years appears in the third volume of Kolakowski's *Main Currents of Marxism*, entitled *The Breakdown*. Unfortunately, his account is marred by historical inaccuracy;

---

* A variation on this theme is taken up in: S. Tagliagambe, *Scienzia, Filosofia, Politica in Unione Sovietica 1924–1939* (Milan, 1978), which interprets the implications differently from Graham. Ludovico Geymonat reviewed the work: "A New Interpretation of the Cultural Conflict in Post-Revolutionary Russia" in *Scientia* (vol. 114, nos. 1–4, 1979, pp. 221–236). Tagliagambe describes mechanists as reducing historical materialism to a general conception that tightly bound the progress of the superstructures to the development of the productive forces and as tending to resolve all historical explanation on the basis of unidirectional causation. He characterizes the dialecticians as emphasizing the relative autonomy of the superstructures and the specificity of their manifestations. The dialecticians, he contends, were guided by a greater respect for the autonomy of scientific research and a more genetic conception of knowledge. He contrasts the intellectual pluralism of the 1920s with the ideological pragmatism of the 1930s that brought the sphere of culture to bear upon the forced march being imposed upon the development of industrialization and collectivization of agriculture. Disagreeing with those who take Marxist philosophy to task for the distortion of Soviet science, Tagliagambe argues that it is not philosophy, but the philosophical vacuum produced by Stalinist pragmatism that is to blame for the violence inflicted on science. Tagliagambe openly defends the dialecticians. He denies any accord between the Stalinist conception of science and the philosophy of dialectical materialism, but affirms a strong connection between Stalinist pragmatism and both Bogdanovian subjectivism and mechanistic materialism.

he reports, for example, that Deborin was dismissed from the editorial board of *Pod znamenem marksizma*, which he was not. More important, Kolakowski's interpretation of the philosophical outcome of the controversies is somewhat twisted. After commenting that the mechanist had a good idea of what science was about, but knew nothing about the history of philosophy, whereas the dialecticians were ignoramuses in the sphere of science, but knew more about the history of philosophy, he concludes: "Eventually, the party condemned both camps and created a dialectical synthesis of both forms of ignorance." He asserts that there was no philosophical issue at stake in the 1930 controversy and that the official "canonical" version of dialectical materialism that was adopted was virtually indistinguishable from Deborin's. He feels that it was simply a power struggle and that in 1931 brought to power "a younger generation of careerists, informers and ignoramuses" who made a career out of betraying colleagues, parrotted the classics and party slogans and presided over the extinction of philosophical studies in the Soviet Union."[70]

Turning from the foreign literature to the Soviet literature on the subject, it is, within the Soviet Union, acknowledged as a shortcoming of Soviet philosophers that they have not written very fully of their own philosophical history.* In the *Dictionary of Philosophy*, edited by Rosenthal and Yudin, Deborin is not even mentioned. However, the major secondary sources in the Soviet Union have been large-scale collective works, such as the six-volume *History of Philosophy* and the *Philosophical Encyclopedia*. In the *History of Philosophy*, book one of volume six is devoted to the history of Soviet philosophy, as is the article on "Philosophical Science in the USSR" in volume five of the *Philosophical Encyclopedia*, but both give only a brief, sketchy, and schematic account of the developments of the 1920s. Mechanism is described as substituting materialist dialectics by vulgar evolutionism and replacing materialism by positivism, objectively hindering the penetration of dialectical materialist methodology into the field of science. Deborin and his associates are mentioned as having criticized the mechanists, but themselves having made mistakes of a formalist and idealist character, expressed in over-valuing the idealist dialectics of Hegel and underestimating the philosophical heritage of Lenin.[71]

In recent years, however, several full-scale studies on Soviet philosophy in the 1920s have emerged, as well as historical articles in such journals as *Filosofski nauki (Philosophical Sciences)*. V.I. Ksenofontov's *Leninist Ideas in Soviet Science in the 1920s*, published in 1975, characterizes the mechanists as suspicious of philosophy and as exaggerating the significance of

---

* This point was made to me on several occasions. Nevertheless, the inhibiting factors against forthright discussion among communists, which I have discussed elsewhere, are operative here and continue to prevent them from fully coming to terms with their history.

the methods of the natural sciences, and the Deborinites as too uncritically enamored of the Hegelian dialectic and too ready to dictate to natural scientists. What is particularly interesting about this work, however, is that the author makes an effort to understand the rise of such trends in their connection with the social circumstances of the times, instead of simply branding them as "mistakes" or correlating them too crudely with political trends. Before judging the mechanists too harshly for hyperbolizing the natural sciences as a sufficient basis for an understanding of social reality, he insists that it is important to remember that religious, mystical, and idealist views were still widespread and influential. As to the Deborinites, while it is true, he admits, that they were too one-sidedly engrossed in Hegelianism, it is necessary to see as well that they did not yet know of the dangers of dictating to science, a lesson made clear only by later Soviet history.[72]

It is indeed vital to assess these controversies within the movements of the times, not to see them as simple direct reflections of political positions, but to understand them as giving expression to the sort of exaggerated positions that come to the surface in periods of social ferment representing sincere efforts to bring into play forces that needed to find their expression vis-à-vis other forces within the new social order.

It is also correct to see the different trends as reflecting the tension between philosophy and natural science, between dialectics and materialism. It is a mistake, however, to believe after the fashion of the neo-Thomist Sovietologists that these tensions could not be resolved creatively. It was vital to the future of Marxism that a synthesis emerge in which neither philosophy nor science, neither dialectics nor materialism, was emphasized at the expense of the other, as they were with the mechanists and dialecticians respectively. A return to the more integrative approach that had characterized Marx, Engels, and Lenin would have been constructive enough, if not for the authoritarianism and suppression of debate that came with it. It was at least potentially a dialectical synthesis of two forms of knowledge, and not of two forms of ignorance, as Kolakowski so bitterly asserts.

Also reflected was the tension between such a theoretical synthesis and the concrete level of political practice. The resolution here involved at least an orientation toward bringing the integrative approach of the founders of Marxism to a new stage by seeing it within a new political context, the new stage of socialist construction. It was still unspecified just how philosophy and science would be brought to bear in the stage of socialist construction. It was not yet clear that those with the crudest and most opportunist manner of drawing the connections would prevail and that it would culminate in such disaster. The issues at stake in these debates were very real and important ones. The most fruitful form of interaction between philosophy and the natural sciences and between both of these and their social context remain challenging subjects of discussion and debate to this day.

The young philosophers of the "new turn" did effect a significant and important reorientation of Soviet philosophy along lines that set it again in the mainstream course flowing from the original inspiration of the Marxist tradition, even if it was based more on appeal to authority than philosophical argument. They paved the way for the systematization of dialectical materialism as a philosophy of science and emphasized the social and political importance of philosophy in a socialist society, even if their own mode of systematization was a least-common-denominator vulgarization, even if an orientation towards sociopolitical relevance was turned into an instrument for a ruthless witch-hunt. The reorientation cannot be reduced to a simple struggle for power, even if such shifts often do carry opportunist elements with the tide, even if in this case the most prominent exponents of the reorientation were perhaps engaged in a simple struggle for power.

The young philosophers did have their shortcomings, and rather glaring ones at that. Their rough edges were perhaps attributable, in some degree at least, to their youth, to the rapidity of their rise to the top, and to the backwardness of their social background, but also to their own ruthlessness, to a taste more for power than for knowledge. They lacked the detailed knowledge of either the natural sciences or the history of philosophy that had characterized their predecessors. They had come up under the revolution. They had never been abroad, nor were they much acquainted with foreign trends or with the level of discussion reached by their foreign counterparts. They were inclined to a simplistic practicalism, to an unhealthy authoritarianism, and to extreme crudity in argument. But, however brash, unsophisticated, and heavy-handed those most to the forefront may have been, the whole of the new generation of Soviet philosophers were not simply "careerists, informers and ignoramuses," nor were they all simply "court philosophers." No doubt many were, particularly those who rose to greatest prominence at this time, but the entire generation of Soviet philosophers formed in the creative intensity of the early years of the revolution cannot be dismissed so easily. There were others, whose names were not on everyone's lips, who were not being proposed as academicians, members of editorial boards and directors of institutes, who quietly and seriously continued their work.

It would be a great mistake to believe that creative philosophical work was cut off by political decree in 1931, as so many of the commentators contend. It was a complex society and it was not the case that all forms of intellectual life were strangled by various forms of administrative fiat at this time. Creative work continued, particularly in the early 1930s, although it was coming increasingly under pressure all the time. Indeed the face that Soviet society

presented to the world in 1931 though its delegation's contribution to the
history and philosophy of science at the History of Science Congress in
London, was a particularly impressive one. The fate of the delegation,
however, constituted a dramatic illustration of the intensification of pressure
and the tragedy resulting from it.

Soviet philosophy was at every stage along the way inextricably tied to
Soviet politics and on many levels. Specific developments in Soviet intellectual
life can only be understood against the background of specific parallel
developments in Soviet political and economic life. The goals of the Five Year
Plan were being pursued with great vigor and determination, but the way
forward proved to be far from smooth and uncomplicated. There was massive
and violent resistance to collectivization, as peasants burned crops and
slaughtered livestock rather than surrender. There was disaster after disaster in
the drive towards industrialization, as unskilled hands and hastily trained
engineers fumbled with sophisticated machinery imported from abroad.
Breakdowns, fires, famine, and unfulfilled plans were put down to sabotage
and espionage and conspiracy with foreign powers. But the evidence for
sabotage and espionage and conspiracy with foreign powers was of a very
dubious sort. Yet, such charges were made in a ruthless attempt to find
scapegoats to blame for problems actually resulting from the backwardness of
a workforce that lacked the extraordinarily advanced levels of expertise in
science, technology, engineering, and agronomy required for such a dramatic
push forward. It was sometimes extremely difficult to know how to distinguish
between bungling and wrecking, between advocates of discredited policies and
sabotage, between contact with foreign colleagues and espionage. In this
atmosphere the commitment of the intelligentsia formed before the revolution
was brought more and more into question. Stalin spoke of the connections
between the bourgeois intelligentsia and international capital, which was
massing its forces against the Soviet régime.*

Bolshevization

As the 1920s passed into the 1930s, there was a sharp turn in the direction of
Soviet intellectual life. Gone were the days when specialists were told that
working conscientiously in their own special fields was enough. Every area

---

* To reinforce the point, there were, between 1928 and 1931, a series of public political trials, opening in 1928 with the "Shakhty Affair," involving mining engineers
accused of wrecking. In 1929, there was a trial of Ukrainian nationalists charged with
forming a secret alliance with Poland. In 1930 came trials of the "Toiling Peasant
Party," on charges of sabotage and espionage in the food supply system, and of the
"Industrial Party," of technical specialists accused of wrecking, subversion, espionage, and sabotage in industry. (Note continued on page 451.)

was to be overhauled and "bolshevized." Every science was to be thoroughly reconstructed from a dialectical materialist point of view. The idea of two sciences—bourgeois science and proletarian science—came increasingly into play. In sciences in which there were contending theories, there was a growing tendency for the proponents of one to label that of their opponents as "bourgeois" and their own as "proletarian" and to become correspondingly cavalier in relation to standards of evidence. In history, and by implication in all other fields, Stalin set the tone in an angry letter to *Proletarskaya revolutsiya* in 1931 denouncing "archives rats."[73] Not only did such historians as Slutsky, Pokrovsky, and Yaroslavsky come severely under attack, while others were dismissed from their jobs and expelled from the party, but still others in other disciplines felt the impact as well in a great rash of denunciation of all forms of "rotten liberalism" and "bourgeois objectivism." There were new, though unspecified, criteria. There were special, though elusive, proletarian standards of evidence to be sharply distinguished from the bourgeois ones and there were the most shocking barbarities, such as assertions that "where class instinct speaks, proof is unnecessary." More vaguely, others simply somehow believed that the very affirmation of dialectical materialism conferred upon young bolsheviks an insight into science denied to their empiricist contemporaries. The stage was set and the Lysenkos were on the rise.

It was the day of the bolshevizing youth in every discipline. They were given enormous responsibilities, sometimes incommensurate with their skills, training, or experience. They suddenly found themselves in positions of authority over far more highly qualified specialists in their fields, indeed often over extremely distinguished scholars of international reputation. In philosophy, the young P.F. Yudin was made head of the Philosophical Section of the Institute of Red Professors within a year of graduating from it. In a few short years, he was Academician Yudin, director of the Institute of Philosophy of the Academy of Sciences. Indeed, there were a number of new academicians in the 1930s, the standards were suddenly sharply different from anything before or since. The tone of Soviet intellectual life had changed drastically. The most serious subjects were discussed in a sort of "agitprop pidgin" and learned institutions became penetrated by a provocative and combative presence. The party organization of the Academy of Sciences set up a newspaper, *For the Socialist Reconstruction of the Academy of Sciences*, calling upon young workers in the academy to criticize the work of their seniors and condemning the survivals of political neutrality in the academy. The Komsomol organization formed an organization called the "Light Cavalry" to intensify class struggle in the Academy of Sciences. So it was in all quarters. A meeting of the Leningrad Writers' Union was actually called upon by some of its members to form "mopping-up squads" to remove from bookshops works that Stalin had just condemned.

But the traditional intelligentsia was still there and working, as was the older communist intelligentsia. Bukharin, although expelled from the politbureau, was still a central committee member and an academician. He held a number of responsible positions in the 1930s and was very much involved in organizing the scientific work of the Academy of Sciences. He was editor of the *Large Soviet Encyclopedia* and a series of volumes on the history of science and technology. He had moved beyond the mechanist approach that had marked his thinking in the 1920s and, in the 1930s, displayed a new integration in his philosophical writings.

Deborin also changed his mind to some extent and continued to hold important academic posts; he was especially active within the Academy of Sciences and was eventually elected to its presidium. His early statements of self-criticism were criticized somewhat ungraciously by some of the younger philosophers. In 1933, he gave a long self-critical speech at the Institute of Philosophy:

> Our fundamental mistake lay in separating theory and practice. I must admit that for a long time I did not understand that accusation, because, after all, I was constantly speaking of practice as the criterion of truth. I was speaking about it and writing about it. Only later did I understand that this was not all that was at stake. The essence of the matter was that we failed to establish a connection between our theoretical and methodological investigations and the concrete tasks of socialist construction, that we separated theory from life. This led to a state where some of the menshevizing idealists developed to a certain degree into "inner emigrants" and later into political opportunists.[74]

He then urged on his followers "complete and unconditional surrender" as the only possible position for an honest communist.[75*] Some followed his advice, while others did not. Luppol and Asmus engaged in self-criticism, while Sten refused to do so, but for the moment they all continued their work. So did the mechanists, although major figures like Timiriazev and Axelrod persistently refused to recant. Axelrod, at a conference organized at the Institute of Philosophy of the Communist Academy devoted to the twenty-fifth anniversary of *Materialism and Empirio-Criticism*, did modify her views with regard to this work that she had once viewed negatively, but she remained an unregenerate mechanist to the end.

There were, however, several early casualties among the older generation of communist intellectuals, most notably David Riazanov, founder and director

---

* He elaborated on these ideas on a number of occasions. In an interview with John Somerville in 1937, he remarked that it was a false conception of the relation of theory and practice that led him to misconceive the relation between Lenin and Plekhanov (cf. Somerville, *Soviet Philosophy*, New York, 1946, pp. 233–234). In 1961, he published a collection of his old articles with a new foreword analyzing what he saw as the errors in his early work.

of the Marx-Engels Institute, a crusty old Marxist scholar, not inclined to hold his caustic tongue, no matter how powerful the opposition. He did not hesitate to speak against particular proposals supported by Lenin in his day, and he continued in the same fashion under Stalin. In 1931, he was abruptly dismissed from his institute, expelled from the party and sent into internal exile. He was charged with treason: in general, with trying to establish betwen the party and its enemies a "third force" and, in particular, with sheltering mensheviks in his institute and with honoring agreements with foreign social-democrats to suppress publication of certain letters of Marx and Engels. The evidence was of an extremely dubious sort.* He was replaced by Adoratsky who had shown himself particularly adept at flowing with the tide, transforming himself within a few short years from mechanist to Deborinite to militant bolshevizer. Meanwhile, Riazanov, the greatest living authority on the life and work of Marx and Engels, after a time in prison, was sent to Saratov where he worked as a librarian until he died in 1938. With the permission of Kirov, he was, for a short time, allowed to continue his research in Leningrad. However, upon the assassination of Kirov, he was forced to return to Saratov.

By 1933, the philosopher Ian Sten, who had been engaged by Stalin to tutor him in philosophy from 1925 to 1928, was also arrested and sent into internal exile. In his case, there seemed to be evidence linking him to actual political opposition.**

Nevertheless, work in the philosophy of science proceeded in the early 1930s, even if the storm clouds were gathering all around it. The rallying cry for the proletarianization of science brought Soviet science to a level of acute philosophical and political self-consciousness unprecedented in the history of science anywhere. This had its positive, as well as its negative, aspects. There were, it is true, the clumsy and hasty reconstructions that justified eccentric scientific theories on the basis of crude philosophical notions put forward with scant regard for experimental evidence. There were the vulgar and superficial

---

* The circumstances surrounding this case were somewhat complex. In the trial of the "Union Bureau of the Central Committee of the Menshevik Party," I.I. Rubin, a research worker at Riazanov's institute, testified against Riazanov. Rubin's sister later testified that her brother had been forced to give false testimony against him and Riazanov regarding Menshevik documents being kept at the institute (cf. R. Medvedev, *Let History Judge,* London, 1976, pp. 132–136). It seems clear that Stalin had a strong antipathy to Riazanov (cf. Medvedev, *Let History Judge,* p. 34).
** Sten, earlier of the "Young Stalinist left," seems to have become involved with the opposition group that formed around M.N. Riutin in the early 1930s. Riutin, the head of propaganda in the central committee apparatus, issued a manifesto combining demands for democratization within the party, after the fashion of Trotsky, with elements of Bukharin's political program. Former supporters of Trotsky, Zinoviev, Kamenev, and Bukharin rallied to the group. Its leaders were arrested and Stalin demanded that they be shot, but was overruled by the politbureau, which decided that they should be sent into exile.

demands that every line of research justify itself by its immediate practical relevance to the goals of the Five Year Plan. There were also leftist excesses in dealing with the older generation of scientists. Still, in all, the atmosphere of the time brought an awareness of the need to bind together the different spheres of philosophy, science, politics, and, indeed, all else, into a new and revolutionary whole. It brought a new intensity to the exploration of the philosophical implications of the results of scientific research and a new degree of seriousness about the role of science in the process of social transformation. These were all to be welded together in a positive way to build a new world.

It did seem that a new world was being built. It was a time of enormous transformations and great upward movement. This was in dramatic contrast to the situation in the west, which was undergoing unprecedented crisis. In 1929 came the "great crash" and the sharp downturn that followed. The Soviets were not reticent about pointing to the contrast and noticing that the economic crisis had brought in its wake a deep and profound crisis in bourgeois consciousness that made its impact felt in every aspect of bourgeois culture, including its science and its philosophy. Science in the Soviet Union was moving irresistibly forward, they argued, as all scientific disciplines were being penetrated by a single method through the philosophy of dialectical materialism and as being concretely harnessed in the construction of the socialist future. Science under capitalism, on the contrary, was in crisis, threatened on all sides, not only by economic collapse, but also by the reactionary philosophical trends that were emerging from the bourgeoisie's ever more acute disillusionment with all forms of material culture. Fascism was on the rise, growing on the confusion, and the Soviet Union stood as a bulwark against it, as a source of clarity and enlightenment. This was how the communist intelligentsia saw it. The contrasts grew sharper by the day. They knew which side they were on and they felt glad to be alive and part of such a titanic and noble struggle.

London 1931: the Soviet Delegation and Science at the Crossroads

As if to put the contrast on display, a large Soviet delegation unexpectedly appeared at the Second International Congress of the History of Science and Technology held in London in the summer of 1931.* It turned out to be an extraordinarily important event, not so much in the history of Soviet Marxism, as in the history of Marxism abroad.** The Soviet papers read at this congress published by the end of the week in a volume entitled *Science at the Crossroads*, gave a good indication of the spirit permeating the best of Soviet thinking in the philosophy of science at this time.

---

* It was organized by the International Academy of the History of Science (founded in 1928) and held from June 29 to July 3. A special Saturday session was added to accommodate the Soviet delegation. The first congress, in Paris, was in 1929.
** Its impact on the course of foreign Marxist thought, particularly its impact on the development of British Marxism, will be dealt with in the next chapter.

The leader of the delegation was Bukharin, who concentrated on showing how philosophy was transformed within a socialist society. The new culture, the new science, the whole new style of life formed a new conjuncture between theory and practice. Bukharin used the term "theoretical practice,"* explaining that theory was accumulated and condensed practice and that practice was itself theoretical. Nearly all modern schools of philosophy, other than Marxism, failed to realize this and began from individualist, anti-historical and quietist presuppositions, making "epistemological Robinson Crusoes" the starting point for their philosophies. He put forward dialectical materialism, not only as rejecting all species of idealism and agnosticism, but also as overcoming the narrowness of mechanistic materialism, that is, overcoming its ahistoricism, its antidialectical character, its failure to understand problems of quality, its contemplative objectivism.[76] This obviously represented a significant evolution from Bukharin's earlier philosophical thinking.

M. Rubinstein, an economist, noted the tendency of all sciences to be transformed into a single system of science on the basis of dialectical materialism. Science, in the process of socialist construction, while modifying the whole of life, also modified itself, undermining existing divisions of the sciences and bringing a new scientific monism into being. Reaching beyond the boundaries of Soviet science, Rubinstein also cited the work of Einstein and the Cavendish Laboratory as evidence of the growing unity of science. Relativity had overthrown traditional notions of gravity, space, and time and radium had turned old views about immobile and immutable elements upside down.[77]

A sweeping paper on the range of conflicting trends represented in current biological thinking was given by the geneticist, B.M. Zavadovsky. The two basic, and seemingly mutually exclusive, tendencies in bourgeois science were mechanism and vitalism. Mechanism was based on the complete identification of the physical and biological sciences, and reduced biological phenomena to laws of a physical character. Vitalism, on the contrary, drew the sharpest possible contrast between them as two different orders of being. Zavadovsky drew attention to the predominance of mechanistic materialism in the period when capitalism was in its prime and to the rebirth of idealist, vitalist, and mystical ideas in the period of the general decline and decay of capitalism. Both positions, Zavadovsky insisted, were caught in insoluble internal contradictions, making it impossible for either position to maintain itself and causing each to slide into the very position it was intended to refute.

The way beyond the whole morass of crises and contradictions encountered in the present state of biological science was to be found, Zavadovsky insisted, in the Marxist philosophy of dialectical materialism. It transcended the equally barren attempts to embrace all the complexity and multiformity of the world through either a single mathematical formula of the mechanical

---

* The term was not invented by Louis Althusser. In fact, it was used by Marx.

movement of molecules or through the vitalist idea of a single "principle of perfection," in effect representing an attempt to know and explain the world through the unknowable and the inexplicable. It overcame both reductionism and mysticism, both mechanism and vitalism. Biological phenomena, although historically connected with physicochemical phenomena, were not reducible to physicochemical laws; they displayed qualitatively distinct laws and required distinctive methods of research, without thereby losing their materiality and cognizability.

Summing up the dialectical materialist position and showing how it broke through all the seemingly irreconcilable antitheses, through its emphasis on the extreme diversity and qualitative distinctness of the various forms of the motion of matter, while still holding to an overriding continuity, Zavadovsky declared:

> Affirming the unity of the universe and the qualitative multiformity of its expression in different forms of motion of matter, it is necessary to renounce both the simplified identification and reduction of some sciences to others and the sharp demarcation and drawing of absolute watersheds between the physical, biological and socio-historical sciences—which frequently takes the form of admitting the existence of causal determinateness of phenomena only in the sphere of physical science, while proposing to seek in the biological sphere for teleological solutions and in the sphere of socio-historical phenomena completely abandoning the search for any order and explanation in the course of historical processes at all.[78]

So it was that the common antitheses were, for Marxists, false ones. The unity of science was still an ideal to be pursued, but without reductionism and without obscurantism. It was a matter of bringing Engels's conception of the dialectics of nature into play in the new situation and letting it illuminate the confusion and the darkness.

The paper that made the greatest impact at the congress and that has gone down in the history of science as a classical statement of the externalist position in the historiography of science was the one by Boris Hessen. Joseph Needham, who was present at the congress, recalls Hessen tripping over proper names, making mistakes of detail, and speaking with unsophisticated bluntness, but nevertheless coming forth with a great "trumpet blast," a veritable externalist manifesto in the history of science that was enormously influential.[79]

Hessen defined his task as applying the method of dialectical materialism to an analysis of the genesis and development of Newton's scientific thought in connection with the period in which he lived and worked. Where, he asked, should the source of Newton's creative genius be sought. Marxism, he answered, eliminated subjectivism and arbitrariness in the interpretation of the history of science by seeing the roots of ideas in the state of the productive forces. Newton was a typical representative of the rising bourgeoisie, a typical son of the class compromise of 1688. Whereas on French soil, materialism was the standard of republicans, in England, religion was their standard. Thus, the materialist germs in Newton's thought were intermingled with theological belief, giving rise to a division of labor between mechanistic causation and God in his pattern of scientific explanation. Newton's matter was essentially inert and required an outside impulse to bring it into movement or to alter or end any movement.[80] In his other work in the philosophy of science, Hessen, a prominent and firm supporter of nonclassical views in physics, showed how other historical circumstances had come to bear to supersede Newton's classical view of physics.

Other members of the Soviet delegation to give important papers included the physicist A.F. Joffe, the geneticist N.I. Vavilov, the engineer N.T. Mitkevich, and the Czech mathematician A. Kolman. On the issues under debate within the Soviet Union regarding the philosophy of science the members of the Soviet delegation held to very different points of view, for example, Hessen and Mitkevich were on opposite sides in the controversy over dialectical materialism and relativity theory. These differences would continue to intensify and the members of the delegation would meet with very different fates.

Marxism and Modern Thought

In 1933 a set of papers illustrating the thinking of the time appeared in the volume *Marxism and Modern Thought*, prepared by the Academy of Sciences to commemorate the fiftieth anniversary of the death of Marx. The overall theme of the book was the stark contrast between Marxism and all other contending schools of thought in the new epoch. The ideas of Marx, developed and enriched over the years, represented, according to the authors, the most progressive outlook in science, philosophy, and history, whereas the others were reflections of crisis, abandonment of science, and denial of progress.

In grand style, Bukharin opened by describing revolutionary Marxism, in the harsh and terrible epoch of the collapse of capitalism, of wars, of revolutions, of proletarian *Sturm und Drang*, as emerging from the chaos, powerful, energetic, and at once destructive and creative. Long gone were the days when a conspiracy of silence had surrounded the ideas of Marx. His influence, even over official bourgeois science, had become great, and was reflected in Marxism's becoming the object of violent criticism, the cause of antithetic systems, and the basis of theories that took over many of its arguments, while planing off its revolutionary edges. This latter operation was performed by the theoreticians of social democracy, whom Bukharin characterized as agents of bourgeois influence over the proletariat, who distorted and castrated Marx and corroded Marxism's revolutionary content.

Supporting the historical course that had been taken by Marxist philosophy in the Soviet Union, Bukharin observed that the "collective worker" was also the "collective philosopher." The ideas of Marx, the all-embracing genius of the centuries, were being transformed into theoretical practice under socialism. Through the influence of Soviet Marxist literature, the penetration of dialectical materialism, into the sphere of the natural sciences, particularly biology and physics, was undertaken. Marxism, Bukharin said, was the only continuer of all the progressive tendencies of the ages that were being strangled by capitalism. While the bourgeoisie, frightened and fearful, was going over to mysticism and inventing its own counterrevolutionary eschatology, the proletariat was confidently marching under the banner of revolutionary science.

Bukharin now took greater note of the Hegelian roots of Marxism and of the whole way it took up the rational elements of the preceding thousand years of philosophical development. He put great emphasis on the difference between mechanistic materialism and dialectical materialism. He criticized the former for its lack of historicity, its abstract and qualityless conception of matter, its positivity, and its contemplativeness. The overcoming of this static, unspecific, and undialectical type of conception had raised materialism to unparalleled heights just as the materialist refashioning of dialectics had raised the dialectical method to the highest degree. But he still repudiated what he called "Hegelian panalogy."

The synthesis of dialectics and materialism, Bukharin argued, was not to be mistaken for a mechanical juxtaposition and it was to be clear that matter was primary. But dialectics having become materialist was the crucial "algebra of revolution," showing the transitory character of every form, the absence of absolute limits, the zig-zag nature of development, inner contradictoriness as the immanent law of motion, the interrelatedness of things, the many-sidedness and universality of connections, the negation of the old form and its presence in the new in sublated form. In the logic of contradictory processes and universal connections, abstractions became concrete, analysis and synthesis were indivisible, and all conceptions became flexible to the maximum degree.

Bukharin went on to make his case for Marxism as the most perfect of all philosophical systems, contrasting its historicism with the eternal categories of the bourgeoisie; its dialectics with their fear of leaps; its strict determinism, infinitely broadening the scope of science, with the indeterminism of the idealist doctrines expressing fear of science; its activist theory of knowledge as higher, truer, and deeper than their superficial empiricism; its confidence in the possibilities of human knowledge as opposed to their many varieties of agnosticism. On these points, he criticized various trends current at the time: axiology, gestalt, neo-Hegelianism, pragmatism, and logical positivism, all the while acknowledging the grains of truth in all these systems.[81]

Deborin's contribution took up some of the same themes. He also pointed to the increase of sharp attacks on modern science, singling out such critics as Sombart, Spann, Jaspers, and Jeans. Parallel with the new political super-structure of monopoly capitalism, he said, a reactionary ideology was developing that was threatening the very foundations of science and culture. Bourgeois liberalism had given way to a reactionary political ideology that was indissolubly connected with the most reactionary tendencies in the spheres of philosophy, science, literature, and art. Classical German idealism, the philosophy of the bourgeoisie in the period of its rise, defended the idea of development, albeit in idealist form. However, in the new period, there was widespread denial of the idea of development, even in the form of bourgeois evolutionism. There was much learned criticism of Marx's "crude" materialism and "mystical" dialectic. The social democratic philosophers, like Marck, Markus, De Man, and Max Adler, here played their part, either simply renouncing the Marxist dialectic or distorting it to make it acceptable to the bourgeoisie.

Deborin took note of the fact that there was a great cry to return to God, to bow down before the holy, to restore the kingdom of the spirit. But it would not be, Deborin asserted, for the whole of natural science, from Darwin via Virchow and Haeckel to Planck and Einstein, had killed God with its matter. It was science that gave men vision, lifted the curtain of the future before them, and allowed them to act consciously in a definite historical direction.[82]

The contribution from Y.M. Uranovsky, a philosopher and historian of science, was devoted specifically to the relation of Marxism to natural science. He strongly repudiated the position that had been put forward by Kautsky in the period of the Second International, denying the integral connection between Marxism and philosophy and natural science. Marx and Engels, Uranovsky stated, had stood on the shoulders of the philosophers of the past and critically accepted everything of value that the history of philosophy could give them, integrating it with the whole past development of the natural sciences as well. A deep necessity penetrated the apparently accidental character of Marx's scientific studies.

In Uranovsky's view, the Marxist conception of the dialectics of nature represented the overcoming of two false conceptions of the relation between philosophy and science. On the one hand, it stood against starting from the activity of pure reason, engaging in the abuse of deduction and the thrusting of artificial connections onto nature. On the other hand, it did not stand on the data of science alone, as a mere aggregate of observations, apart from philosophical thinking. The conception of the dialectics of nature was not meant to solve scientific problems by substituting itself for the data of science. Its purpose was to help in critically interpreting and connecting facts already obtained, to point out paths of further investigation, and to pose uninvestigated questions. The dialectics of nature, Uranovsky agreed, was inseparable from the dialectics of history, with which it was connected by a unity of method. In fact, a real knowledge of nature and a conception of it as a developing whole was only possible with a knowledge of the laws of the history and the development of human society that formed a specific part of nature.

It was extremely important, Uranovsky insisted, to understand correctly the question of the relation of science to philosophy at a time when there were so many reactionary solutions being proposed not only to it, but also to particular philosophical problems being posed by the natural sciences. In biology, there had been a wave of reaction that went to the extreme of repudiating Darwinism. Vitalism was the inevitable shadow of mechanism and its necessary complement. Neo-Lamarckism was appealing to shades of Kant, Schelling, and Mach. But the recent achievements of biology—the mechanics of development, the theories of ferments, vitamins, the facts of endocrinology, genetics, conditioned reflexes—were seen as a complete refutation of vitalism and a confirmation of the materialist dialectics of nature. In physics as well, all was changing. The basic conceptions of Newtonian mechanics were being undermined. No longer could mass, energy, space, and time be conceived of as separate and independent, for it had turned out that these were interconnected and united. Quantum theory had shown the importance of interruption in nature and wave mechanics had demonstrated the unity of continuity and discontinuity. It was impossible, however, to accept the conclusions being drawn by such as Bohr, Jordan, Franck, Reichenbach, Schlick, and Heisenberg. It was necessary to repudiate once again the Machist idea of no object without subject and the renunciation of the category of causality. Social democracy, for its part, had turned from mechanistic materialism to Machism, from Darwinism to neo-Lamarckism and was attempting to refashion Marxism in this light.

It was obvious to Uranovsky that the social conditions of the bourgeois world were not favorable for the development of science. The division of labor had developed one-sidedly. The anarchy of production did not allow proper planning of research. Romanticism, mysticism, intuitionism, indeterminism, pessimism, and fatalism were growing everywhere under capitalism as symptoms of the fact that it was losing its way.[83]

Other contributions to the volume concentrated on the state of Marxist thinking in specific sciences. These included, most notably, those by V.L. Komarov, an older botanist who was at this time vice-president of the Academy of Sciences and shortly to become the president, and by S.I. Vavilov, a young physicist also to become president of the Academy of

Sciences at a later date. As both of these papers bore on ongoing discussions of the philosophical implications of specific scientific theories, it is best to review here the background of controversy in which they found their place.

Biology and physics, along with psychology, were the sciences around which most of such discussion revolved. In each, there were contending theories and different points of view about which was more correct from the point of view of dialectical materialism. Such debates began in the 1920s and proceeded to develop and take new forms in the 1930s.

The Psychology Debate

Soviet psychology in the 1920s was characterized by calls to liberate the discipline of psychology from departments of philosophy, to develop psychology as a natural science, and to do so within a Marxist framework. Aside from specifications that it be materialist and dialectical, that it be neither introspective nor mechanistic, there was at first little idea of what a distinctively Marxist psychology would look like. Non-Marxist trends, such as behaviorism, gestalt and psychoanalysis, found considerable support among Soviet psychologists in the early days. Various schools of psychology vied with one another for recognition as *the* Marxist psychology. Dominating the field were reflexology, stemming from Pavlov and modified in a Marxist direction to become collective reflexology by Bekhterev, and reactology, associated with Kornilov and dealing with larger entities and social context. Nevertheless, both were basically behaviorist trends reducing psychic phenomena to physiological events and remaining aloof from the flow of consciousness. In contrast, was the developmental, cultural, historicist approach associated preeminently with Vygotsky and also with Leontov and Luria, which attended to the data of consciousness, analyzed higher mental processes in relation to the labor process and emphasized the role of language in human development.*

At the First All-Russian Psychoneurological Congress in Moscow in 1923 and at the Second Congress in Leningrad in 1924, the various schools contended with one another. However, at the First All-Union Congress on Human Behavior in 1930, all existing schools of Soviet psychology were criticized as inadequate to serve the needs of socialist construction. It condemned idealism, mechanism, reductionism, biologism, dualism, eclecti-

---

* L.S. Vygotsky was a particularly important thinker. He believed that psychology was in crisis and he attributed this to the division of psychology between two approaches: an atomistic approach, reducing all to the study of elementary processes, such as conditioned reflexes, and a phenomenological approach, describing complex processes in subjective, impressionistic terms, but offering no explanation of them. Both, he argued, were based on Cartesian dualism. Opposing this dichotomy of mind and body, he proposed a psychological monism. He also stressed a materialist explanation of higher mental processes and the necessity of understanding such processes as the product of both biological development and socio-historical evolution.

cism and abstractness. Included under these headings were Freudianism, reflexology, reactology. The Vygotsky School was taken to task as too reliant on foreign trends, as were the others as well. The congress set out certain principles of Marxist psychology: materialism, i.e., mind as a function of highly organized matter, the role of consciousness in human behavior, the irreducibility of the psychic to the physical within the framework of psycho-physical monism.

As the grip tightened on Soviet intellectual life, whole areas of psychological inquiry were ruled out of bounds. The intensification of labor with the inauguration of the Five Year Plan, the return to piecework, the Stakhanovite movement created a climate extremely inhospitable to the pursuit of industrial psychology. The new Soviet constitution of 1936, with its declaration that socialism had been achieved in the Soviet Union, made it no longer admissible to explain individual differences or deficiencies on the basis of social environment. Thus the 1936 decree on pedology, which virtually liquidated the discipline. It condemned two-factor theories (both Blonsky who held biological factors to be primary and Zahlkind who held social-environmental factors to be primary) in favor of a three-factor theory, the factors specified heredity, environment and training. The emphasis on training was to replace an allegedly passive approach with an active, even voluntarist, one.

As was becoming the custom, the philosophers were called in to sort things out. A meeting was called under the auspices of *Pod znamenem marksizma* and the leading psychologists were all taken to task by the philosophers. The events of these years seem to have thrown psychology and the psychologists into utter disarray and confusion. During the purges, psychologists disappeared, and those who remained were in paralysis wondering what to write, what to teach, what to do. It was much clearer what was forbidden than what was wanted. Rubenstein's 1935 text *Fundamentals of Psychology* ventured to set out basic principles, and won him the Stalin Prize in 1942. For the most part, the discipline was characterized by vague enunciation of general principles and quotes from the classics and party decrees, but little basic research. The discipline also suffered from the premature death of Vygotsky in 1934 and the death of Pavlov in 1936.[84]

The Physics Debate

In physics, controversy centered around the theory of relativity and the Copenhagen School of quantum mechanics. There was in the Soviet Union active resistance to the new physics, but, in assessing it, one should realize that there was also much resistance elsewhere. Einstein's theories shattered the conventional views of space, time, mass, energy and motion, and it was to be expected that many whose ideas had been formed on the classical concepts would find it exceedingly difficult to break with them so radically. The Copenhagen School, in particular, was calling into question the objectivity of science and the determinism of nature in a way that could not but provoke heated and troubled debate.

The main opposition to relativity theory in the 1920s came from the physicist A.K. Timiriazev, who insisted that Einsteinian relativity was inseparable from Machism. He was aided in this in considerable measure by Einstein's own acknowledgement of his debt to Mach and by the Russian Machists' continued insistence on the connection between Machist philosophy and relativistic physics. Timiriazev believed that the essence of scientific explanation lay in the formation of visual models. In the 1920s, this position found very little support. A few, such as I.E. Orlov, who, like Timiriazev, was also involved in the philosophical controversy on the mechanistic side, did come to his defence.

In 1922 Lenin had objected to the many attempts to draw Machist conclusions from Einstein's theories, but he was not inclined to question Einstein's physics. Nor was Nevsky, who had been commissioned by Lenin to comment on the relevance of *Materialism and Empirio-Criticism* to developments that had come subsequent to its first appearance. Nevsky held that Einstein was sound in the realm of physics, but less sound when he ventured beyond it into the realm of philosophy.

As the debate over physical theory became intertwined with the debate over fundamental philosophical orientation, there was a growing tendency to equate mechanism with hostility to relativity and Deborinism with enthusiastic acceptance of it. There was a definite logic to this, as the mechanists were indeed fighting to hold onto outmoded physical theories as well as onto the philosophy that was intrinsically bound up with them, while the Deborinites perceived "dialectical" elements in the new theories as representing another step away from mechanist concepts within science itself. There was discussion of the philosophical questions posed by the natural sciences along these lines at the April 1929 conference that endorsed Deborin's views and repudiated those of the mechanists. Otto Shmidt, for example, proclaimed his belief that relativity theory was thoroughly dialectical. Among those who were at this time supporters of Deborin's philosophy, there were such militant defenders of relativity theory as Hessen and Semkovsky.

In the 1930s, the tone of the discussion changed somewhat, as the militant bolshevizers, led by A.A. Maksimov, voiced the demand for the dialectical materialist reconstruction of physics. They did not, however, seem terribly clear about how to go about achieving such a reconstruction. Beyond declaring their hostility to "bourgeois" physics and insisting on an end to philosophical and political neutrality on the part of Soviet physicists, they were not sure how to proceed. Mitkevich was inclined to defend classical theory and insist on the necessity of visual models. Maksimov was more disposed to adopt a critical attitude to both classical theory and relativity and to hold that the dialectical materialist reconstruction somehow transcended both. Most physicists condemned both of these approaches: Joffe, Fok, Vavilov, Tamm, Landau, Rumer, and Bronshtein all defended relativity, disapproving both of the reticence to accept real scientific advance and of the amateurishness of the reconstructions. Tamm, in the pages of the philosopher's own journal, accused the philosophers of flagrant ignorance and an appalling level of scientific illiteracy.

There were varying degrees of scientific literacy among the philosophers concerned. Whether rejecting the new physics like Maksimov or attempting to steer a middle course between outright rejection and wholehearted acceptance like Mitin, most philosophers were groping in the dark and well deserved Tamm's rebuke. There were a few, however, who had a better grounding in the scientific issues and were better equipped to make a constructive contribution to the debate. Most notable among these was Kolman. With great attention to the scientific writings of the pioneers of modern physics, Kolman consistently emphasized Lenin's distinction between the substantial scientific achievements of the new physics and the dubious philosophical conclusions that were drawn from them. Kolman applied this technique in his analysis of the issues in the debates over both relativity and quantum mechanics. Kolman defended relativity, but took issue with Einstein's views on a finite universe, on his denial of general simultaneity as a negation of the objectivity of time, and on his denial of the empirical basis of scientific abstractions. Kolman also defended quantum mechanics, but expressed reservations as to the Copenhagen School's views of complementarity and causality, and accused its leading theorists of extrapolating transitory methodological difficulties into a permanent justification for ontological agnosticism.[85]

Most physicists simply wanted to get on with their science without having to justify themselves to the philosophers, but a few did make the effort to enter into such discussions with the philosophers and to provide a philosophical defence of the new physics. S.I. Vavilov, for example, argued in favor of the method of mathematical extrapolation in physics that was being resisted by those who were clinging to classical concepts. Science, he said, had decisively entered a nonclassical period. Before it stretched a realm full of the unexpected, the unusual, phenomena quite unlike those known in the ordinary world of immediate human experience. Physics had unfolded a world of new scales, deprived of concrete images and models. Relativity theory and quantum mechanics had demonstrated the power of the new methods. Working alternatively from empirical experiment and mathematical hypothesis, the scientist was able to express the vital dialectic of natural processes, which completely failed to fit on the mechanical straight rails of classical physics. Physics needed to move forward, terrified by no scales, however remote from human custom. It was unrealistic to expect immediately the former close bond between human thought and the forces of nature. It was necessary for man to change biologically in order to reach again such harmony and comprehension. It was analogous to the human eye which, after a long period of darkness, assumed a greater sensitivity. Vavilov argued resolutely that Marxism must stand with the new physics. The revolution in physics over the past thirty years had proceeded along three paths: atomic structure, wave mechanics, and relativistic space-time. All three paths, he insisted, led to dialectical materialism. Nevertheless, there were still enormous obstacles, for these three paths had not yet fused into one. The fusion would bring much that was unexpected, but there could be no other way.[86]

To such arguments, Timiriazev responded by calling Vavilov a Machist. He persisted in identifying the defence of relativity with the defence of idealism and declared that it must be rejected in the name of dialectical materialism. Semkovsky and Hessen, for their part, continued to insist that relativity was the realization of dialectical materialism in physics. So it went into the mid-1930s, with the defenders of the new physics under attack, but maintaining the upper hand. Most Soviet physicists held their own, as part of the international context in which physics was developing so dramatically, operated on the assumption that the laws of nature were the same for all, although some admitted that different class interests could indeed put different interpretations on them. In deciding such questions as the validity of relativity, however, they believed that the appropriate answers were in terms of truth or falsity, not in terms of their being proletarian or bourgeois. Although defenders of the new physics were among those swept up in the purges, Soviet physicists were at this time relatively successful in protecting their science from the inhibiting effect of philosophical concepts that were tied to superseded science and from hasty and ill-conceived reconstructions based on philosophical concepts superficially understood and applied a priori.

The Biology Debate

Things went very differently, however, for Soviet biologists. In the 1920s, the debate was focused on the conflicting claims of genetics, as elaborated by such scientists as Mendel and Morgan, and Lamarckism, which was experiencing something of a resurgence.* At first, many Soviet biologists were favorably disposed towards both; they discerned in genetics the material basis of

---

* Genetics was in its formative stages as a science in the early twentieth century and was based on the experiments of scientists including Gregor Mendel, August Weissmann, and T.H. Morgan. It located the transmission of heredity in the genes—particulate, material, self-reproducing, intracellular units. Mutations occurred spontaneously by random selection. It thus accounted for both the relative stability and the variability of species. Lamarckism came to stand for the belief that the evolution of species occurred as the result of many individuals simultaneously adapting to common environmental stimuli.

individual heredity, but they believed that it was inadequate in explaining the
evolution of the species or the influence of the environment in the process of
evolution. Lamarckism was viewed sympathetically as answering these
supposed inadequacies.* Research was carried on by experimenters of both
orientations. The Communist Academy even offered a laboratory to the
Austrian Lamarckist Paul Kammerer in 1925** and the Timiriazev Institute
was a center of Lamarckist research. Meanwhile, the geneticists pursued their
research as well. Some, such as B.M. Zavadovsky and K.A. Timiriazev
argued for the compatibility of Morganist and Lamarckist assumptions and
urged a reconciliation of the two points of view. In a series of discussions held
at the Communist Academy in the late 1920s, militant Morganists such as
A.S. Serebrovsky, I.I. Agol, and N.P. Dubinin argued forcefully against any
such reconciliation. They insisted on the absolute irreconcilability of Morganism and Lamarckism.

The key question was that of the inheritance of acquired characteristics.
Lamarckists cited Engels on their behalf. B.M. Zavadovsky and others of his
persuasion replied that they could not be tied to outdated science and had to
break with belief in the inheritance of acquired characteristics, even if this
meant abandoning views held by Marx and Engels or Darwin and Timiriazev.
Lamarckists accused Morganists of undermining scientific determinism by
reducing evolution to accident and chance. Morganists replied with charges of
anthropomorphism and argued that their picture of evolution as random,
undirected mutation was the surest defence of scientific determinism against
the revival of teleology. Zavadovsky noted nonetheless that in the mid-1920s
voices in favor of Lamarckism were growing louder and stronger, as he had
good reason to know, for he had been challenged by his students at Sverdlov
Communist University who considered genetics to be a "bourgeois science."[87]

At the April 1929 Conference of Marxist-Leninist Scientific Institutions,
Morganists linked Lamarckism with mechanism and pushed for a formal
repudiation of Lamarckism, as well as of mechanism. In this they were not
successful. Nevertheless, Morganism most definitely had achieved the edge in
the biological debate. Agol and others asserted confidently that Morganism
was the realization of dialectical materialism in biology. They established in
the minds of many a connection between Morganism in biology and
Deborinism in philosophy. This was to their advantage in the period when

---

* There was a tension between genetics and evolution early in the century, as
genetics seemed to emphasize stability at the expense of development. However, as
the science developed and as the process of mutation became clearer, its explanatory power broadened.
** Kammerer accepted the post. However, returning to Vienna for his books and
equipment, he was confronted with evidence of fraud in one of his crucial experiments, and shot himself. Cf. Arthur Koestler's biography of Kammerer, *The Case
of the Midwife Toad* (London, 1971).

Deborinism was at its moment of victory, but it was not a factor in their favor when the latter came to be labelled "menshevizing idealism." In 1930, however, both Morganism and Deborinism were in the ascendant. At the All-Union Congress of Biologists in that year, Morganists announced that Lamarckism was in conflict with Marxism by virtue of its teleological character, whereas dialectical materialism was implicit in the science of genetics. This group included not only the biologists Agol, Serebrovsky, and Levit, but also the philosopher Prezent, shortly to become one of the harshest and most implacable opponents of genetics.

One somewhat confusing aspect of this debate was the fact that Lamarckism was repeatedly identified with both mechanism and vitalism. Perhaps on the part of some authors this could be put down to overly facile and superficial labeling. Zavadovsky's remarks on the tendency of these two extremes to pass into one another in his 1931 paper at the History of Science Congress in London, however, offered some justification for it. The theory of evolution, Zavadovsky further argued, was passing through a crisis that could not be solved by any eclectic reconciliation. The Lamarckian man-in-the-street explanation of heredity had the virtue of simplicity, but it was a simplicity that had been outgrown by science. Heredity could now receive its true explanation only according to the more complex formulae of Mendelism and Morganism. He did, however, take exception to the autogenetic enthusiasm of those geneticists, who altogether ignored the influence of external environment.

Aside from the references to dialectical materialism, the debate among Soviet biologists developed along lines parallel to the debate among biologists elsewhere. Everywhere Lamarckism was losing ground to genetics in response to the mounting experimental evidence. The turning point in the debate came in 1927 with the highlight of the 5th International Congress of Genetics in Berlin, the announcement of H.J. Muller's discovery of methods for artificially producing mutations. This discovery, as Serebrovsky excitedly announced in his September 11, 1927 *Pravda* article entitled "Four Pages That Shook the Scientific World," was thought to be the decisive blow to Lamarckism.

The debate continued, but the balance had shifted. V.L. Komarov, the vice-president of the Academy of Sciences, was typical of his generation of biologists with Lamarckist sympathies who were becoming more and more critical of Lamarckism and more and more favorably disposed towards genetics. He began to remark on the poverty of Lamarck's factual material and on the predominance of deduced conclusions and to stress the importance of starting from the facts and combining inductive and deductive methods.[88]

Both Lamarckists and Morganists were at this time claiming unto themselves the mantle of proletarian science. Lamarckists asserted that their position meant that the working class were not slaves of the past but creators of the future. Morganists replied that the persistence and resisting power of hereditary characteristics was more in the interests of the working class as it explained the survival of their human potentialities thorugh generations of poverty, underfeeding, and, generally, the most unfavorable external conditions. Such arguments, however, were not unique to the Soviet debate, but were a feature of the discussion elsewhere as well.*

However, about this time the Soviet debate was set along an altogether distinctive path by an entirely new factor. The militant bolshevizers were demanding that biology, like every other science, be reconstructed on the basis of dialectical materialism. The tendency was to assume that both Lamarckism and Morganism were foreign and therefore bourgeois and needed to be replaced by a new, distinctively Marxist and thoroughly proletarian biology that would transcend both Lamarckism and Morganism. B.P. Tokin, embryologist and director of the Timiriazev Institute declared that Marxists must stop tailing along behind bourgeois science and create "a single Marxist-Leninist school in biology."[89] As to what such a school would be like, there was only the vaguest idea, but there was a certainty that it would prove itself by its relevance to the tasks of social construction, particularly by its practical service to Soviet agriculture. There were heady proclamations about the indissoluble unity between Soviet biological science and Soviet agriculture, but no new theoretical breakthroughs. There was much groping, but no sign of a specifically proletarian biology. Most working biologists continued to accept the same assumptions as their foreign colleagues and in doing so were in fact conscientiously serving Soviet agriculture. Serious work in genetics was proceeding and compared favorably with the state of research anywhere else. And then onto the stage stepped Lysenko.

Lysenkoism

Trofim Denisovich Lysenko was a young agronomist from the Ukraine, who first came into the limelight in 1927 in connection with an experiment in the winter planting of peas to precede the cotton crop in the Transcaucasus. His results, in his remote station in Azerbaijan, were sensationalized in *Pravda*. The article projected an image of him as a sullen "barefoot scientist" close to his peasant roots. Lysenko subsequently became famous for the discovery of "vernalization," an agricultural technique that allowed winter crops to be obtained from summer planting by soaking and chilling the germinated seed for

---

* In an article entitled "Science and Values: The Eugenics Movement in Germany and Russia in the 1920s," Graham shows that there were in the 1920s Marxist eugenists and Marxist Lamarckists, as well as anti-Marxists among both groups. A decade later, eugenic theories were linked to conservative political views and Lamarckist theories were linked to left-wing political views (*The American Historical Review*, December 1977).

a determinate period of time. He was the first to use the term "vernalization," but not in fact the first to discover this technique, as N.A. Maksimov was quick to point out. Lysenko ignored previous studies of thermal factors in plant development and reacted angrily to Maksimov's claim to scientific priority and to his criticisms of Lysenko's experimental techniques. After being overshadowed by Maksimov at the All-Union Congress of Genetics, Selection, Plant and Animal Breeding held in Leningrad in January 1929, Lysenko organized a boisterous campaign around vernalization, and made extravagant claims based on a modest experiment carried out by his peasant father. The Ukranian Commissariat of Agriculture, in the hope of raising productivity after two years of famine, ordered massive use of the vernalization technique. Lysenko was moved to a newly created department for vernalization at the All-Union Institute of Genetics and Plant Breeding in Odessa. There he began to publish the journal *Yarovizatsiya* (Vernalization) in which he disseminated his ideas on a wide scale and created a mass movement around vernalization.

The next stage in Lysenko's career came when, from 1931 to 1934, he began to advance a theory to explain his technique. According to the idea of the phasic development of plants, a plant underwent various stages of development, during each of which its environmental requirements differed sharply. The conclusion Lysenko drew from this was that knowledge of the different phases of development opened the way for human direction of this development through control of the environment. It was a very vague theory, never to be spelt out very fully, but it provided the link in the evolution of Lysenko's platform from a simple agricultural technique to a full-scale biological theory. The underlying theme was the plasticity of the life cycle. Lysenko came to believe that the crucial factor in determining the length of the vegetation period in a plant was not its genetic constitution, but its interaction with its environment.

Lysenko's theory developed in a pragmatic and intuitive way as a rationalization of agronomic practice and a reflection of the ideological environment surrounding it and not as a response to a problem formulated within the scientific community and pursued according to rigorous scientific methods. But the impression was created that Lysenko achieved results at a time when there was a great demand for immediate results and a growing impatience with the protracted and complicated methods employed by established scientists in achieving them. Lysenko's fame as the sort of man who would achieve results continued to spread. With it came a sympathetic hearing for whatever theoretical views he chose to express, no matter how vague or how unsubstantiated. Lysenko's practical achievements were extremely difficult to assess. His methods were seriously lacking in rigor, to put it mildly. His habit was to report only successes. His results were based on extremely small samples, inaccurate records, and the almost total absence of control groups. An early mistake in calculation, which caused comment among other specialists, made him extremely negative toward the use of mathematics in science.

Contemporaneous with Lysenko's vernalization movement was a growing interest in the work of Michurin, the last in the line of an impoverished aristocratic family in central Russia, who cultivated fruit trees and began experimenting with grafting and hybridization. Michurin worked on the assumption that the environment exercised a crucial influence on the heredity of organisms and he queried the relevance of Mendel's "peas laws" to fruit trees. Michurin's name was soon to be seized upon by Lysenko to designate a whole new theory of biology in opposition to classical genetics, even though Michurin himself had no such theoretical pretensions. Nor was he so anti-Mendelist as Lysenko began to make him out to be, for he did not hold to environmental influence on heredity to the exclusion of a recognition of the internal genetic constitution of the organism. Indeed, before his death in 1935, he began to acknowledge the validity of Mendelism.

Up until 1935, neither the agronomic experiments of Lysenko in vernalization, nor those of Michurin in graft hybridization, were seen to have any direct bearing on the theoretical debate in Soviet biology. Geneticists carried on with their work, although there was constant tension surrounding it. In 1931 and 1932, a number of geneticists were branded as "menshevizing idealists" and lost their positions at the Communist Academy. There was increasing pressure to abandon basic research that was unlikely to lead to immediate practical measures that would advance Soviet agriculture and there were strong implications that research in "pure science" was tantamount to sabotage.

A particularly vicious article that appeared in the influential newspaper *Ekonomicheskaya zhizn* in 1931 was directed against Academician N.I. Vavilov, founder and president of the Lenin Academy of the Agricultural Sciences, director of its All-Union Institute of Plant Breeding, as well as director of the Institute of Genetics of the Academy of Sciences. Vavilov was an internationally eminent plant geneticist and an ardent advocate of the unity of science and socialism. The article, which appeared with editorial endorsement, was written by a rather unsuccessful subordinate of Vavilov's, A.K. Kol, who accused Vavilov of a reactionary separation of theory and practice and advised him to stop collecting exotica and to concentrate on plants that could be introduced directly into farm production.[90] Unrealizable goals were imposed on Vavilov's All-Union Institute of Plant Breeding in 1931 and in 1934 he was called in by the Council of Peoples Commissars to account for the "separation between theory and practice" in the Lenin Academy of the Agricultural Sciences.

Lysenko was very much a part of this campaign, stirring up a negative attitude to basic research and virulently demanding immediate practical results. He was capable of the crudest antiintellectualism, remarking on one occasion: "It is better to know less, but to know just what is necessary for practice."[91] He also was inclined to enunciations of the wildest voluntarism: "In order to obtain a certain result, you must want to obtain precisely that result; if you want to obtain a certain result, you will obtain it. . . . I need only such people as will obtain the results I need."[92] Older scientists were, of course, horrified at such talk, so utterly alien to the habits of mind in which scientific method was grounded.

But Lysenko was the man of the hour, suited as he was to step into the role of the man of the people, the man of the soil, who had come up from humble origins under the revolution and who directed all of his energies into the great tasks of socialist construction. He knew well how to whip up massive peasant support, how to woo journalists, and how to enlist the enthusiasm of party and government officials. He began to be pictured as the model scientist for the new era. He was credited with conscientiously bringing a massive increase in grain yield to the Soviet state, while geneticists idly speculated on eye color in fruit flies. Lysenko made the most of this image and became more and more virulent in attacking geneticists and contrasting their "useless scholasticism" with his own great "practical successes." He began to speak of class struggle in science and declared in his speech at the Second All-Union Congress of Shock Collective Farmers in 1935 that "a class enemy is always an enemy whether he is a scientist or not."[93] Stalin, who was present, exclaimed at the end of his speech "Bravo, Comrade Lysenko, bravo."

Another new stage began for Lysenko in 1935 when, no longer a simple practising agronomist experimenting with a new technique, he came forward as herald of a new biology born out of Soviet agronomic practice. He was assisted in making this leap through his collaboration with Prezent, a party member* and specialist in educational methodology in the natural sciences who had philosophical training and who was extremely adept at the sort of ideological demagoguery that was beginning to flourish among a certain section of the younger intelligentsia. It is likely that Prezent brought Lysenko to see the ideological possibilities of his vernalization movement.** Together they announced a new theory of heredity that rejected the existence of genes and held that the basis of heredity did not lie in some special self-reproducing substance. On the contrary, the cell itself, in their view, developed into an organism, and there was no part of it not subject to evolutionary development. Heredity was based on the interaction between the organism and its environment, through the internalization of external conditions. They thus recognized no distinction between *genotype* and *phenotype*.[94]***

---

* Interestingly, Lysenko was never a member of the Communist Party.

** Not that the new theory of heredity followed logically and necessarily from his agronomic techniques.

*** The *genotype* refers to the complex genes hereditarily transmitted to the individual. The *phenotype* designates the totality of characteristics displayed by an individual and is the result of the interaction between heredity and environment.

The science of genetics was denounced as reactionary, bourgeois, idealist and formalist. It was held to be contrary to the Marxist philosophy of dialectical materialism. Its stress on the relative stability of the gene was supposedly a denial of dialectical development as well as an assault on materialism. Its emphasis on internality was thought to be a rejection of the interconnectedness of every aspect of nature. Its notion of the randomness and indirectness of mutation was held to undercut both the determinism of natural processes and man's ability to shape nature in a purposeful way.

The new biology, with its emphasis on the inheritance of acquired characteristics and the consequent alterability of organisms through directed environment change, was well suited to the extreme voluntarism that accompanied the accelerated development of the drive to industrialize and collectivize. The idea that the same sort of wilfullness could be applied to nature itself was appealing to the mentality of those who wished to stress that Soviet man could transform the world in whatever way he chose to do so. Lysenko's voluntarist approach to experimental results and to the transformation of agriculture was the counterpart of Stalin's voluntarist approach to social processes, undoubtedly a factor in Lysenko's managing to capture Stalin's imagination in this period.

However, other political leaders and scientific administrators were not so easily swayed. There was strong resistance within the Academy of Sciences and Bukharin let it be known that he sided with the geneticists—not that this went very well for them once Bukharin was condemned. But the geneticists fought their corner and had very influential support.

A climactic point in this new debate was reached at a special session of the Lenin Academy of the Agricultural Sciences held December 19 to 27, 1936, and devoted to a discussion of the two trends in Soviet biology, now designated as the Mendelist-Morganist trend and the Darwinist-Michurinist trend. The official goal set for the conference was to achieve a reconciliation of the two schools of thought, some kind of accommodation for genetics within the framework of Lysenko's agrobiology. The outcome was the opposite. The open confrontation of the two trends resulted in drawing the lines more sharply than ever and in highlighting the irreconcilability of the two contrasting lines of approach. There were some compromisers present, such as B.M. Zavadovsky and N.P. Krenke, but the overall mood was a severely uncompromising one. The most intransigent group was that of Lysenko, Prezent, and their followers. The geneticists fought hard and unflinchingly for the future of their science, although it must be said that they were more than willing to concede the value of Lysenko's work in the sphere of agronomy. Vavilov, for example, was favorably disposed to Lysenko's ideas about the phasic development of plants and summer planting of potatoes. But nothing less than the total renunciation of the science of genetics by the geneticists would placate the Lysenkoites. Serebrovsky spoke bitterly against the Lysenkoite attacks on some of the greatest achievements of the twentieth century and charged them with using revolutionary slogans towards reactionary ends. They were attempting to thrust Soviet science backward a half-century and this could only hinder the effort to establish scientific research on a new socialist basis. Dubinin posed the issues in similar terms. The sharpest speech in the defence of genetics came from the American geneticist H.J. Muller, a foreign member of the USSR Academy of Sciences who had come to work in the Soviet Union out of commitment to the possibilities of science under socialism. Muller was also inclined to philosophical reflection on his science and had very definite views as to the place of genetics within the framework of a dialectical materialist philosophy of science. He turned the charge of idealism back against the Lysenkoites and accused them of being Machists, hiding behind the screen of a falsely interpreted dialectical materialism.[95]

In the period following the conference, the Lysenkoites carried on a campaign against the geneticists that became more and more vicious and more and more slanderous. Scientific and philosophical arguments increasingly gave way to political ones. The pursuit of genetics was spoken of as synonymous with adherence to the cause of reaction—and it was identified with racism and fascism. Yakovlev, one of the highest administrators in Soviet agriculture, referred to genetics as the "handmaiden of Goebbels' department."[96] Various geneticists and supporters of genetics were named and accused of sabotage, wrecking, espionage, terrorism, and Trotskyism. Prezent, in a 1937 article, singled out Agol, Uranovsky, and Bukharin as representing the "powers of darkness" opposing the creative direction being taken by Soviet biology. These bandits and Trotskyists had supposedly sold out, wholesale and retail, the interests of Soviet science.[97]

The main target of the campaign was Vavilov, who was becoming ever more resolute and forthright in defending genetics and resisting the forces moving to destroy it. He was identified by the opposition as the main stumbling block standing in the way of complete victory for Lysenko's views. Vavilov believed the situation was becoming intolerable and complained of Lysenko's low level of culture, his outmoded scientific views, but most of all about his intolerance and the reprisals that were taken against those who disagreed with him. Vavilov was defiant, despite the danger, and he declared in 1939:

> We shall go the pyre, we shall burn, but we shall not retreat from our convictions. I tell you, in all frankness, that I believed and still believe and insist on what I think is right. . . . This is a fact, and to retreat from it simply because some occupying high posts desire it is impossible.[98]

Vavilov was not being overdramatic. Already he had lost his position as president of the Lenin Academy of Agricultural Sciences, succeeded first by

A.I. Muralov and then by G.K. Meister. In 1937, each of these in his turn was arrested and in 1938 Lysenko succeeded to the post. In 1938 Vavilov was rebuked by the presidium of the Academy of Sciences for isolating the Institute of Genetics from the trend stemming from Academician Lysenko's scientific work. Prior to this, a campaign against A.K. Koltsov, director of the Institute of Experimental Biology, had cleared the way for the election of Lysenko as academician. In 1940, Vavilov was himself arrested, and Lysenko replaced him as director of the Institute of Genetics of the Academy of Sciences. In 1941, Vavilov stood trial and was found guilty of sabotage in agriculture, belonging to a rightist conspiracy, spying for England, and a string of other charges. Although he denied all accusations and the "evidence" consisted of false testimony, he was sentenced to death. After spending several months in a death cell, Vavilov's sentence was commuted, but he died in prison in 1943 of malnutrition.*

Vavilov was not the only one. The growing ascendancy of Lysenko coincided with the purge that reached into virtually every Soviet institution during 1936 to 1939. Already, before Vavilov's arrest, the losses among Soviet biologists had been staggering. In 1936, Israel Agol, Max Levin, and Solomon Levit, all communists working in the field of biological theory, were publicly denounced as "enemies of the people" and arrested. With regard to Agol and Levin, the charges involved vague references to "menshevizing idealism" and association with a Trotskyist conspiracy. As to Levit, the director of the Institute of Medical Genetics, his studies of human heredity had supposedly made him an abettor of nazi doctrines, or so it was declared at a meeting of the science division of the Moscow party organization, presided over by Arnost Kolman. Levit died in prison and his institute was closed. The other two were shot.** They were followed by a host of others. Many were arrested. Of these some were shot, while others simply died in prison. Others were witch-hunted, lost their jobs, and were forced into other areas of work. Institutes were closed down. Journals ceased to appear. Books were removed from library shelves. Texts were revised. Names became unmentionable. The Seventh International Congress of Genetics, which was scheduled to be held in Moscow in August 1937 was cancelled. When the congress did take place in Edinburgh in 1939, no Soviet scientists were present, not even Vavilov who had been elected its president.

---

* Vavilov was posthumously rehabilitated by the USSR Supreme Court in 1955, before the Twentieth Congress of the CPSU.

** All three were posthumously rehabilitated, as were a number of other biologists and agricultural specialists who perished during this period.

Nevertheless, the opposition was still strong. The effects of the purge had been somewhat uneven. Ironically, in some cases, the most outspoken and defiant survived, while the more compromising elements perished. Serebrovsky, Dubinin, Koltsov, Zhebrak, M.M. Zavadovsky, and others continued to resist Lysenkoism. D.N. Prianishnikov had the audacity to nominate the imprisoned Vavilov for a Stalin Prize. Many of their colleagues, however, gave way under the pressure, engaged in abasing self-criticism, and acknowledged the superior wisdom of Lysenko. The degree of demoralization was fairly overwhelming.

In October 1939, there was another conference called in another effort to achieve some sort of compromise. This time it was organized under the auspices of the journal *Pod znamenem marksizma* by the philosophers, who had been called upon by the presidium of the Academy of Sciences to abandon their neutrality in the struggle between the two trends in biology. Mitin attempted to drive a wedge between Lysenko and Prezent, praising the work of Lysenko, but criticizing the "boundless conceit" of Prezent in trying to fasten his "scholasticism" and "bombast" onto Lysenko's work. Lysenko would have none of it, however much Mitin continued to try to persuade him to preserve his practical results from the tendency toward the scholastic imposition of philosophical categories on concrete material. Lysenko continued to object and brought forward quotes from Engels to prove that the classics of Marxism were on his side. Mitin, however, undercut this line of argument by asserting that there were obsolete ideas even in the classics of Marxism, "the holy of holies of our theory." At the same time, Mitin was not defending the geneticists and he drew them into a rather strained analogy that presented views put forward in the biological debate as parallels to "menshevizing idealism" in the philosophical debate, and to the theories of the "Trotsky-Bukharin-Pashukanis gang" and other such "wreckers" in the political debate. Yudin, for his part, called upon geneticists to clear up the "rubbish and slag" that had accumulated in their science. Nevertheless, the philosophers held back at condemning the science of genetics and backing Lysenko's theories as the one and only dialectical materialist position in biology. They appealed to both sides to be less intransigent, asking the geneticists to concede the supreme importance of agrobiology, and the Lysenkoites to suspend their efforts to suppress genetics.[99] In time Mitin and the rest abandoned all such reserve and came in solidly behind Lysenko.

So it went with Soviet biology. There remained two conflicting theories of heredity, each claiming for itself both scientific validity and philosophical superiority. Lysenko continued to move from strength to strength, while his opponents were hounded, purged, jailed, and shot. Nevertheless, neither the party nor the philosophical leadership had decisively committed itself to

Lysenko's theories or to a repudiation of the science of genetics. But it was early days yet.*

Stalin and the Philosophers

The philosophers, at this time, were inclined to be cautious about most things. They busied themselves writing textbooks, a dictionary of philosophy, a history of philosophy, and articles on philosophy for the Large Soviet Encyclopedia. This work was directed by the Institute of Philosophy, which had been reorganized in 1936 and brought under the Academy of Sciences upon the dissolution of the Communist Academy.** Articles in *Pod znamenem marksizma* for the most part stuck very close to the classics of Marxism and current party decrees. There were still polemics, but increasingly only one side of the argument was represented on its pages and with a harshness and stridency that directly correlated philosophical trends, such as mechanism, "menshevizing idealism," and empirio-criticism with political wrecking.

---

* Lysenkoism peaked in 1948, with the official endorsement of Lysenko's views and repression of geneticists'. Within a few years, the struggle resumed again. It was a protracted episode in Soviet history, with complex political, scientific and philosophical issues coming into play. The most notable accounts are: David Joravsky, *The Lysenko Affair* (Cambridge, Mass., 1970); Zhores Medvedev, *The Rise and Fall of T.D. Lysenko* (New York, 1969); Loren Graham, "Genetics," in *Science and Philosophy in the Soviet Union* (London, 1973); Dominique Lecourt, *Proletarian Science? The Case of Lysenko* (London, 1977); Richard Lewontin and Richard Levins, "The Problem of Lysenkoism," in *The Radicalisation of Science* (London, 1976) and Bob Young, "Getting Started on Lysenkoism," *Radical Science Journal* 6/7. My view is that Lysenkoism cannot be understood simply as a story of personal opportunism and political terror, nor as a cautionary tale against the dangers of bureaucratic interference in intellectual life or of ideological distortion of science. These are elements of an analysis, but it is vital to see the emergence of Lysenkoism as no historical accident, as no imposition of alien elements (philosophy and politics) upon science. It must be understood against the background of the tasks of cultural revolution, the drive to create a socialist intelligentsia, the push to transform every sphere of life and thought (including science and agriculture) in a new social order. Such tasks naturally involved struggling with such issues as the ideological character of science, hereditarianism versus environmentalism, determinism versus voluntarism, the relationship of philosophy to biology, the relationship of biology to agronomy, and so on. The proper procedures for coming to terms with such complex issues were short-circuited by grasping for easy slogans and simplistic solutions and imposing them by administrative fiat. It was a tragedy parallel to other tragedies in Soviet life at this time, rooted in the tensions between the monumentally advanced tasks undertaken in Soviet political life and the persisting cultural underdevelopment of Soviet society – and this in conditions of hostile encirclement. The sorts of conclusions to be drawn are: that there are no shortcuts in dealing with such intricate issues; that a certain cultural level is required to deal with them competently. The sorts of conclusions *not* to be drawn are: that science must be kept free from philosophy and from politics; that science is in essence nonideological and that ideology is necessarily antithetical to science.

** Communists having come decisively into control of the Academy of Sciences,

Nineteen thirty-eight saw the publication of the *History of the Communist Party of the Soviet Union (Bolshevik): Short Course*,[100] prepared by a special commission set up by the central committee and often attributed to Stalin alone. Millions of copies came off the press and it became the basic text for the study of Marxism in the Soviet Union. The section on "Dialectical and Historical Materialism" became the preeminent philosophical work of the period. As philosophy for the masses, it was an extremely clear and concise presentation of very complex ideas. However, it cannot now be denied that it had a highly stultifying effect on philosophical creativity. Although it was schematic and derivative and reduced Marxist philosophy to its barest outlines, it was hailed as a new classic of Marxism and incessantly quoted by the philosophers as embodying the ultimate in philosophical wisdom, so that nothing else need really be said. It was presented as proof of the supremacy of Stalin in philosophy as in all other matters.

As a philosopher, Stalin was most definitely not the genius his contemporaries proclaimed him to be. It must be said that he showed an interest in philosophy rare for a political leader with such monumental matters of state calling for his attention and he did not hesitate to enlist the help of the professional philosophers in his efforts to master Marxist philosophy. It must also be said, however, that his intentions in doing so were not entirely honorable, for there is ample evidence to support the charge that he cynically and selectively manipulated the language of Marxist philosophy in the service of policies already decided on an altogether different basis. Stalin never allowed himself to be constrained by Marxism. Beneath the solemn homage to its forms lay a deep disrespect for its underlying spirit. As political expediencies changed, so did the official version of Marxist philosophy change. The exhortation to "think dialectically" was used to support the wildest irrationality that swept over the Soviet Union during the great purge. During the war, however, Stalin began to worry that entrenched habits of thinking irrationally would inhibit the rational thinking necessary to the war effort. Thus Stalin initiated the process of the rehabilitation of formal logic within the Marxist tradition. Thus, on one day, formal logic was the denial of dialectics and banned as "the arm of the class enemy", and then suddenly Soviet philosophers were instructed to write textbooks on formal logic so that "Soviet men may learn to think effectively." In 1941, on direct orders from Stalin, Kolman wrote the first Soviet textbook on logic to appear in many years.[101] The corresponding downgrading of Hegel, whose philosophy Stalin now categorized as an "aristocratic reaction to the French Revolution,"[102] had more to do with the drive to whip up Russian patriotism and anti-German feeling after the nazi invasion of the Soviet Union than with any considered judgement on the history of philosophy. All in all, even though Stalin's intervention in relation to formal logic had positive effects, on the whole he encouraged the worst in Soviet philosophy. He knew

there was no longer any role for the Communist Academy. It was disbanded in 1936 and its institutions were amalgamated into the structure of the Academy of Sciences.

of, condoned, and even actually incited the darkest forces that were overtaking philosophy, science and, indeed, all else at this time.

Not surprisingly, Stalin has fared rather badly with the commentators. In non-Soviet sources, Stalin's philosophical legacy is almost universally rated negatively, if not contemptuously. In Soviet sources, it is today sharply criticized as having hindered the creative development of Soviet philosophy,* omitted the law of the negation in expounding the laws of dialectics, emphasized the struggle of opposites without showing the unity, and having been excessively negative in evaluating the heritage of classical German philosophy. Most Soviet philosophers today reject Stalin's characterization of Hegel's philosophy.

While that particular remark may have been historically inaccurate and excessively negative towards Hegel, not all Marxists would agree with these sources in holding that Stalin's de-emphasis on the Hegelian heritage of Marxism was such a bad thing. In assessing Stalin's contribution to the history of Marxist philosophy, Schaff, although caustically critical of the Stalinist legacy in virtually all other respects, considers Stalin's treatment of dialectics in the *Short Course* to be in some respects an improvement on Engels. Stalin was right, Schaff believes, to stop "coquetting Hegel" and to omit the negation of the negation. His acceptance of the validity of formal logic, which led him to have a text on logic published during the war and to recommend it for study by the officers of the Red Army, was a positive factor in reorienting Marxist philosophy on this question.[103]**

The secondary literature, such as it is, that comments on Soviet philosophy in the 1930s does so in most cases in an utterly dismissive manner. Bohenski typically writes it off as "more quotology than philosophy"[104] and De George comments on the timidity of the large number of "parrot-like pseudo-philosophers and pedagogues" who slavishly, monotonously, and verbosely repeated Stalin's few statements.[105] Indeed, such was the case. However, there were in the 1930s others still working in the field of philosophy who made it not quite the barren wasteland the Sovietologist network has made it out to be. But even in saying this, there is pathos, for so many of those who worked most fruitfully in the field of philosophy of science in the early 1930s, like Agol, Hessen and Uranovsky, were dead by the late 1930s, long before their rightful time, precisely because they did work so fruitfully. When applied to the prevailing atmosphere by the end of the decade, the criticisms are justified. To

---

* One Soviet philosopher, prominent during the period, remarked that it functioned as a "collection of military rules." (Interview with Professor Y.P. Sitkovsky, Moscow, 3 March 1978).

** The debate revolving around the conflicting claims of formal logic and "dialectical logic" was to become a crucially important discussion in the postwar period.

some degree, Soviet philosophers have themselves recognized this in their successive self-criticisms from 1947 on.* Since 1956, however, there has been a tendency to rely too heavily on the phrase "cult of personality" to explain what happened to Soviet philosophy and everything else during those years. The situation was far more complex, and it is unworthy of Marxists to present one man as the embodiment of all goodness on one day and the source of all evil on the next.

Soviet Intellectual Life, "Intensified Class Battles" and the Great Purge

Philosophy and the natural sciences came under the same pressures during this period as every other discipline did, and indeed as absolutely every other area of Soviet life did. The tension continued to mount. The stupendous gains in achieving industrialization and collectivization had been won at a terrible cost and had left enormous bitterness and disillusionment in their wake. Such pockets of discontent were seen as a persistent threat to the stability of the régime. It was argued that there was an oppositional network organized around Trotsky's *Bulletin of the Opposition*, which was so remarkably well informed that it was obvious that the network was well entrenched within the country, and in crucial positions of power at that, and not just outside the country among a handful of exiles like Trotsky. More genuinely menacing was the rise of fascism in Western Europe and the ominous approach of war, which made political stability a matter of life or death, survival or destruction. The specter of a nazi fifth column greatly intensified the growing paranoia.

As a matter of fact, there is little evidence of any organized internal opposition or foreign espionage on any significant scale. Nevertheless, the press pictured the country as full of spies and wreckers and agents of imperialist powers that were planning to disrupt every aspect of Soviet life in every possible way. The population were urged to revolutionary vigilance, to root out the traitors all around them in order to save the revolution from its enemies. The fact was that in 1934, at the end of the first Five Year Plan, the Soviet Union had stood at a crossroads. It could have taken the path of normalization and democratization or the path of continued coercion and terror. It not only chose the second, but intensified the coercion and terror. The assassination of Kirov at the end of 1934 was used to justify the most far-reaching wave of repression. The purges claimed victims from every stratum of the population. There was no corner in which the NKVD did not reach to unmask spies, wreckers, traitors, and double-dealers. The losses were especially high among party members. Every day brought news of arrests of central committee members, commissars, Red Army officers, trade union

---

* Although it cannot be said that any of these official statements have gone far enough, and there is still great fear of speaking publicly of the role of certain figures who are still alive and in high positions.

officials, Komsomol leaders, old bolsheviks, foreign communists, writers, artists, doctors, philosophers, historians, physicists, geneticists, economists, engineers, agronomists, construction workers, railway signalmen, teachers and even children—and then finally members of the NKVD, of the courts and of the procurator's office, that is, the agents of the purge themselves.

The impression was created that the whole original nucleus of the party consisted of conspirators in the service of foreign powers. There was a series of spectacular, highly publicized political trials in Moscow, in which former party leaders were heard to confess to the most fantastic crimes: to conspiracy to assassinate party leaders, to espionage for foreign intelligence services, to sabotage of industry, to creation of conditions of famine in agriculture, to negotiations to cede Soviet territories to foreign powers, to plots to restore capitalism, and on and on. In August 1936, there was the trial of the "Trotskyite-Zinovievite United Centre" at which Zinoviev, Kamenev, and others were sentenced to death. This was followed in January 1937 by the trial of the "Parallel Center" in which defendants such as Radek, Pyatakov, and Sokolnikov were sentenced to death or long terms of imprisonment. The testimony at the first two trials was put together to form the scenario for the third trial that of the "Right-Trotskyite Center," which came in March 1938. The sentence of death was passed on Bukharin, Rykov, Krestinsky, and others—including Yagoda who had headed the NKVD and presided over the preparation of the first two trials.* Trotsky, who had long been in exile, was sentenced to death *in absentia*, a sentence executed in Mexico in 1940.

The *Short Course*, which appeared shortly after the third trial, presented the trials within the panorama of its brazenly fabricated version of Soviet history, as a struggle between the powers of light and the powers of darkness, between white and black, between good and evil. One the one side stood Lenin, Stalin, and CPSU, the Soviet people and the forces of progress. Arrayed against them were Trotsky, Zinoviev, Kamenev, Bukharin, and all the forces of world reaction. These former party leaders, these "enemies of the people," "dregs of humanity," "whiteguard pigmies," "contemptible lackeys of the facists," had been from the earliest days in conspiracy against Lenin, Stalin, the CPSU, and the Soviet state. They had been responsible for the shot fired at Lenin, for the

---

* At the All-Union Conference of Historians in December 1962, it was announced that a reexamination of the materials relating to the Moscow trials had proven that the accusations were false. Pospelov, a central committee member, stated unequivocally that Bukharin was no terrorist or spy. However, the verdicts have not been formally annulled. There is at the present time an international campaign for the rehabilitation of Bukharin. To this day, such publications as the *Large Soviet Encyclopedia*, and the *Dictionary of Philosophy*, carry no entries on Bukharin. I once inquired about why there was no entry on Bukharin in the *Dictionary of Philosophy* at a meeting with Soviet philosophers at the Institute of Philosophy in Moscow. The reply was, "He has not been rehabilitated."

murder of Kirov*, and for the most unspeakable and innumerable crimes.[106] With justification has the book been described as "a monstrous blend of whitewashing and mudslinging, of panegyrics and slander."[107]

At the same time as history was being so blatantly falsified and reduced to a primitive and simplistic schematization, Stalin was making public denunciations of attempts to falsify, vulgarize or oversimplify history. It may have been black humor or it may have been genuine ambivalence. More probably it was crass and manipulative cynicism. In any case, there was wholesale distortion of historical events. Certain names disappeared from the pages of Soviet history overnight, with all books making reference to such persons' historical contributions removed from library shelves. Photographs in the museums and in the new books were doctored. Documents relating to them in archives were destroyed. In 1935, V.I. Nevsky, the director of the Lenin Library, was arrested for refusing to discard specified holdings, even though he received a written order from Stalin to do so. In 1938, all major archives came under NKVD administration.

It was not only political history that was falsified. So were the history of philosophy and the history of science. In philosophy, Plekhanov was slighted, and so at times were even Marx and Engels, and Stalin's role was greatly exaggerated. Mitin and the others even credited Stalin with having been the one to lead the criticism of both mechanism and "menshevizing idealism," thus denying the historical role played by the Deborinites in the 1920s and even by themselves in the early 1930s. In science, Russian scientists were credited with virtually every important scientific discovery, while the real discoverers were deemed unworthy of mention. For this, the Polish physicist, Leopold Infeld, called Rosenthal and Yudin's dictionary of philosophy "a publication that will remain a monument of shame of the past period." He looked in vain for an article on Einstein, but found the formula $E=mc^2$ attributed to Lebyedyev and S.I. Vavilov. In the article on space and time, Einstein was not mentioned, but instead Butlerov and Fyodorov.[108]

Within every academic discipline, there were "intensified class battles" to be fought. In philosophy, mechanism and "menshevizing idealism" were condemned anew and this time the condemnation had a new edge to it. This time proponents of mechanism and "menshevizing idealism" disappeared— some had past records of political opposition and some had not. Mechanists such as Varjas and Tymyansky disappeared. Deborinites such as Luppol disappeared. They most likely died in prison. Karev and Sten were confronted with their oppositional past as part of the roundup of ex-oppositionists that was

---

* Not until the 20th Party Congress was it revealed that Kirov, and not Stalin, had been elected General Secretary at the 17th Party Congress. Before the end of the decade, not only Kirov, but the majority of delegates to that congress had paid with their lives.

taking place everywhere. Karev seems to have come to trial and to have been shot. Sten was arrested in his place of remote exile and shot.

However, orthodox dialectical materialists such as Razumovsky seem to have perished as well. At the same time, leading figures associated with the discredited trends, such as Deborin and Axelrod, survived. The pattern was a somewhat elusive one. While the charges brought against philosophers were not philosophical but political,* one followed rather easily from the other in a situation in which philosophical tendencies were so directly linked to political deviations. Nevertheless, there was a certain arbitrariness about it that made the arrests of some alongside the survival of others difficult to explain.

Among specialists in the history and philosophy of science, Uranovsky and Hessen perished. Uranovsky had been identified by Prezent as "following the wrecking line in the field of scientific politics." He seems to have played some role in obstructing Prezent's biology course at Leningrad University. Hessen was accused of Trotskyism, a standard charge for anyone arrested at that time. Both Uranovsky and Hessen were outspoken defenders of the new physics, as was Semkovsky who also seems to have disappeared at this time. The new physics came ever more sharply under attack, as idealism that was tantamount to subversion. Physicists such as Tamm and Fok were accused of smuggling in enemy ideas. A number of other such physicists were arrested. Some, such as Bronshtein, were shot. Others, such as Berg and Landau, were sent to prison and later released. Still others, foreign communists, such as Weissberg and Houtermanns, who had come generously out of commitment to the cause of socialist construction, were handed over to the Gestapo. Eminent physicists from abroad, such as Irene and Frederic Joliot-Curie and even Einstein, wrote letters to Stalin on behalf of their imprisoned colleagues, but their letters were unanswered. It is said, however, that the intervention of the Soviet physicist Kapitsa, who approached Stalin on the matter, secured the release of Landau. Other sciences were purged as well, though it was biology that suffered the worst. But not even such abstract disciplines as mathematics escaped the net. Certain mathematical theories were identified with "wrecking on the mathematical front," resulting in the arrests of mathematicians. Discussions

---

* Mitin told me that, although there were repressions, they were not along philosophical lines, but with philosophy tied so directly to politics, this point loses much of its force. Y.P. Sitkovsky, a contemporary of Mitin's at the Institute of Red Professors, told me of an incident in which Mitin, when in Stalin's study and upon noticing on his desk books written by Razumovsky and Raltsevich, remarked on the fact that both writers were in prison. Stalin is said to have drily replied: "Well, it is not because of their philosophy" (Third interview with Professor Y.P. Sitovsky, Moscow, 4 April 1978). Sitovsky was himself a victim. A member of the editorial board of *Podznamenem marksizma*, and orthodox both in philosophical and political terms, he was arrested and sent to a labor camp, because of an editorial mistake made in the production of the journal. He was subsequently released. In those days, the smallest miscalculation by a technician or the slightest misprint in a publication could result in arrests.

that began in the pages of learned journals were often taken up in the interrogation rooms of the NKVD and ended in long, dark corridors late at night when a shot was fired from the back.

The accusations and arrests brought a frantic turmoil into the institutions from which the victims had come. After the exposure of an "enemy of the people," the remaining staff would be summoned to discuss "the liquidation of the consequences of wrecking." They were expected to denounce the accused person and to criticize themselves and/or others for lack of vigilance and for not having unmasked the traitor sooner. This often resulted in further accusations. Some jumped on the bandwagon and used the situation to settle old scores, to win acceptance for their ideas by eliminating the opposition, or to gain control of institutions and journals. Aside from motives of spite, jealousy, and lust for power, there was also fear. Some must have believed it to be a matter of accuse or be accused. Slander and intrigue became a way of life in universities and research institutes. Eminent scholars and scientists often had their academic work and their political loyalty publicly called into question by students and undistinguished junior colleagues. In some faculties and institutes, liars, informers, incompetents, and opportunists came to power, while the most honest and serious and able elements perished or went to prison or into other areas of work. In other faculties and institutes, an extremely intricate and delicate equilibrium was maintained, in which colleagues managed to continue to work seriously and somehow to protect one another. But this was usually almost impossibly difficult. Through unbearable pressure, even torture, false confessions were extracted and close and esteemed colleagues implicated. Few could hold out against it and they named their best friends as recruiters or accomplices in the most fantastic conspiracies.

Condemnation resulted in a ban on all works written by the condemned person. Even books making reference to such authors were seized from bookshops and libraries. There were cases with large collective works, like that of the *Large Soviet Encyclopedia*, in which the arrest of Sten resulted in his article appearing under the name of Mitin, in order to save a whole printing from being destroyed. The names of the condemned became unspeakable and sometimes the most elaborate circumlocutions had to be employed when referring to work for which they were responsible or with which they were in any way associated. There can be no doubt that Soviet science was seriously set back by the repressions of these years. It was sharply cut off from the whole international context within which it had previously functioned, with severe restrictions on travel abroad and on access to foreign publications and with great fear and suspicion attending any attempts even to correspond with foreign colleagues. Not only was it denied knowledge of the progress made by scientists elsewhere, but it was deprived of some of its own best minds. As war approached, it became painfully apparent that technological advance had been hindered in areas crucial to the Soviet war effort. Labor camps and prisons throughout the Soviet Union were combed for experts in such crucial areas as aircraft design, and special prison research centers were established in which research necessary to Soviet military technology was carried out by imprisoned experts.

The times were dark. The future would bring better times for philosophy, for the natural sciences and indeed for all else. But nothing can ever excuse the crimes against science and against humanity that occurred during this period. Nothing can ever annul the tragedy that befell its victims. Nothing can ever compensate for the splendid human material that was so flagrantly wasted, the intelligence that was so carelessly squandered, the commitment that was so callously abused. It may be that, as Victor Hugo said, history has no dustbin, despite NKVD destruction of archives. It may be that the longer run is bringing the perspective denied to it in the short run, and that the martyrs of science are faring better than their accusers, bringing Vavilov to triumph at least posthumously over Lysenko. But things have not yet been set right and it is an episode that is far from over.

But why did it happen? A detailed answer must await the future and the opening of archives still closed, but there are certain things that can and must be said. On the one hand, it would seem that the overwhelmingly hostile forces surrounding the still-endangered Soviet state made a certain defensiveness and suspiciousness inevitable. It would also seem that the cultural backwardness, sometimes so incongruous with such highly advanced goals, made a certain clumsiness in dealing with complex questions inevitable. But, on the other hand, the whole tragic cycle, which gave such an edge to the most ruthless and ignorant elements over the healthier and more enlightened ones, does not seem to have been inevitable. There is every reason to believe that socialism could have been built and could have defended itself against its enemies without such terrible destruction and waste that deprived socialism of so much of the knowledge and commitment it needed most. There is no reason to think that all revolutions necessarily devour their own children, though this one surely in large measure did. It cannot be explained away by the phrase "cult of personality," despite Stalin's crucial part in it all. Others too had blood on their hands when they need not have. At the same time, it cannot be denied that things would have gone very differently had Lenin survived or had he been succeeded by Trotsky or Bukharin (particularly Bukharin). Stalin's personality is obviously a central factor to take into account. It is a cruel fact of life that unscrupulousness confers a decided advantage in struggles for power. Having achieved power through ruthlessness rather than through reason, Stalin exercised it in the same way. Because personal rule came to predominate over policies, power was more jealously and more arbitrarily guarded. In Stalin's case, he was exceedingly vainglorious and hungry for power, indeed to the point of absolutism, to the point of aspiring to eliminate all possible contenders, not only in the present but in the past. Jealous of the whole tradition of pre-Stalinist or non-Stalinist bolshevism, he sought to create a "memory hole," to obliterate the real past and to substitute a fabricated one, to destroy all those who could remember. And it was not only a matter of destroying the basis of the past in terms of the record of events and the historical role of others, but in terms of all standards of rationality and morality, all norms independent of administrative fiat, that had prevailed in the past and could prevail in the present. This required not only the murder of the dead, but the corruption of the living. And it was not enough to physically destroy, to implicate in guilt, to terrorize. It was necessary in all cases to break their character, to eliminate all operative factors but the single will.

The physicist Weissberg has shed light on the mechanism at work in his account of his arrest and three-year imprisonment at the height of the terror. He describes his conversations with a fellow prisoner whom he subsequently discovered had been pressurized by the NKVD to persuade him to submit to making a false confession. Rozhansky spoke of the confessions as "political necessities." It was necessary for a communist, he argued, to abandon all "bourgeois concepts such as truth and lies," to abandon all old ideas of honor, and to subordinate all to the final aim. When asked about what was the final aim, he replied:

> We have lost sight of it at a bend in the road, but Stalin can see it. At the next bend we shall see it too.

Describing the effect of this line of reasoning, Weissberg reveals:

> My discussions with Rozhansky were upsetting all my established notions of good and evil, truth and falsehood, and my criteria of judgement . . . I resisted the poison which his words dripped into my brain.

In terms of powers of resistence, Weissberg distinguished himself, but his testimony indicates the intensity of the pressure that brought him close to the edge. At one point, he confesses:

> I felt my reason was about to break down.

He tells how he felt that he had lost the integrality of his personality, that he had become a sum of unrelated parts. It is a most searing account, showing how the very basis of both personal sanity and human community seemed well on the way to being destroyed.[109]

Those who filled the jails, bewildered, disoriented and demoralized, struggled desperately to understand what had befallen them. Many argued that somehow, in some unknown way it was all somehow necessary and that it was necessary to submit, for the sake of some higher laws, some higher processes, beyond their ken—that even if things were somehow going wrong and they must suffer, there was a higher justification.

Those in prison groped for a theory to explain it all, but every new theory was soon invalidated by subsequent happenings. For those arrested included not only old revolutionaries, oppositionists, and former oppositionists, but loyal Stalinists. Even when NKVD interrogators joined those they had interrogated in the cells, they had no theory to explain what they had done and why they had done it. Many since have also struggled to understand. The range of explanatory theses have been very wide indeed: from "you can't make an omelet without breaking eggs" to the "cult of the individual" to the "Thermidorian negation" of the revolution. One thing at least must become clear to anyone serious about understanding it all: it was a massive and extraordinarily complex social process that took on a life of its own and assumed a form that no one had quite planned for it. No simple formula will do. It did not develop according to some necessary law of history, but neither did it happen out of sheer wilfullness.

But how to balance the historical accounts? How to weigh these events vis-à-vis what went before and what went after? The major question is that of continuity or discontinuity in assessing the relationship between bolshevism and Stalinism. At the one end of the spectrum is the continuity thesis, which has constituted the academic orthodoxy of the western Sovietologists for many years and has been the dominant view among anticommunist authors. Ironically close to this is the CPSU position, expressed in the reply of the Soviet journal *Kommunist* to the French communist authors of *L'URSS et nous* dealing with the purges. The Soviet authors declare:

> Contrary to the allegations of the enemies of socialism, the personality cult was unable to disrupt the operation of the objective laws governing the socialist system of society; it did not alter the profoundly democratic, truly popular character of the system or the leading role played in it by the working class and its vanguard, the Community Party[110]

At the opposite end, there is the discontinuity thesis enunciated most forcefully by Trotsky's 1937 declaration:

> The present purge draws between bolshevism and Stalinism . . . a whole river of blood.[111]

A somewhat intermediate position is that of Victor Serge:

> It is often said that "the germ of all Stalinism was in bolshevism from its beginning." Well, I have no objection. Only bolshevism also contained many

other germs—a mass of other germs—and those who lived through the enthusiasm of the first years of the first victorious revolution ought not to forget it. To judge the living man by the death germs which the autopsy reveals in a corpse—and which he may have carried in him since his birth—is this very sensible?[112]

On this matter, Serge is probably closest to the truth, despite the shallow evasions of those who are afraid to face the full consequences of it.

The discontinuity was enormous, for the most fundamental principles of the revolution had been flagrantly violated and overcome by arbitrariness, ignorance, and incomparable baseness. Those who made the revolution had been cruelly and cynically swept away and replaced by a newer element without traditions, without principles, without standards, without scruples. So much had been built and so much destroyed. The revolution had triumphed and brought equality and enlightenment on a vast scale, where previously there was none. The revolution had not only pursued higher knowledge, but had drawn into the pursuit the masses that had heretofore been excluded. Perhaps they did not in every instance pursue it wisely. Perhaps some among their number were charlatans. Still in all, they created out of backwardness an experiment in the most advanced social forms that had yet been conceived in human history until that time. The experiment may have floundered and gone off the rails, but those who initially undertook it did so for the highest of goals and in the name of a cosmological vision that sought to harness the best possible science toward the highest of social and philosophical purposes. They sought to bring into being a new type of convergence of science, philosophy, and politics.

At the same time, there were aspects of bolshevik revolutionary traditions that spelled danger. An insurrectionist mentality, an intolerance towards those who opposed them in good faith, a tendency to subordinate means to ends. Early on there were mistakes made that prepared the way for the crimes that would come later. To a degree, perhaps, it was the enormity of their efforts that dictated the enormity of their mistakes. Those who dare little make fewer mistakes and those who dare much make many. In the case of the Soviet Union, those in the grip of a revolutionary vision set themselves the task of expounding and living by a philosophical worldview that was both in harmony with the class interests of the proletariat and in accordance with the discoveries of the natural sciences. Along the way, there were many differences of opinion as to how to achieve this, but they were grappling with real and important issues and did so in a way that could still be instructive to later Marxists wrestling with many of the same issues. As to the differences and their somewhat distinctive ways of resolving them, it was to some extent because philosophy was taken so seriously and was considered to be so socially important that there was such a sense of urgency about arriving at a philosophical consensus. The truth of things mattered to them and so they pressed hard to discover what it was and to come to social agreement on it. In so doing, ironically, they instituted procedures that could not but be obstacles to the search for truth and to any meaningful consensus. In the end, socialism can only be built upon consent. In failing to appreciate the creative role of clash of opinions, in short-circuiting the process of discovery, debate, and consensus, in employing inappropriate criteria in making judgements, in allowing only one position on any question Marxist legitimacy and associating all others with political treachery, they set the stage for the tragedy that would engulf them.

And tragedy really did engulf them, all the more so for the brave and decent men and women who were rendered powerless to stand against it. For it was carried on in the name of the revolution and many felt they could not oppose it without turning their backs upon the revolution. Even when they began to suspect the worst, they still felt they were faced with the choice: Stalin or Hitler. They chose Stalin. All during this time, the fascists were aggressively on the march. They had their own philosophy, one that looked backwards and one that threatened to plunge the masses again into darkness and subservience. Their appeal to the masses was irrationalist in the extreme, yet they sought to use advanced science towards their own reactionary ends, while at the same time repudiating the reasoning underlying that science. They left as their legacy the memory of incalculable devastation and desolation. What great minds, with so much to contribute to science and philosophy and other areas of human thought, were incinerated at Auschwitz or fired upon at Stalingrad, no one will ever know.

When in 1941 the armies of Nazi Germany attacked the Soviet Union, those that survived, including many philosophers and scientists, went off to the front. The Soviet people, despite all that had torn them asunder, rose as one and fought heroically. Because they did, the future, with as many complexities as the past, was nevertheless to be far more favorable for the philosophy of science than it would have been otherwise. But nothing in the Soviet war effort and its victorious outcome changed the nature of the regime under which it was achieved. It did not, as has been so often implied or asserted, vindicate its Marxist credentials. Science, philosophy, and human life were still in danger. It remained to be seen whether it would be possible to bridge the revolution and its future across such an era of destruction and desolation.

NOTES

1. V.I. Lenin, "To the Population," November, 1917, *Collected Works*, 4th English ed., 26, p. 297.

2. M.N. Pokrovsky, *Narodnoe obrazovanie*, no. 11–12 (1927), pp. 22–23.

3. V.I. Lenin, "Interview with Klara Zetkin," *Lenin on the Emancipation of Women* (Moscow, 1965), p. 107.

4. Lenin, "On the Significance of Militant Materialism," *Selected Works* 3 (Moscow, 1975), pp. 604–605.

5. Lenin, "Once Again on the Trade Unions," *Selected Works* 3, p. 485.

6. For these cases and for other points of historical background to the cultural revolution and to the development of the bolshevik attitude to science and to the traditional intelligentsia, I am indebted to David Joravsky's, *Soviet Marxism and Natural Science 1917–1932* (London, 1961); Loren Graham's, *The Soviet Academy of Science and the Communist Party 1927–1932* (Princeton, 1967); Kendall Bailes, *Technology and Society under Lenin and Stalin 1917–1941* (Princeton, 1978).

7. "Ob akademii nauk RSFSR," *Pod znamenem marksizma*, no. 1 (1923), p. 191.

8. A.A. Bogdanov, *Tektologiya vseobshchaya organizatsionnaya*, 3 vols. (Moscow, 1925–1928); Vestnik kommunisticheskoi akademii 21 (1927), p. 263.

9. *Voprosy kultury pri diktature proletariata* (Moscow, 1925), pp. 13, 16.

10. V.I. Nevsky, in Lenin, *Sochineniya* (3rd ed., Moscow), vol. 13.

11. Adam Schaff, *Structuralism and Marxism* (London, 1978), p. 105.

12. N.I. Bukharin, cited by Julius Hecker, *Moscow Dialogues* (Moscow, 1933), p. 152.

13. S. Minin, "Filosofiiu za bort!" *Pod znamenem marksizma* (PZM), no. 5–6 (1922); "Kommunizm i filosofiya," *PZM*, no. 11–12 (1922).

14. Joravsky, *Soviet Marxism*, p. 93.

15. Bukharin, "Avtobiografiia," cited by Stephen Cohen in *Bukharin and the Bolshevik Revolution* (New York, 1975), p. 14.

16. Bukharin, speech to Press Dept. of Central Committee, May 1924, cf. *Voprosy Kultury pri diktature proletariata*; *Krasnaya nov* no. 4, 1925, pp. 263–272.

17. Bukharin, *Historical Materialism* (New York, 1928), p. 75.

18. Leon Trotsky, "Marxism and Science," *Labour Review* 2, no. 2 (July 1978). The text is an English translation of Trotsky's September 1925 speech to the Mendeleyev Congress.

19. Ibid., p. 104.

20. Ibid., p. 111.

21. The exposition of Trotsky's thought is based on speeches and articles of Trotsky's from the 1920s reprinted in *Problems of Everyday Life* (New York, 1973), particularly "Science in the Task of Socialist Construction" (1923), "Culture and Socialism" (1926), "Radio, Science, Technology and Society" (1926), "Attention to Theory" *PZM* (1922); also Trotsky, *Literature and Revolution* (Ann Arbor, 1960, originally published in Russian in 1924), pp. 218–220.

22. M.N. Liadov, *Pravda*, 18 December 1923 and 8 January 1924.

23. S.L. Gonikman, *Pravda*, 10 January 1924; *PZM*, no. 3 (1924), p. 36.

24. V.V. Adoratsky, Preface to *Pisma Marksa i Engelsa* (Moscow, 1923).

25. I.P. Razumovsky, "Sushchnost," *Vestnik sotsialisticheskoi akademii*, no. 4 (1923); "Nashi 'zamvriplekhanovstsy'," *PZM*, no. 12 (1923).

26. I.I. Skvortsov-Stepanov, "Dialektischeskoe ponimanie prirody-mekhanisticheskoe (Moscow, 1924).

27. Skvortsov-Stepanov, "Dialektischeskoe ponimanie prirody-mekhanisticheskoe ponimanie," *PZM*, no. 3 (1925).

28. Skvortsov-Stepanov, "O moikh oshibkakh 'otkrytykh i ispravlennykh' tvo Stenom," *Bolshevik*, no. 14 (1924).

29. A.K. Timiriazev, "Dialektika," *PZM*, no. 4–5 (1923); "Lenin," *PZM*, no. 2 (1924).

30. L.I. Axelrod, "Otvet na 'Nashi raznoglasiya' A. Deborina," *Krasnaya nov*, no. 5 (1927), p. 162.

31. Axelrod's Introduction to *Gruppa osvobozhdeniya truda* 2 (Moscow, 1924), p. 5.

32. A.M. Deborin, *Dialektika i estestvoznanie* (Moscow-Leningrad, 1929).

33. Deborin, "Engels i dialekticheskoe ponimanie prirody," *PZM*, no. 10–11 (1925), p. 18.

34. Deborin in an editorial marking the publication of Lenin's *Philosophical Notebooks* in *PZM*, no. 1–2 (1925), p. 5.

35. Deborin, Introduction to *Materializm i empiriokrititsizm* (Mowdow, 1925), pp. xxii–xxv.

36. I.K. Luppol, cited by John Somerville, *Soviet Philosophy* (New York, 1946), p. 217.

242   MARXISM AND THE PHILOSOPHY OF SCIENCE

37. N.A. Karev, *PZM*, no. 5–6 (1925), p. 261, cited by Joravsky, *Soviet Marxism*, p. 122.
38. A.A. Maksimov, "Ob istochnikakh," *PZM*, no. 1–2 (1926).
39. S.I. Semkovsky quoted by I.S. Rozanov, "Kievskaya," *PZM*, no. 5 (1927), p. 189.
40. Semkovsky, in *Sovremennye problemy filosofii marksizma*, ed. Deborin (Moscow, 1929).
41. Stepanov, "Dialekticheskoe ponimanie prirody—mekhanisticheskoe ponimanie."
42. Luppol, *Na dva fronta* (Moscow, 1930).
43. Stepanov, "Engels i mekhanisticheskoe ponimanie prirody," *PZM*, no. 8–9 (1925), pp. 54–55.
44. Deborin, ed., *Sovremennye problemy filosofii marksizma* pp. 197–198.
45. Shmidt, *Zadachi marksistov v oblasti estestvoznaniya* (Moscow, 1929), pp. 21–25.
46. Z.A. Tseitlin, in Deborin, ed., *Sovremennye problemy*, p. 159–160.
47. Deborin, ibid, pp. 197–198.
48. Stalin, *Sochineniya* 13 (Moscow, 1946), pp. 38–39.
49. Karl Radek, "Po tu ili druguyu storonu barrikady," VARNITSO, no. 7–8 (1930), p. 5.
50. Stalin, *Sochineniya* 13, p. 142.
51. Deborin, "Itogi i zadachi na filosofskom fronte," *PZM*, no. 6 (1930).
52. Mark Borisovich Mitin, V.N. Raltsevich and P.F. Yudin, "O novykh zadachakh," *Pravda* 7 June 1930.
53. Deborin et al., "O borbe na dva fronta v filosofii," *PZM*, no. 5 (1930).
54. *Pravda* 2 August 1930.
55. *Bolshevik*, no. 19–20 (1930).
56. *Raznoglasiya na filosofskom fronte* (Moscow, 1931).
57. Ibid.
58. Mitin, *Boevye voprosy materialisticheskoi dialektiki* (Moscow, 1936).
59. D.S. Mirsky, "The Philosophical Discussion in the CPSU in 1930–31," *Labour Monthly* (1931), pp. 649–656.
60. Kolakowski, *Main Currents of Marxism* 3 (Oxford, 1978), p. 75.
61. Joravsky, *Soviet Marxism*, p. 61.
62. V. Adoratsky, *Dialectical Materialism* (London, 1934), pp. 44–63.
63. *Filosofskaya entsiklopediya* 2 (Moscow, 1962), p. 379.
64. "Postanovlenie Ts. K.V.K.P.(B) o zhurnale *Pod znamenem marksizma*," *Pravda* 26 January 1931.
65. Julius Hecker, *Moscow Dialogues* (London, 1933), pp. 147–189.
66. Somerville, *Soviet Philosophy* (New York, 1946), pp. 213–228.
67. Richard De George, *Patterns of Soviet Thought* (Ann Arbor, 1970), p. 183. Cf. also Gustav Wetter, *Dialectical Materialism: A Historical and Systematic Survey of Philosophy in the Soviet Union* (London, 1958); J.M. Bohenski, *Soviet Russian Dialectical Materialism* (Dordrecht, 1963); T.J. Blakeley, *Soviet Philosophy* (Dordrecht, 1964).
68. Loren Graham, *The Soviet Academy of Sciences and the Communist Party* (Princeton 1967).
69. Joravsky, *Soviet Marxism*, p. 271.
70. Kolakowski, *Main Currents* 3, pp. 64–75.
71. Rosenthal and Yudin, *A Dictionary of Philosophy* (Moscow, 1967); *Istoriya filosofii* 6, bk. 1 (Moscow, 1965), p. 142; *Filosofskaya entsiklopediya* 5, pp. 359–369.
72. V.I. Ksenofontov, *Leninskie idei v sovetskoi nauke 20 godov* (Leningrad, 1975).
73. Stalin, "O nekotorykh voporakh istorii bolshevisma," *Sochineniya* 8, p. 96.
74. Deborin, cited by René Ahlberg, "The Forgotten Philosopher: Abram Deborin," in *Revisionism*, ed. L. Labedz (London, 1962), pp. 140–141.
75. Ibid., p. 141.
76. Bukharin, "Theory and Practice from the Standpoint of Dialectical Materialism," *Science at the Crossroads* (London, 1971). (Originally published in London, 1931.)
77. M. Rubenstein, "Relations of Science, Technology and Economics under Capitalism and in the Soviet Union," ibid.
78. B.M. Zavadovsky, "The Physical and Biological in the Process of Organic Evolution," ibid., p. 80.
79. Joseph Needham, "Forward," ibid., pp. viii–ix.

80. B.M. Hessen, "The Social and Economic Roots of Newton's *Principia*," ibid.

81. Bukharin, "Marx's Teaching and Its Historical Importance," in *Marxism and Modern Thought* (London, 1935).

82. Deborin, "Karl Marx and the Present," ibid.

83. Y.M. Uranovsky, "Marxism and Natural Science," ibid.

84. There are a number of accounts of the development of Soviet psychology during this period. Cf. R.A. Bauer *The New Man in Soviet Psychology* (Cambridge, 1952); J. McLeish *Soviet Psychology: History, Theory and Content* (London, 1975); L. Rahmani *Soviet Psychology* (New York, 1973); J. Wortis *Soviet Psychiatry* (Baltimore, 1950). I am also grateful to Elizabeth Angus for her unpublished paper "An Outline History of Psychology in the Soviet Union."

85. I.E. Tamm PZM No. 2, 1933, p. 220; Mitin *Boevye voprosy*, pp. 252–379; Kolman PZM No. 6, 1939, p. 120 and No. 10, 1939, pp. 129–145. There are several accounts of the Soviet debate on relativity in the 1920s and 1930s, e.g. Siegfried Muller-Markus's *Einstein und die Sowjetphilosophie* 1 (Dordrecht, 1960), and Joravsky's chapter on "The Crisis in Physics," *Soviet Marxism*. An interesting Soviet account, which takes up certain questions of interpretation found in such foreign works as K.Ch. Delokarov's *Relativitätstheorie und Materialismus* (Berlin, 1977). See also Alexander Vucinich, "Soviet Physicists and Philosophers in the 1930s: Dynamics of a Conflict," in *ISIS* 71, No. 257 (1980), pp. 236–250.

86. S.I. Vavilov "The Old and the New in Physics" *Marxism and Modern Thought* op. cit.; "The New Physics and Dialectical Materialism," *Modern Quarterly*, No. 2 (1939). (Originally in *PZM*, 1936.)

87. Zavadovsky "The Physical and Biological"; also "Darvinizm i lamarkizm i problema nasledovaniya priobretennykh priznakov" *PZM* No. 10–11, 1925.

88. V.L. Komarov "Marx and Engels on Biology" *Marxism and Modern Through*; A recent article giving an account of the early biological debates is A.E. Gaissinovitch, "The Origins of Soviet Genetics and the Struggle with Lamarckism 1922–1929", *Journal of the History of Biology*, Spring 1980, pp. 1–51.

89. B.P. Tokin, in *Protiv mekhanisticheskogo materializma* (Moscow, 1931).

90. A.K. Kol, *Ekonomicheskaya zhizn* 29 January 1931.

91. T.D. Lysenko, February 1934, from the Archives of the Lenin Academy of the Agricultural Sciences, cited by Joravsky, *The Lysenko Affair* (Cambridge, Mass., 1970), p. 96.

92. Lysenko, reported in *PZM*, no. 11 (1939), p. 95.

93. Lysenko, reported in *Pravda* 15 February 1935.

94. Lysenko, *Heredity and Its Variability* (New York, 1946).

95. *Spornye voprosy genetiki i selektsii: roboty IV sessii VASKhNILa 19–27 dekabriya 1936*, Moscow, 1937.

96. P.P. Yakovlev, *Yarovizatsiya*, no. 2 (1937), p. 15.

97. I.I. Prezent, *Yarovizatsiya*, no. 3 (1937), pp. 49–66.

98. N.I. Vavilov, quoted by Zhores Medvedev, *The Rise and Fall of T.D. Lysenko* (London, 1969), p. 58.

99. Mitin, Lysenko et al., "Za peredovuyu sovetskuyu geneticheskuyu," *PZM*, no. 10 and 11 (1939).

100. *History of the Communist Party of the Soviet Union (Bolshevik): Short Course* (Moscow, 1939).

101. This episode is mentioned by Mihailo Markovic in *The Contemporary Marx* (Nottingham, 1974), p. 70. The fact that the text was written by Kolman and ordered by Stalin was brought to my attention by Marković in his comments on my manuscript.

102. On the critique of Stalin's characterization of Hegel, cf. *Filosofskaya entsiklopedia*, vol. 2, p. 383. On the critique of the Stalinist treatment of the laws of dialectics, cf. vol. 5, pp. 359–369. Cf. also *Istoriya filosofii* 6, bk. 1, p. 146.

103. Adam Schaff, "Stalinouski wklad w filozofie marksistowska," in *Mysl Filozoficzna* 2, 8 (1953), pp. 43–85, cited by Z.A. Jordan, *Philosophy and Ideology* (Dordrecht, 1963), p. 130. Schaff also made these points to me in a discussion in Vienna on June 16, 1979.

104. Bohenski, in *Philosophy in the Soviet Union*, ed. E. Laszlo (Dordrecht, 1967).

105. De George, *Patterns of Soviet Thought*, pp. 200–201.

106. *Short Course*.

107. Cited by Roy Medvedev, *Let History Judge* (London, 1976), p. 500. The original source is not altogether clear from Medvedev's citation. Joravsky, the editor, believes it possibly refers to an unpublished manuscript of M.I. Gefter from a session of the Institute of History of the USSR Academy of Sciences.

108. Leopold Infeld *Przeglad Kulturalny* June 21–27, 1956, in an English translation in *Bitter Harvest*, ed. E. Stillman (London, 1959), pp. 240–241.

109. Alex Weissberg *Conspiracy of Silence* (London, 1952), pp. 175, 179, 180, 217, 308.

110. Y. Amburtsumov, F. Burlatsky, Y. Krasin, Y. Pietnyev *Kommunist*, extract "The Missing 10,000,000: CPSU Response" in Comment 17 March, 1979.

111. Trotsky *Bolshevism and Stalinism* (New York, 1972), p. 17.

112. Victor Serge *New International* February 1939.

CHAPTER 5

THE COMINTERN PERIOD:
The Dialectics of Nature Debate

The Formation of the Communist International

It seemed in 1919 as if the whole of Europe was about to erupt into one vast revolutionary conflagration. Soviet Russia was fighting for its life in a hostile world, but its severe isolation was being broken by proclamations from groups of revolutionaries from far and near pledging their loyalty to the new republic and expressing their determination to follow its example. From 1914 on, Lenin had been speaking of the need for the formation of a new Third International, one that would not betray the workers' interests as had the Second, which lay in shambles from the onset of war. The October Revolution forced on the whole of the working class movement the sharp choice of whether to stand with it or against it, and the further choice of whether to work to extend it in their own countries or to seek a less radical accommodation within the existing structures of power. In 1918, communist parties were formed in Germany, Poland, Hungary, Austria, Holland, Finland, Latvia and Greece, and soon after in many other countries, as the movement decisively split between communists and social democrats.

In January of 1919, the Russian Communist Party issued a manifesto calling for the creation of a new International and in March, delegates assembled in Moscow. Despite the difficulties in communication and transportation and despite the absence of real unanimity, the Communist International was inaugurated. The congress, which declared itself the founding congress of the Communist International, sent forth a manifesto "To the Proletarians of the Whole World" in the hope of rallying workers everywhere to side with the cause of revolution and to break with the false promise of bourgeois democracy.

Events were moving very fast. Everywhere workers inspired by the October Revolution were rising up against their oppressors and soviets sprang up from the Dombrowa coal basin in Poland to Limerick and Tipperary in the faraway west of Ireland.

In March, a Soviet Republic was proclaimed in Hungary with Béla Kun at its head. Kun had been converted to bolshevism while a prisoner of war in Russia and, upon his return to Hungary in November 1918, had founded the Communist Party amid the fluid political situation caused by the dissolution of the Austro-Hungarian empire. The new government, a coalition of communists and socialists, quickly nationalized everything from industry and land to children's sweets. The Hungarian philosopher Gyorgy Lukács tried to seize the short period of time given to the régime to "revolutionize souls," to sweep aside the prejudices of the ages, to make way for a new morality and a new culture. As deputy commissar of education, he brought in sweeping measures for educational reform at every level. Arthur Koestler, who was a Budapest schoolboy during the days of the Hungarian Commune, recalls it as a hundred-day spring, when the whole town seemed to have been turned upside down, when new teachers spoke in new voices addressing students as citizens of a strange new world, when colorful, futuristic posters transformed the streets into art galleries, when everyone was eating vanilla ice cream and when people on the streets, who didn't expect it to last, kept saying with surprise "it goes on and on."[1] However, it did not last. By August it was over. The fragile republic, weakened by internal dissension, was overcome by Czech and Rumanian armies and Admiral Horthy's "white terror" prevailed. Those communists who survived went into exile.

Another episode that ended in tragedy took place in Bavaria. There, too, a Soviet Republic was declared that spring, though it had from the beginning something of a farcical quality about it. In the confusion following the assassination of Kurt Eisner, head of government, the Munich Workers' and Soldiers' Council took power in April, against the opposition of the communists. However, when the government, dominated by anarchist-inclined intellectuals, came under attack, the communists came to its defence and subsequently took the helm. On the first of May, however, Munich was encircled by the army and the soviet fell. A brutal massacre followed and the leaders, including the anarchist Gustav Landauer and the communist Eugene Levine, were killed. The Vienna Circle philosopher, Otto Neurath, who had taken up a position as head of the central planning office under the social democrats and had stayed on during the period of the soviet, was arrested and sentenced to prison, until the Austrian government negotiated an exchange that brought him back to

Vienna. Munich subsequently became a center of reactionary opposition to Weimar republicanism.

In June, an attempted communist uprising in Vienna was crushed and it became clear that the first wave of the postwar revolutionary tide was beginning to ebb. The mood was one of retreat. The stresses and strains accompanying it brought crisis to the Comintern in its first uncertain months. Hungarian emigrés in Vienna, engaging in a post mortem on the defeat of the commune, embarked on a bitter factional dispute that was to continue through the 1920s. Most serious, however, were clashes within the German Communist Party (KPD) that resulted in a formal split at their second congress in Heidelberg in October 1919. After the murder of so many of its leaders within the first few months of its foundation—Rosa Luxemburg, Karl Liebknecht, Leo Jogiches—Paul Levi emerged as party leader. Levi not only pursued a policy of participation in elections and in the existing trade unions, but ensured the expulsion of all who opposed this policy, resulting in the formation of the Communist Workers' Party of Germany (KAPD).

During its first chaotic and unsettled year of existence, the Comintern elicited support from the most varied quarters. Adherents of the most diverse revolutionist tendencies pledged their support—from such syndicalist and quasi-syndicalist groups as the American "Wobblies" (the Industrial Workers of the World) to such sophisticated Marxists as those who formed the communist parties in Poland and Germany. In some countries sections of the Comintern consisted of small sectarian groups, like the Dutch Communist Party, formed for the purpose, while in others, already existing mass parties, such as the Norwegian Labour Party and the Italian Socialist Party, came over to the Comintern. At this time, Comintern leaders embraced the variety and diversity and preferred to keep their options open, negotiating in Germany, not only with the KPD, but with the KAPD and the USPD* as well.

However, by the time the Second World Congress took place in July 1920, it had been decided to put the house in somewhat better order. In preparation, Lenin wrote *Left Wing Communism: An Infantile Disorder* and the West European bureau of the Comintern centered in Amsterdam and controlled by the Dutch leftists was dissolved. Things were now to be tightened up, both organizationally and ideologically. The Third International, in a radical departure from the precedents set by both the First and Second Internationals, was no longer to be a series of national parties, but a single Communist Party with branches in different countries. A party line would be laid down for all and would be enforced by iron discipline according to the principles of democratic centralism. Between congresses, the highest authority was to be the executive

---

* Independent Social Democratic Party of Germany.

committee, which would have powers parallel to and superseding the powers of the central committees of the individual parties. It was to be a directive center of a world revolution, a far cry from the "mailbox" concept that had shaped the secretariat of the Second International.

The Second World Congress adopted a list of twenty-one conditions to determine the admission of parties to the Comintern. Henceforth, each party was required to carry out systematic propaganda, including within the army and in the countryside, in favor of proletarian revolution; to remove reformists and centrists from all positions in the working-class movement and to replace them by communists; to combine legal and illegal methods of work; to supervise the activities of its members in parliament; to denounce pacifism; to support colonial liberation movements; to secure the adherence of all sections of the labor movement to the Red Trade Union International as opposed to the "Yellow" Amsterdam Trade Union International; to organize on the basis of democratic centralism and to conduct periodical purges of its membership; to support all existing Soviet republics by all possible means; to revise its party program in accordance with the policies of the International; to accept all decisions of the Comintern as binding; to take the name of "Communist Party"; and to expel all members who voted against acceptance of the twenty-one conditions at a congress called for the purpose.[2] The congress marked a sharp breach not only between communists and social democrats, but between communists and those who were still seeking a basis for compromise, such as the Austro-Marxists who still wished to find a third way between "terrorist Moscow" and "impotent Bern."[3]

The seat of the Communist International was to be in Moscow. It seemed only natural that it should be located in the one socialist country that existed. Indeed, the structure of the Comintern was modeled on that of the Russian party, not because of any sinister design to ensure Russian domination, but simply because the Russian party was the only one to have carried out a successful proletarian revolution. From the beginning, there had been uneasiness about this situation, particularly among the KPD leadership, and Rosa Luxemburg had early warned against the potential subjection of the international movement to the "Russian model." However, it seemed possible that this danger would be circumvented. The Russians at this time fully expected that their preeminence would be superseded as soon as a proletarian revolution triumphed in an advanced industrialized society. Indeed, bolshevik leaders were inclined to state the matter quite sharply, pointing to their own backwardness and to the necessity of a more advanced country taking the lead. However, as E.H. Carr has pointed out, it was only when the revolution obstinately stood still at the Russian frontier and the bright hopes of the summer of 1920 faded, that the gap in authority between those who had succeeded in making their revolution and those who had failed widened, leaving the Comintern shaped in a Russian mold and ensuring Russian dominance.[4]

The congress was followed by bitter debates within the parties on acceptance of the twenty-one conditions and the period between the second and third congresses saw a series of splits and amalgamations based on the new policies. The lines were drawn and redrawn in the turmoil of sorting out who stood where in the new situation that had emerged. Following extremely acrimonious proceedings at an extraordinary congress of the USPD in Halle in October 1920, the majority voted to join the Comintern. This majority then amalgamated in December 1920 with the KPD, giving Germany a mass Communist Party. The minority went back to the SPD. Also in December 1920, the French Socialist Party met in Tours and split, as did the Italian Socialist Party when they met in Leghorn in January 1921. In France, it was the majority that became the Communist Party, whereas in Italy it was the minority, resulting in the secession of the mass Italian Socialist Party from the Comintern. The Czechoslovak party, also split, whereas the Bulgarian, Norwegian, Dutch, Hungarian, and Austrian ones accepted the twenty-one conditions without splitting. The Communist Party of Great Britain (CPGB), formed in London of diverse groups that came together in August 1920, while the second congress was still in session, held another congress in Leeds in January 1921, at which it accepted the twenty-one conditions and adhered to the Comintern. In March 1921 the Independent Labour Party rejected the conditions of adherence to the Comintern, although the minority, which argued in favor, resigned to join the new CPGB. The Socialist Party of Ireland became the Communist Party and duly expelled its few members not in favor of accepting the conditions.

At the same time, things began to flare up again in Germany. After police moved in to disarm the Mansfeld strikers in central Germany, the KPD proclaimed open armed insurrection against the government and announced a general strike. It ended in fiasco, however, as communist strikers fought not only the police, but the mass of workers who did not see their way clear to supporting the strike. The party lost many members and much support. The failure of the "March action" resulted in another wave of recriminations and Paul Levi, although vindicated in his criticism of the "March action," was expelled from the party for having published his criticism.

Black clouds were darkening the revolutionary horizon everywhere. In Italy, faced with an upsurge of working-class militancy expressed in a wave of strikes, Italian industrialists turned to Mussolini, pushing forward the realization of his fascist fantasies. The increasing unity on the right was in contrast to the growing disunity of the Italian left, split not only between communists and socialists, but between rival factions within both the communist and socialist camps.

Such developments had their effect in the change of mood reflected at the Third World Congress of the Comintern in June–July 1921 in Moscow. The prevailing atmosphere, which contrasted sharply with the heady optimism and militancy of the year before, was one of moderation and restraint. The congress signaled a tactical retreat on various fronts. As the Soviet Union shifted its policy from war communism to the New Economic Policy, so did the Comintern move from its policy of revolutionary offensive to the tactics of the united front. Fire was now directed not against social democracy but against ultra-leftism. The idea was to seek limited cooperation with the Second and Second and a Half Internationals, the Amsterdam International, and anarcho-syndicalist organizations. The emphasis was to be on winning over their rank and file and achieving "unity from below." Thus the day of "front organizations" and "fellow travelers" arrived, masterminded in time by the German communist Willi Münzenberg, of whom it was said that he "used kings as pawns and made pawns feel like kings."[5]

During these years, the Comintern was often preoccupied with the affairs of the KPD, as Germany was seen as the nerve center for future prospects of world revolution. The left wing, headed by Arkadi Maslow, Ruth Fischer, and Ernst Thälmann, believed that Germany was still ripe for imminent proletarian revolution and wished to set the date for the final insurrection, whereas the right wing, led by Heinrich Brandler and August Thalheimer, were more inclined to caution and opposed to insurrectionist tactics. In 1923, the fragile Weimar Republic was again in crisis. Some said it was a republic without republicans. Karl von Ossietzky of the antimilitarist review *Die Weltbühne* said the situation was just the reverse: the republicans were without a republic.[6] In any case, liberalism was constantly losing ground. The ranks of the KPD were beginning to swell again, but so were the forces of the right, including the rising nazi movement. More and more the failure of the successive governments of the Weimar Republic seemed to force a choice between looking eastwards to the new revolutionary order in Russia or backwards to a mythical German past. In October 1923 the KPD, under Comintern direction, undertook a complex strategy for taking power, centered in Saxony and Thuringia. Communists, including Brandler in Saxony and Korsch in Thuringia, entered coalition government. The insurrection that was scheduled to follow was a complete fiasco. Again, party membership dropped and the party was declared illegal. Again, bitter quarreling attended the defeat and Brandler and Radek, outspoken opponents of insurrectionism, were made scapegoats for the defeat of the insurrection. Ironically, the leadership of the party passed to the leftist advocates of insurrection.

The parties of the Comintern were at this time torn by factional struggles, centered for the most part on political debates regarding strategies to be employed in achieving power. These coincided with and were exacerbated by the intense factional struggle going on within the Russian party, which was then in the process of moving against Trotsky. On top of this, various parties were in open conflict with the leadership of the Comintern, and directives condemning the "right wing" leadership of the Polish party, the "right deviations" of the British party, and the "ultraleftism" of the Italian party constantly emanated from Moscow.

Philosophy and the Communists: Minimalism versus Maximalism

During this period, there was, on the whole, a great diversity of opinion on questions of what was to be done by the communist parties and of what it meant to be a communist.* Many issues were raised—from straightforward tactical ones to far-reaching and deep-seated theoretical ones. Everything from parliamentarianism and trade unionism, to atheism and materialism were deemed up for discussion. It was by no means clear to the great majority of members of the communist parties what exactly being a communist entailed.

An early Comintern statement on the philosophical dimension of communism came from the executive committee (ECCI) in 1923:

> Communism represents a complete outlook on life, which excludes religion and logically involves atheism.[7]

This set off debate in various quarters, as the pages of such journals as *The Communist Review*, published by the CPGB, testify. One member of the British party, F. Baldwin, disturbed by the ECCI declaration, requested that the party's own executive committee instruct its delegates to the next world congress to question the propriety of this decision. Taking a minimalist approach to the theoretical implications of party membership himself, he wrote:

> I cannot agree that communism represents a complete outlook on life—it seems to me to deal mainly with one set of human activities. . . . On all other subjects, it

---

* In his *Memoirs of a Revolutionary* (London, 1963, p. 177), Victor Serge conveys the atmosphere in the Comintern in those days and of what its activists, particularly in Central Europe, felt about what it meant to be a communist:

Events continued to overwhelm us. Even where they took place at a distance, I find it hard to separate them from my personal memories. All we lived for was activity integrated into history; we were interchangeable... None of us had, in the bourgeois sense of the word, any personal existence: we changed our names, our posting, and our work at the party's need... We were not interested in making money, or following a career, or producing a literary heritage, or leaving a name behind us, we were interested solely in the difficult business of reaching socialism.

> seems to me that we differ wildly, and if we tried to have a party that expressed
> every individual opinion, we might end by having as many parties as there are
> individuals.[8]

This, and other contributions in subsequent issues, showed the degree to which
the British left, including the communists, was still under the influence of
English empiricism.* Others, however, contested this view and affirmed their
belief that communism did indeed imply a complete outlook on life. The debate
resurfaced several times throughout the 1920s, with many on the British left,
sometimes communists, making it clear that they agreed with Bertrand Russell
who thought it a mistake to base a political theory on a philosophical one,** and
with others insisting on the necessity of a philosophical grounding for their
politics. The agitrop department of the Comintern saw fit to chastize the
British party in 1925 for its "aversion to theory" and in 1929 for its omission
of "the Marxist world conception" from its training manual. On the other
hand, there was keen interest in the philosophical foundations of Marxism in
workers' educational circles in which the party was fully involved. The culture
of these working-class autodidacts was highly systematic and speculative, and
militantly atheist, very much in the tradition of the popular materialism of the
nineteenth century secularist movement.*** It was, however, highly derivative.
It was not until the 1930s that Britain became a center for original
philosophical thinking within the Marxist tradition.[9]

In the 1920s, one needed to look further to the east for the vital centers of
philosophical debate. Central Europe was bursting with it. The intellectual
culture of the Weimar Republic was sophisticated and fiery. Amid the speeches
and the street fighting that created an atmosphere of "chaos mingling with
apocalypse," the cafes of Berlin vibrated with a spirit of intense intellectuality.
Left-wing ideas, avant-garde art forms, and contending philosophical theories
were discussed with excitement and with life-or-death earnestness. Austria too
was alive with it. Vienna was at this time the home of Freud and
psychoanalysis, the Vienna Circle and logical positivism, and Austro-

---

* Another factor was the tendency to "workerism" in the labor movement, leading to
periodic outbursts of antiintellectualism and bullying proletarian chauvinism, typi-
fied by "Clydeside Rivetter" in *Sunday Worker*. Fortunately, there has always been a
countervailing tendency in the movement, displayed by a correspondent who
protested that this workerism amounted to "acquiescence in the cultural disinheri-
tance which the bourgeoisie had imposed on the working class" (7 April, 1929, p.4).
** Russell denied that there was any necessary connection between philosophy
and politics, insisting that "the mixture damages both philosophy and politics." He
argued that there was no logical connection between philosophical materialism and
historical materialism.
*** This culture tended to a maximalist approach and had affinities with the
impossibilist trend in France and the Proletkult movement in Russia.

Marxism, as well as being an international crossroads where Europeans of other nations and other trends constantly came and went or else tarried in exile, as did the Hungarian communist György Lukács and the Italian communist Antonio Gramsci.

A lively forum for theoretical debate within the communist movement in its formative days was *Kommunismus*, a Comintern theoretical journal published in Central Europe in 1920–1921,* of which Lukács was an editor. German, Austrian, Polish, Dutch, Italian and Hungarian Marxists, of a decidedly leftist tendency within the political spectrum of the time, dominated its pages. The politics of the journal was hostile to trade-union and electoral activity and supportive of the primacy of workers' councils within the strategy of the revolutionary movement. In philosophy, it signaled the first notes of a neo-Hegelian revival within Marxism, accompanied by a critical approach to the legacy of Engels and an identification of it with a positivist and mechanist interpretation of Marxism.[10]

In this environment, a far more maximalist approach to the theoretical implications of party membership prevailed. The sort of Marxists who gathered around *Kommunismus* firmly believed that Marxism was a complete outlook on life and put great stress on the philosophical dimension of Marxism. Ironically, though, they did so in a way that undercut their own basic intentions, for they took up philosophical positions that implied a severe restriction of the scope of Marxism and actually made it anything but a complete outlook on life. There was a growing tendency to exclude the philosophy of nature from the horizon of Marxism or to assimilate it within the concept of revolutionary praxis as to undercut any meaningful understanding of Marxism as a philosophy of science.

The Dutch Leftists

The major theoreticians of the Dutch party, Anton Pannekoek, Henriette Roland-Holst and Hermann Gorter, represented the "left communist" trend against which Lenin directed his fire in "Left Wing Communism: An Infantile Disorder." Responding to Lenin's criticism, Gorter wrote an "Open Letter to Comrade Lenin" in which he argued that the Comintern was neglecting the impact of bourgeois thought on the consciousness of the proletariat. The reason why crisis after crisis came and yet attempts at revolution misfired was that the proletariat were still under the spell of bourgeois ideology. This was the secret source of the power of the bourgeoisie over the proletariat.[11] From this, the Dutch leftists concluded that it was necessary to emphasize the

---

* The journal was closed down by order of the Comintern Executive in 1921.

philosophical side of Marxism. It was essential to break decisively with all premises of bourgeois thought and all bourgeois organizational forms.

The group was, however, much clearer about the implications of the latter than of the former. Organizationally, their radicalism committed them to abstention from parliaments and trade unions and to the creation of workers' councils. Theoretically, it involved a determination to break away from fatalistic and positivistic conceptions of the world, leading in practice to a certain attenuation of materialism, although it must be said that they were never very clear about what their stance implied in philosophical terms. Gorter seemed to confine Marxism within the limits of historical materialism and to leave the realm of natural science outside its scope. [12] Pannekoek, who was an astronomer by profession and who had earlier written a book on the relation between Marxism and Darwinism, was somewhat more interested in the natural sciences. His philosophy of science involved a denunciation of "bourgeois materialism" and defence of empirio-criticism. He sided with Mach and Avenarius against Lenin, arguing that Lenin took up the matter from the wrong end, by identifying the real world with physical matter. For Pannekoek, the concept of physical matter did not suffice to explain the experienced world and there was a need for more and other concepts: energy, mind, consciousness. [13] Roland-Holst, who so ardently sought within Marxism a unity of rationality and emotion, a deepening of the moral side of man, a new relation between man and nature, became disheartened at ever finding these there and turned again to religion.

Both politically and philosophically, the Dutch council communists* found themselves increasingly isolated. They left the Communist Party in the autumn of 1921 and formed the Dutch Communist Labour Party, analogous to the German KAPD, and yet another new International, a Fourth International (preceding Trotsky's), and Communist Workers' International. Henriette Roland-Holst remained in the Communist Party, but left later in 1927, when the strain of factional struggle throughout the communist movement had simply become more than she could bear.

Lukács, Korsch, and the Neo-Hegelian Revival

What caused the greatest stir on the philosophical front, however, was the publication in 1923 of two influential and controversial books by two influential and controversial authors: György Lukács, *History and Class Consciousness*, and Karl Korsch's, *Marxism and Philosophy*. These two books, linked as

---

* "Council Communism" was an antiparliamentarian and anti-trade unionist movement, advocating direct workers' control and denouncing the socialist and communist parties as counterrevolutionary.

representing a common trend emerging within Marxism at this time, had been written independently, although the authors had come together with a number of other Marxist intellectuals in the "Summer Academy" held in a Black Forest resort in 1922. This gathering, also called the "First Marxist Work-week" (although there was never a second), in time led to the formation of the Institute for Social Research or the Frankfurt School. Although there were greater divergencies between the two than seemed apparent to anyone at this time or than commentators then or now have taken into account, there were indeed certain common themes. And by their contemporaries, they were praised together or damned together. Both emphasized, as did the Dutch leftists, the crucial importance of revolutionary consciousness and the necessity of breaking decisively from bourgeois patterns of thought, but went further than them in unfolding this in philosophical terms. For both, the main target of attack was positivism and what they saw as positivist incrustations within Marxism. Both engaged in extended criticism of the Marxism of the Second International, in both its revisionist and so-called orthodox forms, and expressed concern over the continuation of forms of "vulgar Marxism" within the Third International. They were opposed to the separation of theory and practice, which they saw as following from the conception of Marxism as a science, and they defined Marxism as the revolutionary consciousness of the proletariat, a historical consciousness that overcame all such dualisms as that between theory and practice, science and ethics. They spoke against the fascination of Marxists with the materialism of the natural sciences and proposed a renewed attention to the dialectic. They tended to see the dialectical method as limited in application to the sociohistorical realm and as standing in the sharpest possible contrast to the scientific method, which applied only to nature. Equating the theory of reflection with "naive realism" and substituting for it an emphasis on revolutionary praxis, a type of historicism that led to a radical epistemological relativism, they counterposed a revolutionist, antideterminist activism to a quietist, determinist materialism. The debate as they saw it was between a passive, contemplative acceptance of the world as it was and an active, critical negation of the world in order to transform it. They regretted what Korsch called "Hegel amnesia" and sought the recovery of the Hegelian heritage within Marxism.

This new neo-Hegelian trend within Marxism took from Hegel something very different from what Engels had taken. It was not the grandiose system embracing the whole of history and nature that appealed to them, but the dialectical unfolding of world historical consciousness. In its hostility to the natural sciences and its subjectivist renunciation of the cosmological dimension of Marxism, this trend represented a radical departure from Engels's conception of the dialectics of nature. It was therefore a head-on challenge to the mainstream Marxist position in the philosophy of science.

History and Class Consciousness

Lukács's book made the confrontation quite explicit. Taking a critical view of Engels and rejecting the concept of the dialectics of nature, he wrote:

> It is of the first importance to realize that the method is limited here to the realms of history and society. The misunderstandings that arise from Engels' account of dialectics can be put down to the fact that Engels—following Hegel's mistaken lead—extended the method to apply also to nature. However, the crucial determinants of dialectics—the interaction of subject and object, the unity of theory and practice, the historical changes in the reality underlying the categories as the root cause of changes in thought, etc.—are absent from our knowledge of nature.[14]

However Lukács exempted Marx from this error. In an earlier work, *Tactics and Ethics*, when expounding on dialectics, he wrote: "Marx was altogether too sober and profound a thinker to apply this method to the investigation of nature."[15]

Concerned to break with a contemplative view of reality and to reemphasize the dimension of subjectivity within Marxism, Lukács ruled out the possibility of knowing of natural processes occurring independently of man and declared nature itself to be a societal category. Knowledge was not a reflection of some preexisting reality, but the unity of subject and object, thought and being, theory and practice, in the process of revolutionary transformation. Within this framework, it became impossible to know of the world except in and through human praxis.

In *History and Class Consciousness*, Lukács persistently identified science and analytical rationality with bourgeois consciousness, with a passive, reified and fragmented conception of the world that was endemic to capitalism. Against it, he set the revolutionary class consciousness of the proletariat, which was able to rise to an organic comprehension of the whole, to a realization of totality in the process of revolutionary transformation. Thus the struggle between capitalism and socialism was set in terms of scientific method versus intuition of totality. The epistemologically privileged proletariat, in whom the subject and object of history were destined to coincide, was not, however, to be confused with the empirically existing proletariat. The party was the necessary mediator between the actual proletariat and the total historical vision that constituted its true class consciousness, between what the proletariat actually thought and what it would think if it could rise to a comprehension of the totality of implications of its social position. Only the proletariat, in its ascribed class consciousness, was able to achieve that full historical subjectivity that would negate all reified objectivity.

Running through Lukács's work at this time was a transcendent disdain for the natural sciences and indeed for the whole realm of empirical investigation that often took very extreme forms. Against such as Bernstein, who argued that dialectics violated scientific method and led to conclusions that ran contrary to the facts, his attitude was: so much the worse for scientific method, so much the worse for facts.[16] Again in addressing himself to the question "What is orthodox Marxism?", he actually argued:

> Let us assume for the sake of argument that recent research had disproved once and for all every one of Marx's individual theses. Even if this were to be proven, every serious orthodox Marxist would still be able to accept all such modern findings without reservation and hence dismiss all of Marx's theses in toto—without having to renounce his orthodoxy for a single moment. Orthodox Marxism, therefore, does not imply the uncritical acceptance of this or that thesis, nor the exegesis of a "sacred" book. On the contrary, orthodoxy refers exclusively to method. It is the scientific conviction that dialectical materialism is the road to truth and that its methods can be developed, expanded and deepened only along the lines laid down by its founders. It is the conviction, moreover, that all attempts to surpass or "improve" it have led and must lead to over-simplification, triviality and eclecticism.[17]

While this was a manifestation of a healthy desire to keep Marxism from rigidity and ossification, it presupposed the very separation of method and reality that Lukács proclaimed must be overcome and it expressed the sort of cavalier disregard for the concrete facts that would most surely keep Marxism from developing. To show such indifference to the actual empirical studies that led Marx to the formulation of his method and to make it immune to the process of continual empirical validation would be to deprive it of the very grounding in concrete reality that had given birth to it and that would keep it alive and relevant. However laudable his critique of positivism and his insistence that facts did not interpret themselves, but only became meaningful in the framework of a system, he was courting danger in his scorn for facts, and in his severing of the very connection between scientific knowledge of the world and revolutionary action to transform it that constituted the very essence of Marxism.

But whatever his shortcomings at this time, Lukács, was an important philosopher. His importance in the history of Marxism lay in his passionate commitment to overcoming the numerous dualisms at the heart of bourgeois consciousness, his insistent emphasis on vision of totality as constituting the core of revolutionary consciousness and his connection of philosophical integrality with the social role of the proletariat. However, at this early stage in his long career, he was unable to carry through with his task in an effective way, because he was leaving out of account too much that was too crucial to any true vision of totality. In striving to resolve the dichotomy between nature and history, he simply left out nature. In endeavoring to transcend the dichotomy between science and ethics, he dismissed the relevance of science. In seeking to supersede the dichotomy between object and subject, he denied the claims of objectivity altogether. Ironically, all he was able to achieve at this time was the mirror image of the fragmented consciousness of positivism that he was trying to combat.

The sources of the one-sidedness that blocked him from achieving the sort of comprehensiveness of vision to which he aspired must be sought in the intellectual milieu in which he had been formed. The story of the complex philosophical and political evolution of György Lukács in the years both. preceding and following upon the events of 1923 constitutes one of the most interesting chapters in the history of Marxism. Educated in Germany at a time when university life was dominated by neo-Kantianism, phenomenology, and various irrationalist and romanticist reactions to positivism, Lukács came particularly under the influence of Dilthey and Simmel. In his neo-Kantian phase, he sided with the Heidelberg School rather than the Marburg School, that is, he believed the truth was ascertainable through phenomenological intuition of essences, instead of adopting a more orthodox neo-Kantian epistemological agnosticism. The Heidelberg School did, however, put great emphasis on the traditional Kantian distinction between nature and culture. It was hostile to the naturalistic search for general laws of development applicable to nature and history alike, for history was the realm of the unique and unrepeatable event that could never be subject to causal laws. It made a sharp distinction between scientific and historical knowledge, between causal analysis and hermeneutic intuition, and it considered the latter to be a superior form of knowledge.

Soon, Lukács moved forward from Kant to Hegel and from Hegel to Marx, that is, from subjective idealism to objective idealism to dialectical materialism. It cannot be denied, however, that at the time of writing *History and Class Consciousness* Lukács was still inclined to read Marx through neo-Kantian and neo-Hegelian spectacles. It was the Heidelberg School rift between nature and history that Lukács had brought with him unquestioned into Marxism that formed the basis of his rejection of the dialectics of nature and that made *History and Class Consciousness* as much a neo-Kantian book as a neo-Hegelian one.

Looking back on this period after many years Lukács recalled it as a time when momentous, world-historical changes were struggling to find theoretical expression, when he united within himself conflicting intellectual trends, finding himself changing from one class to another in the middle of a world crisis and not yet able to forge the various elements into a new, homogeneous world outlook.[18]

Lukács saw the October Revolution as the opening of a window into the future and he joined the Hungarian Communist Party in 1918 soon after its inception. He was an extremely active and committed party member, playing a leading role in the Hungarian Commune and, after its defeat, identifying with the Landler faction in opposition to Béla Kún's leadership of the party. From his exile in Vienna, Lukács went to Moscow as a delegate to the Third World Congress of the Comintern and was also very close to the affairs of the KPD. But, despite his concrete activity in the working-class movement, the proletariat of *History and Class Consciousness* bore far more resemblance to Hegel's *Weltgeist* than to the living, breathing working class of his historical experience. Looking at the extraordinarily abstract and ethereal role assigned to the proletariat by the early Lukács, Gareth Stedman Jones aptly comments that its role was not so much that of a concrete historical force, but that of a hitherto missing term in a geometrical proof. The proletariat was the *deus ex machina* whose timely appearance resolved the antinomies of *Geistegeschichte*.[19]

Another aspect of Lukács's thought that showed his remoteness from the concrete during these years was his attitude to the natural sciences. A typical representative of the German academic milieu from which he came, concerned with the critique of positivism and the vindication of philosophy, he failed to realize that science had outgrown positivism and had reached a new postpositivist stage. Lukács functioned with an extremely primitive concept of science as fact-gathering, and he accepted uncritically the positivist interpretation of scientific method. In rejecting positivism, he rejected science as well. It was not, however, actually existing science, about which he knew almost nothing, but a specific academic stereotype of science that he was rejecting.

But Lukács was alive and developing and his continuing political experience and openness to new knowledge eventually was brought to bear on his high-flown philosophical style, bringing him in due course to change his mind regarding many of the propositions in *History and Class Consciousness*.

Marxism and Philosophy

Korsch also was an active party member. He joined the USPD in 1917 and, when it split in 1920, he went with its majority into the KPD. In 1923, he was elected to the Thuringian parliament and went into government as minister of justice in Thuringia. In 1924 he was also elected a member of the Reichstag. A supporter of the Maslow-Fischer line in the KPD, he became editor of the party's theoretical journal *Die Internationale*, when the left wing succeeded to the leadership of the party in 1923.

Lukács's *History and Class Consciousness* appeared just as Korsch's *Marxism and Philosophy* was going to press, giving Korsch time to add an "Afterword instead of a foreword" in which he declared himself to be in fundamental agreement with Lukács's book and calling attention to its relation to his own book. He stated that in so far as there were differences between the two, he preferred to leave them for future discussion.[20]

Indeed, there were differences and in coming years they were to be intensified. But in 1923 what mattered was the common ground: the political utopianism, the epistemological activism, the critique of positivism and "vulgar Marxism," the emphasis on revolutionary consciousness, the revival of Hegelian themes within Marxism. Korsch considered the central issues to be their common critical attitude towards both social democratic orthodoxy and the new communist orthodoxy. Like Lukács, Korsch declared himself a dialectical materialist, indeed he declared *Marxism and Philosophy* to be "materialist in the strictest sense of the word."[21]

On the question of the dialectics of nature, Korsch posed the issue in a very different way from Lukács. Although many of his critics accused him of doing so, he did not explicitly reject the dialectics of nature, nor did he criticize Engels's philosophical writings, or draw an essential distinction between Marx and Engels on this point. His position was that knowledge of nature, no less than knowledge of history, came within the scope of human praxis and was therefore no less dialectical. Seen in this way, there seemed to him to be no ground for rejection of the dialectics of nature.

However, when looked at more closely in light of the way he expressed his rejection of the theory of reflection, Korsch's stance can be seen to have been in reality far closer to Lukács's than to Engels's. Within the framework of his epistemological relativism, nature was so radically historicized as to make it impossible to speak of natural processes as dialectical independently of interaction with human processes. Nature could only be known in and through human praxis. Thus, only human thought, only human historical activity, could be dialectical. A few years later, when clarifying the issues he believed were at stake in *Marxism and Philosophy*, Korsch declared himself on this more sharply, not however in terms of a gulf between Marx and Engels, in the manner of Lukács, but in terms of a gulf between Marx and Engels on the one hand and Lenin and the Leninists on the other:

> Lenin and his followers unilaterally transfer the dialectic into object, nature and history and they present knowledge merely as a passive mirror and reflection of this objective being in the subjective consciousness. In so doing, they destroy both the dialectical interrelation between being and consciousness and, as a necessary consequence, the dialectical interrelation of theory and practice.[22]

Korsch's position on dialectics was a more definitive one than that of Lukács, for knowledge of nature bridged to some extent that gap that Lukács left yawning between history and nature. However, although he did not express it so forcefully as did Lukács, there was in his work as well an aversion for the natural sciences and for the materialism of the natural sciences that was as alien to the whole spirit underlying the philosophy of Marx and Engels as to that of Lenin.

The main theme of *Marxism and Philosophy* was the status of philosophy within Marxism. His argument here was a very confused one. After castigating the Marxism of the Second International for neglect of the philosophical side of Marxism and declaring the aim of *Marxism and Philosophy* to be reemphasis of the philosophical side of Marxism, he turned around and proclaimed that Marxism was not a new philosophy but the abolition of philosophy. In one passage, he would say that "the dialectical materialism of Marx and Engels is by its very nature a philosophy through and through"[23] and in another that it was an "anti-philosophy."[24] The convoluted reasoning that it was an anti-philosophy that was itself philosophical, a surpassal of philosophy that retained its philosophical character, a realization of philosophy aimed at its destruction, could not annul the elementary logical contradiction that lay at the heart of this argument. This particular confusion did have some basis in the writings of Marx and Engels however.* Going even further than Marx and Engels, who unfortunately spoke of the transcendence of philosophy by the positive sciences, Korsch declared that the revolutionary process was a total attack on bourgeois society that brought the abolition not only of its philosophy but of all of its sciences. Just how the new revolutionary man, endowed with mystical proletarian class consciousness, was to come to terms with the natural world without the positive sciences and without philosophical interpretations of the results of positive sciences was something that was never quite explained.

The Reviews: A Storm of Controversy

At all events, the two books unleashed a storm of controversy. The reviews came flooding in, bringing high praise from some quarters and virulent condemnation from others.

In social democratic circles the reaction was mixed. Siegfried Marck hailed Lukács's book and in his own book, *Die Dialektik in der Philosophie der Gegenwart*, published a few years later, he did raise various criticisms of Lukács. He felt, for example, that Lukács's concept of proletarian class consciousness shot beyond the boundaries of historicity. He hinted that there was something dangerous in Lukács's ascription of a sort of scholastic

---

* I argued in chapter 1 that the pronouncements of Marx and Engels on the "end of philosophy" ran counter to the basic thrust of their thinking on the status of philosophy.

infallability to the dialectical method and to his "mythological" conception of the proletariat, and indicated that such a way of thinking could lead to "inquisition."[25]

Kautsky, on the other hand, took a dim view of it all, not surprisingly, as he was a major target of attack in both books. Reviewing Korsch's book, Kautsky accused him of distorting the history of Marxism and of emphasizing revolutionary consciousness to the point of forgetting that revolution is possible only under specific historical conditions.[26]

Among independent Marxists, of whom there were a few, although not nearly so many as in years to come, both books were warmly received. The Hungarian Marxist László Radványi believed Korsch was correct in his emphasis on revolutionary consciousness. Whereas the working-class movement had liberated itself from the hegemony of the bourgeois conception of the world in political and economic terms, it had not done so in philosophical terms. As Radványi saw it, it continued to be in the grip of "the crude, natural-scientific, materialist world-view of the post-revolutionary bourgeoisie." What was of central importance about Korsch's book, Radványi argued, was its break with the view of Marxism as "pan-economism," its proclamation that the economic structure was not the only realm that was fully real, its recognition of the intellectual sphere as a constitutive part of the reality of social life.[27]

The German Marxist Ernst Bloch greeted Lukács's book with great enthusiasm, although he predicted that others would not, singling out the Russians as not understanding the German philosophical heritage and as thinking "like uncultured dogs." He welcomed *History and Class Consciousness* in a situation he saw as one in which the revolutionary process was advancing only with "crippled strides," seeing all its ideas shooting beyond reality, bearing its endless crises without the strength to transform them. The need was to press beyond the fragmentations, beyond the specializations, beyond the separations. The need was to become "whole men," not only in the process of the construction of communism, but in the very formation of the revolutionary forces to achieve it. He saw Lukács's reconstruction of the subject-object dialectic as opening the way to achieve this. He believed Lukács had liberated thought by bringing it into the sociohistorical process, by breaking free of bourgeois reification. Bloch felt himself very much in tune with Lukács's utopian boldness, if anything he thought Lukács not quite bold enough. Bloch, more oriented to the cosmological dimension than Lukács, criticized *History and Class Consciousness* for its purely sociological view of the world.[28]

There was also Herbert Marcuse, who, a few years later, specifically defended Lukács in his polemic against Engels. Marcuse supported Lukács's critique of the dialectics of nature and commended him for seeing through the dualism of nature as the ahistorical object of physics, and nature as the historical environment of human activity. He was, however, critical of Lukács's concept of "correct" class consciousness, arguing that it amounted to a fixation outside the course of events and could be linked to history only in an artificial and abstract manner.[29]

The most intensive discussion of the books took place within the Comintern. Here, too, there were sympathetic reviews, but, on the whole, the reception was overwhelmingly hostile. Most communist critics saw the books as representing a trend that was reneging on materialism, reverting instead to a combination of the subjectivist epistemology of Kant with the idealist dialectic of Hegel.

Among Lukács's fellow Hungarians, there was first of all László Rudas, who had been a close associate of Lukács until this time, but after *History and Class Consciousness* would never be so again. His negative response to Lukács's book seems to have been the cause of his departure from the ranks of the Landler faction within the Hungarian party. His review was a lengthy one, carried in three installments of *Arbeiterliteratur*. At the center of Rudas's critique was his rejection of Lukács's position on the dialectics of nature. He affirmed vigorously that the dialectic extended to the whole of reality, to nature as much as to society. The Marxist dialectic was a science of objective laws of nature and of history, the validity of which was not contingent upon human consciousness. Lukács's rejection of the dialectics of nature, along with his view that proletarian class consciousness made the revolution, constituted in Rudas's view, a lapse into subjective idealism. He advised Lukács to subject his philosophy to a thoroughgoing materialist self-criticism. Comrade Lukács, said Rudas, may have broken with his social past, but not with his philosophical past. Rudas also registered his strenuous objection to the way Lukács and others saw Engels as "the first vulgar Marxist." Another theme Rudas took up was the relation between political and philosophical deviations, asserting a logical connection between political ultra-leftism and philosophical idealism.[30]

The book was given short shrift by Béla Kún, who denounced Lukács and all others who attempted to revise dialectical materialism or "to put it more accurately, to emasculate it by expunging materialism."[31]

In sharp contrast was the review of another Hungarian communist, József Révai, who hailed *History and Class Consciousness* as a major breakthrough in the in the history of Marxism. He saw it as "the first systematic attempt to make the Hegelian movement in Marxism, the dialectic, philosophically conscious." Révai took exception to the views of Engels and Plekhanov who, by putting the dialectic into nature, ended up by naturalizing the dialectic. They had reduced the revolutionary dialectic of consciousness and being to a naturalistic metaphysics that simply stated certain general characteristics of objects. Against this trend stood Lukács who had returned to Marx. Marx, according to Révai, had introduced the future in to the domain of the revolutionary dialectic, not as positing a goal or an end, or as the necessary advent of a natural law, but as an active reality dwelling in the present. As to the past object, the mission of the proletariat was to comprehend itself as the identical subject-object of history.[32]

Yet another Hungarian, Béla Fogarasi, wrote enthusiastically on *Marxism and Philosophy*, praising it as a decisive step towards the elimination of the philosophical residues of bourgeois consciousness from Marxist thought. Fogarasi identified the remnants of bourgeois thought within Marxism as "naive realism" and mechanistic materialism, rather than idealism. He agreed with Lukács about the dialectics of nature.[33]

There was also divided opinion within the ranks of the KPD. Prominent KDP theoretician August Thalheimer noted the appearance of a new edition of Hegel's *Logic* with a remark in the pages of the newspaper *Die Rote Fahne* about the dangers of reading Hegel without a thorough knowledge of the Marxist classics and the natural sciences.[34] A week later, in *Die Rote Fahne*, Hermann Duncker, another prominent KPD theoretician, referred to *History and Class Consciousness* as a "dangerous" book. In his opinion, it amounted to advocacy of idealism, which could only serve to dilute the materialist essence of Marxism.[35] Two years later, expanding on his critique of Lukács, Duncker argued against his separation of method and results in dealing with Marx, and insisted it was impermissible because:

> the Marxian system is so homogeneously built up, all its ideas and lessons stand in such organisational connection that the withdrawal of one little stone would bring the rest of it toppling down.[36]

Both Thalheimer and Duncker were very critical of the views of Korsch as well as of those of Lukács.

Korsch jumped into the fray himself, using the pages of *Die Internationale*, the party's theoretical journal, of which he was himself the editor, to review four books: Lukács's *History and Class Consciousness*, Hegel's *Logic*, Bukharin's *Theory of Historical Materialism* and his own *Marxism and Philosophy*. He made his review a reply to the critics. He remarked on Thalheimer's "fear of Hegel" and contrasted it with Lenin's emphasis on the importance of studying Hegel and his proposal to Marxists to form a sort of society of materialist friends of the Hegelian dialectic. The hostility shown towards Marxists who wished to return to the Hegelian dialectical core constituted a definite retreat from the revolutionary essence of Marxism. He complained that "vulgar Marxism" was coming to prevail in the Third International just as it had in the Second. The objections advanced against Lukács and himself were based on "the empirical method of the natural sciences and the corresponding positive historical method of the social sciences," which was a bourgeois mode of thought.[37] In another article in the same issue entitled "Lenin und die Komintern," Korsch expressed his opposition to the united front line of the Comintern and characterized it a surrender of the revolutionary dialectic to pragmatic expediency. He called for the erection of a "protective wall against the rising flood of communist revisionism" and for the assertion of the independence of European communist parties from the hegemony of Moscow.[38]

In Moscow, all this was creating a stir. Even *Pravda* pronounced on it. The issue of July 25, 1924 declared Korsch, Lukács, Fogarasi, and Révai to be Marxist theoreticians still in need of education in the fundamentals of Marxist theory. It recited for them the "ABCs of Marxist philosophy," asserting a definition of truth as the agreement of a representation with objective reality external to it. Any denial of this was "bourgeois." It also asserted the centrality of the dialectics of nature to Marxism and accused both Lukács and Korsch of repudiating it. The new deviation was described as "an idealist admixture of the philosophy of identity and Machism."[39]

The Soviet academic journals also carried reviews. Somewhat ironically perhaps, it was the dialecticians and not the mechanists who were in the forefront in criticizing Lukács and Korsch. There seems to have been no trace of sympathy for them among the Deborinites. Deborin accused Lukács of going over to Hegelianism. He described Lukács's views as "a colourful mishmash of ideas of orthodox Hegelianism made tasty by an admixture of ideas from Lask, Bergson, Weber, Rickert . . . Marx and Lenin." Deborin characterized Lukács's conception of the dialectic in terms of the identity of subject and object and renounced it as the purest idealism. It amounted, he said, to a reversion to the ideas of Mach and Bogdanov. In line with Lenin's critique of Bogdanov, Deborin held to a realist theory of knowledge. Defending both the theory of reflection and the dialectics of nature, Deborin insisted that thought reflected being, that knowledge reflected a reality existing independently of man, that nature possessed laws of its own independently of human consciousness. From the standpoint of dialectical materialism, nature was in itself dialectical. The same dialectical laws of development applied to both natural and historical reality. Reality was of one piece and the dialectic applied to it universally. There was, moreover, no contradiction between Marx and Engels on this point. Deborin identified Lukács as the leader of a trend, a new current in Marxism that included such "disciples" of Lukács as Korsch, Fogarasi, and Révai and that could not be ignored and needed to be subjected to criticism.[40]

The main Soviet commentator on Korsch was G. Bammel, the author of the critical introduction to the Russian edition of *Marxism and Philosophy* that appeared in 1924,* who accused Korsch of ignorance in questions of episte-

---

* Actually it was the second Russian edition of the book. An earlier translation, published by Kniga without commentary, had appeared in 1924. Soon, however, the "October of the Spirit" publishing house in Moscow came out with another edition, accompanied by Bammel's commentary.

mology and defended the theory of reflection and the concept of the dialectics of nature. Most Soviet critics seemed to believe that Korsch had specifically rejected the dialectics of nature.* Another somewhat odd feature of the reception of Korsch's book was that it included warnings against its supposed positivism.[41]

Throughout 1924, the pages of *Pod znamenem marksizma* and *Vestnik kommunisticheskoi akademii* carried articles on the views of Lukács and Korsch. The leading Deborinites, Luppol, Sten, and others, all joined in, anxious, it would seem, to defend their own neo-Hegelian interpretations of Marxism from this one. Perhaps it was their polemic against the mechanists in the field of philosophy of science that made them so eager to make the point that an emphasis on the Hegelian roots of Marxism did not necessarily imply abandoning the field of the natural sciences in the manner of Lukács and Korsch. They missed no opportunity to defend Engels's dialectics of nature against its critics and to attack Lukács and Korsch for deserting Marxism for Hegelianism, for abandoning materialism in favor of idealism. Nevertheless, despite their best efforts, they too were to go down charged with giving too much ground to Hegelianism and idealism. Mitin, when the time came, declared that the Leninist concept of partisanship in philosophy was "the best antidote against the strong and bold tendencies of revisionism in Marxist philosophy at the present time, i.e., its Hegelian idealist form, beginning with Lukács and ending with the Deborin group."[42] Another interesting footnote to the whole discussion is that Rudas was not the only one to call attention to the connection between philosophical and political deviations. Sten did as as well, asserting that Lukács's book revealed with utter clarity the connection between idealism in philosophy and ultra-leftism in politics. The story has a somewhat ironic twist in light of the fact that Sten was destined to perish for precisely such a combination of idealism and ultra-leftism, whereas Lukács was to survive.

The Bolshevization of the Comintern

In the midst of all the controversy surrounding the two books, the Fifth World Congress of the Comintern was held in Moscow in June-July 1924. Concerned with all the factions, splits, deviations and "ideological crises" that were plaguing the movement, the congress saw the solution in the "bolshevization" of the International. The bolshevization was to be all-embracing. The movement was to be purged of all deviations—political, philosophical, and any other. Radek, Souvarine, Brandler, and Thalheimer were casualties of the campaign against Trotsky. Lukács, Korsch, and Graziadei (an Italian communist who had called into question the theory of surplus value) were

---

* This called into question how many of them had actually read the book.

condemned as "revisionists" in the ideological sphere. In the main report of the congress, Zinoviev went on the attack. It was imperative, he said, not to let "theoretical revisionism" spread and become an international phenomenon. Making it clear whom he meant, he continued:

Comrade Graziadei in Italy published a book containing a reprint of articles attacking Marxism which he wrote when he was a social democratic revisionist. This theoretical revisionism cannot be allowed to pass with impunity. Neither will we tolerate our Hungarian Comrade Lukács doing the same thing in the domain of philosophy and sociology. I have received a letter from Comrade Rudas, one of the leaders of this faction. He explains that he intended to oppose Lukács, but the faction forbade him to do so; thereupon he left the faction because he could not see Marxism watered down. Well done, Rudas! We have a similar tendency in the German party. Comrade Graziadei is a professor, Korsch is also a professor. (Interruption from the floor: "Lukács is also a professor.")* If we get a few more of these professors spinning out their Marxist theories, we shall be lost. We cannot tolerate such theoretical revisionism of this kind in our communist international.[43]

It was shameless demagoguery. There was nothing in Zinoviev's remarks to indicate that he had even read the offending books, let alone had a serious theoretical reason for objecting to them. When Bukharin alluded to them in his speech, he had at least made reference to such ideas as "relapses into the old Hegelianism."[44] In the congress discussions on the affairs of the KPD, Korsch, who was present at the congress as a KPD delegate, was offered the "friendly advice" from Zinoviev that, as the editor of *Die Internationale*, he should study Marxism and Leninism. The KPD was then further advised that it should place its theoretical journal in the hands of Marxists and not people who still needed to study Marxism.

The whole atmosphere was such as not to heighten theoretical discussion, but to close it off, to label dissenting ideas as heretical and push them aside. The adoption of the section of the draft program calling for a "rigorous struggle against idealist philosophy and against all other philosophies other than dialectical materialism"[45] does not seem to have been accompanied by searching philosophical debate. The congress condemned the "philosophical deviation of some of the intellectual centers of the parties of Central Europe aimed at eliminating the materialist essence of dialectical materialism". A disturbing feature of it all was the stirring up of workerist antiintellectualism implied in the derogatory way the term "professor" was being bandied about. Zinoviev was not ranting against the "professors" in the heat of the moment. In

---

* Actually, Lukács was not at this time a professor. Korsch was professor of law at the University of Jena.

a letter to delegates before the congress, he spoke of two time-honored tendencies within the left showing themselves within the KPD: the "devoted workers" who were the best hope of German communism, and the intelligentsia, some of whom were "unripe elements, without Marxist training, without serious revolutionary traditions."[46] Zinoviev was hardly the prototypical honest worker himself, but, as it suited his manipulative purposes to whip up hostility to intellectuals, he did not hesitate to do so. Other Comintern leaders added their voices to his, with Klara Zetkin warning the communist parties against recruiting too many intellectuals, as they could prove inconstant "allies" to the workers. One Comintern leader who refused to go along with it was Bukharin, who once saw fit to remark that the worker was not always right, no matter how black were his hands.

The sort of degradation of intellectuals that began to set in around this time can be seen in an incident involving the Hungarian economist, Jenö Varga, who is said to have consulted Zinoviev prior to the congress about whether he should report the temporary stabilization of capitalism or its imminent collapse.[47] There was a concerted effort in the period after the congress to replace intellectuals with workers in positions of responsibility throughout the Comintern apparatus. At a commission set up to look into the affairs of the KPD, Stalin warned against the ideas of the "petty bourgeois philosopher Korsch" and remarked to the KPD leadership that their lack of theoreticians was "no great unhappiness."[48] The road to revolution was full of hazards for communist intellectuals.

The Parting of Ways

Lukács and Korsch were among those who learned the hard way. Although they took divergent paths, both were perilous, bringing to each his own dilemma, his own torment.

The Way of Korsch

Korsch was defiant. His article "Lenin und die Komintern" published on the eve of the Fifth World Congress, showed him moving into a position of direct political opposition to the political line of the Comintern. When his editorship of *Die Internationale* was criticized by both Zinoviev and Bukharin at the congress, he defended his policy of keeping its pages open to varying points of view. Also at the congress, in the course of a speech by Bukharin, he is reported to have shouted from the floor "Soviet imperialism."[49] The congress made clear his growing rift even with the left-wing Maslow-Fischer leadership of the party, which he thought to be too subservient to the Soviet leadership.

After the congress, however, Korsch continued on as editor of *Die Internationale*, and within its pages for a short time he sang the praises of

Leninism, Stalinism and the process of bolshevization. But not for long. It had already been made clear that bolshevization was not in his favor. The central committee of the KPD received a letter signed by Béla Kún, by now high up within the Comintern apparatus and in charge of its "agitprop" division, referring to favorable reviews of Lukács's book in *Die Internationale* and demanding a tightening of control over the journal. Early in 1925, Korsch was removed as editor. From this time, he openly expressed his opposition to the party leadership, an opposition that intensified with the downfall of Fischer and Maslow and the succession of Ernst Thälmann to the leadership.* During this period, he was attacked in party publications and heckled when he spoke at party meetings. Factional activity at rallies sometimes reached the point of street fighting. In March 1926, Korsch and his associates formed an organized faction within the party, the *Entscheidende Linke* (the decisive left), which published an internal opposition bulletin that noted the Comintern's declared policies of freedom of discussion of all questions within the party. They criticized the "parliamentary cretinism" of the KPD and advocated "clear revolutionary class politics" that included the creation of breakaway trade unions, as a step toward the goal of a socialist state build on workers' councils. In international terms, the group identified with the Soviet Workers' Opposition, criticized the NEP and called the Soviet Union a "dictatorship of the kulaks." They opposed the Comintern line on the temporary stabilization of captialism and argued that the Comintern had been transformed into an instrument of Soviet foreign policy.[50]

A crisis point was reached when the faction opposed the German-Soviet treaty of military cooperation in April 1926. Korsch was presented with an ultimatum: to relinquish his seat in the Reichstag or face expulsion from the KPD. He refused to comply and was expelled from the party amidst a general purge of leftist elements within the KPD. In June 1926, he made a bitter attack on Soviet foreign policy from the floor of the Reichstag. Zinoviev called him "an insane petty bourgeois" and his expulsion was endorsed by the Comintern in June 1926.

After his expulsion from the KPD and the Comintern, Korsch remained politically active for several years. He kept his seat in the Reichstag and maintained connections with like-minded leftist groups abroad, such as the Italian group around Bordiga. He continued publishing his journal *Kommunist-ische Politik*, even after the *Entscheidende Linke* group split; but he found himself increasingly isolated. The leftists split and split and split until each was left with virtually himself alone.

---

* The fall of Fischer and Maslow in Germany coincided with the fall of Zinoviev and Kamenev in the Soviet Union. After Zinoviev was expelled from the politbureau of the Russian party, he was also removed from his post as president of the Comintern. In October 1926, Bukharin succeeded to the post.

In 1930, Korsch published an "Anti-Critique," an analysis of the issues arising from the discussion around *Marxism and Philosophy* and *History and Class Consciousness*, which was appended to the new edition of *Marxism and Philosophy*. Intellectually, as well as politically, he had gone beyond the pale. He had become overtly anti-Leninist. According to Korsch, Lenin's philosophy, which his epigones were following to the letter, was riddled with "grotesque inconsistencies" and "crying contradictions." In an extremely contradictory fashion, he charged Lenin on the one hand with not following the teaching of Marx and Engels on the abolition of philosophy and with trying to turn Marxism into a new philosophical outlook, but on the other hand criticized him for deciding philosophical questions on the basis of nonphilosophical considerations. Complicating things still further, he proceeded to bring his own nonphilosophical considerations into the forefront of his critique of Lenin's philosophy: Leninist theory was, in his opinion, incapable of answering the practical needs of the international class struggle in the present period. One did not need to look further than Korsch to find the "grotesque inconsistencies" and the "crying contradictions." The relationship between philosophy and politics was a question well worth raising, but Korsch did not help much in illuminating it.

He did , however, make one substantive point about Lenin's philosophy that sharply highlighted the difference between Lenin's perspective and his own: the displacement of the accent from dialectics to materialism. Lenin's argument was that there had been a change in the whole intellectual climate that made it necessary to stress materialism against the predominantly anti-materialist trends in bourgeois philosophy, rather than to stress dialectics against vulgar, predialectical, undialectical, and antidialectical trends in bourgeois science. Korsch contended that this was not the case, that there had been no such change. He argued that the dominant basic trend in bourgeois philosophy and natural science was the same as it had been sixty or seventy years before. It was inspired not by an idealist outlook, but by a materialist outlook colored by the natural sciences. He took exception to Lenin's "uncritical approach to the natural scientific materialism of the second half of the nineteenth century." With regard to the relationship of philosophy to the natural sciences, Korsch contrasted Engels's approach with that of Lenin. Engels perceived that the development of the sciences confirmed a materialism that was essentially dialectical, that had no need for philosophy standing above the sciences. Lenin's procedure was supposedly the opposite. His materialism was a philosophy that stood above the sciences as a kind of supreme judicial authority for evaluating the findings of the individual sciences, past, present, and future. This procedure, he claimed, had been taken to absurd lengths by the Leninists, in whom it resulted in a specific kind of ideological dictatorship over the sciences, as well as over the arts, which oscillated between revolutionary progress and the blackest reaction.* In criticizing Lenin for transferring the dialectic from thought to nature and, in so doing, destroying the dialectical interrelation of being and consciousness, theory and practice, Korsch seemed to be making the same argument as Lukács on the dialectics of nature.

About the dialectics of nature, however, Korsch was still denying that he denied it and criticizing his critics who said that he had. It was very difficult to know where he stood with regard to Marxism in relation to the philosophy of science.He was still decrying "vulgar Marxism" for being devoid of any philosophical perspective at the same time as he was deploring it as making Marxism a new philosophy instead of the abolition of philosophy. He was all for the union of dialectical materialism and natural sciences one moment, and against it the next.

He never did clarify what he thought about Lukács either. He railed against the "inquisition" against Lukács, while stating that he had not been sufficiently aware of the extent of the theoretical divergence between them, but without elaboration on it. However, some pages later, he made a point that seemed an implied criticism of Lukács:

> It is therefore completely against the spirit of the dialectic, and especially of the materialist dialectic, to counterpose the dialectical materialist method to the substantive results achieved by applying it to philosophy and the sciences. This procedure has been very fashionable in western Marxism. Nevertheless, behind the exaggeration there lies a correct insight—namely, that dialectical materialism influenced the progress of the empirical study of nature and society in the second half of the nineteenth century above all because of its method.[51]

The criticism of the artificial separation of method and results was fair enough, but it was exceedingly difficult to make anything intelligible out of all his diverse and discordant pronouncements on dialectics, materialism, philosophy, and the natural sciences so as to be able to state what his position actually was.

His "Anti-Critique" was extremely dismissive of his critics, scornfully declaring their criticisms of his views as "not only unjustified but null and void." He simply noted the convergence of the social democratic and communist criticisms of his book and moved on self-righteously. No more than his critics was he reticent to link the philosophical disputes with the political ones. For him, it was all of one piece. The one was the echo of the other.

By 1930, however, Korsch's active political life had all but come to an end. He still gave political lectures in Berlin, where one of his pupils was Bertolt

---

* It might not have been a fair criticism of Lenin, but with regard to later Leninists, his remarks proved very prophetic indeed.

Brecht, but that was about it. In 1933, even that was to end. In February, he gave his last lecture in Germany. Later that night, as the Reichstag burned, he went into exile, as did so many of his contemporaries. After spending some time in Scandinavia and Spain, he came to settle in America.

In 1935, Korsch published an essay entitled "Why I Am a Marxist" in which he continued his onslaught on the orthodox. He criticized the "renaissance of the pseudo-philosophical dialectic," exemplified by such as Rudas, that inflated the materialist dialectic into an external law of cosmic development. He objected to all attempts to force all experience into a monistic construction of the universe in order to build a unified system of knowledge. Coming forth with yet another confused pronouncement on the relationship of Marxism and the natural sciences, he wrote:

> Marxist theory is not interested in everything, nor is it interested to the same degree in all the objects of its interests. Its only concern is with those things which have some bearing upon its objectives. . . . Marxism, notwithstanding its unquestioned acceptance of the genetic priority of external nature to all historical and human events, is primarily interested only in the phenomena and interrelations of historical and social life. That is to say, it is only interested in what, relative to the dimensions of cosmic development, occurs within a short period of time and in whose development it can enter as a practical, influential force. The failure to see this on the part of certain orthodox communist party Marxists accounts for their strenuous attempts to claim the same superiority, undoubtedly possessed by Marxian theory in the field of sociology, for those rather primitive and backward opinions which to this very day are retained by Marxian theorists in the field of natural science. By these unnecessary encroachments, the Marxian theory is exposed to that well-known contempt which is bestowed on its "scientific" character even by those contemporary natural scientists who, as a whole, are not unfriendly to socialism.[52]

He went on, somewhat obscurely, to call for a "less philosophical and more progressive" interpretation of the Marxist conception of the "synthesis of sciences."

In "Why I Am a Marxist" and in his book *Karl Marx* published in 1938, Korsch seemed to be adopting a different attitude toward science. In *Karl Marx* he spoke of Marxist theory as a materialist science of society as well as a practical instrument of the class struggle of the proletariat, and he stressed its empirical and logical character. Some commentators, such as Giuseppe Bedeschi, have maintained that in *Karl Marx*, Korsch had overcome his earlier antimaterialism. This has been disputed by Sebastiano Timpanaro, who argues that Korsch never ceased to be decidedly antimaterialist. If he appeared in *Karl Marx* to have come to a reconciliation with science and to acknowledge a certain analogy between the method of Marxist social science and the method of the natural sciences, it was only a reconciliation with a science already neo-positivized by the prevailing antimaterialist epistemology. According to Timpanaro, the mature Korsch was only more bland.[53] His later texts would seem to bear out Timpanaro's judgement on the character of his intellectual evolution. Gone was the passion, but not the confusion. He never did achieve any clarity about what Marxism had to do with philosophy or the natural sciences.

As time went on, Korsch seemed to cease to care. Until the 1940s, he wrote for such journals as *Living Marxism*, edited by the ex-Spartakist Paul Mattick, but eventually even this sort of involvement dropped off. In the 1950s, living in America during the cold-war years, he renounced Marxism altogether. It brought to mind the surprisingly acidic remark made by his good friend Bertolt Brecht: "Korsch is only a guest in the house of the proletariat. His bags are packed and he is always ready to leave."[54] He died in the United States in 1961, his works having fallen into relative obscurity.

Only later, in the 1960s was there something of a Korsch revival. An Italian translation of *Marxism and Philosophy* appeared in 1966, bringing in its wake commentaries by Italian Marxists such as Colletti, Vacca, Bedeschi, Cerutti, Ceppa, Timpanaro, and others. Acquaintance with the work of Korsch brought Giuseppe Vacca, who assessed it positively, to a turning point in his own intellectual development.[55] Others, such as Colletti, who have had more negative reactions to it, have criticized Korsch for not untying the Gordian knot binding Marx to Hegel and for confusing the critique of bourgeois society with a romanticist critique of science.[56] Timpanaro has seen him as fighting an anachronistic battle against a stage of bourgeois philosophy already superceded by a new stage, which Korsch, unlike Lenin, could not recognize.[57] Similarly, an English translation in 1970 brought forth much discussion of Korsch in Britain and America. Korsch's early works have been enthusiastically received by those of the *New Leftist Review* variety. *Telos*, sympathetic to the neo-Hegelian current within Marxism and instrumental in the revival of it, has devoted sustained attention to Korsch. Many of the articles have been of a very high scholarly standard and of great value in illuminating the intellectual history of the time, but they carry forward the same antimaterialist prejudices against the natural sciences and against the mainstream Marxist tradition as found in Korsch.[58] Despite his dissociation from Marxism, Korsch has come once again into the limelight of Marxist discussion.

Korsch may not have been a great philosopher, but he nevertheless played a constructive role in the context of the 1920s in his emphasis on the importance of revolutionary consciousness in the revolutionary process and in his refusal to conform to a leveling orthodoxy, despite the various forms of intimidation that the communist movement could bring to bear on dissident intellectuals.

The Way of Lukács

Things went very differently for Lukács. His commitment to the proletarian cause within the communist movement was of a different sort, although the price it exacted of him was very high, higher perhaps than it had a right to ask, higher perhaps than he should have paid. The times were hard, so hard that perhaps no alternative could have been an altogether satisfactory one. The history of Lukács's subsequent intellectual development and political involvement is a complex story to unravel, with so many complexities underlying his feigned recantations, his sincere recantations, his partially feigned, partially sincere recantations; Comintern bans on his publications, his own bans on his publications, pirated republications, permitted republications; the days of honor, the days of obscurity, the days of disgrace.

When *History and Class Consciousness* came under attack, Lukács did not defend it. No such "Anti-Critique" as issued by Korsch came forth from him. Unlike Korsch, it would seem that he took the criticisms seriously and reflected on them, although it is doubtful that he was immediately convinced by them. He was weighing things up and he was searching. In 1924, just after Lenin's death, he published a book on Lenin entitled *Lenin: A Study of the Unity of His Thought*. Lukács saw the key to Lenin's greatness as his combination of global vision and concrete analysis, his ability to focus on the nodal points where theory became practice and practice became theory.[59] Indeed, writing this seems to have helped Lukács come closer to achieving such a synthesis himself.* In 1925, he wrote several reviews that showed him grappling with some of the issues raised in *History and Class Consciousness* and still in the grip of a certain prejudice against the natural sciences. This was particularly evident in his critique of Bukharin's manual on historical materialism, which was based on his resistance to Bukharin's "bias towards the natural sciences." Bukharin's "preoccupation with the natural sciences," his use of science as a model, led him to a "false objectivity" that obscured the specific feature of Marxism, that is, that all economic or sociological phenomena derived from the social relations of men to one another. Lukács also felt that Bukharin attributed too determinant a position to technology. Bukharin's philosophy was, in Lukács's view, dangerously close to the bourgeois contemplative materialism rejected by Marx. Bukharin's fault lay in the fact that

> instead of making a historical-materialist critique of the natural sciences and their methods, i.e., revealing them as products of capitalist development, he

---

* In *The Young Lukács and the Origins of Western Marxism* (New York, 1974), Andrew Anato and Paul Breines rather colorfully refer to Lukác's Lenin as "a virtual *Zeitgeist* in a sealed train" and argue that it amounted to an implicit critique of the offical Leninst Lenin (p. 190).

extends these methods to the study of society without hesitation, uncritically, unhistorically, undialectically.[60]

The problem with Bukharin, supposedly, was in his attempt to make a science out of the dialectic. Lukács saw Bukharin's leanings towards the natural sciences as being in conflict with his frequently acute dialectical instinct. He made it clear that he still disagreed with Engels's conception of the dialectic as "the science of the general laws of motion, both of the external world and of human thought." In another review, this time of the book by the German Marxist Karl August Wittfogel, *The Science of Bourgeois Society*, whom he also criticized for an "uncritical" attitude towards the method of the natural sciences, Lukács reiterated his contention that nature was a social category.[61] Lukács was still far from considering Marxism as a philosophy of science, but he was constantly moving his analysis onto a more concrete level. As he later reflected on this period, he was still opposed (and rightly) to technological determinism and historical fatalism, to the elimination of man and social activity, but he was becoming less inclined to make use of voluntaristic counterweights to oppose this mechanistic fatalism.[62]

The political counterpart to Lukács's philosophical evolution away from idealism and voluntarism was a renunciation of utopianism and ultra-leftism. In contrast to what happened to Korsch, it was not Lukács's ultra-leftism, but his turn away from it, that brought him to the point of political confrontation and crisis within the Comintern. Following the death of Landler, it fell to Lukács in 1928 to prepare the political theses for the forthcoming congress of the Hungarian party. The "Blum Theses"* called for the establishment of a democratic dictatorship of the proletariat and peasantry as a transitional stage leading to the dictatorship of the proletariat. The theses were denounced as opportunist by Béla Kún and his faction, who advocated a more maximalist and adventurist line. However, a new turn of events on the international scene put victory in the lap of Kún, a man most observers agreed in characterizing as intellectually inadequate, indecisive, cowardly, opportunist, authoritarian and corrupt.

The Comintern's "Third Period"

The Sixth World Congress of the Comintern was held in August 1928 and marked a sharp leftward turn in the affairs of the International, corresponding with the sharp leftward turn proclaimed in the Soviet Union with the launching of the first Five Year Plan. Thus was inaugurated the famous "third period" in the history of the Comintern. The tactics of the united front were out and

---

* Blum was Lukács's pseudonym within the party.

revolutionary offensive was in. The severe intensification of the general crisis of capitalism and an accompanying radicalization of the masses was predicted. There was no need for alliances. It was "class against class." Social democracy was branded as "social fascism."

There did indeed come an intensification of the general crisis of capitalism with the great crash of 1929 and the depression that followed it. Comintern propaganda contrasted the capitalist gloom and decay with a radiant socialist future. But there was not radicalization of the masses. There was no revolutionary offensive. Instead there was bitter sectarianism between communists and social democrats. May Day of 1929 saw violent clashes between the two groups on the streets of Berlin. Communists not only scorned alliances, eschewed the existing trade unions, and scoffed at parliamentary activity, but they also ruthlessly combed through their own ranks to purge right deviationists.

This was the situation in which the "Blum Theses" came up for discussion. On top of the overwhelming criticism of Lukács that was coming from the Hungarian party, a document from Moscow, entitled "Open Letter from the Executive Committee of the Communist International to the Members of the Hungarian Communist Party," that condemned the "Blum Theses" as right deviationist and liquidationist, arrived. These theses were resoundingly defeated. These theses not only ensured Lukács's defeat in the struggle against Béla Kún for leadership, but brought him the threat of expulsion from the party altogether. To head off this prospect, he recanted and published a self-criticism, attacking the "Blum Theses" as opportunist and right deviationist. The recantation was quite insincere, and Lukács was not convinced. He continued to dissent from Béla Kún's policies and from the Comintern's "third period" policies, but he kept his dissent to himself. One commentator has remarked that Lukács preferred to act out the role to which the logic of *History and Class Consciousness* committed him, a logic that hypostatized the party as the institutionalized will and expression of proletarian class consciousness, endowed with a superior view of total reality. It constituted a built-in veto against opposing his own views to those of the party.[63] He explained in later times that he had been very affected by the fate that had befallen Karl Korsch. Although firmly believing that he was in the right, he believed also that to be expelled from the party would make it impossible for him to participate actively in the struggle against fascism. The self-criticism was his "entry ticket" into the struggle against fascism.[64]

Not that the communist parties seemed very promising instruments for combating fascism at this time. The Comintern analysis of fascism during the "third period," even despite the bitter experience of Italian communists (in Mussolini's jails or in exile or in carrying out underground activity within the circle of the scattered handful of militants that their party had become) was that fascism was a transient episode that could serve to radicalize the masses and prepare the way for communist power. The German communists directed all their fire against the socialists. The socialists, for their part, often seemed to prefer steel barons to bolsheviks, not hesitating to ally themselves with the traditional right against the left. With the socialists and communists at each others throats, Hitler's power was constantly on the rise.

The times were dark and Lukács knew it. His attitude was to hold on and wait for better times. He is reported to have said to Victor Serge:

> The times are bad, and we are at a dark crossroads. Let us reserve out strength: history will summon us in its time.*

Lukács ceased to involve himself in practical politics and devoted himself instead to his scholarly pursuits. In 1929, he was served with an exclusion order in Vienna and went to Moscow where he worked at the Marx-Engels Institute under David Riazanov. There he collaborated in the editing of Marx's *1844 Manuscripts*. Reading these texts, he claimed, marked a landmark in his philosophical evolution, in that Marx's statements regarding objectivity and materiality had the effect of shattering the theoretical foundations of the idealism that found expression in *History and Class Consciousness*.[65] He also witnessed the removal of Riazanov, which was followed by a dramatic decline in the work of the Institute.

In 1931, Lukács emigrated to Berlin, where he worked closely with the KPD, although he concentrated on literary criticism. He wrote for the KPD literary journal *Die Linkskurve* (the left curve), which was permeated with the sectarianism of the "third period." After witnessing Hitler's final accession to power Lukács returned, in 1933, to Moscow, where he worked at the Institute of Philosophy of the Academy of Sciences until the end of the war in 1945.

Criticism and Self-Criticism

Only in 1933 did Lukács break his silence regarding *History and Class Consciousness* and undertake an analysis of the issues involved in the discussion surrounding it. In 1933, he published *Mein Weg zu Marx* (My Road to Marx) in which he gave an account of his intellectual development from the time when he first read Marx as a schoolboy in Hungary, through the 1920s and into the 1930s. He explained that his conversion to socialism did not immediately have any effect on his philosophical thinking:

---

* Serge in his *Memoirs of a Revolutionary* (Oxford, 1963, p. 192) ascribes this conversation to Vienna in or about 1926. Michael Löwy, however—on the basis of textual and historical evidence—convincingly argues that it must have been in Moscow in 1929 (cf. Löwy, "Lukács and Stalinism," *New Left Review*, no. 91 [May–June 1975]).

As is only natural in the case of a bourgeois intellectual, this influence was restricted to economics, and particularly to sociology. Materialist philosophy—I did not then draw any distinction between dialectical and undialectical materialism—I regarded as wholly antiquated in relation to the theory of knowledge. The neo-Kantian theory of the immanence of consciousness corresponded very well to my class position and world view then. I did not subject it to any sort of critical examination and accepted it unquestioningly as the starting point of any kind of epistemological enquiry. I did indeed have reservations concerning the extreme subjective idealism of the Marburg School of neo-Kantianism and of Machism, since I did not see how the question of reality could simply be treated as an immanent category of consciousness. But this did not lead to materialist conclusions, but rather towards those philosophical schools which tried to solve this problem in an irrationalist-relativist fashion, occasionally tending towards mysticism. . . . The influence of Simmel, whose pupil I then was, enabled me to integrate those elements of Marx's thought which I had assimilated during this period into such a *Weltanschauung*.[66]

Lukács viewed the 1920s as the time of his "apprenticeship in Marxism," as a transitional phase from a view of Marxism shaped by neo-idealism to a mature and more authentic Marxism grounded in materialism. He expressed a change of heart with regard to theory of reflection and dialectics of nature. In the following year, on the occasion of a conference organized in 1934 in honor of the twenty-fifth anniversary of Lenin's *Materialism and Empirio-Criticism*, he gave a more detailed autobiographical statement, focusing on the evolution of his philosophical views in the course of his address to the Institute of Philosophy of the Communist Academy. His speech, entitled "The Significance of *Materialism and Empirio-Criticism* for the Bolshevization of the Communist Parties," was published both in the volume containing the proceedings of the conference and as a separate article in *Pod znamenem marksizma*. Here Lukács declared that the errors into which he had fallen in *History and Class Consciousness* corresponded to the deviations criticized by Lenin in *Materialism and Empirio-Criticism*. To explain the background to his own deviations, he recalled:

I began as a student of Simmel and Max Weber . . . and developed, philosophically speaking, from subjective idealism to objective idealism, from Kant to Hegel. At the same time, the syndicalism of Sorel had a great influence on my development, reinforcing my inclinations to romantic anti-capitalism. Thus I entered the Communist Party of Hungary in 1918 with a world outlook that was distinctly syndicalist and idealist. Despite the experience of the Hungarian revolution, I found myself immersed in the ultra-leftist syndicalist opposition to the line of the Comintern in 1920–1921. . . . The book I published in 1923 was a philosophical summation of these tendencies. . . . In the course of my practical party work and in familiarizing myself with the works of Lenin and Stalin, these idealist props of my world outlook lost more and more of their security. Although I did not permit a re-publication of my book . . . , nevertheless, I first came to a full realization of these philosophical problems during my visit to the Soviet Union in 1930–1931, especially through the philosophical discussion in progress at that time. Practical work in the Communist Party of Germany, direct ideological struggle against social fascist and fascist ideology have all the more strengthened my conviction that, in the intellectual sphere, *the front of idealism is the front of fascist counter-revolution and its accomplice, social fascism.* Every concession to idealism, however insignificant, spells danger to the proletarian revolution. Thus I understood not only the *theoretical falsity,* but also the *practical danger* of the book. . . . With the help of the Comintern, of the All-Union Communist Party and of its leader, Comrade Stalin, the sections of the Comintern will struggle for that iron ideological implacability and refusal to compromise with all deviations from Marxism-Leninism, which the All-Union Communist Party achieved long ago. . . . Lenin's *Materialism and Empirio-Criticism* has been and remains the banner under which this struggle is carried forward on the intellectual front.[67]

On a number of occasions, Lukács reaffirmed his critique of *History and Class Consciousness.* In 1938, he referred to the book as "reactionary by virtue of its idealism, its faulty interpretation of reflection, its denial of the dialectic of nature."[68] The repudiation of his earlier work was another "entry ticket" for further ideological struggle. He believed that it was necessary in order for him to carry on. This article, however, was not totally insincere, as his earlier repudiation of the "Blum Theses" had been. He had sincerely changed his mind on the relevant philosophical issues, as he was to reaffirm in later years in the absence of the earlier pressure. He did admit to having acted in a conformist fashion in adopting the current official jargon in the form of his self-criticism.[69]

For years, Lukács did not allow the republication of his book. Officially, it was for a long time a "forbidden book" and had only a sort of underground existence. There were several pirate editions and spurts of renewed discussion of it, much against the wishes of Lukács, who wanted the book forgotten. Lukács responded angrily to Maurice Merleau-Ponty's efforts to resuscitate the book and revive debate on it. But it never was forgotten. It has been argued that Martin Heidegger's *Sein und Zeit,* in 1927, and Karl Mannheim's *Ideologie und Utopie,* in 1929, were conceived as replies to Lukács. His early work has also been seen as a precursor of the sociology of knowledge and a seed of what developed into the Frankfurt School.

In the 1960s, there was a great revival of interest in Lukács's early work and a massive secondary literature has sprung up relating to it. Lukács permitted republication, conceding that it could now be of historical interest, but at the same time objecting strenuously to the tendency to consider certain works of the 1920s as "classics of heresy" and to treat them as offering the answers to the current controversies. Although Lukács was to go on to become an important philosopher in the postwar period, later Lukacsians continued to find inspiration in the work of the early Lukács, whatever he said about it himself. This was the approach of Merleau-Ponty and Lucien Goldmann, of the "Budapest School" group of Agnes Heller and Mihaly Vajda,[70] and of the New Left of the 1960s and 1970s. Paul Piccone refers to the prevailing view that there are two Lukácses, the one "revolutionary dynamite" and the other "a hack." Many, many articles devoted to the young Lukács have appeared in *Telos*, *New Left Review*, and many other such journals. The *Telos* articles have been for the most part enthusiastically favorable, following the tone set by Merleau-Ponty and Goldmann, who regarded *History and Class Consciousness* as a rebirth of dialectical thought after a half-century of eclipse. The New Leftists find attractive the critique of science and its identification with bourgeois reification and the activist emphasis on consciousness and subjectivity. Such qualities have made Lukács's early work "the Grand Central Station of present day Marxism" in Piccone's opinion.[71]

A Marxist of the opposite tendency has also had words of praise for the early Lukács. The anti-Hegelian Colletti, at a far extreme from the neo-Hegelian *Telos*, although seeing Lukács as the major defender of an immediate continuity between Hegel and Marx, nevertheless saw *History and Class Consciousness* as of great value as "the first Marxist book after Marx in which philosophical Marxism ceases to be a cosmological romance and, thus, as a surrogate religion for the lower classes."[72]

Seeing no redeeming features in the work of the early Lukács, Hoffman writes off *History and Class Consciousness* as a product of "praxical fervour" that was empty of any real Marxist content. He characterizes the trend looking to the early Lukács as reflecting all of the weaknesses of Hegelian idealism without any of its strengths. Taking the opposite view of the dialectics of nature, he takes Lukács to task for having reverted to a pre-Darwinian conception of nature, for himself adhering to positivist assumptions, seeing nature as something passive, wholly external to man and without its own dialectical movement. Such an abstract division between the world of nature and the world of man is the "one basis for science and another for life which was rejected by Marx."[73]

A less dismissive critique is that by Mihailo Marković, who notes the revival of the question raised by Lukács in *History and Class Consciousness* in relation to the dialectics of nature. Lukács's position, he points out, was not clear or consistent, there being two different views expressed in the book: the dialectic as a method of understanding society, which was not to be extended to nature, and various types of dialectic to be concretely expounded, with both a dialectic of natural processes and a dialectic of history. The first view, however, was the dominant one, as Lukács was inclined to draw a sharp distinction between the methodology of the natural sciences and that of the social sciences. The social sciences, if based on the dialectic, were revolutionary. The natural sciences were condemned to being nondialectical, bourgeois, contemplative, quantitative, ahistorical, partial, and fragmentary. This view of the dialectic as only a method of understanding society, Marković argues, contains a series of insurmountable difficulties. The question arises of whether the general philosophical conception of overall being is to be renounced or whether the dialectic of society is compatible with a nondialectical method of cognition of nature. In the case of the former, it leads to a lack of methodological coherence that is obviously inadmissible. If the latter be the case, then there are two alternatives: Marxist philosophy is oriented to knowledge of being as a whole, but there are different, even opposite, methods for different spheres of being, or Marxist philosophy is exclusively a social theory, in which case the cognition of nature is or is not a part of social history. If it is, it is possible to speak of dialectics of nature. If it is not, then cognition of nature remains outside history. This indeed was the position of Lukács, who reproached Hegel and Engels for extending the dialectical method to the cognition of nature. But, Marković asks, with regard to Lukács's stated intentions: how can a philosophy deal with totality while parts of that totality remain outside philosophy? In speaking of the dialectics of nature, there have been three different senses of its meaning: the process of nature in itself, the theory of that process, the process by which man changes and comes to know nature. As regards the first, the objective laws of natural processes themselves, there is indeed no interaction of subject and object and Marković holds that Lukács was right in holding that the concept of dialectic in this sense ought not to be used. As regards the second, the cognition of nature and the application of the dialectical method to nature, Lukács argument was mistaken, for there are only differences in degree between the natural and social sciences. Lukács, Marković argues, made the same mistake as those he was criticizing, in considering nature to be pure object without subject.[74]

Another recent commentator to take a critical view of the young Lukács has been Gareth Stedman Jones, representing yet another orientation among contemporary Marxists. Stedman Jones has classified *History and Class Consciousness* as "the first major irruption of the romantic anti-scientific tradition of bourgeois thought into Marxist theory." He is highly critical of such a negative attitude toward the role of science and technology that constituted a radical departure from Marx, who was strikingly free from the tension of either romantic antiindustrialism or utilitarian positivism, free from a mystique of either nature or industry. The attack upon the analytic rationality of modern science, according to Stedman Jones, formed the theoretical core of the whole book and determined all its political errors. The contempt for the concrete facts deriving from it accounted for the disembodied scenario set out in the work and its surreal logic that reduced economics to a shadowy substratum, nearly etherealized out of existence, and contracted the superstructure to a few wooden leitmotifs. Even history itself played a purely spectral role in *History and Class Consciousness*, with the proletariat coming in as a sort of shadowy demiurge. There are, in his view, two main dangers that Marxism must avoid, romanticism and positivism. The early Lukács represents the Scylla and Bukharin and Kautsky the Charybdis.[75]

Lukács himself participated in the whole discussion, of course. Several further autobiographical statements were forthcoming, that gave a less pressured and somewhat more sophisticated analysis of the issues raised by his early work and of his own intellectual development in sorting them out. Most important of these were a 1957 preface to a new edition of *Mein Weg zu Marx* and a 1967 preface to a new edition of *History and Class Consciousness*. In the latter, he made it clear that it was no longer his intention to paint it all in black and white, as if the dynamics of the situation could be confined within the limits of a struggle between revolutionary good and the vestigial evil of bourgeois thought. In retrospect, he believed that the book was right in its revival of Hegelian traditions, although wrong in its failure to subject the Hegelian heritage to a thoroughgoing materialist reinterpretation. He thought that a great achievement of the book was its reinstatement of the category of totality, as well as its attention to the problem of alienation. At the same time, he highlighted what he had sincerely come to believe were the most serious problems with the book:

> The book's most striking feature is that, contrary to the subjective intentions of its author, objectively it falls in with a tendency in the history of Marxism that has taken many different forms. All of them have one thing in common, whether they like it or not and irrespective of their philosophical origin or their political effects: they strike at the very roots of Marxian ontology. I refer to the tendency to view Marxism exclusively as a theory of society, as social philosophy, and hence to ignore it or repudiate it as a theory of nature. . . . My book takes up a very definite stand on this issue. I argue in a number of places that nature is a societal category and the whole drift of the book tends to show that only a knowledge of society and the men who live in it is of relevance to philosophy.[76]

In earlier days, Lukács had sought a total break with every institution and mode of life stemming from the bourgeois world, but he now realized that it was the materialist view of nature that brought about the really radical separation of the bourgeois and socialist outlooks. With the repudiation of the ontological objectivity of nature, the most important pillar of the Marxist world view disappeared. While he had been correct to attack the overvaluation of contemplation, his polemic against it had taken on extravagant overtones and his conception of revolutionary praxis was colored by an overriding subjectivism. This apparent methodological upgrading of societal categories, he said, distorted their true epistemological function. The overextension of the concept of praxis led to its opposite, a relapse into idealist contemplation. Though he believed that the book had a certain historical validity and that it was correct on many points, Lukács expressed regret that it was precisely those aspects of the book that he thought most mistaken that were becoming the most influential.

At any rate, back in the Soviet Union in the 1930s, Lukács was reflecting on it all. It is unlikely that his sojourn in the Soviet Union had such a wholly negative impact on his thought as so many western commentators seem to assume. It is uncertain just how far the Soviet discussions of 1930–1931 affected his thinking. Whether they actually affected him to the degree that he claimed in the situation of 1934, is hard to know. He did say later that he thought much of the criticism of Deborin justified, but he seemed rather detached from it and remarked that its aim was only to establish Stalin's preeminence as a philosopher. Still, whatever its faults, Soviet philosophy was far closer to the natural sciences than was German philosophy in the academic milieu from which Lukács had come and this must have had a beneficial effect on his development.* He was inclined to keep at a distance from political discussion, even to keep at a bit of a distance from philosophical discussion in the 1930s, although he did enter into the Soviet literary debates of the time with a certain vigor. But regarding most other things, he kept his ideas to himself. He did not protest against the purges or the Moscow trials, believing that to attack Stalin was to support Hitler. However, this did not save him from any further political troubles. In 1941, he was arrested in Moscow and accused of having been a Trotskyist agent since the 1920s (notwithstanding the fact that he had at all times been an outspoken anti-Trotskyist). He was subsequently released after personal intervention on the part of Georgi Dimitrov, president of the Comintern. Not even this episode shook his fundamental loyalty. He still knew which side he was on, and was determined to stay within the ranks of the communist movement at any price. Shaken deeply by the triumph of fascism and believing that it marked a deep rupture in the continuity of European culture, he saw the Soviet Union as the only force that could hold out against it and bridge the gap between the progressive culture of the past and the future.

---

* Professor Boris Kuznetsov, a Soviet philosopher and historian of science, famous for his writings on the theme of nonclassical science, recalls an exchange with Lukács at the Communist Academy during Lukács's years in the Soviet Union, in which he argued that Lukács had not grasped the concept of nonclassical science and was trying "to pour new wine into old wineskins." (Interview with Boris Kuznetsov, Moscow, April 3, 1978).

He asserted this even during the Comintern's "third period" and during the Nazi-Soviet Pact, times when it seemed far from the case.

Foreign Communists in the Soviet Union

Other foreign communists felt themselves faced with the same choice. Among their number were various philosophers and scientists who, like Lukács, worked in the Soviet Union. Some came early on, and some came later. Some came fleeing fascism, and some came to be part of the process of socialist construction. Lukács's fellow Hungarian and critic, László Rudas, also worked at the Marx-Engels-Lenin Institute and became a professor at the Institute of Red Professors. The Bulgarian Todor Pavlov, who specialized in theory of reflection, wrote under the name of Dosev during his years in exile. Yet another Hungarian who came to the Soviet Union was Sandor Varjas (known in the Soviet Union as Aleksandr Ignatevich Variash), who became thickly embroiled in the Soviet philosophical discussions. He had been an official in the Hungarian Commune and was imprisoned after its downfall. After coming to Moscow in 1922 in an exchange of prisoners agreed upon between the Soviet government and the Horthy régime, he became a professor of philosophy at Moscow University. He was a leading figure in RANION and the Timiriazev Institute and a participant in the debates held at the Communist Academy, the Institute of Red Professors, and the Institute of Scientific Philosophy. His name appeared constantly in the pages of *Pod znamenem marksizma*. He was one of the leading spokesmen for the mechanist trend and his name appeared in the 1929 resolution denouncing mechanism. The publication of his two-volume history of modern philosophy, attacked by the Deborinites, was an important episode in initiating the whole debate. He responded in the new few years to Deborinite accusations of positivism and hypostatizing logic with counteraccusations of idealism and Hegelian panlogism. His declaration of faith was: "I stand on Marx's point of view that we know one single science."[77] Varjas pursued his great dream of unity of science in the direction of an unabashed and thoroughgoing reductionism. All laws were ultimately physico-chemical. There were no irreducible qualities. After the condemnation of mechanism in 1929, Varjas remained a mechanist and continued to publish his mechanist view. Despite the pressures, he never recanted. Varjas perished in the purges of the late 1930s.

Also prominent in the Soviet philosophical debates, though on the opposite side of the fence, was the Czech philosopher and mathematician, Arnost Kolman. On his release from a Russian prisoner-of-war camp at the end of the first war, he stayed in the Soviet Union and went to Moscow University. He took the side of the Deborinites against the mechanists, but was among the young philosophers who brought Deborin in his turn under criticism. He was involved in all the major discussions in the philosophy of science. Although a learned and intelligent man, his role in this sphere was a complex one. He became head of the science division of the Moscow party organization and presided over the meeting that denounced Levit and linked medical genetics to racism and fascism. He was among the guardians of orthodoxy, who linked philosophical and scientific deviations to political subversion. In the genetics controversy, he was a supporter of Lysenkoism and in the discussion over relativity theory, he was exceedingly vague, although it is likely that he actually knew better than what he was willing to say. Whatever he may have thought of his fellow delegates to the International Congress of the History of Science and Technology in London in 1931, he was not dealing with them very kindly in 1937. He felt obliged, for example, to call the attention of the readers of *Pod znamenem marksizma* to the fact that the "enemy Hessen" had been a student of Joffe, casting a shadow over Joffe as well as Hessen.[78]*

At the 1931 congress, Kolman gave three papers: one on biology and physics, one on mathematics and one giving notice of the existence of Marx's then-unpublished manuscripts on mathematics, the natural sciences, and technology. He maintained that the dialectical materialist understanding of regularity was the key to steering a way through the Scylla of mechanistic fatalism and the Charybdis of indeterminism. He argued that the mechanistic conception of science as formulated most clearly by Helmholtz and the diametrically opposed conception as expressed most sharply by Heisenberg both had their basis in an undialectical conception of the relation between the general and the particular, the dynamic and the statistical. The second tendency, which emphasized particularity, uniqueness and unrepeatability, was exaggerated to the point where generality was transformed into an illusion. This view was gaining ground. Bourgeois science was growing more and more reactionary, moving irresistibly towards unconcealed fideism. Relativity was being taken as evidence of philosophical relativism and subjectivism. Quantum theory was being used to declare the overthrow of causality. Kolman referred to recent articles by such bourgeois scientists as Schlick, Milliken, and Eddington, to illustrate the point. The way beyond the crisis, beyond the extremes of fatalism and indeterminism in physics, beyond mechanism and vitalism in biology, beyond theories absolutizing either continuity or discontinuity in mathematics, was Marxism. The way out of the blind alley was dialectical materialism. This was the experience of the new generation of proletarian investigators of nature giving rise to a new liberated science. Labor and science were joining to give birth to dialectical materialism. The midwife was history itself, the revolution.[79]

---

* In time he would learn what it was like to be an "enemy" and would undergo considerable transformation in his way of linking politics to science and philosophy (cf. Kolman, "A Lifetime in Soviet Science Reconsidered," *Minerva*, Autumn 1978, pp. 416–424).

The Influence of Soviet Marxism Abroad

The Soviet contribution to the 1931 congress had a great influence on Marxist thinking abroad, not only on the British Marxists who were present. References to *Science in the Crossroads* abound in the writings of Marxists from far and wide, from France, Italy, America, and even Japan. The Japanese translation provided a certain inspiration for those who formed the Society for Materialist Studies in 1932 and later the History of Science Society of Japan at a time when communists were driven underground and even those who wished to explore the philosophical implications of Marxism risked arrest for offenses involving "dangerous thought." This was, for example, the fate of Saigusa Hiroto in 1933 and others such as Tosaka Jun and Agura Kinnosuke who pursued their interests in Marxist philosophy and the history of science under considerable pressure.[80] The papers in the volume all made the obligatory references to "social fascism," but showed that the communist movement still had within it searching minds capable of reaching out to others who were searching, even despite the sectarianism of the "third period" when the fortunes of the communist parties were at an extremely low point. Not that the book was received passively and uncritically by foreign Marxists, but, for the most part, they weighed the views of Soviet philosophers very seriously as coming from a land where new paths were being forged, where a gigantic human experiment was underway.

Gramsci's Marxism

Science in the Crossroads was one of the books that found its way into Antonio Gramsci's gloomy prison cell in Italy. He was far from uncritical of it. He weighed the views of Soviet Marxists on various questions, but always he did his own thinking. To have been able to apply his mind to such matters at all in his circumstances was itself a remarkable achievement. At his trial, the public prosecutor had declared: "For twenty years we must keep this brain from functioning." This they failed to do, but they put obstacles in its path that must have seemed at times insurmountable. It was some time before he was even allowed writing materials or books, and, when he was, his writings were scrutinized by the prison censor. In addition to enforced isolation from party and family life, Gramsci's always fragile health was put under severe strain by the harsh prison conditions and he was several times near to complete breakdown.

Indeed, his conditions of life, almost from birth to death, were exceedingly harsh. Nevertheless he made his way from his provincial Sardinian background to become general secretary of the Italian Communist Party. At university in Turin, he and Palmiro Togliatti, another Sardinian to become general secretary of the Communist Party, were caught in the grip of Crocean philosophy, which was at the peak of its influence in Italian intellectual life. They found the breadth and grandeur of Croce's thought inspiring, and it seemed as if the most vital and most progressive current stirring was the neo-idealist critique of positivism. They were also much influenced by Marxism, particularly by the great Italian Marxist of the previous generation, Antonio Labriola.

While in Turin, Gramsci joined the Italian Socialist Party. He worked as a journalist for the socialist paper, *Avanti!*, and soon became one of the city's socialist leaders. He welcomed the October Revolution in a way that highlighted the highly voluntarist nature of his Marxism at this time. He saw the revolutionary will of the bolsheviks as having defied Marx's schema, but as having drawn strength from what was living in Marxism. The life-giving ideas of Marx were those most in continuity with Italian and German idealist philosophy, even if they had become "contaminated by positivistic and materialistic incrustations."[81]

Gramsci was still very much a Crocean, although already a Marxist, and later, as he matured in his Marxism, he undertook to perform the same sort of operation on Croce that Marx had performed on Hegel. He came to see Croce's role in terms of the need for idealist high culture to adapt itself to the impact of Marxism by incorporating certain elements of it, in order to give itself a new elixir. Gramsci wished to write an "Anti-Croce," a sort of new "Anti-Dühring," that would unmask this operation. He came to criticize Croceanism as based on a purely conceptual dialectic devoid of concrete historical content. Even from his earliest days, however, when most under the influence of Croce, he never seems to have adopted the Crocean attitude to the natural sciences, which denied to them any cognitive value. He saw the natural sciences as integral to the great toil of the centuries that had perfected new methods of thought and had freed the human mind of prejudice and of philosophical and religious apriorism.

Gramsci had about him a great intellectual intensity, combining openmindedness with committed partisanship. In a pamphlet entitled "La Città Futura," published by the Young Socialist Federation in 1917, he wrote:

I believe that living means taking sides. . . . I am alive, I take sides. Hence, I detest whoever does not, I hate indifference.[82]

He was speaking in the context of taking the side of the "city of the future," in which socialists would not simply replace one order by another, but aim to satisfy the possibility of the integral fulfillment of the whole human personality as the right of all citizens. There was a great fullness in Gramsci, a disinclination to make artificial distinctions between rationality and emotion. He expressed his belief that it was impossible to know, to really understand, without feeling, without being impassioned. It was a theme he would take up again, and indeed one he would live by, explicitly arguing that strong passions were necessary to sharpen the intellect and to make intuition more penetrating. He continued to despise the shallowness of those who claimed to be impartial and insisted that only the man who willed strongly could rightly identify the elements necessary to the realization of his will.

At the end of the war, Italy erupted in a series of revolutionary strikes and entered a crisis that shook Italian society to its foundations. Gramsci became a leader of the factory council movement in Turin and in 1919 began to publish, along with Togliatti and others, *L'Ordine Nuovo* (the new order), a weekly socialist review that saw the councils as a new form of social organization and as the matrix of a new proletarian culture. Gramsci's revolutionary temperament was aptly expressed in the motto that appeared on every edition of *L'Ordine Nuovo*: "Pessimism of the intellect; optimism of the will."

In 1921, *L'Ordine Nuovo* became a communist daily, following the split in the Socialist Party, in which its editors ranked among the founders of the new Communist Party. Gramsci was a member of its central committee, although in the first years the party was dominated by the ultra-leftist and antiintellectual Amadeo Bordiga. In 1922, the year of Mussolini's March on Rome, Gramsci was sent to Moscow as the party's representative on the Comintern, and there he participated in the Fourth World Congress of the Comintern. At the end of 1923, he was sent to Vienna and in 1924 he was recalled to Italy. Having been elected a member of parliament, he had for a time immunity from arrest. In the Italian party, torn by factual strife, he had emerged as representative of a centrist position against the leftists, who believed that fascism was simply one more form of bourgeois rule and held a maximalist position of no transition between fascism and dictatorship of the proletariat, and the rightists, who took a minimalist approach of limitation of aims to the struggle for bourgeois democracy. The whole of the Comintern was torn by the factional struggle in the Soviet party and Gramsci expressed some disquiet about it all. Under his leadership, the Italian party sided with Stalin and Bukharin against the "United Opposition" of Trotsky and Zinoviev, believing the position of the latter endangered the worker-peasant alliance. Gramsci's letters conveying this position also expressed the view that party unity should be the result of conviction and not coercion, of persuasion and not expulsion. Togliatti, having taken Gramsci's place in Moscow, did not wish to convey these qualifications to the Soviet party.

Civil liberties were being eroded by the hour in Mussolini's Italy. Even as a member of parliament, Gramsci led a semiclandestine life in Rome. A new wave of repression came in October 1926. Although the party had made arrangements for him to go into exile, Gramsci decided to stay and to participate in the debate on the new emergency laws designed to suppress political opposition altogether. In November, he was arrested on his way to parliament. At the travesty of a trial before the Special Tribunal for the Defence of the State, he was sentenced to twenty years. He was sent to a prison in Turi and never again saw his wife or son. His second son, born after his arrest, he was never to see at all. As if the hardships brought by Italian fascism were not enough, there were also the strains brought by the course taken by the Comintern and by Italian communism. Although extremely isolated from political events, he learned of the Comintern policy of the "third period" and of the events ensuing in the Italian party as a result of it. In 1930, there was a purge of those opposed to the new Comintern line. There were violent discussions among the prisoners about the new line, which Gramsci strongly opposed. His brother, however, communicated to Togliatti that Gramsci was fully in agreement with the new policy and with the expulsions, believing this was necessary to save Gramsci himself from expulsion. It would also seem, from the testimony of the Italian Marxist and Cambridge economist Piero Sraffa, that Gramsci was somewhat sceptical about the Moscow trials. By 1937, his health was utterly broken from the prison régime and at the age of forty-six he was dead.*

Gramsci's greatest contribution was made during his difficult prison years. From 1929 to 1935 he filled thirty-two notebooks with his reflections on history, literature, politics, philosophy, science, and a number of other subjects.** Until 1926, his writings had been rather directly tied to day-to-day politics and were more of a journalistic nature. However, in prison, cut off from the world of current political events, his writings assumed a deeper, broader and more theoretical character. He did miss the activity of his former life.

As a thinker, Gramsci has been linked with the trend associated with his contemporaries, Lukács and Korsch. There were indeed a number of parallels. Common themes included the unity of theory and practice, the notion of totality, renewed interest in the Hegelian origins of Marxism, a strong emphasis on revolutionary consciousness, and on the role of ideas and of will in human history, as well as a great stress on the critique of positivism and on positivist incrustations within Marxism. Gramsci, however, dealt with these matters far more astutely and more concretely, although on some issues not

---

* One of the saddest comments about Gramsci has been made by Eric Hobsbawm: that Gramsci was saved from Stalin by Mussolini (cf.,"The Great Gramsci," *New York Review of Books*, April 1974). Others have made the point that Gramsci was saved to the communist movement only by his imprisonment. Had he gone into exile, he could have been an outcast of the Comintern. If he had gone into exile in Moscow, he could have met his death there (cf. Kolakowski's *Main Currents of Marxism* 3 [Oxford, 1978], p. 228).

** In an article on "Gramsci and Marxism" Victor Kiernan makes the point that Gramsci could not know whether anything would come of his prison toil—"a world of ideas hung by a thread, like Catullus's poems surviving for centuries in a single copy in a monastic vault" (cf. *Socialist Register*, 1972).

altogether satisfactorily. He too left some loose ends hanging, though nowhere near so many as Lukács and Korsch. His vision was far more far-reaching and more concretely grounded than theirs came anywhere near to being.*

Gramsci's view of the role of revolutionary consciousness was of a more popular and organic nature than that of the others, and in his political perspective he managed to transcend the extremes of putschism and spontaneism. The experience of the Russian revolution, in his view, could never be repeated in an advanced industrialized society. The tsarist state was not founded upon consensus and its civil society was primordial and gelatinous. However, the nature of bourgeois rule under advanced capitalism was such that its strength was not based on its coercive state apparatus alone, but on its world view having become "common sense." This made the crucial task of the party the construction of a new proletarian world view to replace the bourgeois one and the struggle to win consent for its new ways of thinking and living, its new culture and new morality. This was the meaning of his concept of "hegemony" for which he has now become so renowned and which lies at the heart of his originality and importance as a Marxist thinker: his insistence that Marxists must seek to gain intellectual and moral ascendancy within the institutions of civil society, building the new within the shell of the old, rather than concentrating simply on the seizure of state power. It was necessary for communists to achieve cultural hegemony before taking political power. Only in this way could the liberal consensus really be broken. Only in this way, with revolutionary transformation embracing the whole of life and the result of active popular consent, could socialism really be built.

This way of thinking implied the ascription of a rather important role to intellectuals in the revolutionary process. He followed the logic of this and devoted quite a lot of attention to an analysis of the role of intellectuals. Gramsci believed it was of central importance for the communist movement to overcome both the elitism and political isolation of intellectuals and the antiintellectualism of sections of the working class. In one sense, everyone was an intellectual. In terms of social function, intellectuals were those who brought a new degree of critical elaboration to the intellectual activity existing in everyone. Of these, there were two main types: the traditional intellectuals, whose position in the interstices of society made them seem autonomous and independent of class, but whose position was derived ultimately from past and present class relations; and the organic intellectuals, whose function was to be the articulating and organizing force of the consciousness of a particular social class. The working class was creating its own intelligentsia, a new intelligentsia

---

* This would seem to concur, at least to some extent, with Luckács's own later judgement. In the 1971 *New Left Review* interview devoted to his life and work, he remarked that in the 1920s, he, Korsch, and Gramsci tried to come to grips with the problems facing Marxism at the time but "we all went wrong," though Gramsci, he said, was perhaps the best of them.

arising organically out of the working class and remaining closely tied to it, working out and making coherent the principles and problems raised by its historical movement and striving to win over the traditional intelligentsia. The new intelligentsia would give coherent and systematic expression to popular awareness as a basis for precise and decisive will.

The culmination of this process could be the emergence of a great individual philosopher. The activity of particularly gifted individuals marked the high points of the progress made by common sense. This philosopher of a new type thrown up by the revolutionary movement would have, on the one hand, the flexibility of movement of the individual brain and, on the other hand, the capability of realizing concretely the needs of the mass movement and of being able to address it in the most relevant fashion. Ultimately, however, the new historical synthesis was the product of the whole movement of history, not the individual philosopher. In the end, mass adherence or nonadherence to a philosophy was the critical test of the rationality and historicity of various modes of thought. Arbitrary constructions went by the wayside in the historical process, even if sometimes, through a combination of immediately favorable circumstances, they enjoyed a temporary popularity.

Philosophy loomed very large in Gramsci's view of things. It was inseparable from politics. He saw philosophy as the organized critical reflection of the forms of thought of the old social order and as the organizing principle of the new social order. Philosophical activity was a battle to transform the popular mentality. It was the growing edge of "common sense." The ruling class exercised hegemony through the dominance of its definition of reality of which philosophy was the highest level of elaboration. It was therefore necessary for the working class to struggle on the philosophical front if it were to come to exercise hegemony. Philosophy was rooted in the historical process and integrally connected to the development of the social recognition of the social relations of production. Marxism, in its conscious and explicit recognition of the social relations of production, brought a new intelligibility to the philosophical enterprise. It brought philosophy onto a new terrain. For Gramsci, philosophy had a class character, for there could be no philosophy independent of the social relations in which man lived and thought. It was either bourgeois or proletarian, for there could be no universal philosophy except in a classless society. In elucidating the problem of transition, that is, the trajectory of proletarian class consciousness in its development from the world view of a subaltern class to the world view of a ruling class, Gramsci held that it began as a critical-polemic consciousness and, through its historical activity to transform the social relations of production, it grew into a new synthesis, a new civilization. Marx initiated the critique and discerned the possiblilty of a new phase of historical development as a necessary basis for a new conception of the world, a unitary conception that would overcome all antinomies plaguing human thought for centuries. Lenin, through the October

Revolution, brought Marxism beyond the phase of critique into the new phase, the phase of the historical realization of the philosophical critique, the phase of the positive creation of a new civilization. For Gramsci, no "return to Marx" was possible. Marxism henceforth was to develop out of the innovative historical activity initiated by the revolution.

As Gramsci saw it, everyone was in some sense a philosopher in that everyone functioned with a conception of the world, whether a critical and coherent one or a disjointed and episodic one. Philosophy could not be separated from the history of philosophy. Sometimes a person's pattern of thinking juxtaposed stone age elements with the principles of the most advanced science. The starting point for a critical approach to philosophy was to assess this, to "know thyself" as the product of a historical process that had deposited an infinity of traces without leaving an inventory. It was necessary to order in a systematic, coherent and critical fashion one's own intuitions about the world and to do so in the context of the history of philosophy, within the collective effort of the centuries, to elaborate viable forms of thought. It was necessary for Marxists to achieve a critical equilibrium in relation to the whole of past and present thought, to restructure all knowledge and to bring it to a new historical synthesis.

Needless to say, Gramsci saw Marxist philosophy within the history of philosophy. He situated it not only in terms of the philosophical systems of the past, but also within the spectrum of contemporary philosophical standpoints, as well as within the context of the different philosophical views enunciated within the Marxist tradition itself. He did so in a way that was often illuminating, but on the whole somewhat lacking in perspective. His orientation in this sphere was decidedly one-sided. While affirming that Marxism was the crowning point of all of the previous history of philosophy, he elaborated on this in a very one-sided way, making much of it as a reform and development of Hegelianism and virtually ignoring its roots in French materialism and in the development of modern science. With regard to current philosophies, he was sometimes weak in his criticism of neo-idealism and excessively severe in his criticism of positivism. Within the terms of Marxist debate, he conceded too much ground to idealist tendencies and overstressed the dangers of positivist tendencies. He greatly exaggerated the strength and extent of the mechanist tendency within Marxism. Moreover, he left a yawning gap in his analysis of the state of the question by not really addressing himself at all to the mainstream current in Marxist philosophy. For example, while he considered himself a Leninist and discussed Lenin's views on many topics, he never commented on Lenin's philosophical work. By implication, he seemed to con-sider all trends in Marxism adhering to a realist epistemology and an emphatically materialist ontology to come within the scope of "vulgar Marxism." He never distinguished this trend from the positivist and mechanist trend within Marxism with which he was so preoccupied.

Gramsci's main philosophical adversary was Bukharin. Of all Marxist theoreticians, it was Bukharin whom he singled out for special attention as the very embodiment of "vulgar Marxism". Bukharin for his part had taken a critical attitude to Gramsci's Marxism. In his speech on the problem of the ideological unity of the communist movement at the Fifth World Congress of the Comintern in 1924, he addressed himself not only to the trend represented by Lukács and Korsch, but referred to a "voluntarist idealism" in the Italian party tendential to the rebirth of the "old Hegelianism". Gramsci in time took up a far more elaborate criticism of Bukharin's Marxism.

Gramsci based his attitude not only on Bukharin's earlier manual, *The Theory of Historical Materialsm*, but also on his paper from the 1931 London congress. He expressed his opinion that the 1931 paper did not indicate any significant change in Bukharin's position, in spite of the Soviet debate on mechanism that had taken place in the ensuing years. His judgment of Bukharin was not an altogether fair one, although many of his points with regard to the earlier manual were well-taken.

In general, Gramsci believed that Bukharin had emphasized materialism to the neglect of the dialectic. He took exception to what he saw as a sterile attempt to reduce Marxism to abstract and ahistorical causal laws, to a scheme of flat evolutionism that left no room for will, no room for leaps. This led to fatalism, a delusion undermining political initiative. The revolutionary movement was particularly susceptible to it at times of defeat, when it had lost the initiative in the struggle. Mechanistic determinism could lull the movement into a false sense of security with the belief that, despite temporary defeat, the tide of history was flowing inexorably towards its goals in the long term. It was a form of arid mysticism that became a cause of passivity and idiotic self-sufficiency. Fatalism was the clothing worn by active will when in a weak position.

For Gramsci, the experience on which Marxism was based, that is, historicity in all its infinite variety and multiplicity, could not be schematized in the fashion of Bukharin. Historical events could not be described and classified according to criteria appropriate to the phenomena studied by the natural sciences. The so-called sociological laws resulting from this type of procedure were almost always tautologies and paralogisms with no causal value. They amounted to no more than duplicates of the observed facts themselves. The only novelty was in the collective name given to a series of petty facts. An excessive reliance on statistics favored mental laziness and superficiality. Perhaps the most fertile contribution of Marxism, said Gramsci, was its way of connecting quality to quantity, which Bukharin had reduced to mere wordplay about water changing its state, a mechanical process determined by external agents.

Gramsci's own approach to the Marxist theory of historical materialism was that its canons were only of value *post factum*. With regard to the present and the future, its was necessary to engage in the most detailed analysis of specific factors. Opposed to the linear derivation of superstructure from structure, to the causal explanation of the complex and multiple aspects of historical processes by simple reference to the economic base, he exhorted Marxists to take note of the concreteness and specificity of historical events, the field of possibilities, the multiplicity of factors coming into play, the contrasts, the countervailing forces. Nevertheless, however firm in his critique of economism and his insistence on the complexity of historical processes, he still held to the predominance of the economic in the overall process. The identification of the "cathartic moment," the form of the transition from the economic to the political, was the starting point for Marxism.

Gramsci thought that Bukharin had conceived of the relationship between Marxism and the natural sciences in the wrong way. Bukharin, he said, had failed to give Marxism its proper scientific autonomy and the proper position due to it in relation to natural sciences. Modern atomic theory was a scientific hypothesis that could conceivably be superseded in the historical process of the development of science. Natural science was a form of historical activity. For Marxism, then, it was not atomic theory that explained human history, but human history that explained atomic theory.

Although Gramsci opposed what he considered to be a "near fetishism" with regard to the position of the natural sciences within Marxism, there was in him no trace of the hostility toward the natural sciences that was to be found in Lukács or Korsch. His attitude to the natural sciences was generally a positive one, although at times somewhat ambiguous. His random remarks on physics and relativity theory were exceedingly obscure, revealing his remoteness from the world of the natural sciences and his lack of awareness of the actual state of discussion of scientific theory. Nevertheless, he saw science as necessary to man in order to discern order in the world. He believed that scientific experiment had inaugurated a new form of active union between man and nature, a new type of dialectical mediation between the human and the natural.

The problem of the relation between the sociohistorical world and the natural world was a theme Gramsci raised a number of times. He objected to the reductionist application of the methods of the natural sciences to the realm of the social sciences, but affirmed an underlying continuity. This was the basis of his criticism of Lukács:

> It would appear that Lukács maintains that one can speak of the dialectic only for the history of men and not for nature. He might be right and he might be wrong. If his assertion presupposes a dualism between nature and man, he is wrong, because he is falling into a conception of man proper to religion and to Graeco-Christian philosophy and also to idealism which does not in reality succeed in unifying and relating man and nature to each other except verbally. But if human history should be conceived also as the history of nature (also by means of the history of science) how can the dialectic be separated from nature? Perhaps

Lukács, in reaction to the baroque theories of the *Popular Manual*, has fallen into the opposite error, into a form of idealism.[83]

Not for Gramsci any neo-Kantian chasm between nature and history. So radically had nature been historicized in his thinking that any methodological rift between naturalistic and humanistic knowledge was inconceivable. He did take great pains to distinguish these two forms of knowledge and was extraordinarily anxious to establish the specificity of humanistic knowledge, but it was within an overall continuity. The continuity was perceived, however, from exactly the opposite point of view of such thinkers as Engels and Lenin. For Gramsci, the history of nature was subsumed under the history of man rather than vice versa.

Not surprising then was his opposition to all realist and objectivist epistemologies. He thought Bukharin's polemic against subjectivism at the 1931 London congress to have been badly framed and worse conducted. He considered it futile, superfluous, and even reactionary. It was too tied to uncritical common sense. It was based on a popular belief that was religious in origin. Appealing to it was an implicit return to religious feeling. The fact that the masses found the subjectivist conception fit only for mockery was a reflection of the distance between science and life, between intellectuals and the masses. The purpose of science was to criticize common sense, the immediate product of crude sensation, still often Ptolemaic and anthropocentric. With Bukharin, the most vulgar common sense was imposing itself on science rather than vice versa. The subjectivist conception, proper to modern philosophy, had its usefulness as criticism of transcendence on the one hand and the naiveté of common sense and of philosophical materialism on the other. It had given birth to, and was superseded by, historical materialism. In other words, with Marxism, subjectivism had been transformed into historicism.

What this actually meant in terms of a theory of knowledge was not altogether clear. Gramsci wished to distinguish his position from solipsism, scepticism, or relativism. On the question of the objectivity of the external world, he tended to skirt the issue or to make contradictory assertions about it. He criticized Bukharin for accepting this belief "in its most trivial and uncritical sense" and "in such a mechanical way." Proceeding from this, he went on:

It might seem that there can exist an extra-historical and extra-human objectivity. But who can judge of such objectivity? Who is able to put himself in this kind of "standpoint of the cosmos in itself" and what could such a standpoint mean? It can indeed be maintained that here we are dealing with a hangover of the concept of God, precisely in its mystic form of an unknown God. Engels's formulation that "the unity of the world consists in its materiality demonstrated by the long and laborious development of philosophy and natural science" contains the germ of the correct conception in that it has recourse to history and to man in order to demonstrate objective reality. Objective always means "humanly objective" which can be held to correspond exactly to "historically subjective": in other words, objective would mean "universal subjective."[84]

However, he was equivocal. On the one hand, he asserted that to speak of a reality existing apart from man was to fall into a form of mysticism. On the other hand, he admitted that it was hard to imagine reality changing objectively with changes in ourselves, a belief that not only common sense but science made untenable. He also acknowledged that it was difficult not to think in terms of something real beyond our knowledge, not in the sense of a *noumenon* or an unknown god or an unknowable reality, but in the sense of something still unknown, that will some day become known when the physical and intellectual instruments of mankind have been perfected. Again, in discussing the nature of natural science, he saw it as historical activity and indeed saw nature itself as a historical category. "Up to a certain point," he asked, might it not be said that what nature provided the opportunity for was not discoveries of pre-existing qualities of matter, but creations linked to social interest? Fair enough, perhaps, but when it came to what was beyond the "certain point," he would not not quite say. He did concede that "as an abstract natural force," electricity existed before its reduction to a productive force, but added that it was in a state of historical "nothingness."

In a discussion of the epistemological foundations of science, inspired by the pronouncements of some Italian scientists who were taking up a position similar to Eddington in the philosophy of science, Gramsci repudiated the "grotesque way of thinking of certain scientists, especially English, in connection with the 'new' physics". It was a way of thinking that Gramsci believed meant that science ceased to exist and was changed into a series of acts of faith. If it were true that microscopic phenomena could not be considered as existing independently of the observer, if it were true that such phenomena were not actually "observed" but "created", then, Gramsci said, it was not even a matter of dealing with solipsism but witchcraft. The problem was that the crisis in physics, an initial and transitory phase of a new scientific epoch stemming from the difficulties scientists were encountering in trying to explain microscopic phenomena in macroscopic terms, was converging with a great intellectual and moral crisis and bringing forth a great wave of sophistry. Coming to terms with this was, however, serving to refine the instruments of thought. The most important question to be resolved was whether and how science yielded objective knowledge. Yes, answered Gramsci, despite all the difficulties and complications, phenomena recurred and could be ascertained by observers independently of one another. Science brought forth objectivity, in the sense of establishing what was common to all observers, what was independent of every particular point of view. However, having asserted the universality of science, he backtracked:

> Actually, even this is a particular world view, an ideology. However, this view, as a whole and for the direction it indicates, can be accepted by the philosophy of praxis, whereas one should reject the view of common belief, even though it reaches the same materialist conclusion. . . . All of science is tied to the needs, the life, the activity of man. Without human activity, which creates all values, including scientific ones, what would "objectivity" be? Chaos. . . . For the philosopher of praxis, being cannot be divorced from thinking, man from nature, activity from matter, the subject from the object. . . . To place science at the base of life, to regard science as the world view par excellence, as the one that clears the eyes from every ideological illusion, and which places man in front of reality as it is, means to fall again into the idea that the philosophy of praxis needs philosophical props outside of itself. But in reality, science, too, is a superstructure, an ideology. . . . That science is a superstructure is shown also by the fact that there have been entire periods of eclipse, for example, when it was obscured by another dominant ideology, religion. . . . Moreover, in spite of all the efforts of scientists, science never presents itself as naked objective information; it always appears coated by an ideology and concretely science is the combination of the objective fact with a hypothesis or a system of hypotheses that surpass the mere objective fact. It is true, however, that, in this domain it is relatively easy to distinguish the objective information from the system of hypotheses by a process of abstraction that is inherent in scientific methodology itself, in such a way that we can accept the one and reject the other. That is why a social group can accept the science of another group without accepting its ideology.[85]

One feels Gramsci was groping in the right direction, but such a passage is so riddled with conceptual ambiguity as to serve only to muddy the waters. One is left wondering: Is there or is there not any distinction between science and ideology? Is science to be taken up as integral to Marxism or is it an extraneous element threatening its integrity? Is objectivity reducible to collectivist subjectivity or is it not? It is impossible to state firmly any clear position held by Gramsci on such issues, so often did he shift premises and use the same terms equivocally.

Gramsci was also somewhat equivocal when it came to materialism, using the term in different ways in different contexts. Most often he used it in a negative way. He preferred to see Marxism, not in terms of materialism versus idealism, but in terms of immanence versus transcendence. It was a way of explaining the world from within the world on the basis of the world itself, without recourse to any power outside it to justify the power within it. Marxism was a new form of immanence. It was immanence become historicism, immanence brought onto the concrete terrain of history. Marxism, he insisted, superseded both materialism and idealism, while retaining the vital elements of both. These traditional philosophies were expressions of past societies. It was wrong to consider Marxist philosophy in subordination to other philosophies, as this obscured its originality. It was a new synthesis, embodying a new dialectic, opening up a new road, "renewing from head to toe the whole way of conceiving philosophy itself." Unfortunately, the dialectical unity achieved by Marx had been destroyed by his successors, who had regressed either to philosophical materialism or to philosophical idealism.

Gramsci's term for Marxism was "the philosophy of praxis." Commentators invariably explain that he called it thus to evade the prison censor. Undoubtedly this was so, but the choice of term was by no means arbitrary, for it was well suited to his radically historicist conception of philosophy. As he defined the new conception:

> The philosophy of praxis is absolute historicism, the absolute secularization and earthiness of thought, an absolute humanism of history.[86]

It was a monism, not of materialism or idealism, but of the dialectic embodied in the concrete historical process. The dialectic was not a pre-constituted framework for philosophy, but the very marrow of historiography. It was the logic of connections and mediations, in which all spheres—the philosophical, the economic, the political—were interwoven into an organic unity.

In Gramsci's view, Marxism had about it an integrity and coherence that made it unnecessary to supplement it with any other philosophy. It stood on its own without extraneous additions. He specified tendencies with which Marxism should not enter into combinations. He was hostile to all attempts within the Second International to combine Marxism with such philosophies as neo-Kantianism, and he criticized the philosophical variations in Bernstein and in Max Adler. He took strong exception to Otto Bauer's assertion of the compatibility of Marxist economics with Thomist epistemology.

He particularly warned against the reabsorption of Marxism within the framework of traditional culture or the reemergence within Marxism of tendencies over which it had already triumphed. At the same time, Marxism had a unique capacity to assimilate all that was of value in all other systems of thought, to weave all such elements together in a new synthesis. It was an organic absorption to be sharply distinguished from an eclectic borrowing. Marxism matured and developed always to a higher level through this continuous interaction with other philosophies. It was an ongoing process. Even with philosophies already assimilated within Marxism, Gramsci held this was not a process finished once and for all. Gramsci believed that the Hegelian element in Marxism in particular had not been fully subsumed or resolved into it and retained a vitality of its own. Marxism grew and developed as a historical process still unfolding, in which the necessity for a philosophical-cultural synthesis was constantly renewed.

But its integrity and coherence involved more than not adding on anything extraneous. Even more importantly, it meant not leaving out anything essential. Its constituent parts were so interwoven that none could be negated without destroying its organic unity. For Gramsci, Marxism was an all-embracing *Weltanschauung* that left nothing vital outside its scope. The natural sciences were part of the totality, a part of the structure that could not be omitted without doing violence to the rest. Orthodoxy lay in the conviction that Marxism

> contains in itself all the fundamental elements needed to construct a total and integral conception of the world, a total philosophy and theory of natural science, and not only that, but everything that is needed to give life to an integral practical organisation of society, that is, to become a total integral civilization.[87]

It was this sense of the wholeness of Marxism, the fullness of it as well as the concreteness of it, that more than anything set him off from his contemporaries, Lukács and Korsch. Despite the fact that he did not really achieve clarity with regard to certain of the constituent elements or even see the various aspects in the proper proportions, he nevertheless realized that no philosophy could maintain itself and develop with anything fundamental left out of account. Nothing vital could be outside the scope of Marxism. Despite Gramsci's limitations and ambiguities and blind spots, he was still among the greatest Marxists of his age for his affirmation of the all-embracing character of Marxism and for his highly original and spirited way of highlighting various aspects that were being neglected.

Not until after the war did his *Prison Notebooks* become known and not until the 1960s did they become known outside of Italy. Since this time, the secondary literature has mushroomed. Today congresses are held to discuss his views. The list of books and articles about him is enormous and ever-growing. And it is not only in Italy.

The interpretative literature differs vastly, particularly when it comes to his views on philosophical questions. This is not altogether surprising in light of the fragmentary character of his prison writings, the obscurity of his utterances on materialism, the natural sciences, realism, objectivity, and other such pivotal concepts, as well as the radical differences in orientation among the commentators themselves. Predictably, for example, the Althusserian School has criticized him for historicism, and the Dellavolpean School has taken issue with him for neo-Hegelianism. The English-speaking New Left has tended to read its New Leftist prejudices into him and to embrace him warmly. The Italian New Left has tended to read their antipathy to the Italian Communist Party into him and to have rejected him in disdain.

Despite his own wishes, commentators have read him as either materialist or idealist. Those who argue the former are somewhat nearer the mark than the latter, but the argument on both sides is often weak and evasive, ignoring many inconvenient passages. It is, however, generally correct to contend that Gramsci was not so antimaterialist as various passages taken out of context might indicate. To some degree the problem is a terminological one, in that he had an obvious hostility to the term "materialism," but clearly operated with certain basically materialist assumptions. For Gramsci, the world was to be explained in and through the world itself with no resort to any force outside the world. On this, he was absolutely firm. Moreover, with all his emphasis on consciousness, it was never discussed in a disembodied and disconnected way. Always it was infused with living, concrete historicity. On the other hand, he did undoubtedly give too much ground to idealism, in overstressing the historical roots of Marxism in idealism, in being overzealous in his critique of positivism, and indeed of much that was in reality not positivism at all, and in equivocating in the sphere of epistemology. His philosophical orientation was decidedly one-sided, giving prominence to those features of Marxism that were in continuity with the idealist tradition in the history of philosophy and under-playing those features that were in continuity with the materialist tradition. He could not quite bring himself to acknowledge fully what Timpanaro calls the element of passivity in knowledge, the fact that there is an element in knowledge that is irreducible to praxis, thus conceding too much to idealism and remaining in some respects an unconscious Crocean.[88]

There was moreover the factor that Gramsci's views on the natural sciences within Marxism, so crucial to the historical development of the materialist emphasis in Marxism, were definitely underdeveloped, as his actual knowledge of the natural sciences was not so highly developed as his knowledge of other fields. If he criticized Bukharin for a neglect of the dialectic and of history in favor of concentration on materialism and emphasis on the natural sciences, Gramsci could be criticized for the contrary, a neglect of the natural sciences and of the elements underlying the materialist emphasis in the Marxist tradition in favor of his stress on the dialectic of historical activity. Nevertheless, it cannot be overstressed, in light of the tendentious reading of Gramsci by writers themselves hostile to the natural sciences, that Gramsci's critique of positivism was not a critique of the natural sciences. Paolo Rossi is correct to emphasize the importance of the fact that Gramsci did not identify positivism with scientific method. He never denied the cognitive value of science and, in fact, saw in science, or in a philosophy connected with science, a central element in a *Weltanschauung* capable of overcoming religious superstition. The scientific revolution was part of the heritage of the proletariat.[89]

Although Gramsci did not contribute to the development of Marxist philosophy of science in the sense of elaborating more fully the philosophical implications of the results of the natural sciences, he was nevertheless important to the history of the development of Marxism in relation to the philosophy of science in his way of conceiving the science-philosophy-politics nexus within Marxism. He affirmed against those who denied it the importance of the natural sciences to Marxist philosophy and to socialist politics, seeing it as playing a central and crucial role in the process of the revolutionary transformation of society. His greatness was in drawing attention to the fullness of the revolutionary process, including in it philosophy, science, psychology, and culture, and in analyzing the vital role of these vis-à-vis the struggle for state power. Within this framework, he generated many provocative ideas and proposed many fruitful avenues of approach to many important questions, such as the integrality of Marxist philosophy and its relation to other philosophies, the political character of philosophy, the fusion of rationality and emotion. Gramsci's Marxism may have been in some respects an instrument with rough edges, but it penetrated very deeply.

The Red Decade: Britain and France in the 1930s

Meanwhile, others were coming along who would take up the development of Marxism in relation to the philosophy of science in a new way. The early 1930s saw a revival of interest in Marxism and the birth of a new left-wing intelligentsia in Western Europe that would leave its own distinctive mark on the theory it came to embrace. Although the Comintern policy of the "third period" had been an unmitigated disaster, however reluctant communists were to say so, the communist vision of a new world nevertheless shone through in sharp relief to the darkening reality of crisis-ridden capitalism. As the boom of the 1920s crashed down into the slump of the 1930s, the capitalist world was overshadowed by a harrowing depression that brought massive unemployment and underemployment, cuts in education and social services, and general economic insecurity. Concomitant with the economic crisis and the accompanying struggle for political power came a deep-seated intellectual crisis that made itself felt at every level of culture and in every field of knowledge. It was widely believed that capitalism was in a dying but dangerous state. Indeed, the militant rise of fascism highlighted just how dangerous it could be. So it was that the freshest and ablest minds of this generation set out to analyze the dying culture amidst what they believed to be its death throes and to seek after an alternative vision, one of a living culture that would give full scope to their burning creativity and their bright hopes. Thus began the "Red Decade," with workers rebelling against the dole queues and intellectuals leaving their private dreamlands and coming onto the streets, living and thinking and dreaming in a whole new way. No longer were students in the public eye strikebreaking as in 1926, but welcoming the hunger marchers into Cambridge and striving with the working masses to find a better way.

For their part, the communist parties, most notably in Britain and France, rose to the occasion. Without any official change of policy, they began to back down from the violent sectarianism that had brought communist parties into such severe isolation and so near to extinction. The rising militance of the working class and the growing radicalization of the intelligentsia were met with a broader, somewhat more accommodating attitude toward the role of mass organizations and towards the role of intellectuals in the revolutionary movement. Responding to the new forces that were at work, communists had the clarity and the discipline to spearhead the revitalized left-wing movement emerging and thus managed to bring a steady stream of extremely vital and high-powered intellectuals within the orbit of the communist parties.

In France, the surrealist writers Louis Aragon and Paul Eluard came into the Communist Party. Among scientists, the young physicist Jacques Solomon joined the party, while the most eminent physicists of the day, Paul Langevin and Fréderic Joliot-Curie, drew ever nearer to the party and eventually joined it. Already a member was the eminent biologist Marcel Prenant, who was not only professor of zoology at the Sorbonne, but lectured at the Université Ouvrière as well. The young philosopher Paul Nizan joined the party while still a student at Ecole Normale Supérieure, and, together with those philosophers who formed the Philosophes group, Henri Lefebvre, Georges Politzer, Georges Friedmann, contributed to the development of a school of Marxist philosophy commanding the attention of the French intelligentsia. Philosophers and scientists came together in the *cercle de la Russie neuve* to discuss the implications of Marxism for philosophy and the sciences. Philosophers such as Auguste Cornu and scientists such as the psychologist Henri Wallon and the astronomer Henri Mineur discussed the many contending intellectual trends of the day in the light of dialectical materialism. They all participated in the various conferences at the *maisons de la culture* that drew an increasing number of intellectuals, including philosophers and natural scientists, under communist influence.

In Britain, a Marxist intellectual culture emerged for the first time and the Communist Party poised itself to play a central role in it. For the first decade of its existence, the British party had been much affected by the traditionally pragmatic trade unionism, and even overt antiintellectualism, of the British labor movement. The movement had not only been negligent with regard to underlying theory, but its leaders were even prone to boast that it had no philosophy. Suddenly, however, it all changed and Britain not only became cognizant of the value of Marxist theory, but became a most fruitful and vigorous center for its creative development in the 1930s. Among the many who felt the power of communist ideas at this time—though not all were members of the Communist Party—were writers such as W.H. Auden, Stephen Spender, Christopher Isherwood, Cecil Day Lewis, and Louis McNeice; critics such as Margot Heinemann, Jack Lindsay, Arnold Kettle, and Ralph Fox; historians such as Christopher Hill and Eric Hobsbawm; classicists such as Benjamin Farrington and George Thomson; lawyers such as D.N. Pritt; journalists such as Claud Cockburn; future party functionaries such as James Klugmann; and future spies such as Guy Burgess, Donald Maclean, and Kim Philby. There were even clergymen. The Dean of Canterbury, Hewlett Johnson, was a fellow-traveller, as was the Red Vicar of Thaxted, Conrad Noel. The Reverend Dr. John Lewis, organizer of the Left Book Club, eventually became a party member and a Marxist philosopher to boot. Among philosophers, Maurice Cornforth and David Guest, who were among the privileged circle who sat at Wittgenstein's feet at Cambridge, turned to Marxism and to the Communist Party. There was also the brilliant and many-sided Christopher Caudwell—poet, philosopher, critic, journalist, and much more. Among scientists the turn to the left was truly spectacular. Most prominent were J.D. Bernal and J.B.S. Haldane, both Fellows of the Royal Society, who became communists and looked to Marxism in matters of philosophy of science. Others who formed a part of the radical science movement of the 1930s were P.M.S. Blackett, E.H.S. Burhop, J.G. Crowther, Hyman Levy, Sam Lilley, Joseph Needham, N.W. Pirie, C.H. Waddington, W.A. Wooster, and Lancelot Hogben.

Not only was the alliance of philosophers and natural scientists recommended by Lenin being achieved, but the development of Marxism was giving rise to a new type in men such as Bernal, Haldane, Langevin, and Prenant, who combined within themselves scientist, philosopher, and political activist all in one. The shifting nexus of science, philosophy, and politics was brought to a new level as such men, who were among the outstanding scientists of their day and on the cutting edge of their disciplines, began to proclaim that only in dialectical materialism could their science find an adequate philosophical grounding and that only under socialism could science achieve its full social potential.

Philosophy of Science in the 1930s

In turning to socialism, scientists were most definitely going against the grain, particularly those in Britain, living and working among colleagues who thought it most unseemly for scientists to involve themselves in philosophy and in politics, most of all in Marxist philosophy and in communist politics. But reality was pressing hard at scientists who did not wish to consider the implications of their science beyond the details of their laboratory results. Science was pushing ahead dramatically. Cambridge was buzzing with news of the neutron, the positron, and the splitting of the atom. Still unsettled were the controversies over the meaning of quantum theory and relativity theory. The world wanted to know what it all meant and there was no shortage of theologians or chancers leaping in to fill the gap and tell it.

The widespread discussion of the methodological implications of quantum mechanics and of relativity were not confined to the groves of academe. These discoveries had set off the wildest speculations in the mass media and in the 1930s the air was still alive with it. Much of it was centered around Heisenberg's principle of uncertainty, which stated that it was impossible to determine simultaneously with total accuracy both the position and the velocity of a particle. This principle was founded on a hypothetical experiment involving the gamma-ray microscope, in which the very act of observing a particle drove it from the position it would have occupied unobserved. This emphasis on the role of the observer seemed to call into question the existence of reality independent of the observation of it and to favor Machism and all forms of subjective idealism. The impossibility of predicting exactly the path of a particle raised questions about the existence of determinism, about the very operation of causal laws in science. To some, it seemed that the new discoveries proved there was a radical indeterminism at the heart of the universe; that the principle of causality had been utterly discredited; that all phenomena were of a strictly subjective nature and that there could be no grounds for asserting the objective existence of an external world; that it was completely beyond the scope of any form of materialism to account for such developments.

There was a veritable barrage of popular books, radio broadcasts, and Sunday sermons announcing that science itself, once the iconoclastic destroyer of all that was holy, had come to its limits and there found God again. Concepts applicable at the microscopic level were amateurishly extended far beyond their domain of applicability to demonstrate that there was neither rhyme nor reason to the way things happened in the universe; that all was vanity; that, therefore, God and mystery and free will reigned supreme. Loudly and triumphantly, the obituaries of materialism were once again published for all the world. Indeed, religion, whose death knell had also been previously sounded, did receive a new lease of life. Never had the idea of progress fallen upon deafer ears. The social fabric was ripping apart, the ground was crumbling under one's feet, and science seemed to be raising more questions than it answered.

To make matters worse, there were influential scientists joining the chorus, lending credence to those who were seizing upon the crisis in physics to justify their own philosophical prejudices and suggesting that natural processes must once again seek their explanation in supernatural ones. Arthur Eddington boldly declared: "Something unknown is doing we don't know what—that is what our theory amounts to," and argued from there that religion had therefore become acceptable to the scientific mind. The scientific world was built simply out of mathematical symbols drawn from the human mind. The inference from this was the existence of a universal mind or *logos*.[90] Sir James Jeans was on much the same tack. Science no longer rested on causality and determinism, but on free will and the mind of a mathematical God.[91] It was the sure way to a best seller.

Other scientists, in contrast, thought it a disedifying spectacle: scientists utilizing their scientific knowledge to tell all the world that unreality lay at the basis of science, utilizing their rational powers to lend credence to the view that irrationality was triumphant. Marxists would have none of it and indeed were in the frontlines of opposition to it. They argued that materialism stood on as strong a foundation as ever, and they objected fiercely to such attempts to parasite on every breakdown of traditional concepts in science to bring God back into the picture. While it had certainly become necessary to abandon determinism in the Laplacean sense,* it was impermissible to conclude that a world of absolute indeterminism was the only possible alternative to a world of absolute determinism. In a world of absolute indeterminism, in a world without causality, no science would be possible. All possible outcomes of a given physical state would be equally probable. But this was obviously not the case. It was possible to establish statistical probability for the microscopic phenomena described by Heisenberg and so there was causality. Probablistic causality was a far cry from indeterminism.

The debate was at its most intense in Britain. From here came the most prominent theoreticians of the various philosophies of science that stood in opposition to Marxism: not only Jeans and Eddington, but also such as J.S. Haldane, Alfred North Whitehead, and Bertrand Russell. Here too emerged, in close combat, the most vigorous pursuit of Marxist philosophy of science from such powerful minds as those of J.D. Bernal, J.B.S. Haldane, Christopher Caudwell, and a host of others as well.

British Marxism: A Turning Point

Nineteen thirty-one marked a turning point for British Marxists. In their history of Britain in the 1930s, Margot Heinemann and Noreen Branson call it

---

* Laplace believed that if a calculator were supplied with the position and speed of all the particles in the universe, it could predict the entire future with utter exactitude.

a "watershed year," a year in which the ordered society suddenly turned out to be disordered, immoral, and dangerous.[92] The Labour government fell and Mosley's New Party was formed. Mussolini was already in power and Hitler was close to it. The choice seemed to be between fascism and communism. The middle ground was slipping away by the hour. John Strachey, after resigning in turn from the Labour Party and then the New Party and then allying himself with the Communist Party, saw the choice between communism on the side of the defence of culture, science, and civilization itself, and fascism, representing "mental and moral suicide." The title of his book, *The Coming Struggle for Power*, gave an indication of the prevailing mood.[93] Amidst this situation, the scientists were "at once the most fundamentally hopeful and the most frustrated."[94]

Nineteen thirty-one was also the year in which British scientists gathered in the lecture hall of the Science Museum in South Kensington on a Saturday morning in July at the special session of the Second International Congress of the History of Science and Technology. The session had been organized for participants to hear their Soviet colleagues explain their approach to the history and philosophy of science. Earlier in the week, the unexpected arrival of a large Soviet delegation had created a great stir in Britain. To the majority of scientists present, the Soviet viewpoint was something of a curiosity. To a minority, those in whom it crystallized something that had already been stirring, it had a profound impact. It was an experience that gave the initial impetus to the development of a distinctive school of Marxist thought in Britain.

The various participants have left most interesting accounts of the atmosphere at the session. The science journalist J.G. Crowther commented on the unprecedented enthusiasm for the history of science displayed by the Soviet delegation. The organizers of the congress had been modestly hoping to do a little to remove the neglect of their subject. Among the participants, a few were professional philosophers and historians of science, but most were amateurs or elderly scientists with an antiquarian interest in science, who discussed the subject in a leisurely way, as if a matter of secondary importance. They were truly astonished by the Russians, who discussed the history of science as if it were a matter of unsurpassed importance. As Crowther saw it, the movement, of which Hessen's paper was the most stimulating expression, transformed the history of science from a minor to a major subject.[95]

The mathematician Hyman Levy recollected the discomfort of the audience. The ideas were too novel to be absorbed by the majority and the minority were temporarily tongue-tied at perceiving the width of the gap that had been opened by the speakers and their audience. The long and awkward silence following Hessen's famous "trumphet blast" was finally broken by the young Cambridge student of mathematics David Guest, who stepped to the rostrum and expanded on the theme, drawing out the implications of Hessen's remarks by analyzing other British men of science, such as Karl Pearson and Bertrand Russell.[96] Two British historians of science expressed opposition, but most refused to react. Bernal later remarked that the probable consensus was that "anything so ungentlemanly and doctrinaire had best be politely ignored."[97] For himself, he was struck by the unity, philosophical integrality, and social purpose of the Soviet delegation in contrast with the British colleagues they encountered with their indisciplined array of ill-assorted individual philosophies and remoteness from any social considerations.

The whole thing could not, however, be politely ignored. Within a few days, *Science at the Crossroads*, a book containing the Soviet papers, was published. It had been produced in a flurry in accord with a decision taken earlier in the week (a project described by the *Guardian* as the "Five Days Plan," which had transformed the Soviet embassy in London into a makeshift publishing-house for the occasion. Philosophers and scientists rushed around in rolled-up sleeves and translators and printers worked all through the night). After the book came the stream of reviews and more discussion of the embarrassing and unusual ideas of the exotic foreigners. The reviewer in *Nature* expressed his concern over the possible effect dialectical materialism could have on the direction of research in the Soviet Union. He repudiated the concept of "bourgeois science" and affirmed his belief that the "laws of nature are the same for all of us."[98] The review in the *Times Literary Supplement* expressed unmitigated hostility.[99]

The reviews were the least of it. In their lasting impact on the core of left-wing scientists present, the Soviet delegation had, in Gary Werskey's turn of phrase, "performed a five-day wonder."[100] Bernal, Needham, Hogben, Levy, Crowther, and others all testified to the crucial influence of this event on their future thought and activity. The Soviet delegation might go away, but those they left behind would not. The British establishment would have them on their hands for a long time to come. In the highest gatherings of British science, there was from then on a core of exceedingly able left-wing scientists enthusiastically pursuing a Marxist approach to the history and philosophy of science and high-lighting the multi-faceted social relationships of science. The ideological and sociopolitical assumptions of past and present science were being ruthlessly laid bare. The very raising of such questions aroused considerable discomfort, for, as Bernal realized, the secret of the strength of the spirit of bourgeois science lay in its avoidance of explicit statement.

Nor did the new phenomenon in British science confine itself to the level of discussion and discomfort, for these men were not the sort to be "lecturing on navigation while the ship is going down."[101] Around them emerged a vigorous movement for the defence of science against all forces threatening it and for social responsibility in all areas of scientific endeavor. The movement took many organizational forms, such as the Cambridge Scientists' Anti-War

Group and the revived Association of Scientific Workers* and even in due course the Division for Social and International Relations of Science within the hallowed British Association. It was a broad front in which there was room for a very wide spectrum of opinion and for varying degrees of commitment. Not everyone who participated embraced all the tenets of what became known as "Bernalism," but they did know they stood with the movement for social responsibility in science and not with the movement that sprang up in opposition to it. Its manifesto was in John Baker's "Counterblast to Bernalism"[102] and its organizational form was the Society for Freedom in Science set up by a few scientists such as Baker and by the philosopher of science, Michael Polanyi, which devoted itself to the defence of "pure science" and the absence of any form of social control of science. Even the hostile reaction to the radical science movement testified to the power of its impact.

Science and the Popular Front

This shift to the left among scientists, at a time when the left was virtually identified with the communist parties, was considerably enhanced by a radical policy shift within the communist movement itself. The Seventh (and last) World Congress of the Comintern was held in July-August 1935 and marked the inauguration of the period of the Popular Front. No longer was "class against class" to be the rallying cry nor was there the blurring of the line between fascism and social democracy. The call, signaled clearly and decisively in Dimitrov's key address, was for a united front against fascism. The new policy proved extraordinarily successful and the ranks of the communist parties began to swell, not to speak of the broad organizations that grew up around them. On Bastille Day in Paris, the communist Marcel Cachin embraced the socialist (until then "social fascist") Léon Blum, while the crowd cheered and sang the *Marseillaise* followed by the *Internationale*. In the elections of February 1936, the Popular Front came to power in Spain, and those of May in the same year brought the same in France. In Germany, communists and social democrats only came together again in exile and in Hitler's prisons and concentration camps in a belated, but very real, anti-fascist unity. Comintern agent Willi Munzenberg was once again in his element, directing from his Paris exile a staggering number of organizations and projects that summoned the energies of everyone from communist party apparatchiks to English duchesses. In Britain, the Popular Front spirit

---

* The Association of Scientific Workers was a trade union, successor to the National Union of Scientific Workers founded in 1918. The A.Sc.W. later amalgamated to form ASTMS and then MSF and eventually UNITE.

permeated a number of old and new organizations of the left. The Left Book Club, launched in May 1936, elicited massive support and specialists groups were soon formed among doctors, teachers, poets, scientists, and others. Bernal, a prototypical Popular Front intellectual, was on some sixty committees at one time. He, in accord with many other left-wing scientists, pointed to the Popular Front as the most powerful force for the defense of science.

The movement among left-wing scientists had the dual effect of bringing scientists to the left and bringing the left to science. Nowhere, at least nowhere outside the Soviet Union, did the left so thoroughly integrate science into its political perspective as in Britain. The right-wing of British science might balk at the steady drift to Bernalism, but the CPGB did not. Rajani Palme Dutt, in his report to the 16th National Congress of the party in 1943, clearly stated: "The Communist Party stands with modern science."[103] Although the party was not totally immune from the workerism of the British labor movement, the party only intermittently indulged it.* The party was proud of its intellectuals and believed it was an essential duty of its scientists to be good scientists, even in the eyes of *Nature* and the Royal Society, and it took a dim view of any tendencies of its intellectuals to "drop out" and assimilate themselves into the life-style of the workers and do only routine party work. While many of the party intellectuals were extremely energetic on the ground in all aspects of party work, they were reminded of the crucial importance of their distinctive work of bringing Marxism to bear in their own special fields.

Bernal

At the center of the new movement was John Desmond Bernal. His contemporaries of diverse shades of opinion spoke of him in the most glowing terms, coming forth with superlatives not characteristic of the English. Julian Huxley thought him the wisest man in Britain and most who knew him concurred in thinking him very wise. Indeed, the name by which his friends called him was "Sage." A fascinating fictional portrait of him was drawn by C.P. Snow in his early novel *The Search*, in which Bernal figured as the unusual young scientist Constantine, the bearer of a sort of dazzling and global brilliance.[104] Snow's nonfictional portrait of Bernal, written many years later, is even more fascinating. For Snow, he was "perhaps the last of whom it could be said, with meaning, that 'he knew science.' "[105] Joseph Needham describes him as one of the best minds of their generation.[106]

---

* The same Palme Dutt, in an article "Intellectuals and Communism" wrote: "First and foremost, he should forget that he is an intellectual" (*The Communist*, Sept. 1932). It is interesting to note that workerist antiintellectualism didn't always come from the "honest worker." It came from demagogic party leaders, themselves intellectuals, like Zinoviev and Palme Dutt, when it suited their purposes.

By the mid-1930s (and his mid-thirties), Bernal was already a fellow of the Royal Society and professor of physics at Birkbeck College of the University of London, with important work in X-ray crystallography already behind him at Cambridge's famous Cavendish Laboratory. By the late 1920s, he had become a militant atheist and a communist.* He had come a long way from Tipperary. Born in Ireland and Jesuit-educated, he had, as a schoolboy, combined piety and defiance in an adolecscent mixture, organizing a Society for Perpetual Adoration and fervently supporting the Easter Rising. He had from first to last an intensely philosophical frame of mind and an extraordinary sensitive social conscience. It was his voluminous knowledge, his breadth of vision, and his conscientious activism that most singled him out, rather than his laboratory results. At the level of experiment, he had a tendency to generate seminal ideas and to leave to others the opportunity to pursue the detailed further research.

Bernal did pioneering work, not only in such sciences as X-ray crystallography and molecular biology. He was founder of an altogether new discipline, the "science of science." His book, *The Social Function of Science*, quickly came to be regarded as a classic in this field. Based on a detailed analysis of science, both under capitalism and under socialism, and comparing in particular the state of British science with that of Soviet science, Bernal's dominant theme was that the frustration of science was an inescapable feature of the capitalist mode of production and that science could achieve its full potential only under a new, socialist order. According to Bernal, science was a powerful force in human history, and was destined to become more powerful still. Formidable obstacles were obstructing it, however, in the fulfillment of its destiny. It was outgrowing capitalism. Capitalism was losing its ability to cope with it. British science was severely underfinanced, particularly under the impact of economic crisis. German science had been overcome by barbarity. As the bourgeoisie was losing its ability to use science as it had during its rise and as it lost its ability to rule in the old way, it was inclined to turn on science. Rather than admit the cause of social disorder to be the capitalist system, the ruling class generated a distrust of science that in its most extreme form turned into rebellion against scientific rationality itself. Even in the most unlikely quarters, notably the British Association for the Advancement of Science, voices were being raised calling for the suppression of science. Science itself was in danger and its only hope was in socialism. The cause of science was, for Bernal, inextricably intertwined with the cause of socialism. As he put it in an autobiographical essay, from an early age he saw science as holding the key to the

---

* Bernal was a member of the CPGB only until 1933. After that, it was thought best for him to pursue his broad work outside the party. Whatever the reason, his lack of a party card was only a technicality. He was in every way a communist and not a "fellow traveler."

future and the forces of socialism alone as gathering to turn it.[107] By the time he wrote *The Social Function of Science*, he had come to believe that: "In its endeavor, science is communism."[108] Needless to say, Bernal saw science as a social activity, integrally tied to the whole spectrum of other social activities, economic, social and political.

Science was absolutely central both to Bernal's social thinking and to his philosophical thinking. The scientific method encompassed the whole of life. In considering the relationship between Marxist philosophy and scientific method, Bernal thought that it could not be described simply as scientific method. Nor could it be said to be in any sense an alternative method. It was an extension of scientific method. On the basis of current science, it gave a comprehensive and ordered account of the whole range of phenomena, from nebulae to human society. In Bernal's view: "Marxism transforms science and gives it greater scope and significance."[109] As Bernal conceived the relation between philosophy and science, science was the starting point for philosophy; it was the very basis of philosophy. Marxist social theory emerged within this process. There was no sharp distinction between the natural sciences and the social sciences for Bernal, and the scientific analysis of society was an enterprise continuous with the scientific analysis of nature. For Bernal, there was no philosophy, no social theory, no knowledge independent of science. Science was the foundation of it all.

As Werskey has recently remarked of Bernal: "Never had Frederick Engels' famous notion of 'scientific socialism' been treated so literally."[110] For Bernal the humanistic and the scientific dimensions were one. His vision of the sort of future that science could make possible for mankind was in total contrast to that of Aldoux Huxley's *Brave New World*. Full automation, nuclear energy, and cybernetics could bring a fuller realization of human potential. His futuristic sketches grew increasingly better grounded as his Marxism matured, making the society of the future set out in *The Social Function of Science* far more plausible than the one set out in his earlier work, *The World, the Flesh and the Devil*.[111] His sense of history was sweeping, stretching back into the ancient past and shooting forward into the coming future.

As Bernal saw the transition to the future, scientific and socialist philosophical thinking played a key role. He took issue quite sharply with those, especially those in England, who thought that both science and politics could get on quite well without philosophy. Science, philosophy, and politics were all tightly bound together in Bernal's highly integrated mind. There could be no coherence, no far-seeing vision, without a world view. There could be no adequate and plausible world view not grounded in science. There could be no point in having a scientific world view without living by it and acting on it. The new age was bringing the need for new ideas, new values, new movements. The whole vast body of knowledge accumulated through the ages had to be worked over and revalued. It was a time of profound transformation, bringing deep moral and emotional changes as well as intellectual ones.

Bernal's philosophy of science was in the tradition of Engels. It was time, he thought, for Engels to come into his own in Britain, where he had lived and worked and formulated his exceedingly brilliant and suggestive ideas about the dialectics of nature. Although Engels had suffered complete neglect at the hands of the philosophers and scientists of Victorian England, time was to take its revenge. Whereas the professional philosophers of science of his day were already for the most part completely forgotten, Engels would be remembered if Bernal were to have his way. And he was. Engels would be the inspiration of a new wave of fresh thinking in the philosophy of science. The important thing about Engels's concept of nature, as Bernal saw it, was that he saw it as a whole and as a process. Engels's work could still be of value to scientists in that it was important to carry forward this sort of all-embracing and historical approach to science and to use his methods in pushing forward the solution of further problems. Significant steps that had been taken by science, such as relativity and quantum theory, biochemistry and genetics, confirmed the validity of Engels's approach and made continuing work along the same lines necessary. Engels was a radical thinker. His way of questioning fundamentals was the kind that many years later led to the formulation of quantum theory and relativity theory. In fact, in Bernal's opinion, if Engels's philosophy of science had been more widely known in the scientific world, these theories would probably have been discovered sooner and would be free from the idealistic confusions under which they were still suffering. Engels was in every way a compelling figure for someone like Bernal who saw him as scientist, philosopher, and revolutionary, all in one.[112]

For Bernal, dialectical materialism was the most powerful intellectual current of the time. It provided the basis not only for a revolutionary social movement, but also for the enhancement of science. It was a philosophy derived from science that brought order and perspective to science and illuminated the onward path of science. It was no substitute for science. It was no royal road to knowledge. Induction and proof remained what they were and the hard work still had to be done. It was not a dogma imposed on the findings of science from without, but a method of coordinating the experimental results of science and of pointing the way to new experiments, a method that had been developed in and through the development of science itself. Its role was to clarify and to unify the different branches of science in relation to one another and to other human activities and to suggest directions of thought that were likely to yield further results in the future. It was a science of the sciences. It was a means of overcoming overspecialization and of achieving the unity of science. It placed science within the context of the whole of human and cosmic evolution. Its starting point was the material universe, its central idea was the process of transformation, and its scope was the whole range of human experience.

A resolute monist, Bernal saw the unity of science as grounded in the unity of the universe itself. He affirmed the unity of the universe, not in a hollowly reductionist way, but in a way that recognized the intricacy and complexity of matter that had evolved in such a way that new qualities emerged at higher levels of organization. The origin of the new, however, had to be seen against the ongoing process, so as to avoid the two extremes in which the immediate apprehension of quality was made the basis of mystical speculation on the one hand or mechanically denied altogether on the other.

Bernal's position on the dialectics of nature was quite definite. Any sort of dualism between nature and history was quite foreign to him, as was any tendency to radically historicize nature. There were in the external world processes involving oppositions of actual forces that were constantly giving rise to new and higher syntheses. Dialectical development was not confined to human society or even to living things, but occurred at all stages of the organization of matter, completely independent of human thought. His position here, inspired by Engels, had precisely the same flaws as that of Engels: a huge gap yawning between particular examples of dialectical processes and the assertion of the universality of dialectics, a tendency to use the word *dialectical* equivocally, sometimes simply as synonymous with *developmental* and other times as *development through contradictions*.

Close to the very latest developments in the natural sciences, Bernal saw that science itself was providing the most effective refutation of positivism. Although the work of such writers as Lukács, Korsch, and Gramsci were not known in Britain before the war, Bernal would have been quite unsympathetic to their tendencies either to equate science with positivism or to be so preoccupied with the critique of a discredited positivism as not to see that science itself was under attack. The defense of science and the full incorporation of the history of science into the history of Marxism were of the very essence to Bernal.

Nor was he inclined to put such heavy stress on the Hegelian origins of Marxism. He was not anti-Hegelian and he did acknowledge the role that Hegel had played for Marx and Engels, but he was more inclined to emphasize the continuity of Marxism with the whole development of scientific and historical knowledge. The emergence of Marxism laid the foundations for a whole new way of understanding and changing the universe. Marx was seeking the origins of origins and the new philosophy was inspired less by Hegelianism than by such dramatic developments of time as Darwinism—developments that were giving a new sort of account of the origins of the world and all phenomena in it. With regard to the sources of Marxism, Bernal argued:

True, Marx studied Hegel, but the dialectic of Marx, which neither Hegel nor the Hegelians would accept for a moment, is derived far more from his wide knowledge of the universe, and comes directly from the concrete experience of the economic and political struggles of the nineteenth century than from the philosophy of his youth.[113]

Such remarks, to be sure, reveal far more about Bernal than about Marx. Although historically inaccurate with regard to Marx, it was a legitimate enough position for later Marxists.

This did not, incidentally, keep Bernal from being attacked as overly Hegelian and unscientific. The Oxford philosopher, E.F. Carritt, associated dialectical materialism with the imposition of triadic patterns as a priori laws of nature. Hegelian pedantry, he said, did not become more digestible when it was called dialectical matter rather than dialectical idea. The mind tended to feign symmetrical patterns in nature, whose works were often singular. Patient investigation of one's subject matter could not be avoided by any dialectical shortcuts.[114] Bernal replied that Carritt had a rather perverse view of the Marxist dialectic. Carritt's view, Bernal argued, was actually the Aristotelian derivation of the mean coming to life again in the name of Marxism. Dialectical development was not a matter of swinging between two extremes and settling down to a middle course. The synthesis was not a mean, but something qualitatively new and different. Marxists imposed no patterns on nature, but discerned them there.[115]

While Bernal did not explicitly take on the neo-Hegelian trend within Marxism, he did address himself to some of the trends outside of Marxism that inspired it. He thought the various irrationalist and intuitionist currents of the late nineteenth and early twentieth century represented the backwaters and dead ends of human knowledge. He considered the revolt against reason, embodied in such writers as Sorel and Bergson, to be deeply reactionary. He objected most, however, to those who were scientists and were seeking to bring irrationality into the structure of science itself. Scientists such as Jeans, Eddington, Whitehead, and J.S. Haldane were making science into a modern ally of ancient superstition and trying to create a new scientific mystical religion. Gone were the days filled with talk of the warfare between science and religion. Now, the Archbishop of Canterbury benignly presided over Royal Society dinners, for science had become suave, more respectable, and less materialistic. Jeans and Eddington had assured the bishops and everyone else that real science and real religion were the same thing.

Bernal argued that it was a highly objectionable procedure to make what science could not know, or at least did not know, rather than what it could and did know, the basis for affirmations about the nature of the universe. The fact that the new physics was supposed to have destroyed the older materialism was taken as an excuse for holding any opinion whatsoever. Their technique was to conclude from the fact that science had not demonstrated how the universe had come to be, that it must have been made by an intelligent creator; to go from the circumstance that science had not synthesized life to the assertion that the origin of life was a miracle; to make the uncertainty relation in quantum mechanics into an argument for human free will. Such books as Jeans's *The Mysterious Universe*, ostensibly based on science, were negations of science. Jeans's world was a mythological abstraction from science, an arbitrary reduction of all the concrete universe to a number of abstract categories.

Bernal did realize that such antiscientific philosophies did not spring up out of sheer perversity or willfulness on the part of their exponents, but were symptomatic of a widespread and pervasive confusion. Science was in deep crisis, not only because of external events, such as the economic slump, the rise of fascism, wars, and preparation for wars, but because of the impact of internal forces as well, for many of the traditional assumptions of the scientific enterprise were being undermined from within. Materialism had grown so rapidly that it was temporarily losing its language, but it was acquiring a new one through the connecting of the biophysics of sensation with the ultimate wave mechanics picture of the universe. These new discoveries called for fresh thinking and it seemed to Bernal that only Marxists were doing it in a way that made sense, for only dialectical materialism had the capacity to encompass them without betraying science itself.

The Marxist approach to the philosophy of science was seen by Bernal as still being in the process of being formulated. Marx and Engels and Lenin had only sketched the outlines of it. It was being further developed in the Soviet Union in a lively and sometimes violent process. He was aware of the main outlines of the Soviet debates and saw Soviet science as finding its philosophy in the very course of its revolutionary development. But it was, he remarked, complicated at times by the fact that the older scientists were often hostile to new philosophical ideas, while the younger ones, who were most receptive, often lacked sufficient scientific knowledge. He knew of the clash between Vavilov and Lysenko, but did not seem to realize the gravity of what was taking place in this sphere, seeing it in 1939 as a difference in emphasis rather than a revival of the Weissmann-Lamarck controversy. As he far too sanguinely characterized the debate:

Geneticists were criticised for attributing inherited characters to specific unitary factors in the chromosome, and neglecting cyto-plastic and environmental factors, whose importance was probably exaggerated by their critics.[116]

Bernal himself was firmly committed to the science of genetics and was conducting experiments aimed at discerning the molecular structure of the gene. He was, on the whole, extraordinarily impressed by Soviet science and philosophy of science, at times more so than the situation warranted. When he had first visited the Soviet Union in 1931, he was struck by the overriding sense of purpose there and found the country "grim but great."[117] As time went on, Bernal discovered things that must have disturbed him deeply, particularly things relating to the fate of scientific colleagues in the Soviet Union. He is known to have interceded with the Soviet ambassador in London, Ivan Maisky, in relation to the arrests of the physicists Weissberg and Houtermanns. But in public he said nothing. The great failing of Bernal was in his reluctance to take a critical attitude to the tradition he embraced.

Haldane

In 1928, another of Britain's eminent scientists, John Burton Sanderson Haldane, had also gone to the Soviet Union. There he encountered Marxist philosophy of science for the first time, although it would be several years yet before he would come to embrace it as his own. As he put it himself, "it was hardly love at first sight."[118] In his earliest excursions into the realm of the philosophy of science, Haldane had held that Kantian idealism was the most appropriate philosophy for the science of the Einsteinian era, just as materialism had been for the Newtonian era. In his books written before 1928, that is, in *Daedalus, or Science and the Future* and *Possible Worlds and Other Essays*, he set out the view that the world of physics was reducible to a manifold of transcendental events. Its laws were merely the forms of human perception. The data of modern science was more easily reconcilable with Kant's thought than with any other philosophy. This situation, he predicted, would continue for several centuries. Possibly after another two centuries or so of scientific research, the data of science would support one rather than another of several post-Kantian systems. But Kant had indeed written the prolegomena to any future metaphysics.[119]

Not two centuries were to pass nor even one decade before Haldane would believe this view had been already superseded. Within a few short years, he was himself emphatically post-Kantian and he was no longer playing by the rules set out in Kant's prolegomena. Indeed in every way, the evolution of his thinking through the 1930s was to take him a very long way from the world from which he had come. Exceedingly "well-born," he came in time to give himself over to the cause of the proletariat, though only after going through a number of stages along the way. He was, as Werskey put it, "obliged to conduct class war with himself."[120] Born in Scotland into Britain's intellectual aristocracy and educated at Eton and Oxford, Haldane had every possible advantage. Nevertheless, his mind was open and alert and he set upon the course that was to lead him to Marxist philosophy and to the Communist Party. His commitment was to "reality as such, whether it be bright or dark, mysterious or intelligible."[121] An eminent biologist, fellow of the Royal Society and philosopher of science, he was also the son of an eminent biologist, fellow of the Royal Society and philosopher of science. His father, John Scott

Haldane, devoted himself increasingly to philosophical speculation on the natural sciences. Proceeding on somewhat Hegelian lines, he sought to undermine materialism as a philosophy of science.[122] His son, in taking the further step from Hegel to Marx, saw his work as a continuation of his father's, but there was no getting away from the fact that he was accepting and carrying forward the very philosophy his father had sought to combat. But if he could see some strand of continuity with his father who was a scientist, a philosopher and a liberal, this was impossible with his mother who was an ardent tory who had enlisted him as a child in her activities on behalf of such organizations as Children of the Empire.

Haldane was an extravagant, adventurous, and somewhat larger-than-life character, who was prepared to address himself to almost any subject under the sun to almost any audience. His colorful and futuristic popularizations of science won him a huge following among those open to "advanced views." Sir Peter Medawar, in his preface to Ronald Clark's biography of Haldane, observed that the lives of most academics, considered as lives, almost always made for dull reading, but that this was far from the case with Haldane. His life was fascinating from beginning to end and, unless one was already in the know, there was no way of foretelling what would come next.[123] His mockery of the social conventions of his milieu was notorious, whether it was simply discoursing loudly at High Table on all sorts of unmentionable subjects or his famous battle against Cambridge's Sex Viri (renamed the "Sex Weary") after his dismissal for "gross immorality" stemming from his involvement in a divorce case. J. Maynard Smith tells how Haldane was regarded by the masters of Eton in Smith's day "as a figure of immense wickedness."[124] So great was his notoriety that even novelists found him a rich source of inspiration. It is said that both the physiologist Shearwater in Aldous Huxley's *Antic Hay* and the corrupter of youth, Mr. Codling, in Ronald Fraser's *The Flying Draper* were modeled on Haldane.[125] He was also well known for fearlessly experimenting on himself.

But, despite his wilder exploits, Haldane was a very serious man. As a scientist, his work was outstanding and broke new ground. He undertook to re-found Darwinism upon the concepts of Mendelian genetics and thus to eliminate the seeming contradictions between heredity and evolution. He was the first to estimate the mutation rate in man. In 1929, his investigations into the origin of life had produced a theory giving a materialist explanation for the emergence of living organisms from the inorganic world. His work proceeded parallel to, but independent of, the work of the Soviet biochemist, A.I. Oparin. The connection between the Oparin-Haldane hypothesis and Marxist philosophy has been the subject of controversy among historians of science. It was set off by C.H. Waddington who remarked in the course of a book review:

> In the late twenties and early thirties the basic thinking was done which led to the view that saw life as a natural and perhaps inevitable development from the non-living physical world. Future students of the history of ideas are likely to note that

this new view, which amounts to nothing less than a great revolution in man's philosophical outlook on his own position in the natural world, was first developed by communists.[126]

Disputing this, David Joravsky has maintained that chemical hypotheses concerning the origin of life were in no way the product of Marxist philosophy, as neither Haldane nor Oparin were Marxists at the time of their discoveries.[127] Loren Graham has responded quite differently. He argues that both were already under the influence of Marxist philosophy at the time and underwent an intellectual development in which Marxism played its part. Both subsequently became dialectical materialists and explicitly declared that Marxism was an important influence on their biological thought. Graham maintains that the acknowledgment of some form of connection between Marxism and research into the origin of life is a healthy corrective to the tendency to think of the history of the connection between Marxism and biology only in terms of Lysenko. Wishing to set things in perspective, he claims:

> This tendency to explain an acknowledged calamity in science as a result of Marxist philosophy while assuming that a brilliant page in the history of biology had nothing to do with Marxism is a reflection, at least in part, of the biases and historical selectivity of anti-Marxist journalists and historians.[128]

Whatever the precise timetable of the intellectual development of Haldane and Oparin, it is in any case significant that the sort of men who were in the process of evolving towards a dialectical materialist position in philosophy were in fact the men who tackled the problem of the origin of life and made the crucial breakthrough in this sphere. However, it is of some importance for the history of science to sort out as far as possible exactly how the process unfolded for both Haldane and Oparin.* In Haldane's case, the evidence for Graham's interpretation is even stronger than what Graham has himself set out and even clearer than in the case of Oparin where Graham concentrates his attention (as Oparin is central to the purposes of Graham's study, whereas Haldane is tangential to it).

---

* Regarding Oparin's intellectual development, Joravsky holds that he was a nineteenth-century mechanistic materialist, who later professed dialectical materialism only as a result of political pressure. Graham argues that his shift from his earlier reductionism and his commitment to dialectical materialism was genuine, the evidence being not only in his frequent statements favoring dialectical materialism, but more importantly in the fact that the very method of analysis in his publications was permeated with an assumption of a process philosophy and a concept of differing dialectical levels of regularities in nature. He saw life as a special form of the movement of matter, which arose as a new quality at a definite stage in the historical development of matter. Graham sees Oparin's materialism developing in parallel with the philosophical development of Soviet society and interprets the significance of his discoveries in this context. (Note continued on page 451.)

In delivering the Haldane Memorial Lecture (in memory of his father) at Birkbeck College, London, in 1938, Haldane himself shed light on his own intellectual development in the sphere of philosophy of science. Until going to the Soviet Union, he said, he had been unaware of the existence of Marxism as a philosophy of science, but it had made a deep impression on him, both in virtue of its prevalence and of its connection to concrete scientific research, especially to biological research. Until then, he had no idea that an astronomer, chemist, or biologist might find Marxist principles to be an aid to research. Thereupon, he read Engels's *Anti-Dühring* and *Ludwig Feuerbach* and thought Engels far ahead of his time:

> Had these books been known to my contemporaries, it was clear that we should
> have found it easier to accept relativity and quantum theory, that tautomerism
> would have seemed an obvious hypothesis to organic chemists, and that
> biologists would have seen that the dilemma of mechanism and vitalism was a
> false dilemma.[129]

His immediate attraction to Engels's philosophy was to its dialectical aspects. He had still found difficulty, he revealed, with its materialism. He had continued to read and think about it, however, and he had come in due course to embrace materialism, accepting the priority of matter over mind and the existence of unperceived events. Darwin's work, Haldane observed, had left Marx in no doubt that nature was in existence before mind. His thinking had advanced, not only through realization of the respect of dialectical materialism for science, but also through observation of social tendencies and through finding that these had been predicted by Marxism, though perhaps on too short a time scale. He thus had begun to accept Marxism "as the best available philosophy" and had come to discover that, when a person came to accept Marxism as part of his daily thought, the world became enormously richer in content and fuller of pattern. Not until he began to apply dialectical materialism to concrete problems, however, did he realize its power, he claimed, and then explicitly declared: "I have found Marxism of great value in the planning of biological research."[130]

Haldane further elaborated on this thought process in his 1940 essay "Why I Am a Materialist." Here he said that fifteen years before, that is, in 1925, he had been a materialist in practice but not in theory. Although he had been a strict materialist in the laboratory, he had been a vague sort of idealist outside it. He could not at that time see how knowledge or thought were possible on a materialist basis. He had therefore been compelled to fall back on some kind of idealist explanation, according to which mind, or something like mind, was really prior to matter, and what was called matter was really of the nature of mind, or at least of sensation. He had been, however, too painfully conscious of the weakness in every existing idealist philosophy to fully embrace any of them. His difficulties had been resolved, he claimed, both by reading Engels and Lenin and by the actual progress of science over the previous fifteen years.[131]

As to the relation of Haldane's philosophical evolution to the formulation of his hypothesis regarding the origin of life, that cannot be pinned down precisely. However, certain points are clear. It is true that he said in his 1929 essay, "The Origin of Life," setting forth his discovery, that the hypothesis was compatible not only with materialism but with other philosophical tenets, even with the view that preexistent mind or spirit could associate itself with certain kinds of matter, but it was clear that he did not share this view. Knowing that his hypothesis would cause a stir, as it did, he only wished to circumvent such critics as would consider it sufficient refutation of his position to say that it was materialist.[132] It is also true that in his January 1938 Muirhead Lecture in Birmingham, published in his book *The Marxist Philosophy and the Sciences*, the best known philosophical text of Haldane, he said that he had only been a Marxist for about one year.[133] At the same time, the other texts cited do make it possible to pinpoint the time at which he came under the influence of Marxism, that is, his trip to the Soviet Union in 1928, and testify to the fact that he had been reflecting seriously on Marxist philosophy in relation to the natural sciences, and especially to biology, from 1928 on. Moreover, his statement that he had been a materialist in practice from 1925 and that materialist assumptions circumscribed his thinking in his laboratory is also extremely significant.

At all events, by 1938 Haldane was a committed Marxist, proclaiming forthrightly to all the world, "I think that Marxism is true."[134] He thought it true in every sense. His political evolution had converged with his philosophical evolution and he had become a committed Marxist in his politics as well as in his philosophy of science. His successive transformations had taken him from his mother's toryism through to his father's liberalism, through the Labour Party, then to Republican Spain during the Spanish Civil War and finally to the Communist Party. No longer believing that England had to beware of the danger of communism from the east and Americanism from the west, he now called on his countrymen to look to the east. He joined the CPGB in 1942 and was soon a member of its executive committee.

Haldane, like Bernal, had a highly integrated mind and was intrigued by the interrelations between politics and philosophy, science and politics, philosophy and science. The interrelations were pursued both in his academic work and in his outside political activities. He was not one to confine himself to academic circles, especially during such times of upheaval. As German scientists began to seek exile in England as a result of the rise of fascism, Haldane commented "I began to realize that, even if the professors leave politics alone, politics won't leave the professors alone."[135] Equally critical of scientists and philosophers who looked on the realm of politics with disdain and political activists who regarded science and philosophy as unimportant, he was disturbed, and even baffled, by the existence of Marxists who were indifferent to philosophy of science.

To him, philosophy of science was absolutely vital and it was central to his commitment to Marxism. His Muirhead Lectures show him in the grip of an enormous enthusiasm for what Marxism meant as a philosophy of science. Reviewing the book in which the lectures were published, Andrew Rothstein gave a marvelous characterization of Haldane's mood:

> Like anyone on his first acquaintance with the Marxist method, thunderstruck at the new worlds which that touchstone opens in seemingly familiar things, Haldane hastens to run over the whole range of his knowledge, calling "Open Sesame"—with great profit to himself and his audience.[136]

Unfortunately, Rothstein proceeded to contrast this wonderful freshness of a man who came to Marxism an eminent scientist already, seeing all his scientific knowledge in and through it, with the somewhat truncated philosophical statement of Stalin in the newly publicized *Short Course*—and to do so unfavorably—criticizing Haldane for elaborating dialectical principles more from Hegel than from Marx and advising him to recast his work with Stalin's example in mind.

The problem with Haldane's approach to Marxist philosophy was that he became somewhat overwhelmed by it and was not able to view it critically. Indeed, like Bernal, he accepted wholly the current Soviet interpretation of it, and within a year he was himself praising the section on philosophy in the *Short Course* and calling another Marxist scientist to task for not measuring up to it.[137] This did not necessarily mean that his approach or that of Bernal was a dogmatic or unthinking one. Far from it. Both had undergone a genuine philosophical development and had come to dialectical materialism freely and intelligently. It simply satisfied their minds, especially in the first flush of enthusiasm for it, and they were unable at this time to feel the lack of anything within it or the force of any alternative to it. While they were unable to transcend its framework, they were nevertheless highly creative within it. It was a way of thinking they had genuinely made their own and they brought it to bear on new problems, analyzing new scientific discoveries in light of it, using arguments that were always intelligent and plausible, and modes of expression that were often rich, novel, and colorful—and in Haldane's case, sometimes quite humorous as well. Actually, Medawar's evaluation of Haldane is extremely astute, at least in relation to Haldane's measure as a philosopher at this time. Describing Haldane as the cleverest man he ever knew, especially in his power to connect things in unexpected ways, Medawar judges that he was still not a profoundly original thinker: "His genius was to enrich the soil, not bring new land to cultivation."[138] Clever he surely was and enrich the soil he surely did.

Dialectical materialism, for Haldane, was an all-encompassing *Weltanschauung*. It was "not merely a philosophy of history, but a philosophy which illuminates all events whatever from the falling of a stone to a poet's imaginings."[139] Its pivotal insight was in seeing reality as processes and not things, and in putting particular emphasis on the interconnection of all processes and the artificial character of the distinctions men have drawn. Great was its value for science, Haldane never tired of saying, for it helped to achieve both concreteness and elasticity of thought. It encouraged a scientific approach to all problems without in any serious way limiting the kinds of explanation open to the scientist. He did not, however, want to give a false impression nor to arouse exaggerated expectations. He warned that, in discerning the relation of Marxism to science, it was important to proceed with the greatest caution. At best, Marxism would only tell a scientist what to look for. It would rarely, if ever, tell him what he was going to find. Marxism, in his opinion, had a twofold bearing on science: it was relevant in illuminating the social relations of science as well as in handling problems of "pure" science. It shed light on the history of science by studying science as a human activity, and on the philosophy of science by analyzing the most general patterns of development permeating nature, society, and human thought. Marxism proved of the greatest value to science in highlighting the process of the development of science and the relationship of the different sciences to one another.

Haldane laid great stress on dialectics, which he elaborated, in the manner of Engels, as the "science of the most general laws of change" in society, in consciousness and in the external world, and he made much of the three principles of dialectics that had by this time become formalized in textbooks of dialectical materialism. The principle of the unity of opposites meant that matter was something much richer and more complicated than mechanistic materialists ever dreamed. It was exemplified in quantum mechanics in which the electron displayed properties both of waves and of particles, in metabolism in which living substance was a unity of anabolism and catabolism, in pulsating stars as the result of a conflict between nuclear and gravitational forces. Progress occurred through conflict. Such internal contradictions did not mean that nature was irrational, but did mean that it was unstable. The more man studied nature, the more was it found that what was apparently stable turned out to be a battlefield of opposing tendencies. Haldane insisted that union of opposites was a hard physical fact. There were contradictions in matter. The principle of the passage of quantity into quality meant that transformation was not simply a matter of continuous variation and that there were properties of a system as a whole that could not be located in any particular aspect of it. It was exemplified not only in the classic example of boiling or freezing water, but in such newer discoveries as the thresholds of nerve cells. The principle of the negation of negation meant the sudden emergence of novelty, exemplified in geology in the formation of mountain ranges, in biology in the dependence of evolution on variation and selection, in psychology in the need to pass through guilt, the negation of innocence, in order to aspire to virtue.[140]

With regard to the debate over dialectics of nature, although he did not refer to Lukács and the controversy surrounding his views, indeed may not have known about it, Haldane nevertheless had a very definite position on the question: nature was dialectical. Looking back on the history of philosophy, he argued that it had long been realized that matter as a whole behaved intelligibly, conforming to the laws of logic and arithmetic. The question that arose was whether reason mirrored the behavior of matter or whether matter mirrored the behavior of mind. Kant's view was something intermediate. Hegel's was that logical categories existed eternally. Feuerbach, Marx, and Engels, however, held that these categories were exemplified in nature before they governed thought. Engels treated the Hegelian dialectic as expressing primarily the properties of matter and only secondarily the laws of thought. He held that the principles Hegel had worked out in the realm of thought also applied to material events, not only in the social field, but in such fields as astronomy, physics, and biology. Haldane argued on the side of Engels and, on the basis of such an approach, put forward his own argument for the existence of contradictions in matter: mind was intimately connected with matter and mirrored its behavior; therefore, if there were contradictions in the mind there must be contradictions in matter.[141] Superficially plausible, perhaps, but a woeful muddle on a number of counts: in deriving ontology from logic rather than vice versa; in assuming mind mirrors matter only in a straightforward and uncomplicated way, thus failing to take the possibility of distortion into account; in taking contradiction rather than noncontradiction as a principle of logic with ontological implications; in confusing contradictions with opposing forces; and in failing to consider that in a philosophy of ascending levels there can be specific terms inappropriate to lower levels. It was thus by no means clear that the connection of mind to matter and the priority of matter to mind was dependent on the assertion of contradictions in nature.

Be that as it may, it was not Haldane at his best. Far more worthwhile were his discussions of other related issues and his polemics against alternative views in the philosophy of science. In his discussion of the concept of laws of nature, for example, he set himself against both of the two contradictory views then in vogue: the extreme positivist view enunciated by Vaihinger, that it was only possible to say that phenomena occurred *as if* certain laws held, and the older view that natural law was absolute, even if inaccurately formulated. His epistemology, as it came through in this context, was neither relativist nor instrumentalist nor naive realist. He saw no reason for saying there were no regularities in nature to which statements about laws of nature corresponded. At the same time, influenced by Milne's principle of cosmological relativity, he believed there was no favored point or center in the universe. He also realized that a situation was altered by human knowledge of it. Moreover, he added, the laws of nature were not the same for all time: the laws of nature were changing. But there was nothing arbitrary or haphazard about such change. So far from being laid down by the arbitrary word of a creator, they might prove to be a system as intimately and rationally knit together as the propositions of geometry and yet changing and evolving with time like the forms of plants and animals.[142]

Haldane was an atheist and wished to do his part to clear the many layers of confusion that occasioned human recourse to a God, from the common belief that materialism implied the belief that "a good dinner is better than a good deed" to the new philosophies of science being put forward by Jeans and Eddington. One technique he used was to historicize religious beliefs, sometimes employing ironic metaphor to do so. If Aurelian had reigned as many years as Constantine, he hypothesized, Britain would now be having numerous discussions of "Mithraism and Its Critics." His views on religious liberty were unfolded in the form of a parable about the Republic of Krassnia in which the official doctrine of the state was dialectical materialism, its president was formally anointed as the Chief Materialist and its national anthem was "There Is No God in Krassnia."[143]

As to the new philosophies of science, Haldane insisted that both relativity theory and quantum mechanics were comprehensible in materialist terms, indeed more so than in idealist terms. Quantum mechanics, he argued, raised more difficulties for Jeans and Eddington in their defense of theism than it solved. Quantum mechanics had furnished a great deal of new knowledge not possible before. What it took with one hand, it gave back with the other. It was vital to understand the principle of uncertainty properly. Although it was impossible to predict with certainty the future movement of a given electron, it was possible to predict the distribution of a number of electrons with an accuracy that was very great indeed. Relativity theory, in Haldane's view, became intelligible from a dialectical and materialist point of view once the world was regarded as consisting, not of things, but of processes or events. The classical theory of space and time had to be rejected as postulating something beyond matter, namely an abstract space and time that had properties apart from those of events going on in them. Relativistic space-time therefore reinforced materialism as far as he was concerned.[144]

Haldane's interpretations of contemporary biology, from a Marxist point of view, focused for the most part on a dialectical account of evolution, brought him into conflict with A.P. Lerner of the London School of Economics. Lerner took exception to an article Haldane published on this subject in *Science and*

Society, written incidentally from behind the lines in Madrid in 1937. Haldane was trying to pin the dialectic onto biology from outside it, Lerner asserted, but it was "purely gratuitous." Replying, Haldane insisted that the dialectic was indeed an aid both to the understanding of known biological facts and to the discovery of new ones. He did not go so far as to claim that the results of his research could not have been achieved without a study of Engels and the philosophy of dialectical materialism, but nevertheless thought it significant that they were not achieved without such a study.[145]

But Lerner was the least of his problems in defending the relevance of Marxism to biology. British scientists and philosophers of science, such as Baker, Hill, and Polanyi, were pressing Haldane and Bernal to say what they thought about Lysenko and what was happening to genetics in the Soviet Union. Haldane, a professor of genetics at the University of London, was unwavering in his commitment to the science of genetics, but at the same time hesitant to believe the worst of the situation in the Soviet Union. He took a dim view of geneticists being violently attacked and labeled as anti-Darwinist, and of genetics being denounced as incompatible with dialectical materialism, with the rise of Lysenkoism in the Soviet Union. He was also distressed by the cancellation of the World Genetics Congress, which was to have been held in 1937 in Moscow and the failure of a Soviet delegation to attend the congress when finally held in 1939 in Edinburgh. In public, he bent over backwards to put a brave face on it. To charges of fellow scientists that Soviet science was overladen by fraud and propaganda and that Soviet genetics was being destroyed by the attempt to apply dialectical materialism to science, Haldane tried to shift the focus by throwing back at them the situation at home. Conditions for research in genetics were better in the Soviet Union than in the British Empire, where scientific research was dependent on patronage from wealthy individuals, he argued, and pointed out that the only department of genetics in the University of London was about to fold. In any case, he went on, "hard words break no bones" (!) and remarked that the attacks on genetics had not led to the curtailment of Vavilov's work. He expressed confidence that, as a scientific question, it would be resolved in a scientific fashion. Going even further, he expressed the view that, so long as it did not lead to the suppression of research, such controversies were a sign of healthy scientific thought.[146] For 1938, it was entirely too sanguine. In 1939, Haldane wrote to Vavilov, a long-time friend and the source of his invitation to the Soviet Union in 1928, asking him to write an article for *Modern Quarterly*. Vavilov agreed. However, in 1940 Vavilov was arrested. Yet in 1941, the year of Vavilov's trial, Haldane wrote:

The controversy among Soviet geneticists has been largely one between the academic scientist, represented by Vavilov and interested primarily in the collection of facts, and the man who wants results, represented by Lysenko. It has been conducted not with venom, but in a friendly spirit. Lysenko said (in the October discussions of 1939): "The important thing is not to dispute; let us work in a friendly manner on a plan elaborated scientifically. Let us take up definite problems, receive assignments from the People's Commissariat of Agriculture of the U.S.S.R. and fulfil them scientifically." Soviet genetics, as a whole, is a successful attempt at synthesis of these two contrasted points of view.[147]

It makes rather shocking reading and it makes one wonder what sort of psychic mechanism was at work, what bizarre forces played themselves out in the minds of men such as Haldane and Bernal, as in the minds of so many of the intellectuals of the Comintern.* But one factor surely must have been the severe polarization of the times and the atmosphere of all-or-nothing commitment demanded of all who adhered to the communist movement in such times.

Levy

Another of Britain's scientists who came to Marxist philosophy and to the Communist Party during this period was Hyman Levy, who had come from working-class Edinburgh via Göttingen to be professor of mathematics at Imperial College, London. Already a mature scientist as well as experienced trade unionist and political activist (in the National Union of Scientific Workers and the Labour Party), Levy joined the CPGB in 1930.

A conscientious popularizer of science, Levy believed that all aspects of culture should be permeated by science. He believed in the importance of seeing science as a whole and offered two full-scale "landscape pictures" of modern science: *The Universe of Science* in 1932 and *Modern Science* in 1939. In bringing scientific knowledge and scientific method to bear in transforming the modern mental outlook and in calling forth new conceptions in all spheres, Levy believed, philosophy had a crucially important role to play.

The age of materialism, Levy explained, had reached its zenith in the last century and now the pendulum had swung in the opposite direction. The idealist philosophies of Jeans, Eddington, Milliken, Smuts, and others, represented a reaction against the materialism of a past generation. What was called for, however, was a new approach to materialism. As part of this new approach, scientific method itself had to change and to transform itself for the new age.

---

* His attitude would change quite dramatically in the postwar period. When Lysenkoism reached the peak of its influence in the world communist movement, Haldane's opposition became explicit and resolute.

Levy did not at first find the existing philosophy of dialectical materialism to be quite everything that was called for in such a new approach. In his 1934 essay entitled "A Scientific Worker Looks at Dialectical Materialism," he commented that the terminology of dialectical materialism was "a trifle quaint." With some writers there was no recognition that its language must be fluid. Since the time of Hegel, science had been transformed out of all possible recognition and there was a need to be more adaptable to this. Moreover, the use of terms like "struggle" reeked of animism. Inanimate matter did not struggle. Nor did a "quantity" in any circumstances ever become a "quality," any more than a color ever 'became an inch or the number seventeen. Refreshing, it might seem, to have a Marxist scientist call upon his fellow Marxists to be less bound to Hegelian terminology and more sensitive to science. But where Levy went with this insight was something else, far from the expectations his remarks might have aroused. He retained the term "dialectical" and seemed to have no trouble with such terms as "negation" and "contradiction." He introduced a new term "isolates" that he seemed to believe would solve all problems. The first step in the study of dialectic, Levy argued, was to chip out its isolates, to study them and the laws governing them and to remake the dialectic by seeing them again in their environments.[148]

Thus he proposed to reformulate the three laws of dialectics: The first stated that isolates existed and had a limited range of stability. When the limit was reached, a new phase, a new form of isolate emerged. The second indicated the way in which internal contradiction would arise in any phase and bring it to its limit. The new emerging process was the negation of the original phase. The third asserted that the next phase was attained by negating the factor itself being negated. However, having reformulated these laws, supposedly to take account of developments in science, Levy then cast doubt on their applicability to science at all. The principal application of the laws of dialectics was in the field of social and economic development. They appeared to add little or nothing to the detailed methods of analysis produced by scientific workers. In a sense, they could not be expected to add anything to these, as such laws professed to stand above science. Dialectical materialism was primarily an interpretative method, rather than a method of detailed investigation. The reader was left perplexed on many counts, from asking what counted as "science," with social and economic development evidently left outside its range, to wondering what did dialectical materialism amount to as an interpretative method after all.

Levy's major excursion into the philosophical arena was *A Philosophy for a Modern Man*, widely discussed, as it was a Left Book Club choice in 1938. Still "his own man," he continued with much the same sort of approach, although he was more confident about the applicability of his method to science. He did not explicitly characterize his philosophy for a modern man as dialectical materialism. In fact, the word *dialectical* appeared on only one page out of 281. The central concept of the book is that of "isolates" which, even conceding its possible usefulness as a cognitive tool, nevertheless was made to carry far more weight than it could bear.[149]

The book was reviewed in many quarters and received very varied reviews, from a somewhat dismissive one by Bertrand Russell* to a warmly accepting one by Harold Laski. The book was controversial even within the ranks of the CPGB. Party philosopher and guardian of orthodoxy, Clemens Dutt, gave it a highly critical review in *Labour Monthly*. Levy's philosophy, he contended, diverged from the path of Marxism. It has all the advantages and also all the defects of a self-made philosophy. It represented, not the working over of the whole history of philosophical thinking that Engels had recommended, but the attempt to elaborate a philosophy around a few general principles, sticking as close as possible to the facts of natural science. But his generalizations from natural science, Dutt argued, were no substitute for philosophical thinking. His approach was fundamentally mechanist and preoccupied with discrete objects. In discussing motion, he concentrated almost exclusively on movement in the form of mechanical movement, in the manner of Bukharin.[150]

Levy replied angrily. He summarized the various attacks made on the book from various quarters: It was criticized by logical positivists as simply linguistic confusion. To them, he said, his reply was unprintable. By others, it was criticized for avoiding difficulties by being too abstract, for avoiding difficulties by being too concrete, for not using the language of Marxism, for not going on about the laws of dialectics, and for being mechanistic. He considered Dutt's review unhelpful, inaccurate, shallow, and undialectical. The tone, he said, was not one belonging to a joint enterprise, to which they were all trying to contribute, but one that implied he was an outsider butting in. He spoke of how he saw himself portrayed: "I, poor mechanical devil, am clockworking my way about in the outer darkness." To defend himself against the charge of mechanism, he referred Dutt to his article "The Fallacy of Mechanism," just published in the first issue of *Modern Quarterly*. As to the demand to work over the whole history of philosophical thinking, he answered that he was too busy and suggested that Dutt do it himself. Replying to the reply, Dutt repeated his two main criticisms: Levy was preoccupied with isolation, and his general law of change was a mechanistic scheme in contrast to the dialectical view of the identity of opposites and development by contradictions.[151]

---

* Russell wrote: "Hegelianism becomes bland and sensible in the writings of Caird and Bosanquet, and so does Marxism in the writings of Professor Levy ... The English, when they have been aware of the existence of the Germans, have ... either ... treated them with contempt, or they have edited them and bowdlerized them until their systems seemed compatible with common sense" (*New Statesman* Feb. 12, 1933). There was an ensuing polemic between Levy and Russell in subsequent issues.

Hogben

Another left-wing, philosophically inclined scientist with whom Dutt entered into polemics was Lancelot Hogben, professor of social biology at the London School of Economics. From a Plymouth Brethren background, Hogben had turned into an extremely militant atheist. Although he considered himself a Marxist, he felt no attraction to the CPGB, the Popular Front, or to dialectical materialism. Although he had felt some sympathy for the Communist Party of South Africa when he held a chair at Cape Town in the 1920s, his disillusionment with the communist movement developed with the inauguration of the Five Year Plan, which he felt revived the discredited ideology of industrial capitalism, and reached its peak with the execution of Bukharin and the purges of the late 1930s. Hogben was appalled by the CPGB's support of these policies and felt no enthusiasm for what would become of Britain if they ever came to power. Nor was he drawn to Bernal's "treeless utopia." He preferred England's green and pleasant land to a "beehive city with a single glass roof."[152]

In philosophy of science, Hogben thought dialectical materialism to be "obscurantist rubbish"[153] and wished to disassociate the Marxist conception of historical development through class conflict from the Russians' "mania for dialectical laws," from the "peculiarities of verbiage and historical associations which make it well nigh impossible for an Englishman with a materialistic bias to extract any germs of intelligible meaning from the available writings of communist philosophers."[154]

As to what Hogben proposed to substitute for what he considered to be the mystifying Hegelian jargon of dialectical materialism, the answer was behaviorism. The antinomy of mind and matter was resolved in the concept of behavior, he argued in his major philosophical statement, *The Nature of Living Matter*, published in 1930. The book was directed mainly against the "God-building" philosophies of science of his day, particularly against Eddington, Jeans, Smuts, Whitehead, and J.S. Haldane. These, in Hogben's opinion, represented assaults on the mechanistic-materialist approach that constituted the only sound basis for scientific advance. Science, and its reductionist principles, was the bulwark against the forces of darkness and obscurantism. It was all too cosy, he said, to find a formula to provide a compromise for the conflicting claims of magic and science. Hogben most emphatically would have none of it, even if the alternative was the defiant stoicism of Bertrand Russell (to whom the book was dedicated). It was necessary to have the courage to face the austere neutrality of a universe that mocked human self-importance and to face the ruthlessness of death and decay. Another feature of Hogben's approach was its "publicism," the necessity of drawing a sharp distinction between the private world of individuals and the public world of social knowledge. Science belonged to the latter, and yielded knowledge that was consensual, communicable, verifiable, and ethically neutral. The new philosophers of science, in his view, were trying to collapse the public world of science into the private one of their choice.[155] It left one with the feeling of much left out and many loose ends hanging, about which there was more to be said than an exhortation to face the worst.

A polemic ensued between Hogben and Dutt, as a result of Dutt's review of the book in *Labour Monthly*. Dutt thought Hogben's publicism to be in essence a Machist evasion of the basic philosophical issue of materialism versus idealism, and protested against his reduction of matter, life, and consciousness to the framework of external behavior. Both Machism and behaviorism were corrupt philosophies of science, Dutt contended, and they reflected the decay and crisis of the capitalist world. It was the duty of Marxists to fight for the purity of Marxist theory against mechanist distortions and concessions to idealism, to show the class nature of the groupings and tendencies in philosophy and natural science and to evaluate the positive achievements of bourgeois science on the basis of dialectical materialism. Dialectical materialism represented the scientific outcome of philosophy and indicated a way out of the morass of bewildered conclusions and idealist and religious explanations. Dutt insisted that Hogben was wrong to contend that dialectical materialism was not yet sufficiently developed for its adherents to contribute much to contemporary problems. Dutt characterized it as, not a finished system, but a developing system expressing the interconnectedness of knowledge at a given stage of development of scientific inquiry. It was a system and not merely a method, Dutt emphasized. Dialectics was not a mere form of thinking, but something objectively occurring in nature. The theoretical thinking that reflected the dialectical character of natural processes must itself be subject to the same dialectic.[156] Dutt's reply, to be sure, brought back in much of what Hogben left out, though perhaps somewhat too easily.

Needham

A scientist who conceived of the relevance of Marxism to biology in a very different way from Hogben was another fellow of the Royal Society, the Cambridge biochemist Joseph Needham. He shared Hogben's opposition to vitalism in the mechanism-vitalism debate, but believed his own philosophy of integrative levels to be superior to behaviorism for coming to terms with problems of life and consciousness. He was at first somewhat sceptical about the relevance of the Hegelian dialectic to science, but experienced something of a *volte-face* on this issue upon coming into contact with the Soviet delegation at the London History of Science Congress in 1931. What had the greatest impact on him was the striking parallel between the philosophy outlined in Zavadovsky's paper and the philosophy he had himself been developing. As he recollected in the course of giving the Herbert Spencer Lecture at Oxford in 1937, he had discovered that "in Russia, under the guidance of an elaborate philosophy at that time almost unknown here, a new organicism had been growing up."[157] Like Zavadovsky, Needham had come to hold that biological phenomena were in continuity with but not reducible to physicochemical or mechanical phenomena.

From an Anglo-Catholic background which he never repudiated, Needham nevertheless turned to socialism in his politics and to dialectical materialism in his philosophy of science. Describing himself as from his youngest days "a person who absolutely could not do without a world view,"[158] he always looked upon science in an intensely philosophical way. His philosophical interests made him something of a rarity among biochemists of the era, who thought that scientists should spend their time outside the laboratory in such innocent occupations as golf or fishing, rather than with dubious studies of the history and philosophy of science. His encyclopedism also singled him out, and on one occasion the journal *Brighter Biochemistry* teasingly carried a notice of a new book by Needham entitled *Eggs: From Aristotle to the Present* (in 27 volumes).[159]

In politics, he passed from being on "the wrong side" in the general strike of 1926 to participation in the various activities of the scientific left, such as the CSAWG and the AScW, to support for the Soviet Union and the Popular Front. He did not, however, joint the Communist Party, nor did he subscribe to all the tenets of "Bernalism." The basic difference was that he did not believe that scientific method could or should encompass the whole of life. It was one thing to hold that science could only come to its true fulfillment under socialism, and Needham did hold that science and capitalism were irreconcilable, but it was another to believe that a socialist society should be completely circumscribed by the forms of scientific rationality.

Needham's Marxism was obviously of a very distinctive sort. He sought an all-encompassing vision, and Marxism found its place within in, but Marxism was not itself all-encompassing for him, as it was for Bernal and Haldane. Of a mystical bent, Needham was never inclined to atheism and always retained his "sense of the holy." His faith was in the tradition of Otto's *The Idea of the Holy*, a fideist view of religion based on the experience of the numinous, that spurned proofs of the existence of God and the whole tradition of theological rationalism. He did not, however, allow his God to interefere with his science, his philosophy of science, or his politics. He warned specifically of the dangers of scientists carrying their religion into the laboratory, and of philosophers responding to the breakdown of traditional scientific conceptions by calling upon God to solve problems of philosophy of science.

His position, as he set it out in the 1930s, was that each of the great forms of human experience, that is, science, philosophy, religion, history, and art, was equally valid. Each seemed to lead to a characteristic world view, incompatible with and sometimes even contradictory to those of the others. The proper appreciation of the world by man could not arise from the pursuit of any one by itself, but only in the experience of them all, with little hope of uniting them into a coherent view of the universe. The harmonious man was one to whom science and religion, though always antagonistic, were equally necessary methods of attaining contact with what lay at the core of the world.[160]

By the early 1940s, Needham had modified his position somewhat. Earlier, he had characterized science as abstract, quantitative, deterministic, mechanical, analytic, impersonal, ethically and aesthetically neutral, and religion as concrete, qualitative, paradoxical, alogical and irrational (despite the rational cloak thrown over it by theologians), and normative. Looking back, he considered such a description of science to be too narrow and such a description of religion to be too neo-Platonic. The Achilles heel of his former position had been ethics, which he could never assimilate into one or the other forms of experience. As politics more and more forced itself on his attention, he came to see that his fundamental limitation had been to envisage the experiencing being as a solitary unit. Ethics and politics had proved to be the cement necessary for the unification of the divergent forms of experience. As soon as one began to consider man under his social aspect, the germs of a unified world view began to appear. Ethics and politics corresponded to the valency bonds. He considered that the dividing process had been succeeded by a uniting one and that an integrated world view had emerged from the differentiations he had made. In analyzing this process, he mentioned the creativeness of contradictions giving rise to higher syntheses.

This was where Marxism came in for Needham. Marx and Engels, building on both the dialectical process in idealist philosophy and on the new understanding of evolution that was then dawning on men, influenced by both Hegel and Darwin, took the revolutionary step of placing the resolution of contradiction within the historical and prehistorical process itself. Contradictions were not to be resolved in heaven, but right here on earth. This was the meaning of the dialectical materialist way of expounding cosmic development, biological evolution, and social evolution. As Needham explained it, it seemed that the Marxist philosophy of dialectical materialism at least tied together four of the five great forms of human experience, and ethics and politics besides:

> We cannot consider nature otherwise than as a series of dialectical syntheses. From ultimate physical particle to atom, from atom to molecule to colloidal aggregate, from aggregate to living cell, from cell to organ, from organ to body, from animal body to social association, the series of organisational levels is complete. Nothing but energy (as we now call matter and motion) and the levels of organisation (or the stabilised dialectical syntheses) at different levels have

been required for the building of our world. The consequences of this point of view are boundless. Social evolution is continuous with biological evolution, and the higher stages of social organisation, embodied in advanced ethics and in socialism, are not a pious hope based on optimistic ideas about human nature, but the necessary consequence of all foregoing evolution.[161]

As to the place of ethics in the new synthesis, which had previously caused him problems, his evolutionary morality has been aptly characterized by Werskey as "an unblushing example of the 'naturalistic fallacy.' "[162]

Needham saw dialectical materialism as the solution to the most vexing problems in the history and philosophy of science. He spoke of it as "the quintessence of scientific method,"[163] as "the natural methodology of science itself."[164] The concept of the dialectics of nature was crucial to his understanding of scientific method. Marxism had brought to light the true nature of the dialectical process by bringing it into the sphere of nature:

> Marx and Engels were bold enough to assert that it happens in evolving nature itself, and the undoubted fact that it happens in our thought about nature is because we and our thought are a part of nature.[165]

The facts spoke for themselves, Needham insisted, for nothing was more dialectical than nature. The existence of dialectical tension of opposites explained a lot for Needham. The presence of chaos in nature was necessary for the development of scientific order. The essence of dialectical materialism, for Needham, was its insistence on successive dialectical levels in nature within the framework of emergent evolution. The higher forms of the organization of nature, such as life, consciousness, and human society, were continuous with, but not reducible to, the lower forms. In these successive forms of being in the scale of complexity and organization, it was precisely the complexity and organization at each level that constituted its special quality.

This way of thinking was of inestimable value in dealing with the concrete problems of biology. Most importantly, it solved the problem issuing in the long-debated controversy between mechanism and vitalism. In earlier days, because of ignorance of other forms of materialism, he had veered towards neo-mechanism, but in time he conceived of the controversy as a dialectical deadlock that was resolved by a judicious organicism. Mechanists had been wrong in being overly enamored of oversimplified physicochemical explanations of biological processes, but right in opposing hypotheses of vital forces, entelechies, and so on. Vitalists had been right in their eagerness to safeguard complexity, in their attention to the phenomena of organization, in their insistence that living phenomena would always escape total physicochemical analysis, but wrong in confusing the organizing relations of a given system with its supposed anima. Life constituted a new level of the organization of matter, not inscrutable, but not to be forced into the framework of laws operative at lower levels. There was no place in biology for traces of animism, for souls and vital forces. Living things differed from nonliving things in degree not in kind. There were no sharp and fixed boundary lines in nature. The attack on vitalism was necessary to the scientific study of living things, but mechanist methods of doing so needed to be ruled out, as organisms could not be exhaustively explained in terms of laws governing the behavior of atoms and molecules.[166]

In many ways, dialectical materialism could be of service to the working biologist: by pointing the way towards the most promising hypotheses and by indicating which questions were meaningful and answerable. However, Needham warned, there were dangers. Dialectical materialism was so sharp an instrument that, although there could be no question of its value as a general system, the detailed application of it must always be a delicate and difficult matter. Dogmatism was at all cost to be avoided. The year was 1938 and he was thinking of Lysenko. Speaking briefly of the state of genetics in the USSR, he said that the criticism of gene theory was not well based. He was still, like other British Marxists, hoping for the best, and he also added that the USSR offered greater material support for science than any other human community.[167]

Needham felt that other problems, too, were resolved in and through Marxist philosophy. Dialectical materialism, he believed, was itself the synthesis in the age-old contradiction between materialism and idealism in the history of philosophy. It also illuminated the pattern underlying the history of science in showing that science progressed through new hypotheses that combined all that was truest in previous hypotheses, not by mere compromise, but by creative synthesis. Moreover it cleared up what Needham considered to be his own earlier epistemological confusions. He had earlier been influenced by Mach, but reading Lenin's *Materialism and Empirio-Criticism* had helped him to see that scientific truth was relative but still objective.

Needham had come to be very critical of the various philosophies of science formulated by Mach, Le Roy, and Poincaré, as well as by Jeans and Eddington. For them, theoretical physics, the most fundamental of all the sciences, turned out to be a sort of circular construction of conceptions each of which could only be defined in terms of the others. Needham wondered whether the Victorians would have fought as they did for the place of science in education and in public life if they had believed the pale-blooded doctrine that was now being put forward.

Another problem to which Needham addressed himself was that involving the second law of thermodynamics, according to which free energy was constantly decreasing while entropy was increasing. The development of modern science had led to a curious divergence of world views. For the astronomers and the physicists, the world was continually running down. For biologists and sociologists, a part of the world at any rate was undergoing progressive development. Needham was not so inclined to pessimism as most; he saw the process as the inevitable concomitant of rising levels of physical, biological, and social organization, while at the same time he hinted that the victory of socialism on an international scale could develop new techniques for the conservation of energy. Adding a light note to the discussion, he told the story of a lecturer giving a popular talk on the subject of implications of the second law of thermodynamics. When the meeting was thrown open to discussion from the floor, a member of the audience asked: "How long, sir, did you say it would be before the universe was completely run down?" The lecturer replied that he had said seven hundred million years. His questioner heaved a deep sigh of relief and said: "Thank God. I thought you said seventy million years."[168]

Needham's various personalized essays on these topics were inevitably learned, engaging, and probing. The persistent theme was the desire to encompass the whole range of human experience, to discover the connections as well as the distinctions, to omit nothing that was vital and of value. He sought integrality and was aware of the consequences of the widespread fragmentation he saw all around him. He noticed that those of his colleagues with the narrowest specialized interests tended to be the most reactionary. Without history and without philosophy, a scientist was likely to fall into all sorts of fantasies. The more highly developed his knowledge of social evolution and the more highly developed his basic world view, the better his scientific work was likely to be. Just so, the historian without science was a donnish period prisoner. Without philosophy, he was a pedantic purveyor of meaningless facts. At the same time, he thought that, without religion, the scientist, philosopher, or historian would fall into intellectual pride. He found his ideal expressed by Comenius:

> Can any man be a good Naturalist, that is not seen in the metaphysic? Or a good Moralist, who is not a Naturalist? Or a logician, who is ignorant of reall Sciences? Or a Divine, a Lawyer, or a Physician, that is no Philosopher? Or an Oratour or Poet, who is not accomplished with them all?[169]

Needham knew that he had fallen short of a thoroughgoing theoretical integration. He had brought together science, philosophy, art, history, ethics, and politics, but there was still a great theoretical chasm between these spheres and that of religion. There was no theory to bridge the gap between dialectical materialism and the "sense of the holy." It could, he asserted, only be bridged in practice. The only unification that was possible here, as Needham saw it, was an existential one, unification in the life of a man, unification in action. The answer was to view the world as a whole and to see man's place in it and to exercise the soul in conformity with virtue, without hope of fully unifying the products of its exercise.

Needham's answer, however noble, was nevertheless far from satisfactory to those who could not rest content with such a significant and yawning gap that made it impossible in the end to view the world as a whole or to see truly man's place in it. Needham's commitment to Marxism, as far as it went, was respected, and rightly so. It was, however, riddled with inconsistencies, so long as such a significant area of human experience was excluded from the unity of method that bound together the rest. He did not achieve the integrality he sought, although he did overcome the prevailing fragmentation in many ways. This was no small achievement. Even those who most assuredly affirmed the integrality and all-embracing character of Marxism were not always able to clarify the specific lines of connection in such a vital way as Needham had with those connections he did affirm.

The Carritt-Rudas Debate

One form of connection that Needham took for granted at this time, that between dialectical materialism and communism, was disputed by others. E.F. Carritt, fellow of University College, Oxford, and a professional philosopher, began a discussion on this question in the pages of *Labour Monthly* in 1933 that extended into 1935. He expressed his sympathy with the objectives of communists (though not with their strategy and tactics), but was disturbed to hear both from communists and their bitterest opponents that communism was inseparable from dialectical materialism. It was a philosophy that he found difficult to accept on philosophical grounds. It was, he contended, a Protean theory to interpret, referring to the different positions of Bukharin and Deborin. He also felt that the dialectic could not escape the sceptical consequences of a materialism without immanent teleology. In any case, he did not see why it should have any necessary connection with the socioeconomic doctrines of communism. What Engels's views on motion, heat, light, electricity, and evolution of species had to do with communism, he could not understand.[170]

There was a whole series of replies, nearly all arguing for a necessary connection between dialectical materialism and communism. J.M. Hay argued that it was matter of an indissoluble connection of the theoretical whole with one of its parts. T.A. Jackson contended that it was impossible to be a communist without being a dialectical materialist and that it was impossible to hold to dialectical materialism with alternative views of social development and political practice with logical consistency. From faraway Moscow, Rudas came into the discussion and insisted that it was impossible to be a dialectical materialist without being a communist, for one could not ignore the consequences of one's philosophy. Going further, he actually stated that outside the communist party it was impossible to be a communist, a dialectical materialist or a Marxist. Only a revolutionary could revolutionize his thought. Only those who took part in the struggle to change the world could rightly understand the theory of the world's dialectical development. Communism was a revolutionary movement that arose and grew up on the basis of certain laws of development. Dialectial materialism was the consciousness of these laws of development. It was, Rudas explained, the relation of objective dialectics (communism) to subjective dialectics, the consciousness of it in the heads of its adherents (dialectical materialism). The dialectics of society was only a special case of the general dialectics of the world. Whoever did not recognize the dialectics of nature could not, without illogicality, recognize the dialectics of society either. It was impossible to act correctly with an incorrect consciousness. Without dialectics, both scientists and revolutionaries were lost. The sciences were condemned to grope in the dark, as was the proletarian struggle. In science, according to Rudas, all correct results were dialectical results, even if their discoverers did not know it. Nevertheless, they would arrive at much more correct results if they consciously applied dialectical methods. In one of his articles in the protracted controversy, Rudas had referred to Edward Conze, who came into the discussion and mentioned that dialectical materialism in the Soviet Union had every possible facility at its disposal, including protection from rival philosophical schools through administrative measures and the police. Rudas's reply was both abusive and evasive.[171]

The discussion brought to the surface a serious and important issue, that of the relation of philosophy to politics, but the heavy-handedness and sectarianism of Rudas, who dominated the discussion, did not help much in sorting it out. It did show, however, where British communists stood on the question of dialectics of nature. In the dialectics of nature debate as it unfolded in Britain, members of the CPGB who addressed themselves to the question seemed to be on the same side,* whereas those who held that Marxist socioeconomic tenets were compatible with philosophical trends other than dialectical materialism were outside the party. There had come to be a very great emphasis on dialectical materialism in the party and there was no longer significant support in the party for the minimalist approach to what it meant to be a communist that had been there in the 1920s.

---

* Caudwell, however, had a somewhat more complex position than the rest, neither in favor of an objectivist dialectics of nature, nor on the side of the usual arguments against it. His distinctiveness vis-à-vis this question seems to have gone unnoticed both then and since.

British Marxism and the Soviet Union

Although British Marxism was much influenced by Soviet Marxism, even deferential towards it, it was by the late 1930s in a far healthier state. It did, of course, owe a great debt to Soviet Marxism in 1931, but it did not register the changes that had taken place by the end of the decade. British Marxists, at this time, showed few signs of understanding what was really happening in the Soviet Union and tended to overrate their Soviet counterparts and to underrate themselves. Though affected by the prevailing atmosphere of the Comintern, there was far greater freedom of discussion in the CPGB than in other parties of the Comintern, and its members were much more open to dialogue with those of differing shades of philosophical and political opinion than were communists elsewhere. Arthur Koestler, after his tour of Britain on the Left Book Club circuit, in which he had deviated from the Comintern line on the Spanish Civil War, remarked that the CPGB was notoriously lax in reporting deviations to higher quarters. He found Britain to be "a country neither of yogis nor of commissars" and was constantly bewildered by British communists who "put decency before dialectics."[172]

Still, some were more vigilant than others. When *Modern Quarterly* was set up as a Marxist theoretical journal in 1938 with an editorial board that was both highly distinguished and broadly based, it set a tone that was fresh, lively, serious, and sophisticated. It established and maintained close connections with *Science and Society* in America and *La Pensée* in France, both journals of the same type. Yet Rajani Palme Dutt, editor of *Labour Monthly* and chairman of the CPGB, in the context of an article welcoming the appearance of *Modern Quarterly* and *Science and Society* as showing the growth of Marxism in the English-speaking world, felt compelled to remark that neither was yet the basis for a journal of the standard of *Pod znamenem marksizma*; this, despite the fact that it was the time of the deepest degeneration of that journal. Palme Dutt also took the new journal to task for mentioning only its connections in the US and France and for failing to mention the Soviet Union. At the same time, after stating the importance of the alliance of philosophers and natural scientists, he criticized *Science and Society* for publishing an article by Hogben (one of its foreign editors) without adequate counterstatement (although there had been a reply from Dirk Struik, one of its founding editors.)[173] Interesting, and significant as well, was the response of *Modern Quarterly*. In an editorial in its next issue, it noted the criticism of the new journal by *Nature* for its inclination to identify the progress of science with one form of the organization of society, that is, "Marx, Lenin, and Russia," and by *Labour Monthly* for its lack of authoritative theoretical discussion of Marxism and its failure to use existing Marxist sources. These criticisms, it stated dismissively, cancelled each other out.[174]

Pod znamenem marksizma itself kept a watchful eye on all such publications and never hesitated to pronounce on the work of foreign Marxists. One example of the sort of pattern this followed was to be seen following the publication of T.A. Jackson's *Dialectics: The Logic of Marxism and Its Critics* in 1936.

Jackson

Tommy Jackson, unlike so many other British Marxists writing on philosophy in those days, was not an academic high flyer, but a blacklisted printer turned street-corner orator and Labour College/Secular Society lecturer. Throughout his life of poverty and homelessness, he read voraciously and championed many causes, being particularly devoted to the causes of Irish republicanism, atheism, and socialism. A founder-member of the CPGB and member of its executive committee, he lost his position of leadership because of his opposition to the "class against class" policy of the Comintern's "third period." This represented a significant political evolution on Jackson's part from the days in which he declared at an early party congress that communists should take Labour Party leaders by the hand as a preliminary to taking them by the throat. Jackson had also become quite critical of the unintelligible jargon and the heresy-hunting that was coming to prevail in the communist movement. In an article in *Communist Review* in February 1932, he appealed to the party "to 'liquidate' this process of 'Inprecorization'* and thereby restore plain English and commonsense to their rightful place in the party's esteem."[175] The Comintern may have taken a dim view of such statements and the British party might remove those who made them from its EC, but it did so without the expulsions and anathemas so rife in the other parties of the Comintern. The implications of critical dissent for Jackson meant retirement from full-time work, but it did not mean being cast out into the outer darkness. He used the time to write books, including *Dialectics*. He had a great love for philosophy ever since as a young lad he had picked up G.H. Lewes's *Biographical History of Philosophy:*

> I devoured it like a novel, thrilled again and again by the greatest adventure story ever written—the adventure of the mind of man. I realised with the crushing force of an instant conviction that I had never known what it was to *think*. I had remembered, I had wondered, and I had guessed. But like a navigator without a compass, on a night when the stars are veiled, I had been helplessly at the mercy

---

* "Inprecor" was the popular colloquialism for referring to the Comintern publication *International Press Correspondence*.

of the winds and the waves. . . . I gathered in my book-hunts all the works he
named which came within reach. . . . I was . . . acquiring a grasp of the
universe . . . as a unity in multiplicity in perpetual process of self-transformation.
I was, though I would not have known what you meant if you had told me so,
preparing myself for Marx![176]

When he came to Marxism, Jackson pursued Marxist theory with the same
enthusiasm and with the autodidact's thirst for knowledge. His Marxism was
richer for having emerged within this sort of quest for general knowledge and
culture. Realizing this himself, he pointedly asked:

What do they know of Marxism who only Marxism know?[177]

It was a healthy attitude, making him an entirely different species from those
educated only on textbook Marxism, thereby lacking any basis upon which to
judge it.

Dialectics was a lively and highly polemical book, its style owing to
Jackson's years of experience of speaking on street corners, which had taught
him that a "thousand people would gather for a dog fight who would be
scattered by a sermon." Jackson made no claim of originality for the book, but
it was a competent and engaging exposition of what the Comintern considered
to be "orthodox" Marxism. It put heavy emphasis on the integrality of
Marxism. Marxism, Jackson said, was often presented in England as a loosely
aggregated bundle of separate and distinct theories that had no connection with
each other beyond the fortuitous fact that they all originated with the one man.
Marxism thus became an Old Curiosity Shop in which political amateurs and
literary dilettanti could rummage for decorative oddments. Professor Laski,
for example, believed it was possible to reject Marx's economic system, while
accepting large parts of his social theory. It wasn't logically possible, any more
than it was to reject dialectical materialism, while accepting Marxist socioeconomic theory (as Jackson had also argued in the *Labour Monthly* debate).
What gave Marxism its living unity, Jackson asserted, was its dialectical
materialist method. Not that method could be separated from world outlook,
for neither could be understood apart from the other: method was revealed in
the outlook and outlook was implicit in the method. This meant that:

Dialectical materialism as a logically united outlook and method must be
accepted or rejected as a whole. . . . Especially in Britain, it is common to attempt
an eclectic revision of Marxism (as distinct from an organic development . . .) a
revision which takes the form of proposing the rejection of one of its integral
phases and the retention of the remainder in conjunction with some alien theory
supposed to be more "up-to-date."[178]

The book was a staunch defense of the dialectics of nature, and put great stress on the connection between the Marxist conception of understanding (its subjective dialectics) and its conceptions of nature, history, and revolution (its objective dialectics). Nature and history constituted a unity of the dialectical kind. History developed from nature, without being a simple continuation of nature. Before Darwin, it was thought that there was necessity in nature and freedom in man. However, Darwinism destroyed all notions of fixity of species, all conceptions of insurmountable divisions in nature. Darwinism brought an end to the exclusion of historical progression from nature, showed the origin of species in a union of opposites, in the active interrelation between an organism and its environment, and demonstrated that evolutionary development inevitably included not only gradual progression but mutational leaps. Moreover, moving from biology to chemistry, Mendeleyev's discovery had shown that qualitative differences between the chemical elements pointed to quantitative differences in their composition. In physics, atoms had been discovered to be a structure of electricity, a unity of opposites.

Jackson also saw the new developments in physics as the final vindication of atheism and materialism. Modern physics, he said, had killed the last excuse for the argument for design in the universe. Nineteenth-century mechanical materialism had eliminated God, but kept the rest of the Newtonian universe, with the result that under various disguises the need for some form or other of the God hypothesis, as the original "push off," was constantly reappearing. Newton had postulated as primary an absolute stillness, which it took a miracle, God's primal motion, to overcome. Einstein's universe postulated the opposite: that everything was changing of itself and only confined its changes in particular directions under pressure of counterchanges in everything else. Einstein's problem was how to measure motion from a point which was itself in motion. Einstein showed that human knowledge is knowledge of things from the human point of view, that space and time were qualities of the universe, not its preconditions. It was a thoroughly materialist conception. If the universe was everything and everywhere, and *that* all the time, there was nowhere from which to come or to which to go. Jeans and Eddington, faced with the crisis in modern science, tried to get out of the difficulty by postulating a God who designed the universe as an objective demonstration of higher mathematics. They only made the confusion worse, for their position involved the logical absurdity of attempting to revive Newton's primary postulate as an explanation of phenomena that had admittedly long since transcended all possibility of Newtonian explanation.

Much of *Dialectics* was devoted to taking on the various critics of "CP Marxism" in the English-speaking world. For example, Jackson tried to show that Ramsay MacDonald's aversion to the dialectic was related to his preference for a more gradualist model of social evolution. His major opponents came under three categories: the "Marxism minus Marx" of John Macmurray, the "Modernist Marxism" of Raymond Postgate and Max Eastman, and the "neo-Dietzgenian Marxism" of Fred Casey. Macmurray was professor of philosophy at the University of London and Jackson thought his book *The Philosophy of Communism* represented a relapse from Marxism into a species of modified Hegelianism. Macmurray's method, Jackson thought, made the distinction between nature and history relatively absolute and their connection therefore mystical. Postgate's book *Karl Marx* and Eastman's *Marx, Lenin and the Science of Revolution* were characterized by the identification of Marxist philosophy with cosmic determinism, the repudiation of economic determinism in favor of modern psychology, the detestation of the very term "dialectic," the treatment of Marxism as a system of mechanical materialism upon which, as an afterthought, a mystical-idealist dialectic had been superimposed. Against the mechanical materialism which Postgate and Eastman, to Jackson's way of thinking, fraudulently substituted for Marxism, they paraded all the arguments of the idealists. Against the spurious metaphysic they substituted for the dialectic, they paraded all the arguments of the mechanical materialists. As to Casey, Jackson was particularly worried about his influence as a lecturer in the National Council of Labour Colleges. Against his assertion that dialectical materialism was not part of Marxism, Jackson argued that Marxism was wholly and entirely dialectical and materialist. Casey's position, he asserted, was one of solipsist idealism and his notion of dialectic, limited to one phrase, "opposites are identical," was nothing more than an extension of the homely proverb, "it takes two to make a quarrel."

All these ideas, Jackson insisted, were dangerous. The only difference between revisionism and counterrevolution was that counterrevolutionaries fought to destroy Marxism frankly, as a living whole, while revisionists sought to convert it into a dead aggregate of abstract theories, only mechanically interconnected and capable of every sort of eclectic substitution and replacement in the name of bringing it "up-to-date." In either case, it was the living movement of Marxism against which it was aimed. The end was the same: the conversion of Marxism into a corpse, of interest only to anatomists and museum curators.

Heavy handed towards opponents the book surely was, as was the style with Comintern polemics against "deviations," but it was reflective and reasonable in a way that many such polemics were not and it was motivated by a sincere concern for the integrity of Marxism and not by any of the meaner motives that inspired other such polemics elsewhere. Jackson also made the requisite references to the four great classic authors: Marx, Engels, Lenin, and Stalin, though not constantly and not hackishly. The book was written without the dreadful jargon, what Jackson called the "Babylon dialect," that was over-taking communist writing those days.

This, it seemed, was not quite good enough for *Pod znamenem marksizma*. In their review of the book in 1938, they took Jackson to task for failing to explain the "Leninist stage" in the development of dialectical materialism, for putting inadequate stress on the work of Lenin and Stalin, for not making use of the philosophical work of the USSR in its "fight on two fronts," directed against menshevizing idealism and mechanism, for not considering the effect of the victory of socialism on the development of dialectical materialism. They were particularly critical of his failure to criticize modern mechanists, especially Bukharin.[179] It was the time of the Moscow trials.

Cornforth

Maurice Cornforth was, in the 1930s, one of the younger British communists to concern himself with philosophical matters, and later the author of many books on Marxist philosophy. Looking back on these years, he commented that they all at that time had an unquestioning attitude to the Marxist classics and to the Soviet Union. The communist parties affiliated to the Comintern claimed proprietary rights over Marxist theory, with the final say-so in theoretical matters vested in the CPSU.[180] In the prewar period, British communists were not inclined to break from this, though within this framework they were still creative and they still thought for themselves.

Cornforth, as remembered from his 1920s schooldays by his schoolmate, the poet Stephen Spender, dominated school debates and covered reams of paper with his plays, poems, and letters. He was a vegetarian, taken with Buddhism, walked thirty to forty miles a day. He was tousle-headed and had a tousle-headed dog.[181] Schooldays over, he went on to the University of London and then to Cambridge to study philosophy. Along with David Guest, Cornforth was among the circle that gathered at the feet of Ludwig Wittgenstein upon his arrival at Cambridge in 1929. They argued furiously with Wittgenstein and with each other. On one occasion, Cornforth wrote a paper expounding various idealist notions, and Guest, already left-wing and atheist, responded with uproarious laughter. In 1930, Guest went off to Göttingen to study mathematics, but found himself studying fascism as well, and in 1931 he came back to Cambridge as a communist. He erupted into a meeting of the Moral Science Club with a copy of *Materialism and Empirio-Criticism*, bubbling over with enthusiasm about it and reading out passages about the class basis of philosophy. Some thought he had gone crazy. Cornforth did not. The scene had made a great impression on him. He went home, read the book and decided to join the CPGB. Together in 1931, Cornforth and Guest formed the first communist cells in Cambridge, first one in the university, and then another in the town. The communist cell in Cambridge would be enduringly famous for the galaxy of eminent scholars, as well as eminent spies, associated with it in those heady days. Cornforth and Guest sold the *Daily Worker*, threw themselves into their political work, and managed to finish with first-class degrees as well.[182]

Cornforth believed that his tutors in philosophy had sought clarity without finding it and that their philosophy was irrelevant to social problems. He saw Marxist philosophy as providing that clarity and as illuminating the most pressing social questions. Although in the prewar years, he went to work as a district organizer for the CPGB, rather than as a professional philosopher, Cornforth's Marxist philosophy was nevertheless crucial to his commitment as a communist. He seems to have given much thought to settling his accounts with the philosophy in which he had been educated. On occasion, he tried to bridge the gap between the CPGB and the world of professional philosophers. In 1934, he participated in a symposium with Max Black and John Wisdom on the topic: "Is Analysis a Useful Method in Philosophy?", with the paper afterwards published in the *Proceedings of the Aristotelian Society*. Cornforth's paper was a Marxist critique of analytical philosophy in which he set out to show the historical significance of analytical philosophy in light of the theory of historical materialism and to examine the theoretical weakness of the analytic method from the point of view of dialectical materialism. He began by asserting the importance of history and of epistemology over against the attitudes of analytical philosophers who were indifferent to the historical significance of their theories and unconcerned with the theory of knowledge.

This history of philosophy, Cornforth argued, as part of the history of society, was a history of class struggle. Idealism held to the creation and governing of the world by forces that were spiritual, whereas materialism held that the material world was primary and uncreated, existing and changing by its own laws. The bourgeoisie, as a ruling class, found itself in an increasingly contradictory position. As a dominant class, it clung to idealist theories as instruments of class oppression, but needed science, which supported materialism. For 200 years, many philosophical solutions to this contradiction had been attempted, the most satisfactory being that worked out by Berkeley. It accepted science, but said it was not about what it seemed to be about. If science were actually about predicting the order of human sensations and nothing more, then it couldn't contradict religion. It provided a way in which the bourgeoisie could assimilate science without forfeiting idealism. Variations on such a position were to be found in Mill, Mach, Avenarius, Vaihinger, Husserl, Russell, Poincaré, Wittgenstein, and Carnap. In the end, it meant that philosophical questions demanded extrascientific answers, as witnessed, for example, in Wittgenstein's mystical conclusion to the *Tractatus*. The logico-analytical method was the most highly developed form of the philosophy of the bourgeoisie. This did not mean, Cornforth hastened to point out, that analytical philosophers had deliberately set out to rehabilitate religion and reconcile it with science. It meant that the method came as the culmination of a philosophical development whose objective historical role has been to interpret science in such a way that it did not clash with religion. Nor did it imply that such philosophers themselves drew theistic conclusions, even if the objective historical role of their method had been to offset the danger of atheism.

The validity of the method of analytical philosophy depended on the validity of its key distinctions, such as its rigid distinction between discovery and analysis which corresponded to the distinction between science and philosophy. But this, Cornforth asserted, was epistemologically unsound. It assumed that the process of knowing could be treated in terms of the experience of the individual human subject and that the process of knowing was a purely theoretical activity. However, in Cornforth's way of looking at it, verification was a social activity. Knowledge grew through generations interpreting the world and acting on it. It was the social division of labor that left theoretical enquiries to one section and practical tasks to another, but theoretical problems nevertheless had their roots in practical problems, Cornforth claimed, referring to the arguments put forward in *Science at the Crossroads*. Discovery was not enclosed within the bounds of individual experience. Knowledge could only be adequately understood in the context of human social practice and in relation to the external world. When the connection of discovery to practice was not grasped, it seemed that philosophy had to step in. Cornforth opposed any notion of philosophy superadded to science. The only way to knowledge was by the hard road of science. Analytical philosophy meant a departure from the methods of science in its failure to perceive that discovery was also analysis and that discovery was connected to the external world. In the logical scheme of the analytical philosophers, all transition and transformation was denied and the world was reduced to the form of discrete atomic facts. Thus, he concluded, analysis was not a useful method in philosophy, in so far as it endeavored to elucidate the nature of reality by extra-scientific methods. Its objective function was to cloak the real character of human knowledge, obscuring the real import of discovery, in the interests of the bourgeoisie and in reaction to revolutionary materialism.[183]

Guest

Cornforth's friend David Guest was also giving considerable attention at this time to analyzing contemporary British philosophy from a Marxist point of view. From his earliest years, those around him all seem to have found David Guest, son of a Labour M.P., to be a precocious, provocative, and extraordinarily striking person. He was always fermenting with ideas and projects. His early effrontery was said to take people's breath away. It was thought that after becoming a Marxist his personality somehow changed, as if some inner tension had been resolved. After his various experiences in Göttingen, where he not only studied mathematics and witnessed the rise of fascism, but spent a fortnight in prison, and after his further experiences in Cambridge, where he studied philosophy and mathematics and played such a vital part in the left-wing ferment that was to sweep over Cambridge in the 1930s, Guest went off to Moscow in 1933. There he taught physics and mathematics in a Moscow school for a year and wrote an article in *Pod znamenem marksizma* on modern British philosophy. Upon his return to Britain, he threw himself into lecturing, organizing, and library cataloguing at Marx House, founded in 1933, running a "People's Bookshop" in Battersea, activity in the Young Communist League, as well as research at the University of London. Guest was enormously enthusiastic about the philosophy of dialectical materialism, which harmonized with all his intellectual and political interests. Harry Pollitt, general secretary of the CPGB, spoke of the life of David Guest during these years as an irrefutable answer to the charge leveled at communism that it was rigid and inflexible and had a deadening effect on those who came to embrace it. Speaking of the effect Marxism had on Guest, Pollitt wrote:

> It stimulated manifold energies and brought them into a synthesis that kept him from frittering away his gifts. This firm grounding in the principles of communism unleashed new creative thoughts, brought out his qualities of leadership, and enabled him to express his personality to the full. . . . His was a living Marxism, permeating all his work and thought, not artificaly grafted on to it. . . . His restless inquiring mind, never satisfied, made him determined to test out his ideas and put his ideals into practice. . . . It is very significant that it is not the embittered failures, not the careerists and reckless political adventurers, but the flower of youth who turn to communism and make the best communists.[184]

In 1937, Guest was appointed lecturer in mathematics at University College, Southhampton, but left after two terms to go off to Spain to volunteer for the International Brigade. Harry Pollitt saw him during Easter week 1938, and Guest took him through the olive groves to brigade headquarters. Across the Ebro, on the sides of the mountains, the camp fires of the Moorish troops could be seen, and their strange songs came over with the wind, intermittent with the roar of heavy guns. Pollitt last saw him against the full moon walking back through the fields to join his comrades. A few months later, while writing him a letter his mother was interrupted to answer the telephone. It was Reuters inquiring whether her son was the David Guest who had been shot through the heart on the Ebro front.

Guest's high-voltage personality and his searching mind had left a deep impression on all who knew him. The intensity of his intellectual enthusiasms can be seen in his letters, written at various periods of his life:[185]

From school to Oundle:
> If truth and happiness were incompatible, I should choose truth out of sheer inner necessity.

From Göttingen:
> My head is going round in a whirl of politics, philosophy and mathematics. At the moment, I am absolutely tormented by a desire to express my views and to clarify them. . . . The desire for truth is my strongest impulse.

From Spain:
> I have never felt so much the value of abstract things, of theory seen in its proper relation with practice, than just now. I think I can see things in their proper proportions. I have myself a lively and intense desire to explore whole fields of theoretical work, mathematical, physical, logical and far beyond these, when the conditions for this will become again possible.

It would never be. Dead at 27, only his youthful writings, indicating only his promise, would survive the gunfire of Spain.

Guest's notes from his lectures at Marx House, edited by T.A. Jackson, were published as *A Textbook of Dialectical Materialism* in 1939. His vision of Marxism was expounded as a comprehensive world view: consistent, many-sided, scientific. Guest had felt that without a general view of the world, rooted in the facts of science, socialism was without adequate grounding. The present crisis had shaken the English indifference to theory, the preference for rule-of-thumb methods, the widespread mental inertia. Thus, the growing popularity of all kinds of philosophy, psychology, and pseudoscience, that presented the danger of lapsing into the opposite extreme of embracing any kind of fantasy. Guest was anxious to distinguish the philosophy of dialectical materialism from other contenders in this situation by which those who were seeking could be sidetracked. He especially set it against what he characterized as the new varieties of philosophical idealism, with roots in theology, but supported by arguments of a logico-methodological character. In these philosophies, materialism was made to seem a dogmatic belief in the outer world that would not stand the test of critical thinking. Thus, the transition was made to the philosopher in his study doubting the existence of his writing table, to the philosophizing bishop who rejected the dogma of matter only to substitute the dogmas of religion. Modern idealism adopted a special attitude to science, one of obscurantist misdescription. It recommended itself as the only truly scientific method, yet it was destructive of all science. It denied the existing world, only to replace it with the arbitrary fantasies of the philosopher.

As Guest had seen it, it was an epoch of theoretical confusion, in which fragments of the old conception of the world were mixed up with the new. The need was to dialectically master the new results of the special sciences and so bring them to a unified whole, to show the most general features of process common to all fields (laws of dialectics) as aspects of a single world process. The task of dialectical materialism was to show the way to a complete reconstruction of science, to break down the barriers between the sciences and to pave the way for a unified picture of the world to replace the series of mutually conflicting pictures.

To do so required a "fight on two fronts": against rigidity and against relativism. These two extremes, Guest had remarked, were eclectically combined in Bertrand Russell. To him applied the words Marx spoke of Proudhon: he wanted to be the synthesis, but he was the composite error. He also thought that Lenin's arguments against Mach applied to Russell word for word.[186]

Guest had elaborated on his critique of Russell in his article published in *Pod znamenem marksizma* in 1934. In Russell's early work, Guest contended, his subjective idealist bias was clearer than in his later work. His self-confessed aim had been to render solipsism scientifically satisfactory, and he had said that it would give him the greatest satisfaction to dispense with other people's minds and thus to reestablish physics on a solipsistic basis. He had noted, however, that for the majority human affections were stronger than the desire for logical economy. Later, in his *Analysis of Matter*, Russell had come to hold that matter was less material and mind less mental than commonly supposed. To gain a hearing among scientists, Russell had been forced to adopt a more materialist language, and had admitted that a causal theory of perception, in contrast to phenomenalism, was necessary to science. However, he had done this with typical agnostic half-heartedness and with British love of compromise. Solipsism, being an impossible doctrine to maintain, was often the antechamber to fideism. If Russell's philosophy had never been directly used as a prop for religion, Guest remarked, the dull-wittedness of the clergymen of the Church of England could be the only reason. Russell himself, of course, would have none of it, as Guest conceded. He considered the high points in Russell's work came when he was aroused to anger by some priestly impertinence or by some particularly clumsy attempt to reconcile science with religion. Russell argued against Eddington, who deduced religion from the premise that atoms did not obey the laws of mathematics, and against Jeans, who deduced religion from the premise that they did, while both were accepted with equal enthusiasm by the theologians. Still, Russell insisted he was not a materialist, even though he considered himself still further from idealism. Russell's *Sceptical Essays*, Guest went on, reflected the growth of intellectual discouragement among the bourgeois intelligentsia, their inability to bridge the gulf between theory and practice, the intolerable contradiction between the theoretical confusion of modern science with the growth of its practical achievements.

In the same article, Guest had also addressed himself to the views of Ramsey, Wittgenstein, and Whitehead. Ramsey's subjectivism pictured the world all out of proportion. As to his former teacher Wittgenstein, his extreme subjectivism led to its own *reductio ad absurdum*. His antihistoricism, his contemptuous dismissal of all past and present philosophy led to irrationalist consequences. After pushing philosophy in its rational form out the door, Wittgenstein felt constrained to admit it again in an ultra-mystical shape through the window. Whitehead was a "God-builder" and his criticism of scientific materialism was a "fog of words" that did not take into account the fact that for dialectical materialism there was no "irreducible brute matter."[187] While there was much insight in Guest's criticisms of the leading non-Marxist philosophers of the times, he often failed to take due account of their subtleties and their own positive insights. He was young and the world was in turmoil and he tended to see the battle lines to be much more clearly and sharply drawn than they actually were.

In another article from 1935, inspired by the Rudas-Carritt controversy, Guest had taken on the commonplace complaint against Marxism: the accusation of Hegelianism, the charge that it applied a priori reasoning to the real world, that it sought to fit the movement of history into arbitrary, prearranged dialectical forms. Dialectics, Guest had answered, by no means enabled anyone to dispense with the empirical facts or to deduce the facts from a logical scheme. Its role was to help to deal with the facts when discovered and to suggest where facts might be found and which might be of importance. The transitions in Hegel's logic were often quite arbitrary, sometimes effected by mere puns and verbal tricks.[188]

Highly trained in mathematics and logic, Guest had not been one to dismiss the claims of formal logic, as were many Soviet Marxists in the late 1930s, however aware he had been of its limitations. Formal logic was necessary for dealing with abstractions formed in the first stage of thinking. The principles of identity and noncontradiction were essential. Nevertheless, Guest had said, the spirit of dialectics was breaking through the hard shell of formal logic. Guest had been very concerned with the crisis in mathematics and the struggle between intuitionism and formalism, and had commented that vast realms of thought still awaited dialectical understanding.[189]

It would have to be left to others. The cause of Republican Spain, the democratically elected Popular Front fighting for its life against Franco's fascist forces, captured the imagination of the most promising, the most gifted, the most generous young men of Guest's generation. Many of them went off to Spain to fight for democracy, where it was most under siege, to fight their way to where the fire burned fiercest. The International Brigades were full of workers, poets, philosophers, scientists, historians, teachers, dreamers—even some professional soldiers. The stakes were high—and so were the losses. Not only was Spain lost, but so too were many of the finest minds of this generation. British Marxists suffered exceedingly heavy losses, losing not only David

Guest, but John Cornford, Ralph Fox, and the most creative of British Marxists, Christopher Caudwell.

Caudwell

Christopher Caudwell was the pen name of Christopher St. John Sprigg, London-born, professional writer, from an English Catholic and journalistic background. He left school, the Benedictine Priory in Ealing, and began his working life as a journalist at the age of fifteen. He worked first as a cub reporter at the *Yorkshire Observer*, where his father was literary editor, and then as editor of *British Malaya*. Upon returning to London, he also ran an aeronautics publishing company with his brother, edited one of its technical journals, and designed gears for motorcars. In addition, he wrote reams of poetry, plays, short stories, detective novels, and aeronautics textbooks. He even edited a volume of ghost stories. On top of all this, he read voluminously in philosophy, sociology, anthropology, psychology, history, politics, linguistics, mathematics, economics, physics, biology, neurology, literature and literary criticism and much more besides. He did so, not as a dilettante, but as one striving to come to terms with the knowledge of the centuries and to comprehend what it meant for himself and his own age. Despite his lack of a university education, he was becoming, through the London Library, a person of very considerable learning, although he had not yet found himself amidst it all. He seems to have been somewhat reserved, even introverted, moving within highly circumscribed social limits, living with his older brother, and having very few friends.

In 1934, at the age of twenty-seven, Caudwell became interested in Marxism and began to study it with extraordinary intensity, discovering quickly that it provided the key to the synthesis he was seeking. In the summer of 1935, he wrote his first Marxist book, which was to have been called *Verse and Mathematics: A Study of the Foundations of Poetry*, writing at the rate of 5000 words a day. The draft was finished in September, sent off as *Illusion and Reality*, and was accepted for publication by Macmillan. Upon the completion of this book, he moved to the East End of London. Soon Caudwell joined the Poplar branch of the Communist Party. He threw himself into all the routine party tasks, fly-posting, street-corner speaking, selling the *Daily Worker*, as well as joining battle with Mosley's Blackshirts, bravely facing consequent batterings and even arrest. The members of the Poplar branch quickly discerned the utter sincerety of his commitment and accepted him fully as one of their own.[190]

Caudwell was at last finding himself, becoming at the same time both more outgoing and more highly integrated. Just after his move to Poplar, he wrote to friends from his former life:

> Seriously, I think my weakness has been the lack of an integrated *Weltanschauung*. I mean one that includes my emotional, scientific and artistic needs. They have been more than usually disintegrated in me, I think, a characteristic of my generation exacerbated by the fact that, as you know, I have strong rationalising as well as artistic tendencies. As long as there was a disintegration, I had necessarily an unsafe and provisional attitude to reality, a somewhat academic superficial attitude, which showed in my writing as what Betty has described as the "lack of baking". The remedy is nothing so simple as a working over and polishing up of prose, but to come to terms with myself and my environment. This, I think, during the last year or two I have begun to do. Naturally, it is a long process (the getting of wisdom) and I don't fancy I am anywhere near the end. But *I and R* represented a milestone on the way, and that, I think, was why it seemed sincere, free from my other faults, and, with its necessary limitations, successful.[191]

But it was not really such a long process. Because of the extraordinary intensity of his intelligence, he did, in a tightly compressed period of time, achieve a rare wisdom. During this period of one year of the most vigorous political activity, he wrote prolifically, in the most astonishing outpouring of creative energy, and produced serious Marxist theoretical works that can only be described as original and breathtakingly brilliant. No matter what he was discussing, whether poetry or physics or philosophy or whatever else, he had a way of penetrating to the very core and of illuminating in a new way the whole vast and complex network of its connections. He saw the world freshly, and called attention to patterns that had scarcely been noticed before, because he saw with a vision that was wider, deeper, warmer, clearer. He had become a communist in every fiber of his being and was determined to make both the artist and scientist in himself come to terms fully with the Marxist. Indeed he was consumed with the drive to bring this new world view to bear upon the whole of past and present knowledge and to fight for its theoretical and practical fulfillment. His intellectual clarity and his political commitment seemed to mutually reinforce each other, continually bringing both to yet higher levels. His rationality and his emotions seemed to fuse, in such a way that the sharpness of an argument brought the heightening of passion and the heightening of passion made the argument sharper still. A quote from Lenin, which was intended as an epigraph for his writings from this period, explained what he was up to in his work at this time:

> Communism becomes a mere empty phrase, a mere facade, and the communist a mere bluffer, if he has not worked over in his consciousness the whole inheritance of human knowledge.[192]

This was what he was about. He had become a communist for real and was determined to work over in his consciousness everything he knew, or could know, as a communist. His *Studies in a Dying Culture* included essays on philosophy, psychology, history, religion, ethics, aesthetics, love, and much more.[193] The one on physics expanded until it turned into a separate book, entitled *The Crisis in Physics*.[194] There were others, one on biology, which remained unpublished.[195] He was looking at everything anew, from a communist point of view, and doing so with striking vitality and profundity.

All of his work in the field of Marxist theory was only published posthumously, under the name of Christopher Caudwell,* a name he first used with the publication of his serious novel *This My Hand*, published in 1936. It was a name he reserved for his serious work, saying he was afraid of spoiling his reputation as a writer of thrillers. During his life, his theoretical works were unknown. *Illusion and Reality*, the one book he had ready for press, hadn't yet appeared. None of the rest did he consider ready for publication, thinking of them only as drafts to be rewritten, needing "refining, balancing, getting in it the movement of time, ripening and humanising."[196] He was unknown even to his contemporaries among British Marxists. He moved only in the circles of the "anonymous proletariat" and had very little contact with either the party intelligentsia or left-wing literary circles, who knew nothing of his serious work until he was dead.

When the Spanish civil war broke out, the Popular branch involved itself in the campaign to organize support for the Spanish Republic. When they had raised enough money for an ambulance, Caudwell volunteered to drive it across France to Spain. He left in December, 1936, and, upon arriving in Spain, he joined the International Brigade. Soon, he was a machine-gun instructor and editor of the *Battalion Wall* newspaper. Meanwhile, his brother had obtained an advance copy of the proofs for *Illusion and Reality*. He showed them to Harry Pollitt in an effort to persuade the CPGB that his brother would be of greater value to them as a writer than as a soldier. Most biographical accounts say that a telegram was sent by King Street to Spain recalling Caudwell to England and that it arrived too late. It is, however, doubtful that such a telegram was ever sent. At all events, Caudwell was soon dead. In the famous valley of Jarama, he was killed in action on his first day in battle. He was last seen firing a machine gun, covering the retreat of his section from a hill about to be taken by the Moors.

So it was that the life of this most brilliant and generous young man was taken from him before he had even reached the age of thirty. Soon British Marxists began to realize the magnitude of their loss. They read Caudwell's books with astonishment and discovered that something very precious had been taken from them without their even knowing it.

---

* Caudwell was his mother's name.

Hyman Levy, who edited and wrote the introduction to *The Crisis in Physics** in 1939, hoped that when the nightmare had passed man would count the cost of his ignorance in terms of the human suffering and social sacrifice it had brought, in the destruction of such men of promise, even amounting to genius. Caudwell, he said, had combined a social and scientific understanding that would have been rare in a scientist of mature experience. But, remarked Levy, to find them in this young man was almost phenomenal.[197]

Reviewing the book, Haldane wrote that it was impossible to read it without realizing the immense loss the world had sustained through its author's death. Caudwell had explained the ways in which physical theories were shadows of economic realities and had done so with brilliant success. He had written about science as a poet and had something to say about science, something very important indeed, although he had only half said it. Haldane took into consideration, of course, the fact that the book had been left unfinished and that the drafts of the later chapters had been very rough indeed. While making his criticisms of the book, Haldane nevertheless called it "a quarry of ideas" for generations to come. He also said he knew of no writer in his time whose analysis of the problem of freedom went deeper.[198]

John Strachey was amazed by the flood of works produced by this young man possessed of such creative energy. In his introduction to *Studies in a Dying Culture*, he remarked that here was a young man who not only warmed his hands at the fire of life, but gave it great hearty pokes; a young man who was so interested in everything, from aviation, to poetry, to detective stories, to quantum mechanics, to Hegel's philosophy, to love, to psychoanalysis, that he felt he simply had to say something about them all. That, in Strachey's view, was what a man in his twenties ought to be like. After a wandering decade, he had been, at 29, finding himself, gaining in precision, in capacity to focus— "and then the Moors came." He had embodied an exquisite unity of theory and practice, Strachey noted. He had moreover possessed qualities tragically rare in the British working-class movement: width of perception, generosity of sympathy, understanding of human motivation.[199]

The left-wing literary circle gathered around *Left Review* were puzzled to learn of the existence of such a person unknown to them. Its editor at the time, Edgell Rickword, later wrote the preface to *Further Studies in a Dying Culture*, and saw Caudwell as a man of insatiable intellectual curiosity, consumed by a Faustian ambition to master all the sciences. Not one to stand dreaming on the edge of strife, nor one to plunge into struggle without thought, the philosopher could not but turn soldier in a struggle in which the forces of enlightenment and of obscurantism were so starkly opposed. Marxism had

---

* Levy received the manuscript from the hands of David Guest. I am indebted to Jean Duparc, whose *doctorat d'état* thesis at the University of Paris was on Caudwell, for this information.

entered into the very fabric of his being, so that he thought *in* it. The warmth of emotion glowed through his argument, constituting true eloquence. Rickword told how he had sent the essay on consciousness to a neurologist, fearing it might have become outmoded by subsequent research. The neurologist replied that Caudwell had brilliantly anticipated a whole trend becoming discernable in neuroanatomy.[200] Levy and Haldane had said much the same about his views on physics.

Levy bluntly put the question others must have been asking when they first heard of the publication of *The Crisis in Physics:* What had the crisis in physics to do with Christopher Caudwell? With the author of *Illusion and Reality* and *Studies in a Dying Culture*? What could this poet have to say about science? How could the problems that were vexing modern physics stir one whose mind appeared to move on such a different plane? What could he contribute to the solution of such complex problems? In what way were these very different realms linked in his mind? In what way were they linked in reality?

The answer had already been given by Caudwell himself, many times over in his two published works. Whether a book was about science or about art, it always dealt with both. His philosophy of science could be found in his aesthetics and his aesthetics in his philosophy of science. Variations on the same theme were pursued again and again on various levels in all of his theoretical works, with the striving for a *Weltanschauung* always the central focus and the driving purpose. The fundamental problematic underlying his discussion of the most diverse topics was the crisis in bourgeois culture in all of its aspects: what was the cause of its theoretical fragmentation? Caudwell had sought to discern the most basic thought patterns and to discover the lines of connection between these and the most basic socioeconomic realities. At the heart of it all was the subject-object dichotomy, that had its basis in the social division of labor, in the separation of the class that generated ideology from the class that actively struggled with nature. This dichotomy distorted all realms of thought and activity. It distorted art, science, psychology, philosophy, economics and all social relations. It was a disease endemic to class society that has become most acute in bourgeois society as the most highly developed form of class society. Only an integrated world view and a classless society could bring to a synthesis what had been severed and had grown pathologically far apart.[201]

In his foreward to *Studies in a Dying Culture*, Caudwell had outlined the problem. It was a epoch of confusion, dissension, pessimism, and bewilderment. He had referred to Max Planck's expression of the crisis in science: there was scarcely a scientific axiom not denied by somebody and at the same time there was almost no nonsensical theory put forward in the name of science that would not be sure to find believers somewhere. Bourgeois culture, Caudwell had noted, had achieved much: relativity, quantum mechanics, genetics, psychology, anthropology, new technology. So why the despair? Why did this strange doom hang over bourgeois culture in such a way that its progress seemed only to hasten the decline? Why was each discovery like a Midas touch that prepared a new disappointment? Why was it that the more men sought to find a common truth, a common faith, a common world view, the more their efforts at ideological construction increased the sum of contradictory and partial views of reality? What was the explanation? Observed Caudwell:

Either the Devil has come amongst us having great power or there is a causal explanation for a disease common to economics, science and art.[202]

He had believed it to be the latter—which evoked the further question: why then had not all the psychoanalysts, philosophers, historians, economists, scientists, and bishops who had surveyed the scene not located the source of infection? His answer was that the Freuds, the Eddingtons, the Spenglers, and the Keynes were not the doctors; they were the disease.

It was for Marxists to perform the necessary tasks—both analytic and synthetic: to analyze the causes that made every discovery go bad upon its inventor's hands, to separate real empirical discoveries from the ideological confusion and to synthesize them into an integral world view. There had been great disillusionment in bourgeois culture, but this culture had, in Caudwell's view, yet to shed the last of its illusions. It had shed all secondary illusions—of God and religion, of teleology and metaphysics. But it had not yet rid itself of the basic bourgeois illusion: that man was born free, but was crippled through social organization. In its illusory separation of the individual consciousness from the natural and social matrix of its existence, the bourgeoisie had brought to a new level the dualism inherent in class society, generating in philosophy an ever-sharper separation of individual from society, of mind from matter, of freedom from necessity, of history from nature, making the fundamental subject-object relation absolutely insoluble.

Caudwell believed that the Renaissance charter of the bourgeoisie—to claim for the "natural man" freedom from all feudal restrictions—had originally been the dynamic force of bourgeois civilization. It had been a progressive idea in its time and the bourgeoisie had been a progressive class in its rise, for the division of labor had been necessary to further human advance. But it was an idea that had outlived its usefulness. With its utmost potentialities accomplished, it had become a brake to further development.

For Caudwell, it was the lie at the heart of bourgeois culture, for freedom was the product, not of instinct, but of social relations. Man could not strip himself of his social relations and remain man. Failure to realize this was the cause of the typically modern unease and neurosis. The bourgeois saw himself as a heroic figure fighting a lone fight for freedom, as the natural man who was born free, but was for some strange reason everywhere in chains. He ejected everything social from his soul, making it deflate, leaving him petty, empty, and insecure. With the alienation of social affection from labor, there gathered at one pole all the unused tenderness of humanity, while the other pole was reduced to the sheer coerciveness of bare economic rights to commodities. This generated the most terrible tension, but, with its source hidden in the shadow of the free market, all efforts to break free of it only accentuated it. The bourgeois was always talking about freedom, because it was always slipping from his grasp.

In his illusion of detachment, the bourgeois imagined himself to be a free man, who could direct the social process without being affected by it, who could determine without being determined. He was thus preoccupied with the epistemological problem of the independent observer, wishing to be cognizant of the laws determining the environment he wished to control without himself being controlled, to determine without himself being caught up in the same web of determinism. This brought him to stand in his own light, able to conceive only of self-determined mind in a one-way relation to its determined environment, a free and active subject contemplating a necessary and passive object, making him oblivious to the two-way relation, in which his environment also determined him. He became particularly unable to perceive the determining power of social relations upon him and therefore blind to the actual character of social relations determining his individual choices. In economics, both producer and consumer, whose freely-willed desires were supposed to determine production in the best possible way, were themselves, in their very desires, in their very consciousness, determined by the productive forces and social relations of their time. In epistemology, the mind of the observer was itself determined by the environment he observed.

The tragedy implicit in his situation lay in the fact that bourgeois social relations inevitably constrained the very hopes they produced, being grounded in such an illusion. Ignorant of the real determinism of social relations, veiled by the free market, he was subject to forces he could neither control nor understand, to wars, slumps, crises, revolutions. His ordered society was becoming more and more disorderly, his economy was breaking down into anarchy: abroad, idle capital wildly searching for profit; at home, idle hands vainly searching for work. The capitalist, and even the poet, became darker figures—first tragic, then pitiful, and finally vicious.

The bourgeois idealized this one freedom, freedom from all social restrictions except that by which the bourgeois class ruled, that is, the restriction of the means of production itself. Freedom was therefore elevated to a vague, ideal plane, for to interpret bourgeois freedom in a materialist way would be to acknowledge openly the claim of one class to monopolize the means of freedom. The social product was the condition of freedom, and to monopolize it was to monopolize such freedom as society had produced. The freedom of the ruling class was based on the unfreedom of the ruled. Nevertheless, the bourgeois had high-flown notions. He spoke of fine and noble things, while doing base and hateful ones. And, as he became threatened, the discrepancy widened. Bourgeois philosophy, inextricably tied to its social matrix and unable to rise above the standpoint of the individual in civil society, reflected the rhythms of this social process. To show this, Caudwell traced the history of modern philosophy in terms of the development of the class consciousness of the bourgeoisie.

In the first stage, that of bourgeois revolt, the idea that "I am free in so far as I throw off all social restraint" gave birth to Shakespearean tragedy, Tudor monarchy, voyages of discovery, Galileo, Newton, and Descartes. The discovery of gravity and of analytical geometry, the exploration of the farthest limits of the world marked the crescendo of the bourgeois explosion into its environment.

In the process, however, the basis of freedom had been separated from the basis of necessity. The bourgeois saw the environment as determined by his free will, but he saw his interaction with nature as a one-way relation. He saw himself as determining, but not as determined. The philosophy of the seventeenth century, therefore, saw the world in terms of inert matter and creative spirit. With Malebranche and Descartes, substance (matter) was so inert that it required creation anew for each moment of time. For Newton, all transactions of matter were ultimately effected by spirit. Newton's world was atomistic and inert, requiring God to be the moving and unifying principle, creating, setting in motion and holding together the separate, independent particles of matter. But physicists stood no nonsense from their God in their own particular domain. He was ruthlessly stripped of mercy, love, an only begotten son, everything but his material determinism. Newton's philosophy was one resolutely turned towards the object—nature. It was a world-grasping experimental philosophy.

The mechanistic materialism of Hobbes, Condillac, and D'Holbach expressed the limit of the upward movement of the bourgeoisie. It separated the object from the subject and believed that matter could be completely explained in terms of itself—and that man, as part of matter, could be explained in the same terms.

The very concept of matter had been evolving with the evolution of the bourgeoisie. To Galileo and Bacon, matter was still full of quality and sensuousness. However, to realize matter as owned by the bourgeoisie, to cut the umbilical cord of mutual dependence between man and nature, to free the individual from all relations except those of the free market, that is, the one-way relation of private property, it was necessary to eliminate the observer. As nature was to be apprehended as by a kind of divine apprehension in which there was no mutually determining relation between the active subject and the contemplated object, it was necessary to strip nature of all activities in which the observer was concerned, to strip the object of all subjective qualities. At first, matter was divested of color, sound, solidity, heat, and taste. Motion, time, space, mass, shape were regarded as objective qualities, but in due course, these too were shown to be relative to the observer.

Matter seemed to disappear, or become unknowable. The downfall of mechanistic materialism was not due to the fact that it was materialist, but to the fact that it was not materialist enough. By excluding its qualities from the reality of matter, it contradicted the basis of itself. At the apogee of the first stage of capitalism, its materialism turned into its opposite, mentalism, as the bourgeoisie passed from its extroverted, dominating, exploring period into its introverted, analytical period. Its philosophy turned away from the object toward the subject, for the object was beginning to slip from its grasp. Hence, Berkeley, Hume and Kant—and finally Hegel. The same dualism of subject and object was reproduced, only now calling the opposite parody into existence, without realizing its source.

The rebirth of idealism came as the philosophy of a ruling class whose environment seemed to obey its free will and whose distance from its environment was increasing with the growing differentiation of labor. Mind became correspondingly dissociated from matter. Thus was mind stripped in the same way as matter had been before it, and the subject was by stages progressively cut loose from the object. Thus, active, sensuous subjectivity was to be divested of all qualities that tied it to objectivity, making it active, but active upon a nothing, until finally subjectivity had been stripped of the subject—man—and began to strangle itself. This development reached its climax in Hegel for whom mind was dissolved into ideas independent of the subject.

Nevertheless, it still bore the impress of material reality. By comparing its formal and unanchored qualities among themselves, it was possible in a confused way to extract the most general patterns of activity and change, just as by comparing the categories of objectivity among themselves, it was possible to get the confused but general physical laws of mechanism. The Hegelian dialectic represented the high point of bourgeois subjectivity, but, for all its logical rigor and world-embracing grandeur, it was subjectivity active upon nothing and therefore became mere mystical mumbo-jumbo. Philosophy had become severely impoverished.

Giving expression to his sense of its history, not only as a philosopher but also as a poet, Caudwell wrote his "Hymn to Philosophy":

> I saw your figure in a Grecian mode:
> A stripling with the quiet wings of death,
> Touching with your long fingers a marble lyre.

I was impressed by your immortal age,
I was seduced by your adventurous strength,
I was relieved by your polite reserve.

A winged Idea down the rainbow sliding,
With steady steps treading the smoky air,
All ranks you visit, courteous swamp-foul.

The world's great engines pound asthmatically
Fed through Time's recurrences.
Man walks to man across a trembling swamp.

The scientific sportsman lifts his gun;
The second barrel blasts your blue pin-feathers
And you fall spluttering, a specific bird.

I see your stuffed breast and boot-button eyes
Preserved in cases for posterity
And lean on my umbrella thoughtfully.

I have caressed your sort, I must confess,
But give me beauty, beauty that must end
And rots upon the taxidermist's hands.[203]

Philosophy for Caudwell had become empty and lifeless with the development of class society and with the sources of vitality drained from it.

Most disastrous for philosophy, in Caudwell's view, was its conflict with science. Philosophy had flown apart from physics, as subjectivity had drifted from objectivity, with capitalism at this stage maintaining two conflicing ideologies: mechanism for the scientists and idealism for the philosophers. Subjectivity became the province of philosophy, while objectivity was the province of science. The philosopher was no longer interested in matter. The physicist was no longer interested in mind. Hegelianism could not be a physicist's creed, for it denied the need for physics.

Whereas Hegelianism reflected the rapid evolutionary expansion of capitalism, the next stage in the development of bourgeois philosophy, phenomenalism, was coincident with its decline. Positivism marked the passing of the bourgeoisie from a progressive class to a reactionary one. The world of physics had become so bare of quality, the object of science had become so stripped of all sensuous reality, that it seemed to become only a ghostly dance of equations. Physics was thrown back upon philosophy. Scientists were thrown back upon those qualities they had banished from objectivity and abandoned to subjectivity and came to seek in the laws of thought a certitude that they could not find in the laws of matter, only to find that the subject of science had also been stripped bare. Thought too had become ghostly. Only phenomena were left.

Whereas mechanism had sacrificed the subject to the object and idealism had sacrificed the object to the subject, positivism sacrificed both. The process in which matter had become so stripped of all material qualities as to evaporate into mind, and mind had become so stripped of all mental qualities that it solidified into matter, duly brought forth Mach, Russell, Jeans, and Eddington.

Phenomenalism was a bastard compromise. It attempted to solve the problem of the subject-object relation by making the relation alone real. The problem was that in positivism all phenomena had equal validity, allowing no means of distinguishing between hallucination and real perception, scientific theory and barbarous logic. The world no longer possessed a unity due to its materiality, nor any unity at all. There was no causality—only sensations. Matter was unknowable and elusive. So too was man. The physicist could go round and round in circles without ever knowing if he encountered anything real and without ever knowing what to expect next. But in fact the positivist would not face up to his premises, and was constantly smuggling in some coordinating principle to organize the system and give it a standard of validity. He presupposed the existence of the very things he insisted could not be presupposed, shifting from one premise to another without realizing it, weaving a mesh of nonsequiturs and excluded middles. He could not reconcile the dualism of subject and object, but always brought either subject or object back in again through the backdoor. In Mach, the subject was brought back as the most economical laws of thought and finally in Jeans and Eddington as a mathematical God. The parallel in aesthetics was "art for art's sake," which then went and smuggled in emotivism.

In every way, theory was drifting away from practice. Art was flying away from experience. Philosophy, even philosophy of science, was becoming increasingly remote from science. Positivism, in the name of science, was a philosophy alien to the realm of scientific experiment, and science without the appeal to experiment was pure scholasticism and Alexandrian futility. A theory of science drifting further and further from scientific practice gave rise to more problems than it solved. Thus the crisis in physics. It was a part of the general crisis in science, which in turn was part of the overall crisis in culture. Caudwell saw a causal connection between the crisis in physics and that in biology, psychology, economics, morality, politics, art and, indeed, life as a whole. In a society in which consciousness had become so separated from its environment, because the thinking class had become so separated from the working class, there was growing intellectual fragmentation and cultural disorientation. Consciousness tended to gather at one pole and activity at the other, causing distortion of both. Theory and practice were sundered in consciousness, because they were divided in social reality. Theory, which had emerged historically out of practice, vested in the same individual, broke apart, as thinking became the prerogative of the exploiting class with the division of the labor process, creating a class that passively and blindly labored and a class that directed those labors. Bourgeois ideology, increasingly remote from the world of social labor, became increasingly detached from its foundations and the bourgeois class became more and more parasitic, more and more idealist.

Physics was the extreme case, as it was the realm seemingly most remote from potential ideological distortion. But the disease permeating bourgeois ideology was spreading and even physics was now infected. It was sometimes supposed, Caudwell observed, that the cause of the crisis in physics was the discrepancy between macroscopic or relativity physics on the one hand and quantum or atomic physics on the other. This was an aspect of it to be sure, but the fundamental cause of the crisis lay deeper.

The problem was the metaphysics of physics, a metaphysics not so much generated by physics as physics had been generated by it, not in a self-contained way, but in interaction with the rest of reality. The concepts of physics had been formed amid the same social process in and through which reality had been apprehended and conceptualized in all other fields of knowledge. Here too the bourgeois aimed at a closed world, independent of the observer, that he could watch from the outside. Relativity was a magnificent attempt to solve the problem by recreating the closed world of physics in a subtler form, making it four-dimensional, by the complicated method of tensors, the common invariant element in the functions of coordinates. In the end, this too failed, and physics turned in on itself in mentalism and self-contradiction.

Physics was rent by the same dualisms as all other disciplines, though in each specific discipline these dualisms played themselves out in specific ways. Physics too was plagued by the flying apart of theory and practice, practice becoming specialized, restricted, narrow, and empirical, with theory becoming abstract, uncoordinated, remote and diffuse. The pursuit of physics, in so far as it was advancing along the practical front of experiment, was generating a growing body of knowledge that could not be fit into its existing theoretical framework. There was an explosive struggle of content with form, with increasing anarchy and disharmony both within and between the different domains of physics.

It was the same with all the other sciences. Bourgeois culture was proving increasingly unable to control the forces it generated and to assimilate the good things it produced.It was unable to contain the discoveries it made. Relativity, quantum mechanics, experimental psychology, evolution, genetics, sociology, anthropology, comparative religion were all disruptive forces in bourgeois culture that gave rise to semidialectical philosophies and incomplete attempts at synthesis within the anarchy of bourgeois thought.

But none of these could succeed. It was necessary to go to the heart of the matter: to expose the dualism at the heart of bourgeois culture that tore every science apart within itself, as well as isolating it from every other science and from the living whole of reality in which it needed to find its place. What it came down to was the lack of an integrated world view that could encompass all the sciences with their dramatically expanding experimental results and do so within the context of the whole, living, blood-warm reality necessary to their vitality and healthy growth. Because there was no such philosophy within which they could be integrated, science gathered itself around its most practical fronts in detached and isolated sciences. Science decomposed into a chaos of highly specialized, mutually repellent sciences, whose growing separation increasingly impoverished each of them and contributed to the overall fragmentation of human thought. Ironically, the very development of each of the sciences in this situation accentuated the general disorientation.

Every science was a closed world unto itself. With the separation of biology from physics, for example, the world of biology seemed to have all quality, all change, all development cooped up within itself, seeing outside it only the Sahara of the closed world of physics—quantitative, changeless, and bare. A biology unanchored in physics inevitably began looking for an uncaused first cause, an entelechy, an *élan vital*, or some such explanation of itself.

Within the framework of bourgeois philosophy, the sciences could not be conceived of except as either confusing or dominating each other. Either the fundamental categories of each of the sciences were held to be exclusive, with nothing resulting from their combination except a mishmash or, alternatively, the categories of one science excluded and suppressed the categories of the other. Thus, in behaviorism, the categories of biology suppressed those proper to psychology. In mechanistic materialism, the categories of physics were allowed to usurp those of all other sciences. Either each sphere was utterly distinct or they were all the same: that was the dilemma in which bourgeois science found itself.

Science began to despair of the possibility of a general theory of science, falling back on eclecticism, reductionism, positivism, and even mysticism. Without a world view in which to fit his empirical discoveries, the scientist was left with two fundamental choices: Either he regarded his discoveries as limited to his particular sphere, adopting an eclectic attitude to reality as a whole; or alternatively, he erected a complete ideology on the basis of his own discoveries, inevitably leaving much out of account. What was left unaccounted for was either forcibly reduced to the level of the facts that were accounted for or mystical explanations were sought for the phenomena left unexplained and inexplicable by such a limited ideology. In psychology, for example, Freud proceeded according to the reductive method, whereas Jung tended to the more mystical approach. In any case, each of the alternatives in its own way only intensified the crisis. The internal contradictions within and between the various disciplines could not be solved within the confines of the separate disciplines. They could only be resolved in a larger synthesis that encompassed them all.

As to the situation in physics, the solution of the contradictions within mechanics raised by relativity, those within wave physics raised by electromagnetism, and those within atomic physics raised by quantum theory, had only led to greater contradictions among the three domains. Moreover, the integrations were becoming increasingly unstable. As a result, the field of physics was being occupied by opposing armies. Einstein and Planck were clinging to the categories of mechanism. Jeans and Eddington were attempting to find substitutes in subjectivism. Dirac and Heisenberg were trying to dispense with categories altogether. Essentially, the situation was that physics had not as yet found any substitute for the categories its own research had revolutionized. The new categories required could not be formed within the bounds of physics alone, however. No real solution was possible unless the most basic and fundamental categories common to all domains were to be radically refashioned. What physics needed was a new philosophy.

Einstein and Planck were the last physicists adhering to the old metaphysics of physics, whereas Jeans and Eddington represented the most extreme swing in the opposite direction, the most extreme tendency for physical theory to fly away from physical experiment. Einstein stood out as a larger figure than the rest in his aspiration to an all-embracing philosophy. Although he did manage to bring together a wide domain of physics, he was still unable to encompass the whole complexity of modern physics. Nor was anyone else able. With the breakdown of traditional categories and no new ones to take their place, physicists were becoming inclined to call God back in again to sanction the physicist's belief in unity, to assure him of a worthwhile end to his labors. But it was a God from the other side. Unlike Newton's God who was Matter, this God was Mind, a mind remarkably like that of the physicist's own. All such introjections of the physicist's mind behind phenomena to take the place of a deleted matter represented a certain falling off and disorientation as compared with earlier physicists' more robust viewpoint.

The state of physics amid all the confusion was disturbing, not only to the physicists, but to the general public as well. Taking up the various problems perplexing to the popular consciousness in relation to physics, Caudwell first examined the problem of the relation of physics to perception. The world of physics seemed to be deviating further and further from the world of perception. The world of relativity physics seemed to be taking physics further and further from reality as directly experienced. In answer to the question of whether the world of physics could be restored to the world of experience, his answer was yes, that it must, for physics was built up and validated from the results of perception, even relativity physics. The discrepancy between Newton and Einstein was settled after all on the basis of the Michelson-Morley experiment. The perceived world, Caudwell insisted, was primary and gave status to only certain of the various self-consistent possible worlds.

An even stickier question in physics was that surrounding the status of concepts of causality and determinism that had become particularly problematic with the development of quantum physics and had brought many physicists to deny causality and to assert a radical indeterminism in nature. Responding particularly to the conclusions being drawn by Jeans and Eddington from Heisenberg's principle of uncertainty, to the effect that causality and determinism were no longer principles of physics and that it was therefore possible to guarantee the freedom of the human will, Caudwell analyzed the issue on a number of levels. Basically, his argument was that the fundamental bourgeois illusion, in its false notion of the nature of freedom, had penetrated even into physics. The bourgeois understanding of causality was equivalent to predeterminism, the only sort of determinism that bourgeois could understand. It was the bourgeois nightmare. It was the dread of a class that did not want to be tied to nature by any relation except that of private property, a relation entered into by the individual by virtue of his own free will. Freedom, for the bourgeois, seemed to lie in arbitrary subjectivity, with all causality concealed, hidden in the shadow of the free market. Those, such as Jeans and Eddington, who seized upon Heisenberg's principle to launch a full-scale attack on determinism in physics, picturing the movements of the particles as indeterminate and the particles themselves as unknowable, supposed that this at last secured the menaced free will of the bourgeois. By this bizarre stratagem, they thought they had freed man from the determinism of nature by eliminating the determinism of nature altogether. Even nature was now seen to exhibit bourgeois free will.

On one level, Caudwell analyzed the way this played itself out within the realm of physical theory, tracing it back to the subject-object dichotomy and the separation of the basis of freedom from the basis of necessity in the seventeenth century. He seemed to have very definite ideas about how the resolution of his dichotomy provided the way out of the anarchy engulfing physical theory. His argument at this level was not fully developed, however, as it came only in the draft notes for the chapters of *The Crisis in Physics*, which were left in a very rough state, far from ready for publication, when he went off to Spain. Roughly, he seemed to be indicating that the apparent antinomies of physical theory between quantum and wave, discontinuity and continuity, freedom and determinism, accident and necessity, would find their resolution when thought ceased to move back and forth between mutually exclusive polar opposites. It was necessary to see freedom within the framework of determinism. Otherwise, each was abstracted from the other, distorted and scarred. Determinism and necessity became crystalline and

incapable of evolution. Freedom and accident floated about without roots. It was the universal interweaving of domains and not the concept of strict determinism as such that made it possible to speak of the universal reign of law. Part of Caudwell's argument seemed to rest on a distinction between causality as an active subject-object relationship and predeterminism as a passive one.*
For Caudwell, then, the crisis in physics was not due to the mystical and contradictory nature of the phenomena discovered, but to the attempt of the bourgeois to keep the world of physics closed and to preserve his own freedom outside it, to keep himself at all costs immune from causality.

Running parallel to the crisis in physics was the crisis in biology. In Caudwell's unpublished work *Heredity and Development: A Study in Bourgeois Biology*, the fundamental argument was that all of the sharp antitheses tearing at modern biology—between genetics and evolution, between innate and acquired characteristics, between mechanism and vitalism—were rooted in the characteristic dualism of bourgeois culture. Once the environment had been stripped bare by physics, biology was given the task of accounting for life within it and protecting the freedom of the bourgeois at any price. But bereft of its roots and deprived of an integrated philosophy, it swung back and forth between extreme, one-sided positions and became increasingly anarchic, expressing in its sphere the decay of the bourgeois world view.

Emerging from his discussion of biology was a certain model of the development of the history of science in relation to the history of ideology. Science, in so far as it was always in fresh contact with the world through experiment, progressively illuminated the network of causality operative in the world, but it also threw up a scaffolding, an ideological gloss on the results of experiment. Such ideologies interlaced with scientific discoveries were not superfluous accidents but articulations of systems of nature reflecting the social relations of the time. Both Mendel and Darwin were scientists, devoted to fact and freshly in contact with the world, but both were scientists with a viewpoint, Mendel with a clerical viewpoint and Darwin with a bourgeois one. The world view appropriate to feudal society was one that saw nature as a stable, ordered, hierarchical system. That appropriate to bourgeois society, while capitalism was on the upgrade, was one with every reason to assert change, competition, and equality of opportunity. Thus Mendel, opposed to all that industrial capitalism was doing in the world, approached the study of variations in a spirit opposed to change, resting on eternal verities, seeing an unchanging set of genes in changing mathematical combinations producing a changing pattern of phenotypes. Thus, on the contrary, Darwin, seeing the world through the eyes of the ascendant industrial bourgeoisie, came forth with

---

* Determinism, as it emerged at the end of *The Crisis in Physics*, is the more general category. Within it, he distinguished between causality and probalistic determinism on the one hand and strict Laplacean predeterminism on the other.

evolution and natural selection. The great value of Darwinism, in Caudwell's eyes, was in seeing change in life, in seeing it as determined by the nature of matter and in conceiving of the world of nature as subject to natural laws. The great weakness of Darwinism, however, was in seeing change through the distorting ideology of class society: in picturing progress as the result of an unrestricted struggle for food (or profit) and in imagining life as an insurgent force in a dead universe (the reflex of the bourgeois producer with unrestricted property rights over inanimate things). Darwinism was therefore unable to explain the emergence of new qualities, the change of environment, the existence of variations, the origin of species. Looking at Darwinism from a perspective other than Darwin's own perspective, from an ideology other than bourgeois ideology, Caudwell thought that something survived that was irreducible to bourgeois ideology. Even though both evolution and natural selection coincided with the class consciousness of the ascendant bourgeoisie, he wanted to affirm evolution and to reject natural selection. Evolution was a matter of natural law. Natural selection was bourgeois property extended "down to hell and *usque ad coelum*." It was scientifically superfluous, as transformation was a result of the structure of matter, not of capitalist survival of the fittest, nor of the efforts of an active organism in a passive environment.

Thus Caudwell showed that the world view of an era was far more than the mere sum of the scientific discoveries of individuals. Scientific discoveries received their form and pressure from the social relations of the age. This was shown in the fate of Mendel's ideas vis-à-vis Darwin's. Darwin's theory was in harmony with the spirit of the times. Mendel's conception, Caudwell said, could only come into its own, in the form of Weissmann's germ plasm theory, when strict Darwinism had given rise to the opposite. The development of genetics was exposing the contradictions of the bourgeois standpoint and bringing a continual transformation of fundamental concepts as a result. However, all such transformations took place within the circle of bourgeois categories, producing not the unity of science, but its disintegration into specialized spheres, each of which represented a specific compromise between bourgeois ideology and a specific group of discoveries. Without breaking through the fundamental dualism at the heart of the whole process, the cycle would continue to reproduce itself, with the same antitheses recurring in new forms all the time, clouding the development of science and diverting it along unproductive paths.

The controversy over inheritance of acquired characteristics demonstrated this. Caudwell argued, in an implicit but clear repudiation of Lysenkoism, that the whole controversy was misconceived. Both positions rested on a false antithesis between organism and environment, holding them to be mutually exclusive opposites, rather than mutually determining aspects within the integral web of becoming. Both artificially separated organism from environment, making the one active and changing and the other inert and changeless. In reality, Caudwell contended, it was impossible to distinguish between innate and acquired characteristics, for all characters were a result of a germinal response to a given environment. Every organism as it has actually existed was a synthesis between internal and external forces. A change in external forces could produce a change in character, but only if the organism had a germinal aptitude for responding to that kind of external force in that kind of way. The emergence of new qualities came in the interaction between organism and environment. To attach it exclusively to either term was to reproduce the old dilemmas.

Interestingly, Caudwell saw the science of genetics itself as posing the alternative, claiming that the research of Morgan had itself revealed the inconsistencies of all formulations built upon a dichotomy between organism and environment. Both the conception of an abstract environment selecting mechanistically and the conception of an abstract organism mutating spontaneously were incompatible with the practical results of genetics. The trend of genetics, as Caudwell saw it developing, was to prove that specific characters were not the result of specific genes on a one-to-one basis, but the result of the interaction of numerous genes within an environment. Genes could only express themselves in an integral, interpenetrating relation between the whole organism and the whole environment.

As well-read and as thoroughly well-informed as he was, Caudwell surely knew that this defence of genetics and declaration of the meaninglessness of the assertion of the inheritance of acquired characteristics set him drastically apart from the dominant trend in the Soviet Union on this question. Yet he never commented on the discrepancy. This was a pattern he would follow on other issues in which he diverged from the communist orthodoxy of the time. Whether he simply couldn't face up just yet to open and explicit polemic with other communists or whether he felt it would be breaking ranks in the face of the enemy, it is hard to know. It is not hard to know, however, that he did his own thinking and was not prepared to compromise his intellectual integrity. The tone of implicit polemic is so strong in certain cases as to make it clear enough that in his deepest intuitions he knew that he was both heretical and right. In other cases, his unorthodoxy seems to have passed unnoticed, but in the case of *Heredity and Development*, it was unmistakable, even without any explicit mention of Lysenko,* and it is no wonder that this work went unpublished.**

---

* Caudwell did, however, make reference to "a promising young biologist" driven to "fraud and suicide" over the problem of transmission of acquired characteristics. He was obviously referring to the Austrian Lamarckist Paul Kammerer who had been given a laboratory to conduct Lamarckist research at the Communist Academy in 1925 and committed suicide upon being confronted with evidence of fraud when he returned to Vienna to collect his books and instruments. Kammerer was at this time glorified as a martyr of science in the Soviet Union. They did not concede the charges of fraud, but Caudwell seemed not to be in any doubt about it.
** I wrote to the editor of *Further Studies in a Dying Culture*, ...
(Note continued on page 451.)

The conclusion Caudwell drew from his analysis of the state of biology, poles apart from Lysenko's locating the problem within the science of genetics, was that the problem was not to be located within science, but in the crippling of science by bourgeois social relations. The synthesis could not emerge simply within biology, for it was just the posing of biology as a closed world disconnected from the worlds of physics on the one hand and psychology on the other that was at the root of the trouble. The problem could only be solved by the return of science to a common world view.[204]

Thus, for Caudwell, the new synthesis was bound up with a new form of society. An integrated world view could only be founded upon a new social matrix. From the point of view of the bourgeoisie, it was impossible to grasp the whole. Because the very mode of its existence as a class was based on the contradiction between private ownership of the means of production and the social organization of labor, between spontaneous desire and laws of nature, between subject and object, its ideology became increasingly detached from its foundations. The bourgeois ceased to have a dominating world view, because he ceased to dominate nature through labor. He was no longer in control of the process of production, but it was in control of him. Active control of the whole process was now incompatible with private ownership of the means of production, and therefore incompatible with the existence of the bourgeoisie as a class. Because he could not see his way into the future, the bourgeois increasingly looked to the past, not to the real past, but to an idealized past, a past that need not have led to the present, deluding himself into believing he could roll history backwards and return to an age when private property was not the means of exploitation, when tools were undeveloped enough and scattered enough to be owned by the man who worked them. But history could only move forward, never backward. The bourgeois could no longer discern the rhythm of the historical process. To do so would be to cease to be bourgeois, to comprehend relations other than relations of property, to connect theory with social practice. Consciousness of the whole could not come in mere contemplation. It could only emerge in an active process in contact with nature through the forms of social labor. For the bourgeois, this was impossible, as he had cut himself off from nature. His hold on the object had broken. He had prepared the ground for his own doom.

Meanwhile, nature had disappeared into the "dark night of the proletariat." The object had been slipping from the grasp of the thinking class and emerging into the hands of the laboring class. Those actively struggling with nature, debarred from consciousness by the conditions of their existence, at least followed the rough grain of concrete reality. With the development of the productive forces, the social organization of nature took place within the boundaries of the proletariat, with the bourgeoisie more and more remote from it, parasitic upon it, and inclined to impose upon it forms of organization that were increasingly arbitrary. The proletariat, even within its acceptance of the bourgeois world view, could not accept the bourgeois concept of freedom, for its strength was in its forms of social organization, in trade unions. Its freedom was only socially achieved and socially realized. This was the starting point of proletarian class consciousness. It was only a partial world view stemming from its limited experience, but it had a shattering effect on the hegemony of bourgeois class consciousness. However, it could not rise as yet to a total world view and sought its freedom still within the framework of bourgeois society.*

At first, with the proletariat, all the odds were against it, except its organization and its numbers. At first, the object in its hands could only be concrete and unconscious. But as it developed, the proletariat was not only active and organized, but became more and more conscious. The improved communication and universal education necessary to a capitalist economy was raising the consciousness of the class that would destroy it. The new consciousness was being formed within the shell of the old. The object could only blossom again into social consciousness in the form in which the proletariat could know it, that is, through the process of social labor. The bourgeoisie simply rode on top of this process, most without comprehending what was growing below it.

But some did comprehend. As well as a rising up, there was also a coming down. A section of the most conscious element perceived the true flow of the historical process. As the class struggle advanced, it cut itself off and came over to the side of the proletariat. As a result, a new consciousness was formed, a revolutionary consciousness, a fusion of the most highly developed consciousness produced by the bourgeoisie with the consciousness built on the life experience of the proletariat. Marxism was this new synthesis. Caudwell had described the process thus:

> Hence when bourgeois subjectivity in the shape of its most advanced development, dialectics, is driven by material conditions into the bosom of the proletariat, it once more encounters the object, and the object is now, as a result of technological advance, in its most highly humanised form. . . . It must not be thought that this is a kind of marriage of long-separate twin souls who suddenly

---

* Caudwell's analysis included various types of "outsiders" to the bourgeois worldview, including members of oppressed races and nationalities. His sensitivity to the experience of women as "aliens" within the bourgeois worldview was extraordinary. He showed how the dominant worldview was based on the male experience of life and how the first stages of rebellion took the form of demanding women's rights in male terms and in a male world. He traced the stages of development of women's consciousness as at first partial, inchoate, and noncognitive, as a result of exclusion from control of the economy and the cognitive apparatus of society. He believed that a fusion of male and female experience would occur in the course of the revolutionary process, resulting in the transformation of both in a new and higher synthesis.

embrace. . . . It is not a case, for example, of bourgeois mechanism (objectivity) being fused with bourgeois idealism (subjectivity). For mechanism loses the object ultimately without developing the subject, and dialectics ultimately loses the subject without developing the object. . . . But this new consciousness is not one in which active subject is parted from contemplated object, and the real activity of society sinks into the night of an unconscious class. In dialectical materialism, subject is restored to object, because in the society which generates it, consciousness is restored to activity and theory to practice.[205]

This new consciousness was formed by a sort of tunneling in from both sides, from the movement of disaffected bourgeois consciousness struggling to transcend itself on the one hand, and the movement of organized labor struggling to transcend itself on the other. In the meeting of the two, both were changed. Once the synthesis emerged, it blossomed and filled out and attracted to itself all the genuine dispersed elements of bourgeois consciousness and integrated into itself such advances in knowledge and technique as had been achieved in bourgeois culture. As the revolutionary movement actively expanded and brought into it the most conscious and most progressive artists, scientists, and philosophers, its world view gathered into itself all that was best in art, in science, in philosophy, transforming each of these spheres and transforming itself in doing so.

But the synthesis of the most healthy elements of bourgeois consciousness with the life experience of the proletariat had to take place from the standpoint of the proletariat, for it was the most healthy, energetic element in society, organized as it was around the process of social labor. All previous knowledge had to be analyzed and reconstructed, broken up and formed into a new pattern, along altogether new lines, along the lines of force opened by the emergence of proletarian class consciousness, as the only class that could end the division of society into classes and aspire to a vision of totality. However, in the process, proletarian class consciousness transcended itself, it became a revolutionary consciousness and in due course a socialist consciousness, and it in turn became no longer a consciousness grounded in a limited experience, but one that drew upon the fullness of social experience and came thus to encompass the whole.

Such a revolution in consciousness was bound up, of course, with a revolution in the social relations of production. The expropriators had to be expropriated, ending the enslavement of the have-nots to the haves and the enslavement of both to wars, slumps, depression, superstition, and disintegration. Communism was the realization of a new sort of freedom, returning to the social solidarity that characterized primitive communism, but at the same time gathering up all the development of the interim, the higher level of individuation, the greater complexity of consciousness, achieved through the development of the productive forces. The social division of labor had been necessary to the development of the productive forces, whereupon the further development of the productive forces had prepared the ground for the abolition of the social division of labor. A higher productive capacity made possible a higher level of human development. It was not a matter of going back to a world of a few beasts and crops and a wandering sun, but a world enriched greatly by the development of science and technology under capitalism.

Caudwell's vision of revolution was total. It was a whole new way of thinking and living. Although the weight of emphasis in his own work was on the development of revolutionary consciousness, it was never disconnected from the necessity of revolution in the relations of production. These were always inextricably bound together. The new world view was grounded in a new social movement, but the social movement could not proceed toward its goals without the development of its world view. Caudwell had found in Marxist theory and in the communist movement the framework that pulled it all together in what seemed quite the right way.

But within this framework, Caudwell did his own thinking. He had been inspired by the classics of Marxism as well as by later Marxist works,* but was not inclined to be constantly quoting them or to be circumscribed by other people's formulations. He worked out his own world view, based on his own knowledge and experience, and, though it converged with the dominant thinking among communists at that time in most respects, it was deeper, fresher, and more developed than most. Where it did not converge, he followed his own instincts, though without calling attention to the discrepancies.It is hard to know just how far he was aware of any such discrepancies, but often an undercurrent of implied polemic indicated that he knew quite well.** This was most notably evident in his theory of knowledge, which was in marked contrast to that of Lenin in *Materialism and Empirio-Criticism*.

Epistemology was a pivotal concern to Caudwell in the process of working out the outlines of the socialist *Weltanschauung*. The crux of the Marxist approach to philosophy was its restoration of subject to object and object to subject. It penetrated to the material basis from which the dichotomy emerged and brought to philosophy a fullness and vitality it had lacked through the ideological distortion grounded in the class division of society.

---

* Caudwell had read Marx, Engels, Lenin, Plekhanov, Stalin, Bukharin, Deborin, Jackson, Levy, Bernal, Needham, Haldane, and probably much else besides.

** His friend Paul Beard told me he knew exactly how heretical he was. Elizabeth Beard (interview in London 10 July, 1980) told me that, when Sprigg spoke of Marxism, he used to say that some of it was outmoded. She stressed quite emphatically that no one could tell him what to think. His comrade from Popular days, Nick Cox (interview in London 4 July, 1980), spoke of discussions into the night about "all the world" in which Sprigg would often take a highly innovative approach. He was always cogent and avoided jargon, cliché and shortcuts. On occasion he expressed hard-hitting differences of opinion with other comrades.

Caudwell's epistemology was interactionist rather than reflectionist. In the conscious field, generated by the interaction of subject and object, neither subject nor object, neither mind nor matter, could ever be found completely "pure." As knowing was a mutually determing relation between subject and object, neither could ever be totally separated out and isolated from the other. Knowing was an active relation, a social product. It was the outcome of the labor process past and present. Truth was realized in action; it came, not so much as an end, but as the color of an act. Feeling tone could never be completely separated from the object in experience. Both object and thought, both response and situation, were given in one conscious glow. Consciousness was not, however a mere irridescence, but real, determining, and determined.

It was vital for Marxists not to fall back into the subject-object dichotomy. The uniqueness of Marxism was that it overcame this dichotomy, not be denying one or the other (as did idealism and mechanistic materialism) or both (as did positivism), but by embracing both. Both were real, however inseparable within the conscious field. Each was a constituent of the other. Always the subject was tied to the object as the object was to the subject. Knowledge bore always the impress of both. Arguing against epistemological objectivism, Caudwell stressed the role of active subjectivity and the importance of breaking with the illusion of the detached observer. It was impossible for the mind to stand outside the universe, to know it without disturbing it or being disturbed by it. Truth might seem to be in the environment, to be objective, independent of the subject, and yet all attempts to extract a completely nonsubjective truth from experience produced only metrics. Objectivism could not be sustained and turned into its opposite; complete objectivity brought one back to complete subjectivity and vice versa.

The act of knowing transformed what was known. It was never possible to detach the thing known from the knowing of it. Caudwell opposed all passivist imagery in describing knowledge. Knowledge was not a matter of copying, mirroring, photographing, reflecting. Although he never remarked on Lenin's use of such imagery in *Materialism and Empirio-Criticism*, he had read the book and his rejection of the reflectionist model was quite explicit and polemically expressed. In no uncertain terms, Caudwell made his point:

> The mirror reflects accurately: it does not know. Each particle in the universe reflects the rest of the universe, but knowledge is only given to human beings as a result of an active and social relation to the rest of reality.[206]

This particular passage came in his argument against Wittgenstein's *Tractatus*, but it was a persistent theme and he could not have failed to realize the position he was opposing was also held by certain of his fellow Marxists.

Because of his emphasis on knowing as an active process, Caudwell was not disturbed in the least by Heisenberg's principle, which was causing such a stir.

As he believed that each act of knowing involved a new determining force that could not have been allowed for in any original Laplacean act of knowledge, he saw Heisenberg's principle as underlying the fact that knowledge of reality produced a change in reality. The idea that men learned about reality in changing it reached its fullest expression in Heisenberg. All laws of science were laws stating what action produced what change in reality.

Implying that a contemplative and passivist epistemology endangered revolutionary activism, Caudwell made the point that social consciousness was not simply the reflection of social being. If it were, he said, it would be useless. It was the fact that the mind could conceive of something other than the existing reality that made progress possible. It was the very disparity between man's being and his consciousness that drove history forward.

On the other hand, Caudwell was also critical of epistemological subjectivism. Knowledge emerged in interaction with the environment. Conceptualization did not take place in a vacuum, but was based upon empirical observations. Science developed through both hypothesis and experiment. A preoccupation with logic or any other variety of mentalist introversion resulted not in truth but in mere consistency. Such a preoccupation myopically scrutinized the conscious field without reference to the pressures upon it from outside it, in spite of the fact that such pressures from the past were what actually formed it. Rather than remaining alive to the developing subject-object relation, it remained fixated on the forms it took in the past. Logical laws were social. Language was a social product.

On the question of epistemological priority, Caudwell came down firmly on the side of realism, as opposed to any form of phenomenalism, instrumentalism, or conventionalism, consistent and firm in his critique of Machism and all varieties of neo-positivism. It was, however, a critical realism, a mediated and historicized realism. In terms of the debate within *Materialism and Empirio-Criticism*, it was neither the position of Lenin nor that of Bogdanov. Nor was it the position of Lukács or Korsch either. It was perhaps the position Gramsci was groping for, but never expressed with such confident clarity as Caudwell. When it came down to it, being preceded knowing, knowing flowed from being and evolved as an extension of being. Decidedly post-Cartesian, Caudwell asserted: "I live therefore I think I am."[207] In a concise statement of the fundamental contours of his theory of knowledge he wrote:

> The question of which is first, mind or matter, is not therefore a question of which is first, subject or object. . . . Going back in the universe along the dialectic of qualities, we reach by inference a state where no human or animal bodies existed and therefore no minds. It is not strictly accurate to say that therefore the object is prior to the subject any more than it is correct to say the opposite. Object and subject, as exhibited by the mind relation, come into being simultaneously. . . . We can say that relations seen by us between qualities in our environment (the

arrangement of the cosmos, energy, mass, all the entities of physics) existed before the subject-object relationship implied in mind. We prove this by the transformations which take place independent of our desires. In this sense, nature is prior to mind and this is the vital sense for science. These qualities produced, as cause and ground produce effect, the synthesis, or particular subject-object relationship which we call knowing. Nature therefore produced mind. But the nature which produced mind was not nature "as seen by us." . . . It is nature. . . . as having indirect not direct relations with us. . . . Such a view reconciles the endless dualism of mentalism and objectivism. It is the universe of dialectical materialism. Unlike previous philosophies, it includes all reality: it includes not only the world of physics,but it includes smells, tastes, colors, the touch of a loved hand, hopes, desires,beauties, death and life, truth and error.[208]

Only such an epistemology, which could hold together these different strands without giving way to either naive realism or total relativism, could encompass the whole. Caudwell held to it with utter consistency. It was essential to the underlying unity of his thought and it accounted for the distinctiveness of his position on various issues.

It had implications, for example, for his position on the dialectics of nature. There is no evidence that he knew of the Comintern debate on dialectics of nature centering around Lukács, but he was obviously acquainted with the standard communist formulations of this time to the effect that the subjective dialectics of the mind reflected the objective dialectics of nature. Although he did not specifically criticize any Marxist authors holding this position, it was clear that he disagreed with it. For Caudwell, what was dialectical was neither subject nor object, neither man or nature, but the interaction between subject and object, the interaction between man and nature. He gave expression to his thinking on this in a way that again suggested the possibility of an underground polemic:

> The external world does not impose dialectic on thought, nor does thought impose it on the external world. The relation between subject and object, ego and universe, is itself dialectic.[209]

It was impossible, to Caudwell's way of thinking, to speak of an objective dialectic, to speak of nature as dialectical in itself. The dialectic emerged in the relation and not in the object or subject in itself. It was impossible to detach subject from object, or to know definitely what either would be in itself without the other. All that man could say about nature was generated by his interaction with it. He could not be the detached observer and know nature apart from his knowledge of it. Even in affirming the priority of nature to mind, he knew it was only possible to do so in and through nature as known by mind. Nature, as known, was a product of human society. Nowhere could the line between the two be precisely laid down.

Knowledge of nature was mediated by the labor process. In its historical development,the labor process not only generated economic systems, but mathematics and the sciences. Emphasizing the full extent of the role played by social labor in the history of knowledge, Caudwell wrote:

> Once established, the labor process, extending as remotely as observation of the stars, as widely as organisation of all human relations, and as abstractly as the invention of numbers, gathers and accumulates truth. Faster and faster it proliferates and moves. The bare organism is today from birth faced with an enormous accumulation of social truth in the form of buildings, laws, books, machines, political forms, tools, engineering works, complete sciences. All these arise from co-operation; all are social and common.[210]

In this sense, all natural things were artificial. Nature could only be for man in and through his interaction with it. Emphasizing the essential role of such interaction, Caudwell stated bluntly:

> The dynamic subject-object relation generates all social products—cities, ships, nations, religions, the cosmos, human values.[211]

It must seem that Caudwell took a position similar to Lukács, and so he did, in so far as the dialectic was identified with the subject-object relation and in so far as nature was recognized to be a social category. The difference was that Caudwell was able to hold together the fullness of the subject-object relation. Unlike Lukács, he did not forget the object. In the interaction between man and nature, he did not neglect to give nature its due.

Actually, it would never have occurred to Caudwell to think that such an emphasis on human subjectivity could somehow negate the cosmic dimension of Marxism. He continued to affirm what had been most crucial to Marxists in the name of dialectics of nature. Although Caudwell located the realm of the dialectic in the sphere of human interaction with nature and not in nature itself, he realized that what came in the interaction bore the impress of nature and yielded reliable knowledge of nature. He believed, like the rest, that nature developed, not only gradually and in a straight line, but proceeded also by jumps, and that genuine novelty emerged. He believed that the universe was one, that everything was related to everything else; he affirmed "the altogetherness of everything."[212] His philosophy of science was based on the antireductionist concept of integrative levels: history began where physiology left off, biology began where chemistry and physics left off. Each level embodied fresh laws, inclusive of, but additional to, the laws of previous levels.

What then was the dialectic for Caudwell? He didn't use the term often, but then he didn't use any of the normal Marxist terminology very often.He did, however, see fit to categorize his philosophy as dialectical materialism.

He saw it as a distinctive form of materialism in its stress on the active mutually determining subject-object relation, in its restoration to matter of an inner activity and capacity for history that had been stripped from it in class society, in its synthesis of concreteness and sensuousness with time and process and development. What was dialectical was the relation, the restoration, the synthesis. Essentially, it was a way of thinking, imbued with a sense of development, of time, of history, that resolved the tension of polar opposites in a higher synthesis. It was the transcendence of dualism in a processive, integrated way of thinking. But it was a way of thinking grounded in a way of being, in the mode of existence of the universe itself. Man was a part of nature and his thought could be integrated, because reality could be integrated. Man could think dialectically, because the universe as known by him was one and in motion through a web of mutually determining relations.

Above all, what Caudwell was about was setting out the outlines of an integrated world view within which all the intellectual chaos would be resolved and showing its relation to a classless society within which all social fragmentation would be overcome. Within this scheme, science was a critical factor. Caudwell was never in the slightest inclined to a romantic rejection of science. He knew science and embraced it fully. Very much in the tradition of British Marxists, he understood completely its role in the development of human knowledge and in the cause of human liberation. In this spirit, he pursued the ideal of the unity of science. All of the sciences had to be sifted through, reintegrated and reconstituted within the framework of a consistent philosophy. Each of the sciences had to open out to the other sciences—indeed all disciplines had to open out to all others—and each revitalized in its interaction with the rest. Specialization would continue to be necessary, but integration was absolutely essential. Only in this way could science—or art or any other discipline for that matter—again become coherent and fruitful.

Here Caudwell brought the notion of proletarian science into play, running parallel to his discussion of the notion of proletarian art. In class society, argued Caudwell, every discipline in its implicit assumptions was colored by class consciousness. There was no neutral world of science or of art, free from categories or determining causes. Both science and art were in this sense social activities. Class society had in its time enriched both science and art by drawing each to one pole to intensify its development, but at the price of severing it from the other pole, with which it now had to be reunited to develop further. Due to the conditions under which they emerged, bourgeois science and bourgeois art were one-sided, produced in a culture in which a yawning gap had opened between theory and practice, between consciousness and social labor. Thought had stiffened and broken apart in its separation from its organic nexus. Bourgeois science and art were in crisis, in so far as they were bound up with a culture that was dying. However, with the emergence of a distinctive proletarian class consciousness and with its fusion with a part of bourgeois consciousness separating itself out and adhering to the pole of the proletariat, the gap began to close as the new consciousness drew to itself the development of science and art, refashioning them and integrating them within a consistent world view and within a healthy social force. At this period, something emerged that could be called proletarian science or proletarian art. It was a transitional stage in the development of science or art. It was the process of fusion in which all past knowledge was sifted through and reconstructed within a new philosophy.

The new philosophy was not, of course, given, but emerged within this process and on the basis of concrete knowledge. There were certain criteria that needed to be fulfilled to make a philosophy adequate to its new tasks:

1. It needed to be able to explain all the scientific discoveries of its era within the one framework. Not one of the contending bourgeois philosophies was competent to do this.
2. It needed to include, as real and unified, all forms of experience—colors, sounds, values, aims, time, space, and change.
3. It needed to account for the historical evolution of all the various arts, sciences, and religions and to explain the evolution of their own explanations of themselves.[213]

The process of striving to overcome intellectual disintegration and to integrate the sciences and all knowledge and experience in a unified philosophy ran parallel to the process of striving to resolve economic anarchy in collectivization of the means of production. The task of the reconstruction of science was hastened and intensified with the achievement of socialist revolution. It would proceed and be brought to a new level in and through the process of socialist construction, as the tunneling in from both sides and accelerated assimilation of scientific knowledge into the new consciousness continued. It demanded both that bourgeois scientists come to identify with the proletariat and its tasks and that the proletariat come to identify with science and its tasks. Both tasks required a complete refashioning of consciousness. The synthesis needed to come from both sides. The assimilation could not be achieved by one side alone. It was this emphasis that most distinguished Caudwell's position from that of Proletkult. With the achievement of a full assimilation, proletarian science became socialist science, in the course of the revolutionary social transformation in which the proletariat became no longer a specific class, having expanded to encompass the whole of concrete living, that is, in the course of the transition to a communist society. There could be classless science only in the achievement of a classless society. Until then, there could only be bourgeois science or proletarian science.

It is an argument needing to be looked at freshly in the aftermath of the dismal and horrific episode surrounding Lysenko, in which the notion of "proletarian science" has left a bitter taste in everyone's mouth. It must be said emphatically that the notion of proletarian science does not stand or fall with Lysenko. Indeed, Lysenko represented the violation of the very process Caudwell was envisioning. For Caudwell, proletarian science was the integration of the sciences (in which the science of genetics incidentally played a crucial part) within an integrated world view. Caudwell said quite firmly that it was not a matter of imposing the dictatorship of the proletariat on science. It was not a matter of the honest worker telling the scientist what was what in his laboratory or in his theory. Nothing was to be imposed on science. Nothing was to be imposed on the scientist, not even by himself. It was a matter of the assimilation of the scientist to the cause of the proletariat, to the construction of a new society, in which he played his full part *within* the process and *as* a scientist. Science was to be developed by *scientists*, but a new type of scientist, with his feet more firmly on the ground, with his mind more opened to the whole, with his life and work more organically connected to the society of which he formed a part.

The notion of proletarian science was, of course, fraught with dangers, as subsequent history has revealed in a most tragic way. Caudwell had no way of knowing what was happening in the Soviet Union even as he wrote, but it is clear what he would have thought. He believed that there was something terribly wrong with a society when it became coercive, rigid, and "tremblingly alive to heresy."[214] The denunciation of genetics and the arrests of geneticists were the very negation of the process he insisted must be undertaken. So too were the forced and opportunist reconstructions of the sciences in terms of a superficial and dogmatic interpretation of dialectical materialism. Caudwell did anticipate that a certain crudity and clumsiness might be inevitable in the early stages, rather like what resulted when the proletariat occupied for the first time a role in administration, which hitherto had been peculiarly the prerogative of the bourgeoisie. At first, every mistake would be made, except the fatal bourgeois mistakes. Clearly, he did not see just how crude and how clumsy it could actually be, nor did he realize just how fatal proletarian mistakes could be. However, he firmly stated that what was required was a laborious refashioning, a careful, complex and protracted process, a process that was in fact disastrously short-circuited in the Soviet Union.

The dangers Caudwell saw were more from the other side. He worried about scientists not committing themselves fully enough. His concern was over the bourgeois scientist, accustomed to the role of "lone wolf,"[215] who would be prepared to merge with the proletariat and accept its theory and organization in every field except his own. If he remained unintegrated and kept his science in a preserve apart from the development of Marxist theory, the result could only be a distortion of both science and Marxist theory. With the two in essentially separate compartments, either there was no interaction or politics burst into his science in the form of crude and grotesque scraps of Marxist terminology and his science burst into his politics in the form of gratuitous outbursts of bourgeois indiscipline, assertion of the independence of science or distortions of Marxist theory through the retention of bourgeois categories in his own discipline.If scientists, engineers, teachers, historians, economists, artists, soldiers, administrators all indulged the impulse to keep their own sphere separate and to assert its independence from the rest, there would be no revolutionary society at all. The independence of disciplines was bourgeois, the result of the fragmentation of bourgeois ideology, and it must not be carried over into the proletarian state.

It was vital that the scientist not settle for the easy road and mechanically reshuffle the categories of bourgeois science or mechanically import categories from other spheres of proletarian thinking into science. What the proletariat asked of the scientist was that he take the difficult creative road, that he do his own thinking, that he refashion the categories of science, so that science could play its full part in the realization of the new world view and indeed the new world. The scientist was not being asked to play fast and loose with experimental results, nor to "apply" dialectics to science, nor to "fit" the results of science into the categories of dialectics. He was being asked to look at his results rigorously but freshly and to take a broader view, that is, to examine their implications in terms of a world view grounded in the most advanced science.

The same applied to all intellectuals in the revolutionary movement and in the new society. Caudwell, much like Gramsci, envisioned the emergence of a new type of intelligentsia, an active intelligentsia, no longer remote from production, from labor, from the whole vast social process. The new intellectuals would be rooted in the working-class movement and develop the new consciousness that would express and carry forward the realization of its destiny.

It is obvious that Caudwell was trying to do much the same as what earlier Comintern theoreticians were trying to do. Although he was unacquainted with the efforts of Lukács, Korsch, and Gramsci, as they were unknown in Britain at this time, he was much like them in his emphasis on revolutionary consciousness, that is, on the necessity of breaking radically with bourgeois thought patterns and evolving totally new ones. Like them, too, he underlined this enterprise with an epistemology emphasizing active subjectivity. Caudwell, however, was the most integrated, in so far as he understood the role of science within this enterprise. It is true that his achievement was largely programmatic and that much remained to be done. Nevertheless, given the chaos of conflicting theories on matters of fundamental orientation and given the fact that so little could be taken for granted at the programmatic level, even among the Marxists, his achievement was no small one.

Korsch's merit was in calling attention to the neglect of the dimension of human subjectivity within Marxism and in defying authoritarianism in the communist movement, but in his constructive efforts he was woefully unsuccessful. Lukács went further along the way. Indeed, the parallels between himself and Caudwell are striking: the emphasis on proletarian class consciousness as the only force capable of achieving vision of totality, the determination to overcome the subject-object dichotomy, the stress on the active side of the knowing process. But the contrasts are just as striking. Caudwell was not only far clearer about the nature of class consciousness and the process of its emergence and development, but he was actually able to achieve a vision of totality. Lukács was not. He left far too much out of account, most notably nature, science, objectivity, determinism. Caudwell was able to hold together in one vision both man and nature, both revolution and science, both subjectivity and objectivity, both freedom and determinism.

As for Gramsci, he came closest to a full synthesis, though he was epistemologically vague and remote from science. In Caudwell, the synthesis was more total, more tightly integrated, more firmly grounded in science. In terms of political strategy, on the other hand, Gramsci's analysis was more detailed and more refined. Naturally so, as he was far more politically experienced than Caudwell. It is the affinities that are most significant, however: most importantly, the realization of the fullness of the revolutionary process. Both saw revolution in terms of total transformation of thought and culture, a struggle for men's hearts and minds that was connected with, but not reducible to, the struggle for state power. Also, both were competent to come to terms with non-Marxist ideas in an appreciative but critical way, well able to move within a wider intellectual world, with a simultaneous stress on the integrity of Marxism and opposition to eclectic combinations of Marxism with non-Marxist ideas alien to its fundamental structure.

Caudwell's importance in the history of Marxist philosophy has never been recognized. Such secondary literature as exists, and it is not much, has tended to concentrate almost exclusively on his contribution in terms of aesthetics and literary criticism. Such surveys of the history of Marxist philosophy as exist either ignore him altogether or give him only the briefest of passing mention. In Kolakowski's *Main Currents of Marxism*, for example, he is discussed in one paragraph out of three volumes, and *The Crisis in Physics* is rather unperceptively described as "a Leninist attack on idealism, empiricism and indeterminism in modern scientific theory,"[216] whereas Korsch is given an entire chapter. In most other surveys, Caudwell is not even mentioned.

Even among British Marxists Caudwell has been virtually forgotten and never really subjected to full-scale assessment in the first place. The major occasion on which it was attempted, the "Caudwell Discussion" in *Modern Quarterly* in 1951, was a rather disgraceful affair, in which his philosophy was subjected to heavy-handed and ill-founded attack. Cornforth and Bernal in particular were extremely severe and even called his credentials as a Marxist into question, accusing him of a too favorable attitude to bourgeois intellectual trends, most notably those associated with Freud, Einstein, and modern geneticists. The discussion, which took place during the high tide of Lysenkoism in the world communist movement and the Soviet campaign against "cosmopolitanism," revealed more about the state of British Marxism in its time under the pressure of these events than about Caudwell. Caudwell's emphasis on genetics was one of the reasons why he posthumously ran into so much trouble at this time. Even his most severe critics, however, could not help but acknowledge his originality and brilliance. In the course of the discussion, there were those who came to his defence. Various contributors, particularly George Thomson and Alick West, answered the attacks fairly capably.[217]

However, even Caudwell's most devoted defenders, however appreciative of his insights and aware of his profundity, have tended to see his importance for them primarily as a "home product" and haven't fully grasped his importance within the development of Marxist philosophy as a whole. Those on both sides have made far too much of Haldane's "quarry of ideas" metaphor, probably far more than Haldane intended by it, missing Caudwell's essential contribution, which was a tightly constructed and unified analysis of the nature of bourgeois culture and the socialist alternative. He has been interpreted too randomly—as a source of insights—rather than as the bearer of a coherent world view.

The years have passed with his philosophical contribution falling into oblivion. Even in an article giving a rather astute analysis of the tradition of British Marxism in relation to science, which actually raps the knuckles of present-day British Marxists for their neglect of this tradition, Hilary and Steven Rose never mention Caudwell. Calling for a revival of discussion of the theme of bourgeois versus socialist science, opened in Britain by Hessen and effectively closed by Lysenko, they actually state that British Marxists in the 1930s never unraveled the argument, obviously ignorant of Caudwell's contribution in this domain.[218]

Now and then, someone remembers, but almost entirely in terms of literary theory. David Margolies's study of Caudwell's aesthetics is appreciative, though somewhat out of focus in giving the impression that the central focus of Caudwell was a theory of the social function of literature.[219] His central focus in fact was *Weltanschauung* and his interest in literature fell into place around this. Francis Mulhern's analysis of Caudwell tries to see his literary theory more in terms of its underlying philosophy. Something of Caudwell manages sporadically to break though a screen of Althusserian misinterpretation, and what basically comes out of it is: "Christopher Caudwell is an historicist, but not every historicist is Christopher Caudwell."[220] The *New Left Review* was to swing even wider off the mark with Terry Eagleton's passing dismissal of Caudwell, summed up in the conclusion that there is "little, except negatively, to be learned from him." Caudwell's work is written off as Stalinist, idealist, speculative, erratic, "bereft of a theory of the superstructure," "punctuated by hectic forays into and out of alien territories," and "strewn with hair-raising theoretical vulgarities."[221]

A breath of fresh air has come with E.P. Thompson's attempt to settle his intellectual accounts with Caudwell. It is an extremely perceptive and constructive analysis of Caudwell's work as a whole, which says many things that have been crying out to be said. He seeks to highlight what were Caudwell's most central and creative preoccupations, which he rightly notes have been misunderstood by younger Marxists. Insisting that Caudwell's work was more significant than had been realized, with its impulse not yet exhausted, Thompson suggests a new way into the evaluation of Caudwell. He proposes seeing Caudwell as an anatomist of ideologies, as brilliantly laying bare the deep structure of the characteristic illusions of the epoch, as analyzing the generation of modes of intellectual self-mystification. It was, Thompson asserts, a brave, even Promethean, venture: to effect a rupture with a whole received world view. It was "the most heroic effort of any British Marxist to think his own intellectual time." Caudwell was walking abroad in his world, encountering the largest ideas and issues and not retreating into the introverted security where Marxists existed in a universe of self-validating texts. What resulted from his efforts was, in Thompson's opinion, a body of thought that was more interesting, more complex, and more heretical than has been supposed. In refusing the orthodox closures offered by reflection theory and the base/superstructure model, Caudwell was opening the door to a more creative tradition.

Thompson is far from uncritical of Caudwell, however, considering him to be impatient of mediations, to be overready to see everything in terms of binary oppositons, to give way to lesions of logic, shifts, and jumps, to be uncritical of the Soviet Union. While these criticisms are not without some basis, Thompson perhaps exaggerates their extent and significance, which brings him to draw conclusions out of harmony with the overall tone of his analysis. In the end, he says, nothing in Caudwell's work was of a maturity or consistency to merit consideration as a classic and perhaps ninety percent of Caudwell's work must be set aside. He does admit that there is a residue of ten percent that is of such extraordinary vitality and relevance that it presents itself for a renewed interrogation among today's Marxists.[222]

To appreciate Caudwell's full stature as a Marxist thinker, he should not be seen simply as a source for random insights, but as a philosopher who made an original and highly integrated contribution to the development of Marxist philosophy that has yet to be grasped and assimilated by other Marxist philosophers. In terms of the problematic of this book, that is, the conjuncture of science, philosophy, and politics within Marxism, Caudwell offered a most highly developed position illuminating the interrelations. His was a most tightly constructed and plausible argument, analyzing the variety of ways these spheres have converged historically, focusing on the chaos of the later bourgeois period and on the new integration that must be effected by Marxists.

This is not to deny his faults. He was sometimes in too much of a rush and settled for neat antitheses that were a bit off the mark, but these were not such as to distort his overall argument ordered around his most basic underlying themes. Most of the problems emerging from Caudwell's work have to do with the fact that there has been much water under the bridge since. Caudwell, as did his contemporaries of the 1930s left, saw capitalism as undergoing its final crisis and the bourgeoisie as convulsing in its death throes. Events have proved this wrong, calling for a more complex analysis and a more protracted time scale. Subsequent events in the history of Marxism, particularly in the Soviet Union, have also complicated matters in a way that Caudwell could not foresee. His view of Soviet society seems euphoric now and, as Thompson rightly says, it "leaves a dusty taste in our mouths today."[223]

British Marxism and its Successors: Discontinuity with the Past

It is not only Caudwell's work that has fallen into neglect, however much his constitutes the most striking case, but the whole tradition of prewar British Marxism. Present-day British Marxists, particularly those gathered around *New Left Review*, never tire of saying that there is no native British Marxist tradition and that it has been necessary to look abroad for sources of a creative development of Marxism, to the Frankfurt School, to Lukács, Gramsci, Colletti, and Althusser. Perry Anderson seems to let no occasion go by without remarking that there has never been a Marxist culture in Britain or stating that Britain has not produced a single major Marxist theorist in the twentieth century. Ordinarily not considering such Marxists as Bernal, Haldane, Needham, or Caudwell worthy of mention, when he does bother to cast a passing glance at British Marxism in the 1930s, it is only to dismiss it with a condescending reference to "the fantasies of Bernal."[224] David McLellan's *Marxism After Marx* includes a chapter entitled "British Marxism" that comes to a grand total of three pages, with only one short paragraph to bridge the gap between Sylvia Pankhurst and Raymond Williams. There is little to be learned from it, except that the Left Book Club came after the Social Democratic Federation and before *New Left Review*. There is no trace of Bernal, Haldane, Needham, Jackson, Cornforth, Guest or many others.[225]

Not much better can be expected from the rest of the world by way of recognition when British Marxism is treated thus in Britain. Among foreigners, there is Kolakowski, who has even gone to settle in Britain. He also seems to consider it negligible. In three volumes, there is not a single mention of Bernal, and Haldane gets only passing reference as endeavoring "to prove the affinity of Marxism with modern science." For Haldane, as well as for the American Marxist H.J. Muller, Kolakowski asserts, "Marxism figured in aspects that were not specifically Marxist; in biology it appeared chiefly in the form of a general opposition to vitalism and finalism."[226]

But there have been occasional voices piercing the forgetfulness. Hilary and Steven Rose have expressed regret that the concerns of earlier British Marxists with science have found no echo among the present generation, and they have set a certain value on the contribution of such as Bernal, Haldane, and Needham (though not Caudwell). Not that they have been uncritical. For example, they have expressed their regret that earlier Marxist scientists did not adequately take up Hessen's challenge and indeed spent their time and theoretical strength "in a relatively fruitless endeavor to demonstrate the negation of the negation, the interpenetration of opposites and the transformation of quantity into quality in a variety of scientific developments."[227]

Gary Werskey has made the most elaborate attempt to assess the contribution of earlier British Marxists.* His book *The Visible College* is a collective biography of Bernal, Haldane, Needham, Levy, and Hogben. Although it is well researched and full of interesting incidents and observations, it suffers from Werskey's failure to understand properly the role philosophy played for his subjects. Observing that today's left has never heard of Bernal, Werskey makes the point that earlier British Marxists were the authors of a large and distinctive body of socialist thought. In developing a Marxist cultural tradition in Britain, they not only fashioned their own novel, coherent, and contrasting interpretations of Marxism, but closely linked their theoretical perspectives to a set of highly effective political practices as well.[228]

This represented something of a modification in Wersesky's judgements on the earlier generation. A few years before, he took the position that the "thirties movement" took its stand on a "a stultifying, scientistic variant of Marxism, allied to an uncritical defence of Stalinism," which he blames on the reformist and opportunist policies of the CPGB and the Popular Front:

> What Marxism remained in the scientists' movement was either harmless—in respect of British politics—natural philosophy or propaganda for Soviet science. And in the quest for new adherents the older radicalised scientists therefore began to grow less sensitive politically and intellectually to the full-blown

---

* Since writing this, Stuart Macintyre's book *A Proletarian Science* (Cambridge, 1980) has appeared. It is a well-researched account and extremely fair assessment —both critical and sympathetic—of British Marxism in the period between 1917 and 1933.

> emergence of Stalinism in Russia. They failed to see that the very concepts they
> were borrowing from the Russians were, in the Soviet context itself, being used to
> rationalise the power of Stalin: the naturalistic world view of dialectical
> materialism, which introduced into Soviet Marxism determinism in thought
> making for dogmatism in action, the criterion of practice, which meant that Stalin
> had the right to arbitrate on all policy disputes, and the programme of
> proletarianising science which allowed Stalin to replace dangerous bourgeois
> henchmen with unquestioningly loyal peasants and proletarians.[229]

While they may have been blind, perhaps to some degree wilfully so, to what was happening in the Soviet Union, Werseky is perhaps just as blind to the nature of the CPGB and the Popular Front. The severity of his judgement is also based on a severe failure to understand the philosophical issues at stake, that is, in asserting a necessary identification of the concept of determinism with political passivity and of the notion of proletarian science with Lysenkoism. Such identification arbitrarily rules many alternative possibilities out, namely the very possibilities that Caudwell so eloquently set out. Moreover, his rejection of their philosophy of science as scientism proposes no alternative relation of science to Marxist philosophy.

The severity of Werseky's judgement had been moderated somewhat by the time his collective biography was finished. While the book is not so full of acidic remarks about the CPGB and the Popular Front, and Werksey is more appreciative of some of the Marxist scientists' ways of connecting science to Marxism, as representing a "novel extension of Marxism," he still is lacking in sympathy for his subjects in varying degrees. He has particularly serious reservations about "Bernalism." Bernal's Marxism, he insists, was always in danger of becoming like that of his Soviet counterparts—"history with the people left out." He argues that the concept of science occupied a pivotal position in the expositions of dialectical materialism from this period because persons and classes did not. The general laws of motion were substituted for the working class. Werskey also believes that the concept of dialectics of nature or a Marxist approach to biology is peripheral to their real thinking. Their philosophy of science, their "excursions into dialectics," according to Werskey, remained external to their scientific research. He insists that they in no way made their science and politics dependent on belief in dialectical materialism. As the conjuncture of science, politics, and philosophy was absolutely crucial to what they were about, it is a rather serious flaw in what is in many respects a very fine piece of scholarship that brings this trend in Marxism up for discussion again.

French Marxism: A Parallel Development

French Marxism underwent a parallel development in the prewar period. Only in the early 1930s did the Communist Party begin to overcome the obstacles to the spread of Marxism among the French intelligentsia. The Comintern's policy of bolshevization, which discouraged emphasis on national characteristics and national intellectual traditions, repulsed the French intelligentsia, which was more than most inclined to a national traditionalism, even chauvinism, and resistant to foreign ideas in the best of circumstances. The centralized traditionalist and highly bureaucratic university system reinforced the wall of resistance. The situation was not helped either by the antiintellectualist outbursts of the Comintern leadership on such occasions as the Fifth World Congress in 1924, which heard Zinoviev's condemnation of the "professors" and Klara Zetkin's warning against admitting too many intellectuals—"inconstant allies"—into the communist parties. The congress also condemned Boris Souvarine, editor of *L'Humanité*. *L'Humanité*, naturally enough, was thrown into confusion and underwent a sharp transformation in which the "La vie intellectuelle" page abruptly disappeared. The onset of the "third period" saw the party rocked by violent purges. The succession of Maurice Thorez to the party leadership in 1930 coincided with something of a restoration of equilibrium. Thorez was one of the earliest in the communist movement to call for a united front, reaching out to antifascist intellectuals in a common cause. From then on, intellectuals were actively recruited by the party and encouraged to think through the problems of their own disciplines in the light of Marxism. The first generation of communist intellectuals, which had been for the most part lost in the convulsive sectarianism of the 1920s, began to be replaced by a new generation.

The Philosophies Group

In the 1920s, a group of younger philosophers, the Philosophies group, made their way into the party via their own serious study of Marxism. At first, under the influence of such trends as surrealism and psychoanalysis, they turned to Marxism as best able to restore the "sense of being" capitalism had destroyed in modern man. Philosophy they asserted to be fully a part of the revolutionary process.

Politzer

Georges Politzer, Hungarian-born agrégé in philosophy at the Sorbonne, concerned himself with many issues; his interests included not only philosophy, but psychology, physics, and politics. At first determined to introduce Freudian psychology into French academic life, he came to criticize its "false psychologism." Upon his conversion to Marxism, he wrote a critique of the foundations of psychology and strove to set psychology upon new foundations that underlined the social character of human personality.

Politzer devoted himself to polemics, not only against Freudianism, but against a variety of other intellectual trends vying with Marxism at the time, particularly Bergsonism. He fully identified Marxism with modern rationalism, characterizing it as the only vehicle capable of carrying forward the best of French intellectual traditions and the surest bulwark against the threat of all forms of modern irrationalism.

In addition to his wide range of activities, such as creating and directing the center of documentation of the French Communist Party (PCF) and writing a column in *L'Humanité*, Politzer also taught the extremely popular course on dialectical materialism at the Workers' University in Paris. In this course, Politzer stressed that Marxism must not be presented only as a political doctrine, for it embodied a general conception, not only of society, but of the universe itself. He stressed the intimate relationship between dialectical materialism and modern science. Philosophical questions concerning matter were to be answered by science. Philosophy was dependent on science and the state of empirical knowledge at any given time. Dialectical materialism had developed, not only on the basis of the results of science, but under the influence of the spirit of science. As science had achieved a more profound knowledge of nature, the interconnections of the sciences had come to be seen with greater clarity. Dialectics, with its understanding of change, reciprocal interaction, contradiction, and progress by leaps, represented a new spirit of science.[230]

Lefebvre

Henri Lefebvre, another of the Philosophies group, also held in the 1930s to this interpretation of Marxism, aspects of which he was to repudiate later. Explaining his position vis-à-vis the "dialectics of nature," he wrote:

> The dialectical analysis is valid for any context. By incorporating the experimental sciences (physical, biological, etc.) and using them to verify itself, it can therefore discover, even in nature, quality and quantity, quantity turning into quality, reciprocal actions, polarities and discontinuities, the complex but still analyzable becoming . . . the laws of the human reality cannot be entirely different from the laws of nature. . . . *Capital* shows how, in Marxian thought, the concrete dialectical materialism is made universal and acquires the full dimensions of a philosophy: it becomes a general conception of the world.[231]

The book was unusual for its time in its attention to Marx's *1844 Manuscripts* and its emphasis on the concept of alienation. Although at this juncture, the weight given to the concept of alienation did not seem to him to entail rejection of the concept of the dialectics of nature, Lefebvre's stress on Hegel and on the importance of Hegel for Marx set him even at this time within the neo-Hegelian trend within Marxism. He was to evolve and expand his views on all these matters and to produce a formidable and insightful body of work in the postwar period.

Nizan

Another young philosopher to come forward in the same period was Paul Nizan, a normalien agrégé in philosophy, who called for the development of a purely proletarian philosophy, one that would see the history of philosophy as involving far more than simply the development of ideas. The blindness of bourgeois philosophy caused it to work against the great universal aims it claimed to pursue.[232]

Friedmann

Among the Philosophies group was one who early became disillusioned with the state of Marxist philosophy. After three trips to the Soviet Union, Georges Friedmann denounced in 1938 what he diagnosed there as an atmosphere of sterilizing polemic that drowned science in merciless partisan struggle and indulged a lazy taste for facile formulas.[233]

Cercle de la Russie Neuve

Most of the French Marxists interested in philosophy and science continued, however, to see the Soviet Union as a source of great inspiration to them in their own efforts, as the surest safeguard of the values of science and reason against the barbaric forces on the march to destroy everything connected with scientific rationality.

This attitude among those concerned with history and philosophy of science was generated particularly by the influence of *Science at the Crossroads*. The scientific commission of the *Cercle de la Russie neuve* was a focal point for discussion of the relevance of Marxism for the sciences. The first volume of essays, published by various of its members, *A la lumière du marxisme*, testified to the importance that was beginning to be attached to Marxism as a philosophy of science. References to Vavilov, Zavadovsky, Joffe, and Hessen once again indicated the great influence of the Soviet delegation to the 1931 History of Science Congress upon their foreign contemporaries. The authors of this volume expressed their alarm at the situation of science in the bourgeois world. The economic anarchy of capitalism gave rise to narrow scholasticism in science. This tendency to exclusivism, overspecialization, intellectual cloistering, and severe isolation of disciplines brought about a profound disequilibrium that could not be solved by Bergsonian mysticism or any of the other variants of the bourgeois struggle against science. It was only dialectical materialism that showed the way out of the various crises in the bourgeois world.

The authors interpreted the various crises in terms of class. The wave of materialism in the nineteenth century, associated with French anticlericalism, represented the final struggle waged by the capitalist class against the last remnants of feudalism. The rebirth of idealism represented the philosophical peace desired by the middle class, which had gained their objectives against the older class and had begun to feel the first signs of their own decay with a new class on the rise. The class presently in power, they believed, had reason to dread the subversive conclusions of a synthetic science.[234]

There was quite an impressive array of eminent scientists drawn to Marxism, both because of its politics and its philosophy of science, and entering into such discussions. Among these were the biologist Marcel Prenant, the astronomer Henri Mineur, the psychologist Henri Wallon, the physicists Paul Langevin and Jacques Solomon. Among the various vehicles for extending the influence of Marxism among French scientists during this period were the "houses of culture." A 1939 conference on "Revolution and the Sciences" attracted wide attention, with the various prominent Marxist scientists present maintaining that the philosophy of dialectical materialism aided them in various and significant ways in coming to terms with their particular sciences.

Langevin

The eminent physicist, Paul Langevin, was particularly influential within this movement. Director of the Ecole de Physique and a physicist of the most prestigious international reputation, he did not hesitate to give open expression to his views during the process of his gradual conversion to Marxism. In earlier days, he had been a member of the Ligue des Droits de l'Homme and a supporter of Dreyfus. Later, he was a supporter of the October Revolution. Through his trips to the Soviet Union in 1928 and 1931, he came to extol Soviet science as a science that had put itself in the service of the intellectual and moral liberation of mankind.

On a number of occasions, Langevin articulated the relationship of his growing commitment to Marxism to his thinking as a physicist. In 1938, he explicitly asserted: "The more I progress in my own science, the more I feel myself becoming a Marxist." In Marxism, in the writings of Marx and Engels and Lenin, Langevin said, "I have found the clarification of things that I would never have understood in my own science."[235] A few years earlier he had announced to his fellow physicists at an international scientific conference, "Whether you like it or not, there is no other way of understanding the development of the physics of the atomic nucleus in the twentieth century than the way of dialectical materialism."[236] He also claimed that, upon grasping the fundamentals of dialectical materialism, he came to understand the history of science in a new way, to see that dialectical processes marked out all its essential moments.[237]

In the first issue of *La Pensée*, a Marxist journal founded in 1939 to take up in France the tradition begun by *Science and Society* in the United States and by *Modern Quarterly* in Britain, Langevin undertook a philosophical analysis of the debate surrounding the concept of physical determinism. He took on those who had seized upon the overthrow of Laplacean determinism to speak of the "free will" of the electron, those like Eddington, who were proclaiming that religion had thus become acceptable to the scientific mind, those like Jordan, who were declaring the "liquidation of materialism." What science was witnessing, he argued, was not a crisis in determinism, but an insufficiency in the microscopic realm of the conceptions that had succeeded in the macroscopic realm. The world was proving to be infinitely richer than Pascal had imagined. It was necessary to forge new notions, new instruments. Physics was emerging into a world of neutrons, mesotrons, and neutrinos. There had been many surprises and there were many surprises yet to come. However, none of the new discoveries gave any warrant for the proclamations of indeterminism. To renounce the concept of determinism, Langevin insisted, would be to abdicate. It would be to deprive science of what had always been its strength and its success: confidence in the intelligibility of the world. Nothing in the present difficulties called for such irrationalism as was coming forth. Indeed, man was standing at a particularly crucial point in the development of reason. Reason was not given a priori, nor had it the rigid limits some were setting on it. As knowledge progressed, so did human reason develop and so did the conception of determinism. Those who were proclaiming the bankruptcy of this concept might have been referring to modern science, but it was not from science that their ideas came. It was from an old philosophy hostile to science, which they then sought to reintroduce into science. When idealist philosophers referred to the ideas of idealist physicists, they were simply taking back from them the very ideas they themselves had given to them.[238]

Solomon

Langevin's student and son-in-law, the precocious young physicist Jacques Solomon, was particularly ardent in pursuing the connections between politics and philosophy and physics. He had become a committed communist and was extremely active both within the houses of culture and with Langevin's materialist study groups. During a holiday in the Alps in 1938, he and his friend Georges Politzer translated *Dialectics of Nature* into French. A dominant theme with Solomon, as with other French Marxists, was the commitment to human reason against all forces threatening to destroy it. Communism, to them, was the surest defence of rationalism.[239]

As a physicist, Solomon had studied in Cambridge, Copenhagen, Vienna, and Moscow as well as in Paris. Those who knew him as a physicist unfailingly remarked on his preoccupations, which were both encyclopedic and eschatological. He wanted to scale the heights, to study the depths, and also to know every facet in its intimate details. It was such an insatiable need to understand that led him to so many ideas about field theory, nuclear physics, and cosmic radiation, and which also led him to Marxism. His colleagues noted a profound transformation in him when he became a communist. His friend Léon Rosenfeld saw him as burning with a sombre fire, with the fierce resolution of a man engaged in a merciless struggle. Years later, Rosenfeld admitted that, observing the maturity and outstanding productivity of Solomon's scientific thought, he had secretly regretted the overwhelming ascendancy of his political commitment. Rosenfeld observed that Solomon's humanity was as broad as it had ever been, but that it had lost some of its former gentleness. Even on a scientific or philosophical plane, Rosenfeld saw him as becoming harder and more cutting.[240]

Prenant

Perhaps the most full-scale attempt to come to terms with the problems of a particular science on the part of a French Marxist was Marcel Prenant's *Biology and Marxism*. Prenant, professor of zoology at the Sorbonne, was a founder member of the French Communist Party and lectured weekly at the Workers' University in Paris. Prenant began his book by addressing himself to those who thought dialectical materialism to be separable from communism, strongly asserting their inseparability. He went on to note that dialectical materialism was associated in the minds of some with intellectual tyranny and was accused of destroying the objectivity of knowledge. But, Prenant argued, far from placing tyrannical restrictions on science, dialectical materialism was of the nature of science itself. It was the experimental method continued without a break, not afraid to face its own consequences. The best empirical biologists were materialists and found themselves thinking dialectically in spite of themselves whenever they aimed at synthesis. But without a consciously dialectical and materialist philosophy, they did so only in flashes and failed to keep it up.

Dialectical materialism, according to Prenant, was a comprehensive philosophy that strove to draw into its synthesis the totality of knowledge. It not only fit the facts of modern science—as did other varieties of enlightened organicism—but it provided fruitful working hypotheses likely to lead to fresh advances in science. Laboratory experiments were the living sources from which Marxist science flowed. He strongly emphasized that the conclusions in his book were founded upon experiment. He warned against the danger of Marxists depending too heavily on quotes from the classics. In something of an unusual touch, he added an appendix with quotes from the classics to make the point that the Marxist classics were about more than the social sciences, but was at pains to make the point that his ideas were based on experiments.

As to the mechanist-vitalist debate tearing at modern biology, Prenant argued that it must be transcended in a higher synthesis. In the field, there were two extreme points of view: one referred all phenomena of life to properties intrinsic to itself and ended up invoking some vital principle or entelechy; whereas the other invoked the action of environment with its mechanical, physical, and chemical forces and tried to reduce the whole of life to phenomena of one order. Prenant agreed with Bergson against mechanism, but argued that an élan vital was an easy way out. It was a mere phrase and an utterly sterile one. Vitalism attempted to establish indeterminism in biology, but determinism was indispensable to scientific progress.

It was a lack of acquaintance with dialectics, in Prenant's opinion, that caused biologists to come to one-sided solutions. In a dialectical view of living matter, that comprehended the complex interrelations of the living being with its environment, there was no unbridgeable gulf between cause and effect, between quantity and quality. It was necessary not to deny causality, but to modify and extend it, adopting statistical or probabilistic methods. The recognition of the phenomenon of chance did not mean an absence of determinism, but corresponded to a summation of a number of distinct causalities that were too complex to be analyzed by present methods of scientific investigation.

But mechanists were distrustful of probability and statistical methods for fear they opened the way to indeterminism and miracles. Mechanists were unwilling to admit either theoretical or practical limits to determinism. Moreover, they failed to understand the historical character of phenomena. Nor were they able to accept the occurrence of sudden change. The predispositions entailed the complete abandonment of the social sciences to idealism.

Actually, the great majority of biologists were neither mechanist nor vitalist, but empiricist and agnostic. They claimed to stand on experiment alone, but in reality oscillated between one extreme and the other. Somehow, however, through it all, Prenant believed biology was approaching a dialectical and materialist outlook. All of its decisive advances involved the shedding of rigid concepts and diametrical oppositions under the pressure of experimental fact. Those who rose to a higher synthesis, based on experimental results, inevitably approximated dialectical materialism. The wider the synthesis, the more striking the approximation.

An important feature of the book was its discussion of genetics. By 1937, it was clear to foreign communists that genetics was heavily under attack in the Soviet Union. This notwithstanding, Prenant was resolute in his defence of the science of genetics. His explanation of heredity was along uncomprisingly Mendelian-Morganist lines. There could be little doubt, Prenant firmly stated, about the role played by chromosomes in the determination of heredity. He was, however, critical of the two extreme positions on heredity, both of which, he claimed, derived from a lack of dialectical insight. Heredity and evolution seemed to be antitheses, but were resolved in an experimentally based dialectical synthesis. He took issue with those geneticists who regarded the chromosome system as entirely independent of its surroundings and the genes as utterly disconnected particles that fell into juxtaposition. He also opposed extreme Lamarckists for whom the conception of heredity gave way completely before that of evolution and environmental determination. Prenant's position, a far cry from Lysenkoism, was that the inheritance of acquired characteristics had never been proven experimentally in any decisive fashion. Such experiments as had been done had been lacking in sufficiently strict controls. Sometimes what was at stake was really the inheritance of certain specialized genotypes selected by the environment, when the genotypes most frequent in the old environment proved incapable of surviving in the new. The external influences did not decide the mutations, but threw the cell into a state of crisis in which such mutations had a chance of taking place. The basic fact of organic evolution was that living matter changed, not as an expression of an internal law taking place independent of all outside, but as a result of interaction with its environment while retaining its own composition and structure. Mutation of itself could not explain the direction of evolution, but the selection of favorable mutations by the environment could. Thus, Prenant proposed to account for both evolutionary development and the relative autonomy and fixity of heredity.[241] Prenant continued in his opposition to Lysenkoism and argued in *La Pensée* in 1939 that true genetics was anything but racist.[242]

French Marxism and its Successors: A Parallel Discontinuity

All in all, prewar French Marxism generated an interesting literature and a vital movement bringing together science, philosophy, and politics, comparable to that of prewar British Marxism. This has been similiarly forgotten by the present generation of French Marxists, particularly those of the Althusserian strand, who put a positive value on a lack of historical consciousness. Sartre was once stirred to remark that "today's Marxists behave as if Marxism did not exist and as if each one of them, in every intellectual act, reinvented it, finding it each time exactly equal to itself."[243]

Those who have written of prewar French Marxism have often done so quite dismissively. Typically, Jean Duvignaud takes figures such as Nizan, Politzer, and Lefebvre seriously until the moment of their joining the party. At that point, the story goes, they totally delegated to the party their right to think and criticize. As to the scientists, their Marxism was only a vague scientism. They produced specialized work that, in reality, had no significance for Marxism.[244]

Similarly, David Caute insists that neither the scientists nor the philosophers were really inclined to integrate Marxism with their creative work, whatever pious declarations they made to the contrary. He is convinced that Marxism had nothing to do with any actual scientific research. He simply cannot imagine what Marxism could possibly have to do with microbes, the orbit of Venus, calculus, the sexual impulse, or the population patterns of the North Canadian rabbit.[245] Those who have attempted to see the relationship between Marxism and science have always been subjected to such caricatures. While some writings have drawn the connections in a way that has been superficial and forced, the writings of such French Marxists as Prenant and Langevin show that something much more integral has been involved.

The French Marxists of the prewar period believed that Marxism had to do with everything, and those not themselves inclined to such a maximalist approach cannot possibly comprehend those who were. There is no doubt, however, that party membership could create difficult intellectual and moral tensions, but only willful blindness could write off the Marxism of such as Langevin and Prenant as being of negligible value.

Crisis and War: The Fate of French Marxists

But tensions there were. One casualty was Paul Nizan. In 1939, another shift in Comintern policy threw many a communist into a seemingly irresolvable dilemma. The world was once again at war. The armies of the Third Reich were rampaging their way through Europe. Then, at the very moment it mattered most to the world, the Comintern abruptly withdrew its commitment to a united front against fascism. The Nazi-Soviet Pact of 1939 came as a terrible shock to communists everywhere. Support for the "phoney war" policy was rigidly enforced on all foreign communist parties. The PCF in particular was convulsed by the new policy. *L'Humanité* was banned, even while still calling for national defence. With the party toeing the Comintern line, and France at war, communists appeared truly treasonous. Even communist deputies broke with the party in the early days of the war. Numerous other party members left. Paul Nizan, who was mobilized and killed at Dunkirk, was one of them. In disedifying, but typical Comintern style, he was accused of being a police informer.

Times were very bad indeed for French communists. In a purge of the Chamber of Deputies, communists were deprived of their seats for "propagation of slogans of the Communist Third International," which now constituted a crime against the state. In March 1940, forty-two communist deputies were brought to trial in camera. Langevin, even though he had opposed the Nazi-

Soviet Pact, came to their defence. Trying to break what was being claimed as a necessary link between patriotism and anti-communism, Langevin declared that the communists stood for a superior form of life. Nevertheless, thirty-two deputies were sentenced to five years in prison, fined, and deprived of all civil rights, while others, including Maurice Thorez, were similarly sentenced in their absence. The PCF was forced underground.

Meanwhile, the distant gunfire and rumble of cannons came closer and closer. Holland and Belgium had fallen and it became clearer and clearer that soon it would be proud France. Through its roads poured streams of stunned, exhausted, beleagured humanity. In the early hours of June 14, 1940, the Wehrmacht entered Paris. The swastika was draped over the Chamber of Deputies and the Eiffel Tower. In the afternoon, the Reichswehr made its triumphant entry with tanks and goose-stepping troops, heightening the humiliation and despair. The fifth columnists came out of the wood and entered into full collaboration.

The communists began organization of resistance with great heroism. The resistance was carried out on all fronts—from the military to the industrial to the cultural. It was as dangerous to be discovered standing at a clandestine printing press as with arms in hands. Some carried out sabotage. Others considered every poem an act of war. In fact, the resistance brought forth a great cultural renaissance, bursting like a bright rationalist flame, penetrating the irrationalist darkness.

In October 1940, Paul Langevin was arrested in his laboratory, subjected to Gestapo interrogation, and interned. This sparked vigorous protests through the university. Joliot-Curie declared his laboratory closed until Langevin was freed. At a public demonstration, German troops appeared, fired upon the demonstrators, and threw hand grenades into the crowd.

The university resistance led by Georges Politzer, Jacques Solomon, and Jacques Decour, who produced a defiant journal of resistance, *L'Université libre*, from this time on was extremely bold and vigorous. It endeavored to expose all the intrusions of the enemy, especially in university affairs. It protested against the arrests of Jewish professors and students and the retrograde changes in curriculum. It served also to spark active resistance in the schools and colleges.

In May 1941, even before the "phoney war" turned again into an anti-fascist war for the communists and the ambiguities of Comintern policy dissipated with the nazi invasion of the socialist homeland, the central committee of the French Communist Party issued a manifesto calling for the formation of a National Front against fascism and for the liberation of France. In March 1942, amid a huge Gestapo dragnet which rounded up over 140 communists, Jacques Decour, Georges Politzer, and Jacques Solomon were among those arrested. Communist women were not exempted and Hélène

Solomon, Jacques Solomon's wife and Paul Langevin's daughter, and Maie Politzer, Georges Politzer's wife, were also included and deported to Auschwitz. Jacques Decour, Georges Politzer, and Jacques Solomon were tortured in Gestapo custody. Upon refusal to give information or to promise to collaborate, they were executed on May 23, 1942. Once again, politics took a heavy toll on science and philosophy.

The execution of Jacques Solomon and the deportation of Hélène Solomon shook Joliot-Curie and Langevin very deeply. Joliot-Curie had been arrested in 1941. After his release, he became even more active in the resistance as president of the National Front in the occupied zone. He decided to join the Communist Party. "If I should be seized and shot," reasoned Joliot-Curie, "I wish to die a communist."[246] A party member from 1942, his party membership was not, however, made public until after the liberation. Langevin, released from prison in 1940, was arrested again in 1942, and worked during his periods of freedom in the resistance with every ounce of his energy. In 1944, at the age of 72, he too joined the Communist Party. His daughter, Hélène Solomon, survived Auschwitz and became a communist deputy after the liberation.

For his part, Marcel Prenant was a partisan leader during the occupation. His views on genetics were not only an affront to Lysenkoism, but to nazism as well. His insistence that genetics did not support racism and his further statement that there were not pure races made him a most dangerous enemy. He too was arrested and narrowly escaped death in a nazi concentration camp. At the end of the war, he was among those liberated from Neuendamme and returned to Paris.

All in all, fascism took an exceedingly heavy toll of communist lives, among them its most courageous, its most promising, its most gifted theoreticians. Italian fascism had taken Gramsci. Spanish fascism had taken Caudwell and Guest. Now German fascism had taken Politzer and Solomon.

The Exiles to the West:
German and Austrian Marxists after the Rise of Fascism

The war wreaked havoc also in the lives of those who survived. The onward march of fascism scattered the forces of progress to the four winds: underground, into camps, into prisons, and into exile. As the Wehrmacht consolidated and extended itself, everything else dispersed. As Brecht put it:

> All
> Choose the hurricane rather than the Germans.[247]

But there were Germans and there were Germans, as Brecht, who saw both the dramatic contrasts and the disturbing similarities, knew well.

Brecht

Such contrasts. In the homeland of Einstein, the books were burning. From the laboratories,from the libraries, from the lecture halls, the scientists, the philosophers, the poets went forth, "changing our countries more often than our shoes."[248] The brightness of knowledge gave way before a ruthless and rampaging darkness. The blackness seemed to cancel out all else. In Brecht's words:

> Truly I live in dark times.
> A guileless work is folly. A smooth forehead
> Betokens insensitiveness. He who laughs
> Has not yet heard
> The terrible news.[249]

And yet, in exile, Brecht laughed—not lightheartedly, to be sure, but ironically. The irony had its source in his reflections on Marxist philosophy: on dialectics, materialism, and the German character. In his *Flüchtlingsgespräche*[250] (Refugee Dialogues), written in Finland in 1941, Brecht set the scene with a chance meeting of two refugees from Nazi Germany in a railway station in Helsinki. Ziffel, a physicist, and Kalle, a metal worker who had been in a concentration camp, consider the meaning of the term "German," deciding it meant being thorough: in agriculture as well as in the extermination of Jews. The German, it was often said, had a natural bent for a professorship of philosophy. And it was said with a soulful, bloodthirsty expressiveness. Pursuing the theme:

The Germans have no aptitude for materialism. When they do have materialism, they immediately make an idea out of it. A materialist is then someone who believes that ideas come from material circumstances and not the reverse and after that matter does not appear again. One might as well imagine that in Germany there are only two sorts of people, priests and opponents of priests. The upholders of the Here and Now, lean and pale figures, who know all the philosophical systems, the upholders of the Beyond, corpulent gentlemen, who are connoiseurs of wine.

The conclusion: "A good beef stew comports well with humanism." As to dialectics, it was the "wit" in every object and it took humor to understand Hegel, said Ziffel, after summarizing Hegel's *Logic* thus:

It treats of the habits of ideas, those slippery, unstable, irresponsible entities; how they abuse each other, and fight with knives; and they sit down to dinner together as if nothing had happened.

The new situation might bring the Germans to a new level of dialectical understanding, Brecht thought:

> Emigration is the best school of dialectics. Refugees are the keenest dialecticians. They are refugees as a result of changes, and their sole object of study is change. They are able to deduce the greatest events from the smallest hints—i.e., if they have intelligence. When their opponents are winning, they calculate how much their victory has cost them; and they have the sharpest eye for contradictions. Long live dialectics.

But Brecht's ironic Marxism was not the style of that of other of his fellow refugees from Germany, most notably the Marxists of the Frankfurt School, who had already gone into emigration in the United States, where Brecht eventually made his way. Brecht thought much of it to be circuitous and convoluted word-spinning.[251] He undoubtedly had them in mind among the sort who affirmed materialism and then never gave another thought to matter. For their part, they thought Brecht's Marxism tended towards "vulgar materialism." Earthy Brecht's Marxism surely was, but vulgar it was not. His mind was one well attuned to the complexities. There were many differences among the emigrés. One rather important one was the stress put upon the importance of science for the working class, which again set Brecht's orientation apart from that of the Frankfurt School. He was greatly taken with Einstein in particular, whose lectures on the latest developments in physics he had attended in Berlin.

The Frankfurt School

The Frankfurt School was one of the few centers of Marxist thought outside the Comintern. The Institute for Social Research, founded in 1923 and endowed by Felix Weil, a wealthy grain merchant who became a communist, had been formally attached to the University of Frankfurt. In its early days, it had communists,* social democrats, and independent Marxists of various descriptions on its staff and it maintained close links with Riazanov's Marx-Engels Institute in Moscow. There were also significant differences in philosophical orientation.

In the inaugural address of Carl Grünberg, the first director of the institute and the first avowed Marxist to hold a chair at a German university, he stressed the character of Marxism as a social science and announced that "the materialist conception of history neither is, nor aims to be, a philosophical

---

* Among the communists was Richard Sorge, fated to become one of history's famous spies. Head of a Soviet spy ring in Japan, he was executed by the Japanese during the war.

system."[252] The Frankfurt School was from the beginning much influenced by Lukács and Korsch, though some like Karl August Wittfogel were of the way of thinking that Lukács and Korsch labelled as "positivist."

The ascendancy of Max Horkheimer to the directorship in 1931 signaled a reorientation, and set the institute much more firmly within the neo-Hegelian trend within Marxism. Actually, it was both neo-Kantian and neo-Hegelian in the style of Lukács. It took from Hegel the dialectic, but in its emphasis on critique and its rejection of all systems, it was moving in the spirit of Kant and not Hegel.

The new members of the institute who joined it at this time, such as Theodor Adorno and Herbert Marcuse, along with Horkheimer, pursued philosophy in this spirit, laying the foundations of "critical theory." Critical theory involved a radical rejection of system: it was, as Martin Jay put it, "the gadfly of other systems."[253] It was a philosophy that developed itself and defined itself in terms of a series of critiques of other philosophies. At the center of its critique of the Marxism of the Comintern was its opposition to the systematization of Marxism. In its activist epistemology and in its stress on revolutionary praxis, Marxism was not to be seen as a key to the truth, but as the impulse to social change. Critical theory was oriented, not to being, but becoming.

In his "Materialism and Metaphysics," Horkheimer argued that true materialism was not a new type of monistic metaphysics based on the ontological primacy of matter. He criticized Marxists who made a fetish of the material world and argued that materialism could not be extended to epistemology, as it led to a passivist objectivism. True materialism, he contended, was dialectical, and involved an ongoing process between subject and object.[254] It was the very thing Brecht meant. It was hard to know what Horkheimer's materialism had to do with matter.

For the new members of the institute, as for Lukács, there was no dialectic without subjectivity. The dialectic was situated in the realm of the subject-object relation. There was great concern with mediating subjectivity and stress on the social construction of knowledge. It was, however, a one-sided concern and a lopsided stress. Like Lukács, they tended to forget the object. Like Lukács, they were remote from the world of science. There were, all the same, great flashes of insight in their writings, especially in their historical analysis of how all the institutions of capitalist society became expressions of a contradictory inner essence. Within the prevailing division of labor, both the individual capitalist and the individual scientist were acting according to the imperatives of an obscure social mechanism, no matter how independent and free they thought themselves to be.

Their attitude to science was a complex one. Blatantly antiscientific and antitechnological passages could be found in Horkheimer, Adorno, and, most particularly, in Marcuse, but there were also modifying passages, particularly in Horkheimer. Horkheimer was opposed to indiscriminate attacks on scientific method and to the various irrationalist trends in the German academic world that gave way to subjectivism, irrationalism, and hostility to science. But, even at the best of times, they were overly anxious to separate philosophy from science, without a comparable concern to show the relationship of philosophy to science. The constant concern to distinguish scientific rationality from reason itself was inevitably based on a severely stripped notion of scientific rationality. They tended to accept the positivist account of science and to assume that the critique of positivism implied the critique of science. As time went on, the emphasis shifted from a critique of science as embodying a logic of passive contemplation to a critique of science as an instrument of domination. This gave rise to their distinctive interpretation of fascism, not as a revolt against reason, but as the triumph of instrumental rationality.

With Hitler's rise to power, the members of the institute were in extreme danger, being not only Marxists, but many of them Jewish as well. In 1933, Horkheimer was dismissed from the University of Frankfurt and the institute was closed down for "tendencies hostile to the state." Though most of its members managed to escape, Wittfogel was sent to a concentration camp, but later released. In 1934, Horkheimer negotiated the formal transfer of the institute to Columbia University in New York. It adapted itself to the American scene rather more than some of their fellow Marxists (Brecht among them) thought appropriate. To some degree, this trend was fed by their growing pessimism. No longer inspired by Lukács's faith in the revolutionary proletariat, which embodied the vision of totality, they began to lose faith both with the proletariat and with the possibilities of totality.

Marxism and Psychoanalysis

Another tendency among Central European Marxists in exile was to look for a synthesis of Freud and Marx. In the Vienna of the 1920s, Marxism and psychoanalysis intermingled and seemed anything but incompatible.

Hollitscher

The Austrian communist Walter Hollitscher, who was much concerned with philosophy of science, was at the same time a psychoanalyst. His doctoral thesis was on the ideological basis of the various contending positions on the principle of causality in quantum theory. Upon emigration in 1938, he was invited by Carnap to teach philosophy of biology and psychology at the University of Chicago, but went instead to London as a clinical psychoanalyst, where he wrote on the relationship between psychoanalysis and sociology. He admitted later than it took some years "to fathom the full range of incompatibility between Freud and Marx."[255]

Reich

The story of Wilhelm Reich was quite different. Events did not allow him Hollitscher's scope for finding his own way within the framework of the communist movement. Reich too was both a psychoanalyst and a member of the Austrian Communist Party. At the same time, he founded the Socialist Society for Sex Consultation and Sexological Research. In 1929, he visited the Soviet Union, where he published *Dialectical Materialism and Psychoanalysis*. In 1930, he moved to Berlin, where he became a member of the KPD and founded the German Association for Proletarian Sexual Politics. In 1933, he published *The Mass Psychology of Fascism*. His attempted reconciliation of Marxism and psychoanalysis assumed that the oppression of the masses by the ruling classes could not be explained simply in terms of material power (which, he contended, Marxist theory had attempted to do). The ruling classes were successful in control of the masses, because bourgeois society created a submissive character structure in them and trained them to suppress natural instinctual energies. Society was not derived from character structure, but character structure was derived from society. Thus, during the rise of fascism, the masses failed to pursue the economic interests of their own class, but willingly followed the most authoritarian leaders.[256] Reich's vision of the future was relatively optimistic, however. Breaking from the pessimism of Freud, Reich believed that a socialist society would put minimal restrictions on natural, instinctual energies and provide the conditions for a liberated sexuality. He put great stress on orgasm as a central human experience and saw orgastic potency as a function of the whole human personality and of the whole social order.

His synthesis was unsatisfactory both to his fellow Marxists and to his fellow psychoanalysts, and he had the dubious honor in 1933–1934 of being almost simultaneously expelled from both the Communist Party and the International Psychoanalytical Association. The KPD felt he was diverting energies into the campaign for sexual hygiene that should have been directed into other forms of political activity. The psychoanalytical movement felt itself endangered by the rise of fascism and didn't want itself associated with communism. His efforts were praiseworthy in that he attempted to claim the psychological dimension for Marxism, to place the psychological within the framework of the sociological, and to instill confidence in human reason and in science in relation to such issues. The problem was his own lack of discipline in relation to scientific method, an impatience with vigorous experiment and rigorous analysis that allowed for no check upon his speculation and made for a certain lack of proportion in this theories.

Fromm

The Frankfurt School was also highly involved in the enterprise of synthesizing Freud and Marx. Most prominent in the endeavor was Erich Fromm, who believed that Reich had stopped far short of setting out the guidelines for a fully social psychology. Reich had still restricted psychoanalysis to the sphere of individual psychology. For Fromm, psychology dealt with the socialized individual. It was wrong, Fromm thought, to charge Marxism with having a simplistic psychology of acquisitiveness, but he did think that Marxism needed to be enriched by psychoanalytical insights on a number of points. This process could provide a more comprehensive knowledge of the factors operative in the social process, namely the nature of man himself. It could provide the missing link between ideology and economics. Lacking a satisfactory psychology, Marx and Engels could not explain how the material basis was reflected in man's head and heart, how ideologies were produced. Psychoanalysis, Fromm felt, was a fully materialist psychology, to be classed among the natural sciences, that could introduce a refinement of method in the Marxist theory of historical materialism. Fromm was also critical of such aspects of Freudian theory as the absolutization of the Oedipus complex.[257] Freud's theory was a pioneer effort, he argued, and as such left many loose ends hanging and many years would be spent refining it. Like the rest of the Frankfurt School, and like Reich, Fromm went into emigration in the United States.

Marxism and the Vienna Circle

The philosophers of the Vienna Circle, whose members varied in their views as to the political significance of their philosophy were also in emigration. At one extreme was Moritz Schlick, who denied any relationship between science and social development. At the other extreme was Otto Neurath, who considered himself to be a Marxist and asserted a logical connection between Marxism and logical empiricism. It represented yet another variation on the Machist trend within Marxism. It was at the opposite extremity from that of the Marxism of the Frankfurt School, which was derived from the traditions of classical German philosophy. The Vienna Circle represented a radical break with all such philosophical traditions. It saw reason as being threatened by dark forces and it was a progressive movement. Reason was identified utterly and completely with scientific rationality (and an extremely strict conception of scientific rationality at that) and nothing was considered to be outside its scope. It sought to bring science to bear upon the whole of reality. Its "scientific world conception" admitted only empirical statements of verifiable observation and analytic statements of logic and mathematics. Traditional philosophical questions were ruled out of court. Such questions as the reality or nonreality of the external world led to statements that were unverifiable and therefore meaningless. Its great goal was the unity of science, leading to the search for a neutral system of formulas, and for a formal symbolism freed from the "slag" of historical languages, for a total system of fundamental concepts. Starkly it declared:

> Neatness and clarity are striven for, and dark distances and unfathomable depths rejected. In science, there are no "depths"; there is surface everywhere: all experience forms a complex network, which cannot always be surveyed and can often be grasped only in parts. Everything is accessible to man; and man is the measure of all things. Here is an affinity with the Sophists, not with the Platonists; with the Epicureans, not with the Pythagoreans; with all those who stand for earthly being and the here and now.[258]

Neurath

Neurath saw the philosophy of science as directly connected to politics. The proletariat was the appropriate bearer of the scientific world conception. The bourgeois form of thought was becoming more and more indefinite like a squid—sending out its feelers in all directions, here mathematizing, there psychologizing; here proceeding in a technical manner, there in an occultist one. It was no longer able to hold the wealth of scientific detail together by a unitary approach. In its anxiety to close ranks against the proletariat, the bourgeoisie was making its peace with the powers of yesterday and therefore cultivating all forms of metaphysical and half-theoretical thought. Thus the bourgeoisie, by virtue of its class position, was becoming more and more unscientific. It showed first in the field of social theory, but in the long run it was impossible to prop up a false view of history without adopting unscientific presuppositions all around. The interests of the proletariat, by contrast, coincided with the values of science. It had no reason to hide the truth about history. It was the carrier of a strictly scientific attitude.

For Neurath, the cause of unified science thus coincided with the cause of socialism. Marxism was a world view that started with the recogition that all history was the history of class struggle and predicted the decline of the capitalist order and the coming of the socialist order. Round this center were grouped certain opinions about the way the world hung together, about everything that made a unified picture of the world possible. At first, Neurath defended materialism against idealism as characterized by a closeness to actual life and acceptance of the scientific attitude and as connected with progressive ideas in political and social matters. In time Neurath came to argue for the rejection of both idealism and materialism as unverifiable metaphysical positions. Although he thus argued vigorously for the rejection of all forms of metaphysics, as superseding both mechanistic and dialectical forms of materialism, he continued to consider himself to be more in continuity with the history of materialism than with idealism. This expressed itself particularly in his preference for the physicalist as opposed to the phenomenalist tendency within positivism.[259] Neurath emigrated in 1934 to Holland, where he devoted himself to the Unity of Science Movement, and, pushed further by the nazi advance, went on to England. He died in Oxford in 1945.

Neurath's Marxism, refreshing though it was in its determination to subject all to the bright and purifying light of reason and to leave no corner unpenetrated by the liberating force of science, suffered from a stripped conception of reason and a severely truncated view of science. The story of the rise and fall of logical positivism and its principle of verification is well known. Having tied his conception of Marxism to it, Neurath's Marxism stood or fell with it. A strict monism was a sound enough methodological principle, but it could not maintain itself from an emaciated base. A rather critical appreciation of logical positivism on the part of an American Marxist from this period, V.J. McGill, was to the point:

> Though it must show itself in the end a two-edged sword, and quite incapable of coping with the heavy artillery of its enemies, the resistance it offers to reaction should be recognised.[260]

American Marxism

In America, where many exiles found themselves, there were other debates revolving around the variety of interpretations of Marxist theory and its relation to science, largely outside the confines of the Comintern.

Hook

The discussions of the 1930s revolved largely around the views of Sidney Hook. Hook was a professional philosopher, trained in the tradition of American pragmatism, whose conversion to Marxism naturally led him to look for the common ground between Marxism and pragmatism. After receiving his Ph.D., he went abroad, pursuing his studies in Berlin, Munich, and Moscow (at the Marx-Engels Institute). In 1932, he supported the presidential campaign of the Communist Party in the United States. In 1933, he published *Towards the Understanding of Karl Marx*, hailed as a "revolutionary interpretation of Marx" and in 1934, he published *From Hegel to Marx*, expanding on his distinctive interpretation of Marxism.

Essentially, Hook was counterposing Marxism as critical historicism to Marxism as deterministic materialism. Influenced by Lukács and Korsch as well as by James and Dewey, the center of focus was the concept of revolutionary praxis. Rejecting the legacy of Engels, as based upon a monistic system rather than a unified method, his was a radically antimonist and antideterminist position. His criticism of "orthodox Marxism" centered around its neglect of the activist and critical dimension of Marxism, in favor of a positivist and scientistic *Weltanschauung*. He was, needless to say, highly critical of the objectivist epistemology of *Materialism and Empirio-Criticism*. He echoed Lukács's objections to the dialectics of nature and yet criticized Lukács for his extreme antinaturalism. A characteristic feature of Hook's Marxism of these years was the attempt to balance the dialectical side of Marxism with the naturalistic side of pragmatism.

Hook saw Engels's concept of the dialectics of nature as both undialectical and antinaturalistic. Marxists created great confusion in their ambiguous way of using the term *dialectical*. Sometimes it meant no more than the commonplace fact that change was observable in all fields of thought and activity. Sometimes it meant that every account of physics had to operate with contrasting and complementary principles in order to do justice to the polarities and oppositions in the structure of nature. Hook's own conception of the dialectic, which he identified with Marx's, was that it was historical, restricted only to a consideration of the causes, nature, and effects of human activity to destroy the equilibrium of a polarized society and to determine the direction of a new form of society. The dialectic was the principle of social activity and the medium of class struggle. But there was no need to show that there were sudden leaps and jumps in nature to justify revolution in society. Engels's overextension of the dialectic, as Hook saw it, led to a violation of naturalism. As Hook rather crudely put it, Engels had swallowed more of Hegel than any naturalist could properly digest and it kept coming up throughout his work. However, Hook argued, Galileo's laws of motion and the life history of an insect had nothing to do with the dialectic, except on the assumption that all nature was spirit. The results of physics could be used by the bourgeoisie, Hook went on to say, but there was no such thing as bourgeois physics. To read the class struggle back into nature was to imply that all nature was conscious.[261]

Eastman

Hook came under attack from all sides. The first line of attack was represented by Max Eastman. Also a professional philosopher and a student of John Dewey, Eastman had preceded Hook in the enterprise of reconciling Marxism with pragmatism. In the early 1920s, an enthusiastic supporter of the Russian revolution, he had spent nearly two years in the Soviet Union and became an associate of Trotsky. It was Eastman who made Lenin's testament known abroad. Trotsky, Eastman's source, was still in a position of power and was forced to disavow Eastman's revelations. With the move against Trotskyism in the Comintern, Eastman had been expelled from the Communist Party of the United States of America. By the early 1930s, Eastman had no time for dialectical materialism in any form and, in "The Last Stand of Dialectical Materialism", a 1934 pamphlet directed against the Marxism of Sidney Hook, attacked Hook's 1933 book as a hopeless attempt to defend the indefensible. By the late 1930s, Eastman had rejected socialism altogether.

Hook and Eastman versus the Communists

Meanwhile, both Hook and Eastman were under attack from the other side and from Trotskyists and Communists alike. In 1933, with John Dewey considered a "social fascist" philosopher, the communists were not predisposed to be sympathetic to any attempt to reconcile Marxism with pragmatism. Party leader, Earl Browder, duly came forward in polemic in "The Revisionism of Sidney Hook" published in *The Communist*.[262] A later issue carried an exceedingly harsh polemic against Hook from the pen of Ladislaus Rudas, whom Hook had attacked as the embodiment of Stalinist ideology in his "Why I Am a Communist: Communism Without Dogmatism" in the 1934 symposium sponsored by *The Modern Monthly* in which John Dewey, Bertrand Russell, and Morris Cohen had contributed articles under the heading "Why I Am Not a Communist."[263]

Hook and Eastman versus the Trotskyists

Trotskyists, committed to the defence of dialectical materialism, also polemicized against Hook and Eastman, particulary in the pages of *New International*. From France, the Trotskyist Pierre Naville criticized Hook's critique of dialectical materialism. And from exile in Turkey, Trotsky himself wrote a letter to the *Nation* taking Hook to task for undermining the philosophical foundations of Marxism, and casting doubt on the scientific character of Marxism.[264] Upon his arrival in Mexico, Trotsky expressed to the American Trotskyist George Novack his anxiety over Eastman's repudiation of dialectical materialism and urged Novack to struggle resolutely against pragmatist incursions on Marxist philosophy.[265]

Debates among American Trotskyists

Soon, however, Trotskyists were debating among themselves on the validity of dialectical materialism and its relevance to the revolutionary struggle. In 1939–1940, the U.S. Socialist Workers' Party broke into a bitter factional dispute. The great debate that split the Trotskyist movement worldwide broke out over the nature of the Soviet Union and the necessity of defending it against imperialist attack. The majority, led by James Cannon and supported by Trotsky, held that the Soviet Union was a "degenerate workers' state" and must be defended against its enemies without making any concessions to Stalinism. The breakaway group, led by Max Schachtman and James Burnham, repudiated this position. The Soviet Union was simply a "bureaucratic collectivism" unworthy of such defense.

This debate then brought other issues to the surface, most notably the philosophical one. James Burnham, like Hook, a professor of philosophy at New York University, held that the notion of dialectics was meaningless and was in any case without any necessary connection to politics. Trotsky's "Open Letter to Comrade Burnham" argued vigorously against any move to divorce politics from philosophy and again came to the defense of dialectical materialism as essential to the revolutionary struggle. Marxism without dialectics, Trotsky argued, was like a clock without a spring.[266] This polemic, George Novack said, brought him to appreciate fully the importance of a correct philosophical method and the grave political consequences implicit in departures from it. In the party discussion in New York City that wound up the furious faction fight, Novack debated with Burnham on the philosophical issues involved. From this came the first of his many works on Marxist philosophy, *An Introduction to the Logic of Marxism*, based on talks given to Socialist Workers' Party (SWP) members in 1942 in defence of "dialectical logic" and of the objectivity of contradictions.[267]

Hook and Eastman: the Break with Marxism

Meanwhile, both Hook and Eastman had taken leave of Marxism altogether in an unqualified return to American pragmatism, and accepted not only Dewey's epistemology but his social and political philosophy as well. It was the same eventually with Burnham. Dewey, for his part, didn't take much notice of Marxism, considering that, having broken with Hegelianism, there was no need for him to come to terms with its Marxist extension. Marxism was one more form of the absolutism he had left behind. In presiding over the International Commission of Inquiry investigating the Moscow trials, he had occasion to associate with Trotsky and Trotskyists and to pay closer attention to Marxism. In announcing the commission's verdict that the trials were judicial frame-ups in 1937, he took the opportunity to repudiate the ideas of both Stalin and Trotsky.

Hook's break with Marxism, a break that would harden with the years into a more and more vicious form of anti-Marxism, was marked by his *Reason, Social Myths and Democracy*. Eastman's 1941 backward look *Marxism: Is It Science?* was an explicit attack on dialectical materialism as a philosophy of science. Marxists, he argued, wished to see the world as an escalator. Although Marxists professed to reject religion in favor of science, they cherished the belief that the external universe was evolving in exactly the same direction as they wanted it to go. It was a revival of religious certainty against scientific scepticism. They could not face a struggle to build a communist society in a world indifferent to them. Thus, they saw themselves as traveling on a moving stairway taking them the way they walked. Their enemies were on the same stairway, but walking in the wrong direction. The dialectic was a bold manoeuver in the defense of animism against science. Dialectical materialism was against science. The pretense of Marxism to objective analysis was a bluff. In Lenin's Russia, Eastman said, a whole coterie devoted themselves to the enterprise of translating all modern science into the terms of the Marxist dialectic, producing a continual stream of abstruse volumes of the highest technicality. It was difficult to imagine, Eastman asserted, a more futile employment of the human mind than this restating of every finding in the form of an analogy to the proletarian revolution. All this "incredible professorial hocus pocus" was the result of fear, although Lenin at least had not been inclined to leave to the dialectical universe the tasks that belonged to himself.[268]

Wilson

Another work that engaged in the same sort of critique of dialectical materialism and its relation to science, one inspired by the works of Hook and Eastman, was Edmund Wilson's *To the Finland Station* of 1940. Described on its cover as "the classic study of the origins of communism," its chapter on "The Myth of the Dialectic" took up the theme of the dialectic as a religious surrogate. Marx and Engels, he argued, had been led astray by the abstractions of classical German philosophy, by its great nouns with their capitalized solidity. From the moment that Marx and Engels admitted the dialectic into their semimaterialist system, they had admitted an element of mysticism. There was in the dialectic, he insisted, a large element of pure incantation.

Addressing himself to the examples of dialectics in science given by Bernal and Haldane, Wilson argued that they were dramatic formulas for the dynamics of certain social changes and that they lacked universal application. Mendeleyev's periodic table owed nothing to the antithesis and synthesis. In the Freudian theory of repressed desires, the dialectical cycle was far from inevitable. The transformation of ice into water had nothing to do with the dialectical trinity. The mapping out of processes of mutation and selection as triads in no way proved the relevance of the triad. Referring to the Lerner-Haldane controversy in *Science and Society*, Wilson argued, as did Lerner, that Haldane's biology derived nothing from the dialectic, that the latter was a purely gratuitous addition. The dialectics of Bernal and Haldane came down to "the poetry of imaginative people who think in abstractions instead of images."

The dialectic served as a semidivine principle of history, to which it was possible to shift the human responsibility for thinking, for deciding, for acting. Wilson regarded this period as that of the "decadence of Marxism" and in such a time the dialectic lent itself to tyrannical repressions. In his view, the communists of the Third International, leaving history to the dialectical demiurge, acquiesced in the despotism of Stalin.[269]

Science and Society

Perhaps the most important forum for the discussion of the relation of Marxism to science in the United States from 1936 on was the journal *Science and Society*, a broadly based Marxist theoretical journal formed in the spirit of the Popular Front. In its discussion of philosophy of science, the journal leaned heavily on the British Marxist scientists, who numbered among its foreign editors—Bernal, Haldane, Levy, Hogben, and Needham. Also among its foreign editors were Langevin in France and the American scientist H.J. Muller, then working in the USSR. On the home front, there was a wide range of views among its contributors—from the American communist philosophers Howard Selsam and Harry Wells and the Dutch communist mathematician and historian of science Dirk Struik (who came to settle in the United States as a professor at MIT) to the Dutch breakaway-communist, Anton Pannekoek (who also settled in the United States), to the pragmatist humanist philosopher Corliss Lamont of Columbia University. One of the issues that presented the greatest problems for its editors and readers was Lysenkoism. Bernard J. Stern and J.B.S. Haldane, among others, grappled with it. But none did so more dramatically than H.J. Muller.

Muller

The story of Hermann Muller is one more story of a committed intelligent Marxist pushed away by the far less-intelligent and less-committed powers that were on the rise in the communist movement. A pioneering geneticist, a student of T.H. Morgan early in the century, and discoverer of the mutagenic capacities of X-rays, which later won him a Nobel Prize, Muller was from an early age a Marxist both in his politics and in his science. He saw in socialist society the best hope for science. He first went to the Soviet Union in 1922 and returned there to work in its Institute of Genetics at the invitation of Vavilov in 1933. Muller claimed that his scientific thinking was tightly bound up with his Marxism.

An article published in the Soviet Union in 1934 and suppressed for many years (republished by Loren Graham) shows Muller a devoted exponent of the unity of Marxism and science, especially as it applied to his own science of genetics. He referred enthusiastically to Lenin's defence of materialism and dialectical method. He elaborated in great detail on the various trends in genetics and analyzed them in terms of a range of philosophical assumptions that were holding back the advance of the science of genetics. Dialectical materialism, Muller insisted, was the key to the dilemmas facing geneticists. It kept science close to the evidence, but to the evidence as a whole, not just a restricted section of it, and it opposed the static, cut-and-dried mechanistic attitude that neglected complicated processes and their interrelations with other processes in favor of a more dynamic and synthetic approach. Such an approach would keep scientific energy from being misspent, from being blocked by the veiled antimaterialism that surrounded the gene with vagueness, and by antidialectical antitheses between Mendelism and Darwinism. The article also put forward a hard-hitting attack on the inheritance of acquired characteristics. Muller concluded with a declaration of his faith in the immense potentialities of science when seen with the clear light of Marxism:

> Capitalist society, with its idealist and religious background, would naturally shrink before such revolutionary implications, but for Marxist-Leninist dialectical materialism, in consideration of the facts of genetics, it is one of the logical phases of the world movement, and only in a society freed from the prejudices and conflicts of class, race, and sex, and from the encumbrances of religious superstitions and customs, can an effectual attack be made upon this hitherto impregnable recess of nature, this last stronghold in which the gods of the past still find some refuge. With socialist enthusiasm, however, this hidden way can be opened up to furnish a new avenue of unending victory for the triumphant workers.[270]

But Muller fell foul of Lysenko. His opposition to Lysenkoism at the 1936 conference was resolute, spirited, and vigorous. In 1937, he, like his friend Haldane, went off to Spain during the Spanish Civil War and then back to the United States. He never returned to the Soviet Union. With the increasing assaults on the science of genetics and with the deaths of his friends and colleagues Agol, Levit, and Vavilov, Muller grew quite bitter and became a sharp critic of the Soviet Union. And the Soviet Union became a sharp critic of Muller and all such "Mendelist-Morganists." But Muller remained a Marxist.

Marxism and Pragmatism*

The relationship between Marxism and pragmatism was a recurring theme in the discussions of American Marxism. It is a theme that has by no means played itself out, indeed a theme within a larger theme that has far from played itself out, that is, the relationship of Marxism to non-Marxist philosophies. The debate within American Marxism over pragmatism highlighted the problems. The central and recurring problem was the tendency for both sides to harden so as to prevent a sufficiently fruitful interaction. As on so many other occasions in the history of Marxism, those who sought to open Marxism to another philosophy they had good reason to take seriously, didn't always do so in the most competent fashion, or with sufficient clarity about what was of greatest value in the other philosophy and about what was crucial to the integrality and distinctiveness of Marxism. Ultimately, the criterion had to be what best made sense of experience, but there was also the question of what justified calling any given position Marxist. In Hook's case, for example, a call for the recovery of neglected subjectivity and emphasis on revolutionary praxis was one matter, but whether experience justified the rejection of monism and determinism and whether there was enough left to justify calling such a position Marxist were very different matters. The response of his opponents, reasserting the very formulations objected to, which were often genuinely problematic, did not help in sorting out these questions. It might have been far better for someone to have looked to the radical contextualism, to the processive naturalism of American philosophy, rather than to its pluralism and instrumentalism. It certainly would have been better for other Marxists to have been genuinely open as to what might come of such probing.

Another problem has been the strong element of a caricature in such debate. There was far more involved in the Marxist affirmation of an ontological continuity between the spheres of history and nature than was allowed for in Eastman's and Wilson's identification of dialectical materialism with surrogate religion. There were many alternatives between Eastman's escalator and Wilson's dialectical demiurge on the one hand and a universe unpredictable and utterly alien to human purposes on the other. On the question of the dialectic, there were many possible positions between incantation on the one hand and the objectivity of Hegelian contradictions on the other. Those who affirmed the integral connection between politics and philosophy were surely right to do so, but perhaps without sufficient insight into the complex and subtle character of the various mediations.

---

* See the Introduction for an analysis of the relation between Marxism and pragmatism with regard to the philosophy of science.

Yugoslav Marxism

Another center of particularly sharp debate was Yugoslavia. Intense controversy over reflection theory, socialist realism, and modern science burst into the open in Zagreb in 1938. It began with a group of communist writers and artists associated with the journal *Pecat*, who defended the autonomy of art against political subordination. Closely connected with them was a group of intellectuals, who sought to bring Marxism into accord with quantum mechanics, relativity theory, psychoanalysis, and genetics. In the many articles and several books that emerged in the next two years, more and more left wing intellectuals were drawn into struggle, for or against these trends, which were branded as "revisionism." In the collective work *Knjizevne sveske* published in 1940, various prominent communist leaders castigated the revisionists. Milovan Djilas, in due course (after the war) to be branded a revisionist himself, accused the revisionists of this day of Bukharinism. Edward Kardelj castigated them as attempting to disarm the working class. The present crisis of science, he asserted, could be overcome only by adopting dialectical materialism as the methodological basis of science. This led to the flourishing of science, he claimed, citing the example of Soviet science as his evidence. Todor Pavlov, the Bulgarian philosopher, added to the assault with an expression of contempt for the philosophizing physicists of the day: "Bohr, Heisenberg and their followers in Zagreb." Tito, for his part, took a strong stand against the "rotten liberals." In the journal *Proleter* in 1939, he took issue with those who felt they could improve on Engels, Lenin, and Stalin by resurrecting the "bankrupt theory of Mach."[271] Not unexpectedly, the revisionist faction was duly expelled from the Communist Party of Yugoslavia.*

The Role of the Comintern

There were pockets of Marxist discussion here and there outside the Comintern, but none of the breakaway communist groups between 1919 and 1943 ever achieved great influence and there were very few "independent" Marxists around. The previous generation of social democratic theorists, such as the Austro-Marxists, Kautsky and Krzywicki, were still around, though not

---

* Among those expelled was Ivan Supek, a physicist who had studied under Heisenberg and supported the Copenhagen interpretation of quantum mechanics. In April 1981 in Dubrovnik, Professor Supek gave me an account of the 1938–1940 debate and its aftermath. Expulsion for him entailed repeated attacks from the party and attempts to isolate him within the partisan movement during the Nazi occupation. After liberation, the ideological conflict continued and the revisionists were again condemned. At the 1st Congress of Croatian Cultural Workers in 1944, when Supek delivered a report on science, he was branded as anti-Marxist and counter-revolutionary by the communists.

generating much in the way of new ideas or commanding much enthusiasm. Really the Comintern was at the center of Marxist discussion during the period of its existence—for mixed reasons, perhaps the best and the worst of reasons. There was definitely far more to it than what was implied in de Man's bitter remark: "The ruminant Marxism of the socialists is powerless against the carnivorous Marxism of the communists."[272] The greater vigor of the communist movement cannot be simply marked down to ferociousness.

How to assess the Comintern and the fate of Marxist theory within it? It is a complex matter, for it had its moments both of glory and of shame. But existing accounts of it speak only of the glory or only of the shame. Communists themselves have been singularly reluctant to look fully into the dark recesses. A quick glance, immediately covered by some formula about distortions occurring during the period of the "cult of personality," is generally as much as can be borne.* But the rest is generally left to the anticommunists, who reduce the whole thing to the dark corners of it.

But the darkness weighs heavily upon it. The attempt to assess it both sympathetically and honestly can be the most searing of tasks. For the questions must be squarely put: Did the Comintern justify the fervent hopes it aroused? Or did it amount to a "false dawn of history," "a misfired renaissance"?[273] Was the Comintern "a glittering citadel for outsiders and a deathtrap for those within its walls"?[274] Was Moscow, at the center of it, "a matriarchate which practised infanticide"?[275]

There is, unfortunately, much to support the worst that can be said about it. As other revolutions failed to follow upon the Russian one, the Soviet party came to dominate the Comintern, at first through its prestige as the only party that had successfully carried through a socialist revolution. However, as time went on, this prestige was more and more abused. Soviet domination became more and more institutionalized into the very mode of existence of the Comintern and the other communist parties came to be more and more subordinate to the interests of whatever forces were dominant in the CPSU. The sharp shifts in policy emanating from Moscow, often reflected far more about the nature of the internal power struggles within the CPSU than about the rhythms of the class struggle in the countries where they had to be unquestioningly applied. Not only the internal but the external affairs of the Soviet Union became the overriding factors, with the Comintern becoming more and more the instrument of Soviet foreign policy.

It was not only that the Soviet Union used the Comintern apparatus to recruit Soviet intelligence agents. This at least was understandable. Foreign

---

* Western European communists, however, have in recent years done much to break through this syndrome.

communists felt a great loyalty to the Soviet Union and gladly acted in the interests of the proletarian revolution abroad as opposed to the interests of their own ruling classes at home. What was almost impossible to understand, and foreign communists simply could not see it, was that the Soviet Union did not feel the same loyalty to them. Many aspects of the relation of the Soviet Union to the Comintern involved an astonishing callousness, even manifest betrayal, towards those whose loyalty they inspired. It should have been possible, if perilous, to be a communist party among communist parties, engaged in class struggle, and simultaneously a government among governments, engaged in high diplomacy, without running with the hare and hunting with the hounds. However, whether it was possible or not, it was not the way of the Soviet Union.

It was well enough, for example, during the period of the Popular Front, when the interests of the communist parties and the Soviet state coincided in the struggle against fascism. But when the Soviet Union entered into the pact with Nazi Germany and Britain and France did go to war with Germany, British and French communists were put in the most impossible of situations with the directive of the Comintern being to desist from antifascist propaganda and naming Anglo-French imperialism as the basic aggressive force and as the main target of communist propaganda. The consequences in France, with resistance to German aggression greatly weakened and with communists left to be identified with treason, were particularly disastrous. If the interests of Soviet foreign policy conflicted with the interests of the working class elsewhere, then it was so much the worse for the interests of the working class elsewhere.

The Exiles to the East: The Fate of Foreign Communists in the Soviet Union

Some of the most tragic episodes, however, concerned the fate of foreign communists and antifascists living in the Soviet Union. Fleeing from fascism, some went west and some went east. It was better to have gone west. Communists living under the social order they were pledged to overthrow may not have been greeted with open arms, but they survived. Communists who sought the warm embrace of the socialist motherland in their time of trouble, who were greeted with open arms, perished in large numbers, executed without trial or even notice of execution. Many of them disappeared in the purges, partly as a part of the struggle to render the communist movement finally and fully submissive, to replace those with their own revolutionary traditions and a trace of independence of mind with newer and more pliable elements with no independent histories. But it was only partly that, for the atmosphere of irrationality and paranoia that swept over the Soviet Union in the 1930s allowed of no such clear and semirational explanation for all that happened. As the Austrian communist Ernst Fischer, trying to come to terms with it many years later, asked: "How understand a game in which the devil cuts the cards?"[276]

To look at the Hungarian party, for example, the orthodox and obedient Bela Kun perished, whereas his rival György Lukács survived.* The heretical Sandor Varjas was arrested, but so was the orthodox Ladislaus Rudas, who was sending forth a constant stream of hard-hitting, hard-line polemics against all possible deviations.

The Case of the Communist Party of Poland

A special cse, a particularly horrific case, was that of the Polish party. It was a party with long and militant revolutionary traditions and a sophisticated intellectual heritage. Like all other parties of the Comintern, it suffered from factional struggle, purges, recantations, expulsions, and from following all the abrupt and enforced twists and turns of Comintern policy. It suffered too from having to operate under conditions of illegality through the whole period of its existence, from the bolshevization process aimed at liquidating its Luxemburgist heritage, from involvement in the factional disputes of the Soviet party, from the "May mistake" in which they supported the Pilsudski putsch, under which they continued to suffer severe police persecution, from abrupt changes in leadership decided in Moscow and handed down from above, from the sectarianism of the "third period" that severely isolated and demoralized the party. But through all such trials and tribulations, and despite all its own faults, it was an impressive and heroic party, devoted to the interests of the Polish working class and of the communist movement as a whole.

In 1938, by a resolution of the Comintern, the Polish Communist Party was dissolved, on grounds of being a hotbed of Trotskyism. It was nothing of the kind. Such Trotskyists as had been in the party had long since been expelled. Nevertheless, the party was annihilated. Polish communists, who happened to be in the Soviet Union, were arrested, sent to camps or executed. The fortunate ones, sad to say, were those who were in prison in Poland and not those who escaped to the east.

Not long after the dissolution of the Polish party came the dissolution of Poland. In 1939 came the Nazi-Soviet Pact, containing a secret protocol

---

* Luckács's own arrest in 1941, during the period of the Nazi-Soviet Pact, may have been connected to the emphasis he had always put on the struggle against fascism. Michael Löwy provides evidence for this explanation. A political autobiography covering the period to April 1941 in the Luckács Archives in Budapest—probably written on the order of the NKVD as the basis for his interrogation—made no mention of the struggle against fascism, which he had previously insisted to be central to his political commitment (*New Left Review* 91, 1975).

providing for the partition of Poland between the two parties. Soon after, Nazi Germany invaded Poland from the west and the Soviet Union "liberated" it from the east. It could be that Stalin had been clearing the ground for this eventuality by eliminating any possible opposition to it among Polish communists. Other factors may have played a part as well, for it came at a time when a whole generation of older communists were being eliminated in any case. Isaac Deutscher made the point that "the psychological profile of even the most orthodox Polish communist left much to be desired from the Stalinist point of view."[277] Deutscher, although at odds with the party from the time of his expulsion in 1932 (for "exaggerating the dangers of nazism," against the policy of the third period in which social fascists were the main enemy), testified to the slanderous character of the charges against its leaders and regarded the dissolution of the party as an unparalleled crime.* Whatever the precise explanation, it was a cruel and tragic betrayal of the trust that foreign communists placed in the Communist Party of the Soviet Union.

The Nazi-Soviet Pact: Further Consequences

That was not the end of it. There was another secret protocol, other consequences, and other betrayals. The other secret protocol enjoined each side to suppress agitation against the other. The result was the Comintern directive turning a nonaggression pact between the Soviet Union and Nazi Germany into an accommodation between communism and fascism and allowing French communist deputies to be put on trial for treason. But far more sinister were the implications for German communists. In 1938, certainly a part of the overall process of the massive purge, but also possibly once again to prepare the way for the pact, over 800 German communists and antifascists, who had sought asylum in the Soviet Union, were arrested and imprisoned, including leaders of the KPD. Eberlein, KPD delegate to the founding congress of the Comintern, held a press conference denying reports in the Swiss press that he had been arrested, only to be arrested the next day. Some were actually seized in the House of Political Emigrés in Moscow. Some who had been tortured by the Gestapo were tortured again by the NKVD. Their institutions, such as the Liebknecht School and the Thälmann Club, were closed down. But worst of all was the almost unbelievable, and certainly unforgivable, fact that after the pact, German, Austrian, and Hungarian communists and antifascists were handed over by the NKVD to the Gestapo. Of these many were Jews, who went on to ghetto uprisings, concentration camps, and gas chambers. Among them were the German physicists Alexander

---

* The accusations against Polish communists were never retracted by the CPSU until 1956. The old party and its leaders were posthumously rehabilitated after the 20th Party Congress of the CPSU.

Weissberg and Friedrich Houtermans, who had been working in the Soviet Union. Appeals for their release, after their arrests in 1937, by Irène and Frederic Joliot-Curie, Paul Langevin, Albert Einstein, P.M.S. Blackett, and J.D. Bernal were to no avail. Appeals from Hitler and Ribbentrop obviously carried greater weight. Whatever arguments can be put forward with regard to the necessity of the nonaggression pact, there can be no justification for the secret protocols with their dire consequences for Comintern policy, for Poland, and for communists and antifascists in exile.

And to the fellow-traveling intellectuals in the west, who searched their souls and said that this was simply too much, refusing to parrot the hollow explanations and to abide the hypocritical talk of higher laws of history, the response of the CPUSA was typical:

> Why do many intellectuals retreat at sharp turns in history? They are impotent subjectivists. . . . Don Quixotes. . . .[278]

War and the Dissolution of the Comintern

In 1941, after the nazi invasion of the Soviet Union, communists were free to be antifascists again, although in truth the overwhelming majority of them always were. Communists distinguished themselves in their heroic resistance. Indeed, French communists already had, in the year preceding the new shift in Comintern policy. The Polish party was reconstituted and such Polish communists as were left, in spite of all they had endured, reformed themselves and fought courageously in the underground of nazi-occupied Poland.

In 1943, in the midst of Stalin's negotiations with Churchill and Roosevelt, the Comintern was abruptly dissolved, as it served the interests of Soviet foreign policy to dissociate itself from revolutionary activity in the west.

Marxism and the Comintern: Balancing the Accounts

However, the Comintern must not only be remembered by its official policies, by its intrigues, by its betrayals. It must be remembered too by its brave struggle against the powers of reaction, by the noble commitment of its members to building a new world, by the earnest probing of its theoreticians to work out a new world view. It is true that much of what was noblest and most creative was often stifled and what was most base was most encouraged, but the fact remains that, through it all and in spite of all the obstacles placed in its way, the Comintern was still able to inspire noble commitment and creative thinking. It is true, too, that the figure of Stalin looms uncommonly large over it all and casts a dark shadow over the history of the Comintern, but the Comintern belonged too to the hundreds of thousands of men and women who believed ardently and worked unflinchingly, who knew nothing of the intigues or the cynical manipulation of their efforts. The Comintern may have been the instrument for sending hundreds of communists to their death at the hands of other communists, but it was also the context in which communists went generously to die in the International Brigades, risked their lives daily in the underground of Hitler's Germany, formed the backbone of the resistance to nazi occupation, refused to speak under torture, or simply carried on thinking, writing, struggling as their circumstances allowed, for a new social order.

The communist historian, Eric Hobsbawm, who has not been one to shy away from the difficulties, has nevertheless made the point that it is necessary to resist the temptation to dismiss the Comintern *en bloc* as a failure or as a Russian puppet show. Unless certain factors are taken into account, Hobsbawm insists, it is simply impossible to grasp what the Comintern was about, to understand the sense of total devotion that motivated party members. It is vital to recapture a sense of the immense strength that they drew from the consciousness of being soldiers in a single international army, following a single grand strategy of world revolution. This gave the movement a certain immunity against the terrible collapse of its ideals. About the Soviet Union, only the naive, Hobsbawm explains, believed that it was a workers paradise, but even among the most sophisticated, it enjoyed the sort of general indulgence that the 1960s New Left reserved only for Cuba and Vietnam. Moreover, to be separated from it was to be cut off in their eyes from any possibility of effective revolutionary activity. What individuals and groups did break away were notoriously isolated and unsuccessful.[279]

And what of the development of Marxist theory amidst it all? How to balance the accounts with regard to the conjuncture of politics, philosophy, and science in and through the Comintern? There was, to be sure, much counting against both the healthy development of revolutionary theory and the constructive pursuit of revolutionary practice. For some, it was simply more than could be borne. In bitterness and despair, many turned away and told their stories. Ignatio Silone, the Italian communist leader involved in the high politics of the Comintern, told of various incidents that were symbolic of the prevailing atmosphere that reflected and shaped communist thinking. One took place at a meeting of a special commission of the Comintern executive discussing the ultimatum issued by the British TUC ordering the disbanding of the Communist Party-led minority movement under threat of expulsion. The representative of the CPGB gravely explained the dilemma: liquidation or expulsion. A Russian communist, however, wondered what all the fuss was about and put forward what seemed to him the obvious solution: declare submission and do exactly the opposite, to which the British communist replied "But that would be a lie." Loud laughter rang through the room and then spread all through Moscow, the like of which had perhaps never been heard before. To Silone, the Englishman's short simple sentence outweighed all the long, heavy, oppressive speeches of his years in the Comintern and became a symbol for him. A second incident involved the return of the French communist Jacques Doriot to Moscow from a special mission in China. He gave his friends an extremely disturbing account of the blunders of the Comintern in the far east. The next day, speaking before a full session of the Comintern executive, he said exactly the opposite. He later confided to Silone with a superior smile, "It was an act of political wisdom." A third incident involved the demand of the Soviet communists that the Comintern endorse a document proclaiming the condemnation of Trotsky without reading it. Any remarks as to the inappropriateness of approving a document unread were met with utter hostility. Silone experienced on a number of occasions what he considered to be the utter incapacity of Russian communists to be fair in discussing opinions conflicting with theirs. The adversary, simply for daring to contradict, at once became a traitor, an opportunist, a hireling. An adversary in good faith was something inconceivable to them.[280]

Arthur Koestler, a member of the KPD during the last days of the Weimar Republic and then an associate of Willi Münzenberg in the organization of Popular Front activities in Paris, indicated the way Marxist philosophy was sometimes brought to bear in the day-to-day life of the party. When faced with the sudden, sharp, and seemingly arbitrary shifts of Comintern policy, which often involved an abrupt change of slogans from one day to the next with the new ones completely contradicting those of the day before and with the real reasons for the change concealed, a party member might remark on the contradiction. The reply inevitably was "But you are not thinking dialectically, Comrade." "Gradually," said Koestler, "I learned to distrust any mechanistic preoccupation with facts and to regard the world around me in the light of the dialectical interpretation. It was a satisfactory and blissful state; once you had assimilated the technique, you were no longer disturbed by facts." When someone he recruited had a crisis of conscience and wrote a report of their joint activities, Koestler told of his reaction "I could not face reading in black and white the factual record of actions which I insisted on regarding through a haze of dialectical euphemisms." Koestler regarded the dialectical tight-rope acts of self-deception, sometimes performed by men of good will and intelligence, as more disheartening than the barbarities committed by the simple in spirit. Eventually, he came to see the dialectic as the rationalization of a pathological species that had completely severed relations with the subconscious. It initiated a way of thinking that made the monstrous acceptable: "the necessary lie, the necessary slander, the necessary intimidation of the masses to preserve them from short-sighted errors, the necessary liquidation of oppositional groups and hostile classes, the necessary sacrifice of a whole generation in the interests of the next."[281]

Another set of incidents and observations stemming from experience within the KPD has come from Rosa Levine-Meyer. Hers is very much a widow's account, having seen things very much in and through her two husbands, Eugene Levine and Ernst Meyer, both leaders of the KPD. Nevertheless, some interesting stories are told and some valuable perceptions break through. She told how Karl Radek had justified his cowardice, dissociating himself from his admired friend Rosa Luxemburg during the drive against Luxemburgism, by declaring "It was a historic necessity." She told how a crushed Gerhard Eisler insisted to her:

> If you cannot lie and swindle in favour of the Soviet Union now, you have never been a communist.

He advised her:

> Don't fight. It will lead you nowhere. The time will come when historic truth will be restored. It does not matter now, everything must be done in favour of the Soviet Union.

She testified as to how easily the corruption set in, describing an article on the Five Year Plan she wrote herself as "24 pages of lies which read in retrospect like the ravings of a lunatic" and noting that it was amazing what could be done starting off from a tiny twist of conscience. As to the effect of it all, she judged:

> The moral decline inevitably brought a harrowing decline in theoretical standards.[282]

About this she was surely right.

All of these things infected the prevailing atmosphere in the Comintern: the deliberate deceit, the unconscious self-deception, the high-handed authoritarianism, the degrading servility, justified in terms of a primitive, semimystical schematization of History with a capital H, History with the people left out, that came to overshadow all perceptions of actual facts and experiences of living persons. It often began in innocence, posed as the sacrifice of some perhaps laudable principle for some supposedly necessary gain in the interests of the class struggle. But the erosion of rationality and morality, the entrenchment of the habit of lying, the destruction of trust, and the sheer human degradation outweighed any gains made by such methods. The end did not justify the means. Unfortunately, even the best often reason as did Lukács:

> Marxism knows that small villainies can be performed with impunity when great deeds are being achieved; the error of some comrades is to suppose that one can produce great results simply through the performance of small villainies.[283]

About this, Lukács was quite wrong. The one way of thinking constantly deteriorated into the other. It gave over essential ground. It allowed the rot to set in, and once it did, there seemed to be no way of stopping it. And often the lying, the swindling, the cruelty, the destruction had nothing to do with class struggle or great deeds at all. In any case, the use of the categories of Marxist philosophy to override the demands of rationality and morality could not but have a corrosive effect on the development of Marxist philosophy. So it did. Thus, the stale and dreary textbooks of dialectical materialism that gave the impression that philosophical thinking had reached its finished form and was reliably embodied in the decisions of the party leadership. Thus, too, the heresy-hunting preoccupation with revisionism aimed at those who wished to carry on the process of philosophical thinking and thus the fear of what conclusions might be reached by those proceeding in accordance with considerations of rationality and morality.

But there was more to it. Rigidity, cowardice, cruelty, deceit, and corruption of power may have characterized some, particularly some highly placed, and may have colored much of the high politics of the Comintern. But the men and women who made it a real and living movement were made of finer stuff and acted on very different motives. They were the bearers of the Marxist tradition in its most vigorous form during these years. The power of Marxism, with its invitation to think in a whole new way and to work toward a whole new social order, broke through all the rest. The example of the October Revolution, such a gigantic and daring social experiment, and the call of the Comintern to extend it to the far corners of the earth, brought into the communist movement the most inquiring minds inspired by Marxism and the most energetic personalities seeking to carry it through in political commitment. It was not only the movement of Stalin, but the movement of Gramsci, Caudwell, Bernal, Haldane, Guest, Langevin, Solomon, Politzer, and others, who sought to bring the most advanced science to bear upon what they considered to be the most coherent and most firmly grounded philosophical world view. Whatever obstacles were placed in its way, Marxism developed and extended itself in a most creative and credible way. It also received serious setbacks from which it has never recovered.

NOTES

1. Arthur Koestler, *Arrow in the Blue* (London, 1969), pp. 28–92.
2. *Der zweite Kongress der Kommunistischen Internationale* (Hamburg, 1921), pp. 387–395.
3. Friedrich Adler, cited by Milorad Drachkovitch and Branko Lazitch, "The Communist International," in *The Revolutionary International* (London, 1960), p. 171.
4. E.H. Carr, *The Bolshevik Revolution* 3 (London, 1966), pp. 201–203.
5. David Caute, *The Fellow-Travellers* (London, 1973), p. 57.
6. Karl von Ossietzky, *Das Tage-Buch* (20 September 1924), cited by James Joll *Europe Since 1870* (London, 1976), p. 311.

7. Reported in *Workers' Weekly*, 7 July 1923.

8. F. Baldwin, "Should We Combat Religion?", *The Communist Review*, vol. 4 (1923–24), p. 543.

9. Bertrand Russell *The Practice and Theory of Bolshevism* (London, 1920), pp. 119–121; For communications between the Agitation and Propaganda Dept. of the Comintern and the CPGB, cf. *Communist Papers* (London, 1926), p. 33 and *Communist* 3 (1929), pp. 450–460; for a history of British Marxism from 1917 to 1933, cf. Stuart Macintyre *A Proletarian Science* (Cambridge, 1980).

10. The article by Erwin Ban, "Engels als Theoretiker," *Kommunismus*, 3 December 1920, cited in chapter 1, is an example of an early article drawing a sharp dividing line between Marx and Engels and assessing Engels's role in the development of Marxist philosophy as a negative one.

11. Hermann Gorter, "Offener Brief an den Genossen Lenin," *Organisation and Taktik der Proletarischen Revolution* ed. H. Bock (Frankfurt-am-Main, 1969), p. 192–227.

12. Hermann Gorter, *Der historische Materialismus* (Stuttgart, 1919).

13. Anton Pannekoek, *Lenin as Philosopher* (London, 1975).

14. Georg Lukács, *History and Class Consciousness* (London, 1971), p. 24.

15. Lukács, *Tactics and Ethics: Political Writings 1919–1929* (London, 1972), p. 16.

16. Ibid., pp. 19–27.

17. Lukács, *History and Class Consciousness*, p. 1.

18. Lukács, "Preface to the New Edition" (1967) *History and Class Consciousness*, pp. ix–xlvii.

19. Gareth Stedman Jones, "The Marxism of the Early Lukács, *New Left Review*, no. 70 (November–December 1971), p. 46.

20. Karl Korsch, *Marxismus und Philosphie* (Leipzig, 1923), p. 71.

21. Korsch, *Marxism and Philosophy* (London, 1970), p. 92.

22. Korsch, "The Present State of the Problem of *Marxism and Philosophy*—An Anti-Critique [1930], op. cit., p. 117.

23. Korsch, *Marxism and Philosophy*, p. 67.

24. Ibid., p. 95.

25. Siegfried Marck, "Neukritische und Neuhegelsche Auffassungen der Marxistischen Dialektik," *Die Gesellschaft* 1 (1924), pp. 573–578; *Die Dialektik in der Philosophie der Gegenwart* 1 (Tübingen, 1929).

26. Karl Kautsky, *Die Gesellschaft* 1 (1924), pp. 306–314.

27. László Radványi, *Archiv für Sozialwissenschaft und Sozialpolitik* 53, no. 2 (1925), pp. 527–535.

28. Ernst Bloch, "Aktualität und Utopie zu Lukács, *Geschichte und Klassenbewusstsein,*" *Der neue Merker* (October 1923), pp. 457–477.

29. Herbert Marcuse, "Zum Problem der Dialektik," *Die Gesellschaft* 7, pp. 29–30.

30. László Rudas, "Orthodoxer Marxismus?",*Arbeiterliteratur* no. 9 (1924); "Die Klassen-bewusstseinstheorie von Lukács, 1 and 2, ibid., no. 10 and 12 (1924).

31. Béla Kún, "Die Propaganda des Leninismus," *Die Kommunistische Internationale*, April, 1924.

32. József Révai,*Archiv für die Geschichte des Sozialismus und der Arbeiterbewegung*, no. 11 (1925), pp. 227–236.

33. Béla Fogarasi, *Die Internationale*, 15 June 1924, pp. 414–416; *Vestnik sotsialisticheskoi akademii* no. 6 (1923).

34. August Thalheimer, *Die Rote Fahne*, 20 May 1923.

35. Hermann Duncker, *Die Rote Fahne*, 27 May 1923.

36. Duncker, *International Press Correspondence* no. 55 (July 1925), p. 764.

37. Korsch, "Uber materialistische Dialektik," *Die Internationale*, 2 June 1924, pp. 376–379.

38. Korsch, "Lenin und die Komintern," ibid., pp. 320–327.

39. *Pravda*, 25 July 1924.

40. A.M. Deborin, "Lukács und seine Kritik des Marxismus," *Arbeiterliteratur* no. 10 (1924).

41. Cf. David Joravsky, *Soviet Marxism and the Natural Sciences* (London, 1961), p. 114; and Iring Fetscher, "Der Verhältnis des Marxismus zu Hegel," *Marxismus-studien* (1960), pp. 119–120.

42. M.B. Mitin, "Uber die Ergebnisse der philosophische Diskussion," in Bukharin, Deborin et al., *Kontroversen über dialektischen und mechanistischen Materialismus* (Frankfurt, 1969), pp. 366–367.

43. *Fifth Congress of the Communist International: Abridged Report*, Communist party of Great Britain, n.d., p. 17.

44. Ibid., p. 132

45. *International Press Correspondence*, no. 136 (1924), p. 1796.

46. Cited by E.H. Carr, *Socialism in One Country* 3 (London, 1972), p. 103.

47. Reported by Drachkovitch and Lazitch, "The Communist International" in *The Revolutionary Internationals* London, 1966, p. 193.

48. Cited by Douglas Kellner, "Korsch's Revolutionary Historection," *Telos*, no. 26 (Winter 1976), p. 87.

49. Carr, *Socialism in One Country*, 3 p. 79, n. 1.

50. *Kommunistische Politik: Diskussionsblatt der Linken*, 1926.

51. Korsch, "Anti-Critique," (1930), in *Marxism and Philosophy*, p. 118.

52. Korsch, "Why I Am a Marxist?", originally published in the United States in *Modern Quarterly* in April 1935, reprinted in *Three Essays on Marxism* (London, 1971).

53. Korsch, *Karl Marx* (London, 1938); Timpanaro, *On Materialism* (London, 1975), p. 225.

54. Bertolt Brecht, *Gesammelte Werke* 20 (Frankfurt, 1967).

55. Giusseppe Vacca, *Lukács o Korsch?* (Bari, 1961).

56. Lucio Colletti, *Problemi de Socialismo* no. 10 (1966).

57. Timpanaro, *On Materialism*, pp. 221–254.

58. *New Left Review* has tended more to criticize Korsch in passing, whereas *Telos* has carried many articles on Korsch giving a detailed exposition and assessment of his views. Issue no. 26 (Winter 1975–76) is devoted to Korsch. Two particularly useful articles are those by Paul Breines, "Praxis and Its Theorists: The Impact of Lukács and Korsch in the 1920s," *Telos*, no. 11, (Spring 1972), and by Russell Jacoby, "The Politics of Philosophy from Lukács to the Frankfurt School," *Telos* no. 10 (Winter 1971–72).

59. Lukács, *Lenin: A Study of the Unity of His Thought* (London, 1970).

60. Lukács's review of Bukharin's *Theorie des historischen Materialismus* originally appeared in *Archiv für die Geschichte des Sozialismus und der Arbeiterbewegung* 11 (1925); an English translation is included in *Tactics and Ethics*, pp. 134–142.

61. Lukács's review of Wittfogel's *Die Wissenschaft der bürgerlichen Gesellschaft* appeared in the same issue and also appears in English translation in the above collection, pp. 143–146.

62. Lukács, 1967 Preface, *History and Class Consciousness*, p. xxxiii.

63. Morris Watnik, "Relativism and Class Consciousness," in *Revisionism*, London, 1962, p. 147.

64. Lukács, 1967 Preface, *History and Class Consciousness*, p. xxx; cf. also the interview with Lukács on his life and work published in *New Left Review*, no. 68 (July–August, 1971), pp. 55–56.

65. Lukács, 1967 Preface, *History and Class Consciousness*, p. xxxvi; NLR, no. 68, pp. 56–57.

66. Lukács, "Mein Weg zu Marx," *Internationale Literatur* 3, no. 2 (1933).

67. Lukács, "Znacheniye *Materializma i Empiriokrititsizma* dlya bolshevizatsii kommunisticheskikh partii" *Pod znamenem marksizma*, July–August, 1934, pp. 143–148.

68. Lukács, *Essays über Realismus* (Berlin, 1948), p. 158.

69. Lukács, 1967 Preface, *History and Class Consciousness*, p. xxxviii.

70. Lucien Goldmann, "Reflections on *History and Class Consciousness*" in *Aspects of History and Class Consciousness*, ed. Istvan Mészáros (London, 1971); Maurice Merleau-Ponty, *Adventures of the Dialectic* (London, 1974); Ferenc Feher, Agnes Heller, György Markus, and Mihaly Vajda "Notes on Lukács' Ontology," in *Telos*, no. 29 (Fall, 1975).

71. Paul Piccone, "Dialectic and Materialism in Lukács," *Telos*, no 11 (Spring 1972); cf. also other issues of *Telos*, especially no. 10 and no. 11 devoted entirely to Lukács.

72. Colletti, *Marxism and Hegel* (London, 1973), p. 178.

73. John Hoffman, *Marxism and the Theory of Praxis* (London, 1975), pp. 60, 150, 183–184.

74. Mihailo Markovic, "Dialectic Today," *Praxis: Critical Social Philosophy in Yugoslavia*, ed. Markovic and G. Petrovic, in *Boston Studies in Philosophy of Science* (Dordrecht/Boston, 1979).

75. Stedman Jones, *New Left Review* 70, pp. 27–64.

76. Lukács, 1967 Preface, *History and Class Consciousness*, p. xvi.

77. A.I. Variash, "O tom, kak ne nado pisat kritiku," *PZM* no. 5–6 (1925), p. 218; cf. Joravsky, op. cit. *Soviet Marxism* on Varjas.

78. Ernst Kolman, *PZM*, no. 11–12 (1937), pp. 232–233.

79. Kolman, "Dynamic and Statistical Regularity in Physics and Biology"; "The Present Crisis in the Mathematical Sciences and General Outline for Their Reconstruction," in *Science at the Crossroads* (London, 1931).

80. For an account of the situation of Marxist studies in Japan in the prewar period, cf. Shigeru Nakayama, "History of Science: A Subject for the Frustrated," in *For Dirk Struik*, ed. Robert S. Cohen and Marx W. Wartofsky (Dordrecht-Boston, 1974).

81. Antonio Gramsci, "La revoluzione contro il "Capitale," *Avanti*, November 24, 1917.

82. Gramsci, *La città futura* (Turin, 1917).

83. Gramsci, *Selections from the Prison Notebooks* (London, 1971), p. 448.

84. Ibid., p. 445.

85. Gramsci, *Quaderni del Carcere* 2 (Turin, 1975), pp. 1451–1459, translation in *Telos* 41 (Fall, 1979), pp. 154–155.

86. Gramsci, *Selections*, p. 468.

87. Ibid., p. 462.

88. Timpanaro, *On Materialism*, p. 57, 236–237.

89. Paolo Rossi, "Antonio Gramsci sulla scienza moderna" *Critica marxista* no 2 (1976).

90. Arthur Eddington, *The Nature of the Physical World* (London, 1935), p. 291.

91. James Jeans, *The Mysterious Universe* (Cambridge, 1930).

92. Noreen Branson and Margot Heinemann, *Britain in the Nineteen Thirties* (London, 1971), p. 257.

93. John Strachey, "The Education of a Communist," *Left Review*, no. 3 (December 1934), p. 69; *The Coming Struggle for Power* (London, 1932).

94. Branson and Heinemann, *Britain in the Nineteen Thirties*, p. 259.

95. J.G. Crowther *The Social Relations of Science* (London, 1941), pp. 430–432.

96. Hyman Levy, "The Mathematician in the Struggle," in *David Guest: A Scientist Fights for Freedom*, ed. Carmel Haden Guest (London, 1939), pp. 150–152.

97. J.D. Bernal, "Science and Necessity," *The Spectator*, July 11, 1931, reprinted in Bernal's later *The Freedom of Necessity* (London, 1949). Other accounts of the congress include Joseph Needham's Foreword to the 1971 edition of *Science at the Crossroads*. This edition also contains a well-researched introduction "On the Reception of *Science at the Crossroads* in England" by Gary Werskey. Werskey also gives an account of the congress in his book *The Visible College* (London, 1978), pp. 138–149 as does Neal Wood in *Communism and British Intellectuals* (London, 1959), pp. 123–125.

98. F.S. Marvin, "Soviet Science," *Nature*, August 1931, pp. 170–171.

99. "Science," *Times Literary Supplement*, 10 September, 1931, p. 687.

100. Werskey, *The Visible College*, p. 146.

101. W.H. Auden, *Poems* (London, 1933), p. 79.

102. John R. Baker, "Counterblast to Bernalism," *New Statesman and Nation*, July 29, 1939.

103. Rajani Palme Dutt, in *Unity and Victory: Report to the 16th Congress of the Communist Party* (London, 1943), p. 31.

104. C.P. Snow, *The Search* (London, 1934).

105. Snow, "J.D. Bernal, a Personal Portrait," in *Science and Society*, ed. M. Goldsmith and A. MacKay (New York, 1964), p. 24.

106. Joseph Needham, "Desmond Bernal: A Personal Recollection," *Cambridge Review*, November 19, 1971, pp. 34–36.

107. J.D. Bernal, cited by Werskey, *The Visible College*, p. 75.

108. Bernal, *The Social Function of Science* (London, 1939), p. 415.

109. Bernal, "Dialectical Materialism and Modern Science," *Science and Society*, Winter 1937.

110. Werskey, *The Visible College*, p. 137.

111. Bernal, *The World, the Flesh and the Devil* (London, 1929).

112. Bernal, "Engels and Science," *Labour Monthly*, no. 8 (1935); cf. also his review of *Dialectics of Nature* when first published in 1940, reprinted in *The Freedom of Necessity* (London, 1949).

113. Bernal, "Dialectical Materialism," in *Aspects of Dialectical Materialism* (London, 1934), p. 95.

114. E.F. Carritt, "Dialectical Materialism," ibid., pp. 137–146.

115. Bernal, "Notes in Reply to Mr Carritt's Paper," ibid., p. 153.

116. Bernal, *The Social Function of Science*, p. 237.

117. Bernal, cited by Werskey, *The Visible College*, p. 148.

118. J.B.S. Haldane, "Professor Haldane Replies," *Science and Society*, Spring 1938.

119. J.B.S. Haldane, *Daedalus, or Science and the Future* (London, 1924); "Kant and Scientific Thought," in *Possible Worlds and other Essays* (London, 1927).

120. Werskey, *The Visible College*, p. 160.

121. J.B.S. Haldane, *The Causes of Evolution* (London, 1932), p. 170.

122. J.S. Haldane, *The Philosophy of a Biologist* (Oxford, 1936).

123. Sir Peter Medawar, Preface to *J.B.S.: The Life and Work of J.B.S. Haldane* by Ronald Clark (London, 1968).

124. J. Maynard Smith, Introduction to *Science and Life* (London, 1968), p. viii.

125. Werskey, *The Visible College*, p. 86.

126. C.H. Waddington, "That's Life," *New York Review of Books*, February 29, 1968, p. 19.

127. David Joravsky, *The Lysenko Affair* (Cambridge, Mass., 1970), p. 280.

128. Loren Graham, *Science and Philosophy in the Soviet Union* (London 1973), p. 258.

129. J.B.S. Haldane, *"The Marxist Philosophy,"* the Haldane Memorial Lecture, 1938, at Birkbeck College, University of London, London, pp. 3–4.

130. Ibid., p. 19.

131. J.B.S. Haldane, "Why I am a Materialist," (1940), in *Science and Life: Essays of a Rationalist* (London, 1968), pp. 27–35.

132. J.B.S. Haldane, "The Origin of Life" (1929), ibid., p. 1–11.

133. J.B.S. Haldane, Muirhead Lectures, published in *The Marxist Philosophy and the Sciences* (University of Birmingham, London), 1938.

134. Ibid., p. 16.

135. J.B.S. Haldane, cited by Ronald Clark, *J.B.S.*, p. 118.

136. Andrew Rothstein, "Vindicating Marxism," *Modern Quarterly*, no. 3 (1939), p. 290.

137. J.B.S. Haldane, "Marxism and Science," *Modern Quarterly*, no. 8 (1939).

138. Medawar, op. cit.

139. J.B.S. Haldane, Preface to *Dialectics of Nature* (London, 1940), p. xv.

140. J.B.S. Haldane, *The Marxist Philosophy and the Sciences*, and "Dialectical Materialism and Modern Science" in four parts over four issues of *Labour Monthly*, 1941.

141. J.B.S. Haldane, *The Marxist Philosophy* and *The Marxist Philosophy and the Sciences*.

142. J.B.S. Haldane, "The Laws of Nature" (1941) in *Science and Life*.

143. J.B.S. Haldane, "If . . . " (1934) and "What is Religious Liberty?" (1939) in *Science and Life*.

144. J.B.S. Haldane, *The Marxist Philosophy and the Sciences*.

145. J.B.S. Haldane, "Dialectical Account of Evolution," *Science and Society*, Summer 1937; A.P. Lerner "Is Professor Haldane's Account of Evolution Dialectical?" and "Professor Haldane Replies," *Science and Society*, Spring 1938.

146. J.B.S. Haldane, "A Note on Genetics in the USSR," *Modern Quarterly*, no. 4 (1938).

147. J.B.S. Haldane, "Genetics in the USSR," *University Forward*, no. 6 (1941), p. 21.

148. Hyman Levy, "A Scientific Worker Looks at Dialectical Materialism," in *Aspects of Dialectical Materialism* (London, 1934).

149. Levy, *A Philosophy for a Modern Man* (London, 1938).

150. Clemens Dutt, "The Philosophy of a Natural Scientist," *Labour Monthly*, no. 4 (1938).

151. Levy, "The Philosophy of a Natural Scientist: A Reply," *Labour Monthly*, no. 5 (1938); Clemens Dutt, "A Rejoinder," *Labour Monthly*, no. 6 (1938).

152. Lancelot Hogben, cited by Werskey, *The Visible College*, p. 202.

153. Hogben, in an interview with Werskey, 26 July 1968, cited op. cit., p. 161.

154. Hogben, "Contemporary Philosophy in Soviet Russia" *Psyche*, October 1931.

155. Hogben, *The Nature of Living Matter* (London, 1930).

156. Clemens Dutt, "The Hesitant Materialist," *Labour Monthly*, no. 10 (1932); Hogben, "Materialism and the Concept of Behaviour," *Labour Monthly*, no. 1 (1933); Clemens Dutt, "Dialectical Materialism and Natural Science," *Labour Monthly*, no. 2 (1933).

157. Joseph Needham, "Integrative Levels: A Revaluation of the Idea of Progress," *Modern Quarterly*, no. 1 (1938).

158. Needham, quoted by Werskey in his introduction to Needham's *Moulds of Understanding* (London, 1976), p. 15.

159. Cited by Werskey, *The Visible College*, p. 79.

160. Needham, *The Great Amphibian* (London, 1931); *The Sceptical Biologist* (London, 1929).

161. Needham, "Metamorphoses of Scepticism," (1941) in *Moulds of Understanding* (London, 1976), p. 219.

162. Werskey, Introduction to *Moulds of Understanding*.

163. Needham, Foreword to *Biology and Marxism*, by Marcel Prenant (London, 1938), p. v.

164. Needham, "Integrative Levels," *Modern Quarterly*, no. 1 (1938), p. 34.

165. Needham, "Metamorphoses of Scepticism."

166. Needham, ibid.; "Integrative Levels"; Foreword to *Biology and Marxism*.

167. Needham, Foreword to *Biology and Marxism*.

168. Needham, "Evolution and Thermodyanimics" (1941) in *Moulds of Understanding*.

169. Comenius, *A Reformation of Schools* (1642), quoted by Needham in "Metamorphoses of Scepticism," p. 229.

170. E.F. Carritt, "A Discussion of Dialectical Materialism," *Labour Monthly*, no. 5 and no. 6 (1933).

171. Replies by J.M. Hay and T.A. Jackson in *Labour Monthly*, no. 8 (1933); by L. Rudas in no. 9 and no. 10 (1933); also L. Rudas, "Dialectical Materialism and Communism: A Postscript," in no. 9 and no. 11 (1934); reply by Edward Conze, no. 11 (1934); cf. also Rudas's articles on Sidney Hook in no. 3, no. 4 and no. 5 (1935).

172. Arthur Koestler, In *The God That Failed*, Richard Crossman (New York-London, 1965), p. 61; *The Invisible Writing* (London, 1954), p. 382.

173. Rajani Palme Dutt, *Labour Monthly*, no. 3 (1938).

174. Editorial in *Modern Quarterly*, no. 2 (1938).

175. T.A. Jackson, *Communist Review*, February 1932, cited by Stuart Macintyre, "T.A. Jackson—Working Class Intellectual," in *T.A. Jackson: A Centenary Appreciation* (London, 1979), p. 22.

176. Jackson, Memoirs cited by Vivien Morton "A Memoir," ibid., pp. 5–6.

177. Jackson, unpublished manuscript, cited by Stuart MacIntyre in *A Proletarian Science*, p. 138.

178. Jackson, *Dialectics: The Logic of Marxism and Its Critics* (London, 1936), p. 481.

179. Review in *Pod znamenem marksizma*, translated version published in *Modern Quarterly*, no. 3 (1938).

180. Maurice Cornforth, Preface to *Communism and Philosophy* (London, 1980).

181. Stephen Spender, in *The God That Failed*, p. 211.

182. Maurice Cornforth, in *David Guest: A Scientist Fights For Freedom*, ed. Carmel Halden Guest (London, 1939), pp. 95–101.

183. Maurice Cornforth, "Is Analysis a Useful Method in Philosophy?" *Proceedings of the Aristotelian Society*, supplement vol. 136 (London, 1934), pp. 90–118.

184. Harry Pollitt, "David Guest," in *Lectures on Marxist Philosophy* (Calcutta, 1971), pp. 9–10, originally published as *A Textbook of Dialectical Materialism* (London, 1939).

185. David Guest, in *David Guest: A Scientist Fights for Freedom*, pp. 58, 61, 159.

186. Guest, *Lectures in Marxist Philosophy*.

187. Guest, "The Machian Tendencies in Modern British Philosophy," in *David Guest*, originally in *PZM*, 1934.

188. Guest, "The Marxist and Idealist Conceptions of Dialectic" (1935), in *David Guest*.

189. Guest, review in *Labour Monthly*, no. 10 (1935), reprinted in *David Guest*.

190. My knowledge of the biographical background to Caudwell's work comes from interviews with those who knew him and from unpublished papers held in the Caudwell archives in the Humanities Research Center, University of Texas, Austin. I am grateful to Rosemary Sprigg, Paul Beard, and Elizabeth Beard for permission to quote from unpublished correspondence.

191. Christopher St. John Sprigg in a letter to Paul and Elizabeth Beard, 21 November 1935.

192. Lenin, "The Tasks of the Youth Leagues," *Selected Works* 3, p. 414. Caudwell expressed his intention to make this quote the motto for his *Studies* in a letter to Paul and Elizabeth Beard, 30 November 1935.

193. The bulk of essays intended as *Studies in a Dying Culture* were published in two volumes: *Studies in a Dying Culture* in 1938 and *Further Studies in a Dying Culture* in 1949. *Romance and Realism* remained unpublished until 1970.

194. *The Crisis in Physics* (London, 1939). The manuscript was in a far from finished state when the author left for Spain. Some chapters were in good draft form, whereas some others were only rough notes.

195. *Heredity and Development* and *Communism and Religion*, unpublished.

196. Sprigg to Paul and Elizabeth Beard, December 1936. This letter constitutes Caudwell's literary testament.

197. Hyman Levy, introduction to *The Crisis in Physics*, by Christopher Caudwell (London, 1939).

198. J.B.S. Haldane, "Marxism and Science," *Labour Monthly*, no. 8 (1939).

199. John Strachey, Introduction to *Studies in a Dying Culture*, by Christopher Caudwell (London, 1938).

200. Edgell Rickword, Preface to *Further Studies in a Dying Culture*, by Christopher Caudwell (London, 1949).

201. This is the underlying argument in all of his major works: *Illusion and Reality* (London, 1973); *Studies and Further Studies in a Dying Culture* (New York and London, 1971); *The Crisis in Physics* (London, 1939); *Romance and Realism* (Princeton 1970).

202. Christopher Caudwell, *Studies in a Dying Culture*, p. xix.

203. Caudwell, *Poems* (London, 1965), p. 12.

204. Caudwell, *Heredity and Development: A Study in Bourgeois Biology*, unpublished manuscript.

205. Caudwell *The Crisis in Physics*, pp. 73–75.

206. Caudwell, *Illusion and Reality*, p. 218.

207. Caudwell, *Further Studies in a Dying Culture*, p. 239.

208. Ibid., pp. 228–229.

209. Ibid., p. 227.

210. Ibid., p. 97.

211. Ibid., p. 113.

212. Caudwell, *The Crisis in Physics*, p. 151.

213. Caudwell, *Further Studies in a Dying Culture*, p. 236.

214. Caudwell, *Illusion and Reality*, p. 49.

215. Ibid., p. 316.

216. Kolakowski, *Main Currents of Marxism* 3, p. 115.

217. Maurice Cornforth, "Caudwell and Marxism," *Modern Quarterly*, no. 1 (1951); George Thomson, "In Defence of Poetry," *Modern Quarterly*, no. 2 (1951); "The Caudwell Discussion," *Modern Quarterly*, no. 4 (1951) (contributions by Margot Heinemann, Edward York, Werner Thierry, O. Robb, J.D. Bernal, Edwin S. Smith, Maurice Cornforth).

218. Hilary Rose and Steven Rose, "The Radicalization of Science," *Socialist Register*, 1972, pp. 116, 117, 129.

219. David Margolies, *The Function of Literature: A Study of Christopher Caudwell's Aesthetics* (London, 1969).

220. Francis Mulhern, "The Marxist Aesthetics of Christopher Caudwell," *New Left Review* 85 (May-June 1974), p. 53.

221. Terry Eagleton, "Raymond Williams: An Appraisal," *New Left Review* 95, (Jan.-Feb. 1976), p. 7.

222. E.P. Thompson, "Caudwell." *Socialist Register* 1977, pp. 228–276.

223. Ibid., p. 271.

224. Perry Anderson. For example: *Considerations on Western Marxism* (London, 1976); "Socialism and Pseudo-Empiricism," *New Left Review* 35 (1966); "Components of a National Culture," *New Left Review* 50 (1968).

225. David McLellan, *Marxism After Marx* (London, 1979), pp. 307–309.

226. Kolakowski, *Main Currents of Marxism* 3, p. 114.

227. Rose and Rose, "The Radicalization of Science," p. 110.

228. Werskey, *The Visible College.*

229. Werskey, "Making Socialists of Scientists: Whose Side Is History On?", *Radical Science Journal*, no. 2/3 (1975), p. 35.

230. Georges Politzer, *Elementary Principles of Marxism* (London, 1976).

231. Henri Lefebvre, *Dialectical Materialism* (London, 1968), p. 107.

232. Paul Nizan, *Les chiens de garde* (Paris, 1932).

233. Georges Friedmann, *De la sainte Russie à l'URSS* (Paris, 1938), pp. 129–142.

234. *A la lumière du marxisme* (essais) (Paris, 1935).

235. Paul Langevin, *La Pensée et l'action* (Paris, 1950), p. 301.

236. Langevin, cited by A.F. Joffe "Marxist Philosophy and Modern Physics," *Communist Review*, May 1949, p. 492.

237. Langevin, *La Pensée et l'action*, p. 173.

238. Langevin, *La Pensée*, no. 1 (1939); English translation "Modern Physics and Determinism," in *Modern Quarterly*, no. 3 (1939).

239. Jacques Solomon, "Pour le libre developpement de la science," *Cahiers du bolchevisme*, October, 1938.

240. Léon Rosenfeld, "In Memory of Jacques Solomon," *Selected Papers of Leon Rosenfeld*, ed. R.S. Cohen and J. Stachel; *Boston Studies in the Philosophy of Science* (Dordrecht, 1978), pp. 297–301.

241. Marcel Prenant, *Biology and Marxism* (London, 1938).

242. Prenant, *La Pensée*, July-August 1939.

243. Jean-Paul Sartre, *The Problem of Method* (London, 1963), pp. 50–51.

244. Jean Duvignaud, "France: The Neo-Marxists," in *Revisionism*, pp. 314–316.

245. David Caute, *Communism and the French Intellectuals* (London, 1964), p. 307.

246. Frederic Joliot-Curie, cited by Maurice Goldsmith, *Frederic Joliot-Curie* (London, 1976).

247. Bertolt Brecht, "Der Taifun," translated by Frederick Ewen in *Bertolt Brecht* (London, 1970), p. 380.

248. Brecht, "An die Nachgeborenen," *Gedichte* 4, pp. 143–145.

249. Ibid.

250. Brecht, *Flüchtlingsgespräche* (Berlin, 1961) translated by F. Ewen, *Bertolt Brecht*, pp. 376–378.

251. Brecht's "Turandot oder der Kongress der Weissiwasscher" was a satire directed at the Frankfurt School. Cf. Michael Morley's *Brecht: A Study*. London, 1977.

252. Carl Grünberg, "Festrede gehalten zur Einweihung des Instituts für Sozialforschung an der Universität Frankfurt am Main, 22 Juni 1924," *Frankfurter Universitätsreden* 206 (Frankfurt, 1924).

253. Martin Jay, *The Dialectical Imagination* (London, 1973).

254. Max Horkheimer, "Materialismus und Metaphysik," *Zeitschrifter für Sozialforschung* 2, no. 1 (1933).

255. Walter Hollitscher, in a letter to the author, 25 September 1979.

256. Wilhelm Reich, *The Sexual Revolution* (London, 1969).

257. Erich Fromm, "The Method and Function of an Analytic Social Psychology: Notes on Psychoanalysis and Historical Materialism," in *The Essential Frankfurt School Reader*, ed. A. Arato and E. Gebhardt (New York, 1978), pp. 477–496.

258. *The Scientific Conception of the World: The Vienna Circle* (Dordrecht, 1973), p. 8. (Originally published, 1929)

259. Otto Neurath, *Empiricism and Sociology*, ed. Marie Neurath and Robert Cohen (Dordrecht-Boston, 1973).

260. V.J. McGill, "An Evaluation of Logical Positivism," *Science and Society*, no. 1 (1936), p. 78.

261. Sidney Hook, *Towards the Understanding of Karl Marx* (New York, 1933); *From Hegel to Marx* (Ann Arbor, 1962). (Originally published, 1936.)

262. Earl Browder, "The Revisionism of Sidney Hook," *The Communist* 5, no. 12 (1933).

263. Ladislaus Rudas, "The Meaning of Sidney Hook," *The Communist* 5, no. 14 (1935).

264. Léon Trotsky, *Writings* (New York, 1972), pp. 200–201. (Originally published, 1932–33.)

265. George Novack, "Léon Trotsky's Views on Dialectical Materialism," *Polemics in Marxist Philosophy* (New York, 1978), pp. 269–270.

266. Trotsky, *In Defence of Marxism* (New York, 1965), p. 40.

267. Novack, "My Philosophical Itinerary," ibid., pp. 20–21; *An Introduction to the Logic of Marxism* (New York, 1971).

268. Max Eastman, *Marxism: Is It Science?* (London, 1941).

269. Edmund Wilson, *To The Finland Station* (London, 1960), pp. 181–201. (Originally published, 1940.)

270. Hermann J. Muller, "Lenin's Doctrine in Relation to Genetics," appendix to *Science and Philosophy in the Soviet Union*, pp. 453–469. (Originally published, 1934.)

271. My account of these events is derived from my discussions with Yugoslav philosophers and scientists in Dubrovnik Yugoslavia in April 1981, particularly Professor Ivan Supek, Dr. Srdjan Lelas, Professor Svetozar Stojanović and Professor Mihailo Marković.

272. Hendrik de Man, cited by Drachkovitch and Lazitch, *The Revolutionary Internationals*, p. 199.

273. Arthur Koestler, *The Invisible Writing: An Autobiography* (London, 1954), p. 241.

274. Drachkovitch and Lazitch, *The Revolutionary Internationals*, p. 193.

275. Louis Fischer, cited ibid., p. 195.

276. Ernst Fischer, *An Opposing Man* (London, 1974), p. 283.

277. Isaac Deutscher, "The Tragedy of the Polish CP," in *Marxism in Our Time* (London, 1972), p. 154.

278. V.J. Jerome, *Intellectuals and the War* (New York, 1940), pp. 16, 23.

279. Eric Hobsbawm, "Problems of Communist History," *New Left Review* 54 (March-April 1969), pp. 85–91.

280. Ignatio Silone, in *The God That Failed*, ed. Richard Crossman (London, 1965), pp. 92–93.

281. Koestler, ibid., pp. 29, 31, 84.

282. Rosa Levine-Meyer, *Inside German Communism* (London, 1977).

283. Lukács to Serge, cited by Serge in *Memoirs*, (London, 1963), p. 186.

Afterword (1993)

So much water has flowed under the bridge, so many tides have come in and gone out again, since the story set out here ended. Days of spectacular advance, days of stark defeat, still lay ahead.

After 1945, partisans came down from the mountains and up from the underground and became professors in the academies of the new order, which came after the devastation of war and the defeat of fascism. In Eastern Europe, and soon after in Asia, followed by areas of Latin America and Africa, the institutions of Marxism-in-power multiplied dramatically. Marxism was in the ascendancy in the universities, research institutes, and journals of the new socialist states.

For some, adherence to Marxism was a matter of deepest conviction. For others, it was brazen compulsion or abject conformity. It was the difference between living fire and stagnant water. It was the dissonance between the elation of discovering a pattern in history and the shock of finding show trials to be part of the pattern. It was the distance from the long march through "let a thousand flowers bloom" to the ravages of the red guards. It was the gulf between Paris in May and Prague in August of 1968.

The contradiction between the two forces played itself out over and over in a continuing drama, one that could only end in tragedy. The codes and procedures of the one-party state could never be the dynamism of a vibrant intellectual force. In the end, socialism could only be built upon consent. Such consent could not be enforced by constitutional provision or policed by a state security apparatus. In fact, it proved altogether antithetical to such measures.

However, through all the purges of institutes and editorial boards, through all the politburo decrees and dogmatic denunciations, there was still something vital there struggling for life. There was more even to Zhdanovism than the policing of intellectual life.* It was an assertion of

---

* Zhdanovism refers to the postwar campaign bringing ideological analysis to bear upon the arts and sciences, named after Andrei Zhdanov, a close associate of Stalin, who was the ideological spokesman of the central committee of the Communist Party of the Soviet Union.

the ideological character of all knowledge, all art, all science. It was a belief that philosophy was a fundamental force in the social order. It often had farcical consequences, to be sure: Lysenkoite falsification of experimental results, preparation of delegations to international philosophical congresses as if for Warsaw Pact maneuvers, barbarous editing that made every text sound exactly the same, equally dead and equally deadening.*

So many stories, so many intricate biographies, shaped in such complex ways by this turbulent history, come rushing through me, as I try to convey something of my sense of this period and to break the hold of the clichés of the prevailing view of it. The philosophers and scientists, whom I came to know, were troubled true believers, as well as cynics, cowards, careerists, and conformists.

Their different fates will haunt me to the end of my days: Wolfgang Harich, defending his own imprisonment to me; Radovan Richta, leading reformer, then leading normalizer, fearful, compromised, embodying a tortured ambiguity, knowing better than what he did; Adam Schaff, Svetozar Stojanović and Mihailo Marković, in their quite different ways both loyal and dissident and always lucid.** Others, such as Herbert Hörz, managed to live and work, even high up in the structures of political-academic power, without such sharp conflicts. They were not all of the sort who predominate in any society, that is, those who serve not truth, but power. Some managed to serve truth as well as power ... at least for a time.

There was enlightened and honorable work done in philosophy and

---

* I once got drawn into a Kafkaesque dispute with a Soviet editor in Prague, Ivan Frolov (who emerged later as a top advisor to Gorbachev and was editor of *Pravda* at the time of the August 1991 coup), which resulted in a text (a chapter in *Dialectical Materialism and Modern Science* [Prague, 1978]) published in my name, which bore almost no resemblance to anything I had written, even expressing views contrary to my own in several places. It was a violation of my intellectual and moral standards in procedures that had become routine in such institutions as *World Marxist Review*. It was part of a set of converging circumstances, which made me leave the Communist Party in 1979.

** Wolfgang Harich (GDR), Radovan Richta (Czechoslovakia), Adam Schaff (Poland), Svetozar Stojanović and Mihailo Marković (Yugoslavia) and Herbert Hörz (GDR) were all prominent philosophers and party members, whose fates represent the various problems even the most committed Marxists had with the politics of Marxist orthodoxy. So many sad stories came my way. Once I was torn between attending an official reception or remaining outside to hear a terrible tale of denunciation, purging, suicide, exile. Another time I had a most disturbing encounter with a once prominent philosopher, whom I found living like a forgotten and frightened animal in an attic, still afraid of the light, even after everything had changed.

science. It is important to say so now, as hardly anyone will say it, but it is nonetheless true.

There were controversies continuing from the prewar period, and there were new ones too, all of them rooted in efforts to work out the most appropriate lines of connection between philosophy, politics and science and to come to terms with the times. New interpretations of Marxism, coinciding with contemporary philosophical currents, struggled for assent. There came existentialism, phenomenolgy, neopositivism, post-structuralism, postmodernism ... and finally post-Marxism.

Marxism became a formidable force, not only in countries defining themselves as socialist, but in the most prototypically capitalist ones as well. With the emergence of the new left, Marxism made its influence felt in the academies of the west (more an ideological concept than a geograph-ical one). Although it never took state power, it did seize the intellectual and moral initiative for a time. The new left posed new questions to the old left as well as to the old right and ever shifting center. Eurocommunism represented a merging of old and new left currents, which promised much in its moment.The most vibrant debates of the day were conducted within the arena of Marxism.

This text is a product of that period. Reading it again, I have been glad to rediscover and reappropriate it. I stand over it unreservedly, more than ever confirmed in my rejection of the corrupting pressures bearing upon me at the time I wrote it, whether from east or west, from left or right. I have left the text as it was, except for typographical corrections and refer-ences to a second volume covering the postwar period, which will not now be written by me, because of complex circumstances and shifting priorities in my own work.

However, it is necessary to remark upon the difference between the "now" of the text and the "now" of the publication of this new edition.

In the intervening time, the storm clouds have burst and the dams have broken. Now, after the deluge, we stand amid the ruins of our ravaged utopia and ask hard questions about the tragedy that has engulfed it.

Marxists had built a world view based on a vision of history as moving, in however complicated a way, in the direction of a transition from capi-talism to socialism, only to see the opposite happening before our eyes. This after having our hopes raised that socialism could be reformed, that the promises of glasnost and perestroika could be fulfilled, hopes that have been dashed against the rocks of history.

East and west, Marxism has fallen.

In the west, Marxism is scorned by those swept off their feet in post-modernist pastiche.The new left, overtaken by the new right, has scat-tered to the four winds. Many enthusiasts of Eurocommunism proved more Euro than communist, and there didn't seem much point in continuing to publish *Marxism Today* after its editor and many of its authors had long since ceased to be Marxist, no matter how far the term was stretched.

In the east, Marxism has been transformed from orthodoxy to apostasy overnight. From witch-hunter to witch-hunted, some might say. But those who were the inquisitors then are often the inquisitors now, some transmuting with dizzying speed from obedient apparatchiks to born-again free-marketeers. Honest intellectuals will have as hard a time under the new world order as under the old one, some even harder.

For a brief moment, it was otherwise. Banned books gathered crowds at the window displays of Berlin's bookshops. The theories of Bukharin and Bahro were discussed with urgent intensity and reviewed as if they had just been published.* Now airport best-sellers have replaced them. Those who had breathed the air of the Prague Spring lifted their heads in the sun once more, only to be cast aside again. *World Marxist Review* suddenly went from being mind-numbingly boring to amazingly stimulating, just before the political base sustaining it collapsed.

Everything opened up, just before closing down again.

Marxism has perhaps never been at such a low ebb. It is generally thought at present to have an inglorious past and no future.

This text, however, is testimony to the glory, as well as the terror, of its past. It will not do to dismiss it. Nor will it do to appeal to a pure, unadulterated Marxism untouched by all this tragedy; to touch the totems of its classical texts as if they were Platonic forms; to rescue Marx from Engels, Lenin, Stalin, and finally from Gorbachev. Whatever Marxism is, it is in and through the whole bloody, messy, human process that is its history. At all times, there was intelligence and integrity in it, even if mixed with stupidity and treachery all along the way. It is a rich and complex story, and I am proud to tell it here.

As to its future, we shall see.

Despite such devastating disaffection and desertion and denunciation, I do not believe that it will disappear. Despite everything, it still has an explanatory and ethical power persisting through all its problems and cutting through all the confused and craven chatter surrounding it. Questions of world view cannot be settled by majority vote. The problems that gave rise to it have not yet been solved.

All is not well in contemporary intellectual life, any more than in any

---

* Rudolf Bahro, *The Alternative in Eastern Europe* (London: New Left Books, 1978).

other area of contemporary life. The battle of ideas has not yet been won.*

Arguably, Marxism is still the only mode of thought capable of coming to terms with the complexities of contemporary experience. It is still unsurpassed in its capacity for clarity, coherence, comprehensiveness and credibility.

Within its resources are perhaps the only possibilities for penetrating the meaning even of events that its previous adherents never anticipated. The potency of Marxism is not so much in any of its existing tenets as in its habit of large-scale and deep-rooted thinking, discerning the trajectory of history as it comes, looking for a pattern of interconnections where others see only random chaos, going back further into the past, reaching wider within the present, facing with greater composure into the future.

Marxism may be rejected, but it has not been refuted. It still needs to be seriously studied and critically considered. Books such as this still deserve to be read and the story they tell to be reassessed and reappropriated.

Marxism may be repudiated, but it resonates nevertheless.

Helena Sheehan
Dublin, 1992

---

* Despite Francis Fukuyama, *The End of History and the Last Man* (London: Hamish Hamilton, 1992).

Afterword (2017)

History has continued rushing onward with dramatic and sometimes startling results. In marking the time of the story told in this text and the time of this new edition, there was first a flourishing followed by a fall in the fortunes of Marxism.

The last years of the socialist experiments in Eastern Europe brought renewed interest in the critical traditions of Marxism and raised hopes in a more democratic and efficient socialism. The demise of these experiments had many unforeseen and unintended consequences. Many of those who took to the streets were hoping for a better form of socialism and were shocked to see all that they had so laboriously built being so rapidly and ruthlessly expropriated. Others did seek a capitalist road, but it was a romanticized dream of democracy and prosperity that never came to pass. Instead, public property was privatized, production plummeted and people were plunged into poverty. Compounding the tragedy, people who had lived together in peace and purpose turned on each other as they splintered in murderous mini-ethnicities.

The institutional bases of Marxism were decimated in that part of the world. Some of those who once defended Marxism responded to the new reality by renouncing it. Some seemed to have suffered amnesia regarding their own former lives. Others believed what they said they believed and lost their jobs for it and were moreover marginalized in myriad modes. Every outdated or newfangled belief found desperate and often incoherent adherents. The loss of world view and its sustaining social order manifested many pathological symptoms.

There were some positive phenomena too. Freedom to speak, to trade, to travel and to contest elections were welcome and overdue.

The attendant opening of archives and the revelation of hidden truths has been of great value, although this thrived more in the glasnost-perestroika period than after it. With respect to research relevant to this book, there were sources I would have used if they had been available at the time. However, as far as I am aware, nothing has been revealed to negate, or even to undermine, either the narrative or the arguments of this book. Nevertheless, some interesting and significant material has emerged.

A major discovery came with the revelation that Nikolai Bukharin, major player in the story set out in this book, wrote four book length manuscripts while imprisoned in the Lubyanka awaiting his death. One of these was a remarkable autobiographical novel entitled *How It All Began.** The introduction was written by Stephen Cohen, Bukharin's biographer, whose determination played a part in unearthing these pages from deep in a Kremlin vault. There was also a philosophical text entitled *Philosophical Arabesques.*** I was honoured to be asked to write the introduction to this. The most amazing thing about this text was that it was written at all, that its author was not totally debilitated by despair, seeing what had been so magnificently created being so catastrophically destroyed. The revolution in which he so ardently believed was not only condemning him to death but slandering his life. His name was expunged from the books telling of the history he had participated in making and only reappeared decades later when there was brief Bukharin revival under Gorbachev. In this interval, the story of the revolution was retold in a society where this story crucially mattered just before this society collapsed and was replaced by a society where it was either retold in a hostile way or no longer mattered to many.

Bukharin was the personification of a path not taken, primarily politically and economically, but also philosophically and scientifically. He stood for what he called "socialist humanism," socialism with a human face, socialism with an open mind, socialism with an honest voice, socialism with an outstretched hand. He advocated a more evolutionary path to socialism, an opening of a process where a society would grow into socialism, where those who questioned might be persuaded and not necessarily coerced or executed, where theoretical questions were settled by theoretical debates and not by accusations of treason, purges of editorial boards and disappearances in the night. In philosophy, he was inclined to stress grounding in the natural sciences and to be wary of "Hegelian panology," although he did take seriously the roots of Marxism in the whole history of philosophy.

Philosophical Arabesques was an ambitious systematic work of philosophy exploring the contours of this. This approach to philosophy set Marxism not only within the whole history of philosophy, but also within

---

* N.I. Bukharin, *How It All Began* (1938), Introduction by Stephen Cohen (New York: Columbia University Press, 1998).

** N.I. Bukharin, *Philosophical Arabesques* (1937), Introduction by Helena Sheehan (New York: Monthly Review Press, 2005).

the whole battle of ideas of world culture of his times. It was a highly polemical text, engaging seriously with virtually every major intellectual trend of its times. It displayed an astute knowledge of the intellectual life of the epoch and the world historical context from which it emerged. He saw the grain of truth in every previous philosophy and saw Marxism in continuity with the centuries-long struggle to conceptualize the universe. He acknowledged the partial perspectives in each of the contemporary trends contending with Marxism and argued that Marxism superseded every one-sided view of the world to bring philosophy to a higher synthesis than had ever been achieved. It was an integrative and grounded way of thinking that offered a fresh way into the complex new problems of the era. This was in contrast to the approach to Marxism that prevailed in the Soviet Union at that time, isolating it from all outside forces, shutting down all internal debate, reducing it to a simplistic scheme where canonical formulations were recited repetitively, presenting all philosophical questions as basically settled.

It was in that atmosphere that Lysenko and Lysenkoism were thriving. In all the years since its rise and fall, there have been constant references to it and diverging interpretations of it. This book put much emphasis on this controversy and took a strong interpretative stance. The predominant position of seeing it simply as a morality tale warning against "ideological interference in science" has continued, but more complex analyses have found their place in this literature too. The best have been those addressing it within the whole nexus of debates about nature and nurture, about heredity and environment, about genetics and evolution, and doing so in light of advances in science as well as socio-historical context. The most significant development on this terrain has been a resurgence of belief in inheritance of acquired characteristics, these days scientifically called epigenetic transgenerational inheritance and often referred to as neo-Larmarckism. This has brought the Lysenkoism controversy to the surface again. A knowledgeable examination of the scientific theories and their complex political contexts has appeared in Loren Graham's *Lysenko's Ghost: Epigenetics and Russia*.*

Graham has also published an engaging and experiential account of his years of research into science in the USSR and the Russia succeeding it in *Moscow Stories*.** Perhaps the most interesting of these stories was his

---

* Loren Graham, *Lysenko's Ghost: Epigenetics and Russia* (Cambridge: Harvard University Press, 2016).
** Loren Graham, *Moscow Stories* (Bloomington: Indiana University Press, 2006).

encounter with Lysenko, whom he had tried unsuccessfully to interview on several occasions, but took advantage of a chance encounter when he saw him eating alone in an Academy of Sciences dining room. Lysenko told Graham that he knew his work and believed that it misrepresented him, denying any responsibility for the deaths of Vavilov and other Soviet scientists. In speaking to Graham when I was writing this book, I had remarked on how significant were many of his experiences in the course of writing his books and wrote in my book about an important conversation that he had with A.I. Oparin about the Haldane-Oparin thesis on the origin of life.

I have often wished that more authors would publish the stories behind the stories of their research. Indeed in a book to be published in 2018, I have written the story behind the writing of this book, when I was immersed in my material and had many significant experiences, including during my time in the USSR.* While there, I was based in semi-clandestine International Lenin School, also known as the Institute of Social Sciences, in Moscow. The opening of archives and subsequent research has brought some revelations about this institution, particularly in its earlier years as a Comintern school, where it educated emerging leaders of the communist movement. I had no idea at the time I was there about the fate of Irish communists who had proceeded me. There was not only a purge of Bukharinism but of Larkinism at the school. Dissent on philosophical matters had particularly tragic consequences for Patrick Breslin, who perished in the purges, whose fate was disclosed in Barry McLoughlin's *Left to the Wolves*.** There is still very little published about the institution in the post-Comintern period, so I hope my account when published will do something to fill that gap. During my time there, I encountered the emerging forces of glasnost-perestroika as they were struggling for survival in the previous period.

Other new sources of insight into the history unfolded here have come in various biographies and autobiographies, both adding detail and nuance to the picture of figures directly profiled here and into the wider fate of Marxism in the historical process. There have been several biographies of J.D. Bernal, for instance, a major figure in this history.*** The Verso one,

---

* Helena Sheehan, *Navigating the Zeitgeist* (New York: Monthly Review Press, 2018).
** Barry McLoughlin, *Left to the Wolves: Irish Victims of Stalinist Terror* (Newbridge: Irish Academic Press, 2007).
*** Francis Aprahamian and Brenda Swann, eds., *J.D. Bernal: A Life in Science and Politics* (London: Verso, 1999); Andrew Brown, *J.D. Bernal: The Sage of Science* (Oxford: Oxford University Press, 2005).

consisting of chapters by different authors dealing with different aspects of his work, was sympathetic to his Marxist convictions, whereas the Brown one, while valuable in its massive research, tended to be admiring of his scientific contribution, bemused by his sexuality, condescending to his philosophy and hostile to his politics. It was widely reviewed in a manner that endorsed this view of Bernal. I took issue with this at a conference dedicated to assessing the legacy of Bernal in Limerick in 2006, at which Andrew Brown also spoke, making a critical defence of Bernal's political and philosophical position.*

Subsequent research has also been turned to the role of the 2nd International History of Science Congress outlined in this book. The doctoral thesis of Christopher Chilvers traced the contexts of the congress, analysed the impact of the surprise arrival of the Soviet delegation and followed up on the subsequent fates of those involved.** There were a number of international academic events marking the 75th anniversary of this congress in 2006. I spoke at several of them. The one at Princeton University brought together those who had written about it, including Graham, Werskey, Chilvers and myself to reassess it years later. It was a rich and lively interaction. Gary Werskey, who had been out of the field for some years, came back into it with some gusto for this occasion and delivered a sweeping and engaging paper addressing the whole field of Marxism and science as analogous to a symphony in three movements.***

My own equally sweeping paper traced the influence of Marxism on science studies in socio-historical context.**** I looked back on the tempestuous tides shaping the history of Marxism in its relation to science, not only in the decades examined in this book, but in the ensuing ones and argued that Marxism has made the strongest claims of any intellectual tradition before or since about the socio-historical character of science, yet always affirmed its cognitive achievements. This is still its distinctiveness and its strength.

---

* Helena Sheehan, "J.D. Bernal: Philosophy, Politics and the Science of Science," *Journal of Physics*, vol. 57 (2007), 29–39.

** Christopher Chilvers, *Something Wicked This Way Comes: The Russian Delegation at the 1931 Second International Congress of the History of Science and Technology,* PhD thesis, Oxford University, 2007.

*** Gary Werskey, "The Marxist Critique of Capitalist Science: A History in Three Movements," *Science as Culture*, vol. 16, no. 4 (December 2007), 397–461.

**** Helena Sheehan, "Marxism and Science Studies: A Sweep Through the Decades," *International Studies in the Philosophy of Science*, vol. 21, no. 2 (2007), 197–210.

Subsequent decades have seen the influence of Marxism overshadowed by other forces, from neopositivism to postmodernism to the escalating commercialization of science as part of the intensifying commodification of all knowledge to the current right populist attack on science. In the various types of science wars that have raged, a better knowledge of the Marxist tradition would have brought light and clarity into a dark and muddy terrain.

In this year that marks the centenary of the October revolution, in which we are reassessing this monumental movement in the history of the world, I hope this book may spark an interest in the role played by the debates about philosophy and science in it all and that it may be useful to consider what was at stake. This book testifies to its struggles and its achievements as well as its traumas and its tragedies. There may be fewer centers of institutional power, fewer thinkers and researchers in this field than in times past, but I still believe that any honest assessment makes it clear that there is still nothing to match Marxism to make sense of science and indeed all else.

Index of Names

(Continued from note on page 41.) Engels and many following upon him may have given numerous and fascinating examples of processes following the patterns set out as "laws of dialectics," but there has always been an enormous logical gap between these particular examples and the general claim that these patterns constitute "laws" that govern every instance of change. Such a procedure, moving on the basis of selective induction to claims of universality, cannot be justified. There is also the further problem of the ambiguity surrounding the use of the term "dialectic," with Engels and later Marxists constantly shifting ground and equivocating, sometimes using the word dialectical as simply synonymous with developmental, and other times in a stricter sense as designating development in and through leaps, contradictions, and negations. In any case, whatever Engels's own debt to Hegel, the notion of dialectics is inessential to making a case for the pervasiveness of change, for the interrelatedness of all that exists, for ascending levels of the integration of matter. What is essential is the opposition to static, atomistic, dualist and reductionist modes of thought in the name of developmental, monist and emergentist ones.

(Continued from note on page 202.) In 1931, there was the trial of the "Union Bureau of the Central Committee of the Menshevik Party" for wrecking within the system of economic planning. One lesson that was supposed to be taken from all this, as the prosecutor Vjshinsky made clear at the Shakhty trial, was that the "bourgeois specialists" were no longer to be trusted.

(Continued from note on page 318.) Russian communism of the 1920s provided the kind of atmosphere in which the posing of a materialist answer to the question "what is life?" seemed entirely natural. An interesting, and extremely significant, postscript to the story came to my attention in a discussion on this matter with Loren Graham in Cambridge, Mass., September 29, 1979. In August 1971, Graham left the proofs of his book *Science and Philosophy in the Soviet Union*, in which he put forward the above interpretation of Oparin's intellectual development, with Oparin's office in Moscow. In an interview the next day, Oparin remarked to Graham that in reading the proofs he discovered things about himself that he had scarcely realized before, thus strongly confirming Graham's interpretation.

(Continued from note on page 367.) Edgell Rickword, enquiring about the reasons for the exclusion of the biology study and asking if the conflict between Caudwell's ideas and those of Lysenko was a factor, with the publication of *Further Studies* coinciding with the high tide of Lysenkoism in the world communist movement in 1948–1949. Rickword replied that the essay was quite long and that it had been decided that *Further Studies* should not exceed *Studies* in length. He added that a third series had been envisaged, but gave no explanation of why it never appeared (letter to the author March 12, 1980).